Lecture Notes in Artificial Intelligence 9437

Subseries of Lecture Notes in Computer Science

More information about this series at http://www.springer.com/series/1244

Yiyu Yao · Qinghua Hu · Hong Yu
Jerzy W. Grzymala-Busse (Eds.)

Rough Sets, Fuzzy Sets, Data Mining, and Granular Computing

15th International Conference, RSFDGrC 2015
Tianjin, China, November 20–23, 2015
Proceedings

 Springer

Editors
Yiyu Yao
University of Regina
Regina, SK
Canada

Qinghua Hu
Tianjin University
Tianjin
China

Hong Yu
Chongqing Universtiy of Posts
  and Telecommunications
Chongqing
China

Jerzy W. Grzymala-Busse
University of Kansas
Lawrence, KS
USA

ISSN 0302-9743   ISSN 1611-3349 (electronic)
Lecture Notes in Artificial Intelligence
ISBN 978-3-319-25782-2   ISBN 978-3-319-25783-9 (eBook)
DOI 10.1007/978-3-319-25783-9

Library of Congress Control Number: 2015952762

LNCS Sublibrary: SL7 – Artificial Intelligence

Springer International Publishing AG Switzerland is part of Springer Science+Business Media
(www.springer.com)

Preface

This volume comprises papers accepted for presentation at the 15th International Conference on Rough Sets, Fuzzy Sets, Data Mining and Granular Computing (RSFDGrC 2015), which, along with the 10th International Conference on Rough Sets and Knowledge Technology (RSKT 2015), was held as a major part of the 2015 International Joint Conference on Rough Sets (IJCRS 2015) during November 20–23, 2015, in Tianjin, China. IJCRS is the main annual conference on rough sets. It follows the Joint Rough Set Symposium series, previously organized in 2007 in Toronto, Canada; in 2012 in Chengdu, China; in 2013 in Halifax, Canada; and in 2014 in Granada and Madrid, Spain. In addition to RSFDGrC and RSKT, IJCRS 2015 also hosted the Third International Workshop on Three-way Decisions, Uncertainty, and Granular Computing.

IJCRS 2015 received 97 submissions that were carefully reviewed by two to five Program Committee (PC) members or additional reviewers. After a rigorous process 42 regular papers (acceptance rate 43.3) were accepted for presentation at the conference and publication in the two volumes of proceedings. Subsequently, some authors were selected to submit a revised version of their paper and after a second round of review, 24 more papers were accepted.

At IJCRS 2015 a ceremony was held to honor the new fellows of the International Rough Set Society. The fellows had the possibility to present their contributions in the rough set domain and to propose open questions. This volume contains five of the 11 contributions authored by the fellows Jiye Liang, Dominik Ślęzak, Shusaku Tsumoto, Guoyin Wang, and Ning Zhong.

This volume contains reports of the top six finalists from IJCRS 2015 Data Challenge, which had 90 registered teams with members from 18 different countries and a total of 1,676 submitted solutions. It also contains a summary paper by the challenge organizers Andrzej Janusz, Marek Sikora, Łukasz Wróbel, Sebastian Stawicki, Marek Grzegorowski, Piotr Wojtas, and Dominik Ślęzak.

It is a pleasure to thank all those people who helped this volume come into being and made IJCRS 2015 a successful and exciting event. First of all, we express our thanks to the authors without whom nothing would have been possible. We deeply appreciate the work of the PC members and additional reviewers who assured the high standards of accepted papers.

We deeply acknowledge the precious help of all the IJCRS chairs (Davide Ciucci, Guoyin Wang, Sushmita Mitra, Wei-Zhi Wu) as well as of the Steering Committee members (Duoqian Miao, Andrzej Skowron, Shusaku Tsumoto) for their work and suggestions with respect to the process of proceedings preparation and conference organization. We would like to pay tribute to the honorary chairs, Lotfi Zadeh, Roman Słowiński, Bo Zhang, and Jianwu Dang, whom we deeply respect for their countless contributions to the field.

We gratefully thank our sponsors: the Tianjin Key Laboratory of Cognitive Computing and Application, School of Computer Science and Technology at Tianjin University, for organizing and hosting the conference, the Institute of Innovative Technologies EMAG (Poland) for providing the data and funding the awards for the IJCRS 2015 Data Challenge, and Springer for sponsorship of the best paper awards.

Our immense gratitude goes once again to Qinghua Hu for taking charge of the organization of IJCRS 2015.

We are grateful to Alfred Hofmann and the excellent LNCS team at Springer for their help and cooperation. We would also like to acknowledge the use of EasyChair, a great conference management system.

Finally, it is our sincere hope that the papers in the proceedings may be of interest to the readers and inspire them in their scientific activities.

November 2015

Yiyu Yao
Qinghua Hu
Hong Yu
Jerzy W. Grzymala-Busse

Organization

Honorary General Co-chairs

Jianwu Dang   Tianjin University, Tianjin, China
Roman Słowiński   Poznań University of Technology, Poznań, Poland
Lotfi Zadeh   University of California, Berkeley, CA, USA
Bo Zhang   Tsinghua University, Beijing, China

General Chairs

Qinghua Hu   Tianjin University, Tianjin, China
Guoyin Wang   Chongqing University of Posts and
                              Telecommunications, Chongqing, China

Steering Committee Chairs

Duoqian Miao   Tongji University, Shanghai, China
Andrzej Skowron   University of Warsaw and Polish Academy
                              of Sciences, Warsaw, Poland
Shusaku Tsumoto   Shimane University, Izumo, Japan

Program Chairs

Davide Ciucci   University of Milano-Bicocca, Milano, Italy
Yiyu Yao   University of Regina, Regina, SK, Canada

Program Co-chairs for RSFDGrC 2015

Sushmita Mitra   Indian Statistical Institute, Kolkata, India
Wei-Zhi Wu   Zhejiang Ocean University, Zhejiang, China

Program Co-chairs for RSKT 2015

Jerzy W. Grzymala-Busse   University of Kansas, Lawrence, KS, USA
Hong Yu   Chongqing University of Posts and
                              Telecommunications, Chongqing, China

Workshop Co-chairs for TWDUG 2015

Davide Ciucci   University of Milano-Bicocca, Milano, Italy
Jerzy W. Grzymala-Busse   University of Kansas, Lawrence, KS, USA
Tianrui Li   Southwest Jiaotong University, Chengdu, China

Program Co-chairs for TWDUG 2015

Nouman Azam   National University of Computer and
                                     Emerging Sciences, Peshawar, Pakistan
Fan Min   Southwest Petroleum University, Chengdu, China
Hong Yu   Chongqing University of Posts and
                                     Telecommunications, Chongqing, China
Bing Zhou   Sam Houston State University, Huntsville, TX, USA

Data Mining Contest Organizing Committee

Andrzej Janusz   University of Warsaw, Warsaw, Poland
Marek Sikora   Silesian University of Technology, Gliwice, Poland
Łukasz Wróbel   Silesian University of Technology, Gliwice, Poland
Sebastian Stawicki   University of Warsaw, Warsaw, Poland
Marek Grzegorowski   University of Warsaw, Warsaw, Poland
Dominik Ślęzak   University of Warsaw, Warsaw, Poland

Tutorial Co-chairs

Dominik Ślęzak   University of Warsaw, Warsaw, Poland
Yoo-Sung Kim   Inha University, Incheon, Korea

Special Session Co-chairs

JingTao Yao   University of Regina, Regina, SK, Canada
Tianrui Li   Southwest Jiaotong University, Chengdu, China
Zbigniew Suraj   University of Rzeszow, Rzeszow, Poland

Publicity Co-chairs

Zbigniew Ras   University of North Carolina at Charlotte, Charlotte,
                                     NC, USA and Warsaw University of Technology,
                                     Warsaw, Poland
Georg Peters   Munich University of Applied Sciences, Munich,
                                     Germany and Australian Catholic University,
                                     Sydney, Australia
Aboul Ella Hassanien   Cairo University, Cairo, Egypt

Azizah Abdul Manaf   Universiti Teknologi Malaysia, Kuala Lumpur,
                              Malaysia
Gabriella Pasi   University of Milano-Bicocca, Milano, Italy
Chris Cornelis   University of Granada, Granada, Spain

Local Co-chairs

Jianhua Dai   Tianjin University, Tianjin, China
Pengfei Zhu   Tianjin University, Tianjin, China
Hong Shi   Tianjin University, Tianjin, China

Program Committee

Arun Agarwal	Igor Chikalov	Jun He
Adel Alimi	Wan-Sup Cho	Christopher Henry
Simon Andrews	Soon Ae Chun	Daryl Hepting
Nidhi Arora	Martine de Cock	Joseph Herbert
Piotr Artiemjew	Chris Cornelis	Francisco Herrera
S. Asharaf	Zoltán Csajbók	Chris Hinde
Ahmad Azar	Alfredo Cuzzocrea	Shoji Hirano
Nakhoon Baek	Krzysztof Cyran	Władysław Homenda
Sanghamitra	Jianhua Dai	Feng Hu
Bandyopadhyay	Bijan Davvaz	Xiaohua Hu
Mohua Banerjee	Dayong Deng	Shahid Hussain
Nizar Banu	Thierry Denoeux	Namvan Huynh
Andrzej Bargiela	Jitender Deogun	Dmitry Ignatov
Alan Barton	Lipika Dey	Hannah Inbaran
Jan Bazan	Fernando Diaz	Masahiro Inuiguchi
Theresa Beaubouef	Ivo Düntsch	Yun Jang
Rafael Bello	Zied Elouedi	Ryszard Janicki
Rabi Bhaumik	Francisco Fernandez	Andrzej Jankowski
Jurek Błaszczyński	Jinan Fiaidhi	Andrzej Janusz
Nizar Bouguila	Wojciech Froelich	Jouni Järvinen
Yongzhi Cao	G. Ganesan	Richard Jensen
Gianpiero Cattaneo	Yang Gao	Xiuyi Jia
Salem Chakhar	Günther Gediga	Chaozhe Jiang
Mihir Chakraborty	Neveen Ghali	Na Jiao
Chien-Chung Chan	Anna Gomolińska	Tae-Chang Jo
Chiao-Chen Chang	Salvatore Greco	Manish Joshi
Santanu Chaudhury	Shen-Ming Gu	Hanmin Jung
Degang Chen	Yanyong Guan	Janusz Kacprzyk
Hongmei Chen	Jianchao Han	Byeongho Kang
Mu-Chen Chen	Wang Hao	C. Maria Keet
Mu-Yen Chen	Aboul Hassanien	Md. Aquil Khan

Deok-Hwan Kim
Soohyung Kim
Young-Bong Kim
Keiko Kitagawa
Michiro Kondo
Beata Konikowska
Jacek Koronacki
Bożena Kostek
Adam Krasuski
Vladik Kreinovich
Rudolf Kruse
Marzena Kryszkiewicz
Yasuo Kudo
Yoshifumi Kusunoki
Sergei Kuznetsov
Wookey Lee
Young-Koo Lee
Carson K. Leung
Huaxiong Li
Longshu Li
Wen Li
Decui Liang
Jiye Liang
Churn-Jung Liau
Diego Liberati
Antoni Ligęza
T.Y. Lin
Pawan Lingras
Kathy Liszka
Caihui Liu
Dun Liu
Guilong Liu
Qing Liu
Xiaodong Liu
Dickson Lukose
Neil Mac Parthaláin
Seung-Ryol Maeng
Pradipta Maji
A. Mani
Victor Marek
Barbara Marszał-Paszek
Tshilidzi Marwala
Benedetto Matarazzo
Nikolaos Matsatsinis
Stan Matwin
Jesús Medina-Moreno

Ernestina Menasalvas
Jusheng Mi
Alicja
    Mieszkowicz-Rolka
Tamás Mihálydeák
Pabitra Mitra
Sadaaki Miyamoto
Sabah Mohammed
Mikhail Moshkov
Tetsuya Murai
Kazumi Nakamatsu
Michinori Nakata
Amedeo Napoli
Kanlaya Naruedomkul
Hungson Nguyen
Linhanh Nguyen
Maria Nicoletti
Vilém Novák
Mariusz Nowostawski
Hannu Nurmi
Hala Own
Piero Pagliani
Krzysztof Pancerz
Taeho Park
Piotr Paszek
Alberto Guillén Perales
Georg Peters
James F. Peters
Frederick Petry
Jonas Poelmans
Lech Polkowski
Henri Prade
Jianjun Qi
Jin Qian
Yuhua Qian
Keyun Qin
Guo-Fang Qiu
Taorong Qiu
Mohamed Quafafou
Annamaria Radzikowska
Vijay V. Raghavan
Sheela Ramanna
Zbigniew Ras
Kenneth Revett
Leszek Rolka
Leszek Rutkowski

Henryk Rybiński
Wojciech Rząsa
Hiroshi Sakai
Abdel-Badeeh Salem
Miguel Ángel Sanz-Bobi
Gerald Schaefer
Jeong Seonphil
Noor Setiawan
Sitimariyam Shamsuddin
Lin Shang
Ming-Wen Shao
Marek Sikora
Arul Siromoney
Vaclav Snasel
Urszula Stańczyk
Jarosław Stepaniuk
Kazutoshi Sumiya
Lijun Sun
Piotr Synak
Andrzej Szałas
Marcin Szczuka
Tomasz Szmuc
Marcin Szpyrka
Li-Shiang Tsay
Gwo-Hshiung Tzeng
Changzhong Wang
Chaokun Wang
Hai Wang
Junhong Wang
Ruizhi Wang
Wendyhui Wang
Xin Wang
Yongquan Wang
Piotr Wasilewski
Junzo Watada
Ling Wei
Zhihua Wei
Paul Wen
Arkadiusz Wojna
Karlerich Wolff
Marcin Wolski
Michał Woźniak
Gang Xie
Feifei Xu
Jiucheng Xu
Weihua Xu

Zhan-Ao Xue
Ronald Yager
Hyung-Jeong Yang
Jaedong Yang
Xibei Yang
Yan Yang
Yingjie Yang
Yong Yang
Yubin Yang

Nadezhda G. Yarushkina
Dongyi Ye
Ye Yin
Byounghyun Yoo
Kwan-Hee Yoo
Xiaodong Yue
Sławomir Zadrożny
Hongyun Zhang
Nan Zhang

Qinghua Zhang
Xiaohong Zhang
Yan-Ping Zhang
Cairong Zhao
Shu Zhao
Xian-Zhong Zhou
William Zhu
Wojciech Ziarko
Beata Zielosko

Additional Reviewers

Banerjee, Abhirup
Benítez Caballero,
    María José
Błaszczyński, Jerzy
Chiaselotti, Giampiero
Czołombitko, Michał

D'eer, Lynn
De Clercq, Sofie
Garai, Partha
Hu, Jie
Jie, Chen
Li, Feijiang

Liang, Xinyan
Wang, Jianxin
Wang, Jieting
Xu, Xinying
Zhang, Junbo

Contents

Rough Sets and Graphs

Rough and Fuzzy Hybridization

Granular Computing

Rough Sets: The Experts Speak

Decision-Oriented Rough Set Methods

Jiye Liang[1,2]([⊠])

[1] Department of Computer Science and Technology, Taiyuan Normal University,
Jinzhong 030619, Shanxi, China
`ljy@sxu.edu.cn`
[2] Key Laboratory of Computational Intelligence and Chinese Information Processing
of Ministry of Education, Shanxi University, Taiyuan 030006, Shanxi, China

Abstract. Rough set theory is a very effective multi-attribute decision analysis tool. The paper reviews four decision-oriented rough set models and methods: dominance-based rough set, three-way decisions, multigranulation decision-theoretic rough set and rough set based multi-attribute group decision-making model. We also introduce some of our group's works under these four models. Several future research directions of decision-oriented rough sets are presented in the end of the paper.

Keywords: Rough set · Multi-attribute decision making · Group decision making

1 Introduction

Multi-attribute decision making involves the examination of a discrete set of alternatives that are described along with several attributes. In the past several decades, much attention has been paid to multi-attribute decision making from an expert to a group of experts, determined decision environment to uncertain decision environment, and single method to integrated method. There are mainly four kinds of tasks for decision makers to solve multi-attribute decision-making problems [1]:

- Choice: to identify the best object or select a limited set of the best objects from the set of alternatives.
- Ranking: to construct a rank-ordering of the alternatives from the best to the worst ones.
- Sorting: to classify/sort the alternatives into predefined homogenous groups.
- Description: to identify the major distinguishing attributes of the alternatives, explore the relations of attributes, and build decision rules for illustrating how to make a decision.

Until now, a variety of methods have been suggested to solve multi-attribute decision-making problems. The representative ones are Multi-attribute Utility Theory (MAUT) [2], Analytic Hierarchy Process (AHP) [3], ELimination Et choix Traduisant la REalité (ELECTRE) [4], Preference Ranking Organization

© Springer International Publishing Switzerland 2015
Y. Yao et al. (Eds.): RSFDGrC 2015, LNAI 9437, pp. 3–12, 2015.
DOI: 10.1007/978-3-319-25783-9_1

Method for Enrichment Evaluations (PROMETHEE) [5], Technique for Order Preference by Similarity to an Ideal Solution (TOPSIS) [6], Rough Sets (RS) [7,8], et al..

In practice, decision makers are facing with the following three decision scenarios [8]: selecting the most importance attributes for making the right decisions, giving a prescription on how to make a decision under specific circumstances, explaining a decision in terms of circumstances under which the decision has been made.

Rough set theory proposed by Pawlak in [7,8] is one of the most important methods for solving multi-attribute decision-making problems in the three scenarios mentioned above. It has demonstrated its superior performances via attribute reduction and rule acquisition in multi-attribute decision making [7–11].

Attribute reduction offers a systematic framework for selecting the independent and essential attributes in a decision system. In the past three decades, many reduction techniques have been developed for selecting the important or essential attributes from different aspects. The discernibility matrix reduction [11] and heuristic reduction [12–15] provide effective tools for selecting the suitable attribute set from the aspects of algebra and information. Rule acquisition is another important contribution in decision-oriented rough set methods. Decision rules derived from the decision examples provide an easy-to-comprehend way for illustrating the decision foundation of decision makers. Researchers have designed many rule induction algorithms in different rough set models. LEM2 is one of the most widely used rule induction algorithms [16]. DOMLEM in [17] extracts dominance rules from decision systems under the framework of dominance-based rough set. In 2001, Grzymala [18] proposed a new rule induction algorithm called MODLEM, in which the discretization process and rule induction are performed at the same time. Several variations of these algorithms have been proposed in many other extended rough set models [14,15,19].

For the remarkable performances in selecting important attributes and explaining decisions, rough set has become an important multi-attribute decision model, and the rough decision methods have received special favors of researchers and practitioners from different fields. The paper introduces four decision-closely-related rough set models and some of our group's works under the four models. We also discuss several research directions of the decision-oriented rough set methods.

2   Decision-Oriented Rough Set Models and Methods

Rough set theory is an effective tool for multi-attribute decision analysis [7]. Over the past thirty years, researchers have proposed different extensions of rough set such as fuzzy rough set [20–22], dominance-based rough set [10,17,23–26], probabilistic rough set [27–31], covering-based rough set [32], multigranulation rough set [33–36], and others. All of these models can be applied to aid the multi-attribute decision making for different application demands. This section aims to introduce the following four decision-closely-related rough set models.

(1) Dominance-based rough set model

In multi-attribute/criteria decision-making problems, there in nature exist preference structures between condition attributes and decision attributes. Greco et al. proposed the eminent extended rough set model called dominance-based rough set (DRS) model by replacing the original indiscernibility relation with the dominance relation, which takes into consideration of the **dominance principle** [10, 25, 26], *if the performances of an object are all not worse than the other on all considered attributes, its assignment should not be made to a worse decision class than the others.*

Dominance-based rough set considers an ordered decision information system $S = (U, C \cup d, V, f)$, where d ($f(x, d)$ is single-valued) is the overall preference called the decision and all the elements of C are criteria (attributes with preference structures). Furthermore, $\mathbf{CL} = \{CL_1, CL_2, ..., CL_r\}$ is a set of classes of U, such that an alternative x belongs to one and only one class $CL_s \in \mathbf{CL}$ and each element of CL_r is preferred (strictly or weakly) to each element of CL_s if $s > r$. The dominance relation $xD_P y(P \subseteq U)$ means that x dominates y with respect to each attribute/criterion $p \in P$. Given $P \subseteq C$ and $x \in U$, the dominating and dominated sets of x with respected to P are defined as

$$D_P^+(x) = \{y \in U : yD_P x\} \tag{1}$$

and

$$D_P^-(x) = \{y \in U : xD_P y\}. \tag{2}$$

The upward and downward unions of dominance decision classes are defined as

$$CL_t^{\geq} = \bigcup_{s \geq t} CL_s, CL_t^{\leq} = \bigcup_{s \leq t} CL_s. \tag{3}$$

The P- lower and upper approximations of $CL_t^{\geq}$ are

$$\underline{P}(CL_t^{\geq}) = \{x \in U : D_P^+(x) \subseteq CL_t^{\geq}\} \tag{4}$$

and

$$\overline{P}(CL_t^{\geq}) = \bigcup_{x \in CL_t^{\geq}} D_P^+(x), \tag{5}$$

respectively.

The P- lower and upper approximations of $CL_t^{\leq}$ are defined as

$$\underline{P}(CL_t^{\leq}) = \{x \in U : D_P^-(x) \subseteq CL_t^{\leq}\} \tag{6}$$

and

$$\overline{P}(CL_t^{\leq}) = \bigcup_{x \in CL_t^{\leq}} D_P^-(x), \tag{7}$$

respectively.

For accomplishing the sorting and description tasks, DRS permits to construct the approximations of dominance class set on basis of the dominance preference knowledge bases and then to produce the corresponding decision rules satisfying the dominance principle [10, 24, 26]. It has also been proved that the preference model in the form of dominance rules is more general than the MAUT or the outranking models, and is more understandable for users because of its natural syntax. Researchers have studied the dominance-based rough sets in various of complex decision sceneries [37, 38]. Moreover, the decision rules derived by DRS can also be used to give a recommendation for the choice or ranking problem [23, 39].

Recently, Song and Liang [40] proposed a two-grade DRS-based approach for multi-attribute decision making to obtain a complete rank of alternatives, which are described with interval attributes taking values in the form as $[a^L(x_i), a^U(x_i)]$.

In the first grade, $DDI_a(x_i, x_j)$ and $DDI_A(x_i, x_j)$ are defined as directional distance indexes for the ordered pair (x_i, x_j) under an attribute a and an attribute set A. Then alternatives can be ranked according to the entire directional distance index $DDI_A(x_i)$. It is deserved to point out that there often exit some objects being put into the same place.

In the second grade, a new order mutual information $E(A^{\geq}; B^{\geq})$ is defined to depict the consistence degree between two dominance ordered structures. Accordingly, the importance of attributes is depicted via the order mutual information. Finally, a complete rank is obtained by the whole weighted directed distance indexes $DDI_A^*(x_i, x_j)$.

The DRS-based two-grade multi-attribute decision approach establishes the complete rank for alternatives by analyzing the dominance structures among criteria with interval data. The approach has been successfully applied to solving stock selection problems in China and also shown the effectiveness in risk-averse multi-attribute decision making.

(2) Three-way decision theory

In practical decision making, adequate judgment information brings an immediately "acceptance" or "rejection" decision, and the not-sufficient judgment information usually causes a "deferment" decision. This kind of decision making extends the two-way decision. It is used in everyone's daily life and widely applied in many social works as medical decision making, social judgement theory, hypothesis test in statistics, management sciences, and peering review process. Yao first gave the unified formal description of three-way decisions in [41]. The two-state three-way decision model with an evaluation function is introduced in what follows. Detailed description of the other three-way decision models can be found in [41].

An evaluation function $v : U \to L$ is used to estimate the states of objects in the universe U, where $(L, \preceq)$ is a totally ordered set. By introducing a pair of threshold (α, β), $(\beta \prec \alpha$ i.e. $\beta \prec \alpha$ and $\beta \neq \alpha)$ and the evaluation v, the three regions of three-way decisions are constructed as

$$POS_{(\alpha,\beta)}(v) = \{x \in U | v(x) \succeq \alpha\}, \tag{8}$$

$$NEG_{(\alpha,\beta)}(v) = \{x \in U | v(x) \preceq \beta\} \tag{9}$$

and

$$BND_{(\alpha,\beta)} = \{x \in U | \beta \prec v(x) \prec \alpha\}. \tag{10}$$

Decision-theoretic rough set (DTRS) [42] is a well-formed concrete three-way decision model, in which evaluation function v is interpreted as the possibility of an object belonging to the objective region, L is specialized as the unit interval [0,1] and α, β are the minimum cost thresholds. DTRS gives the thresholds α, β and evaluation function $v(x)$ the clear semantic interpretations by minimizing the decision costs with Bayesian theory [30]. The superiority of DTRS has invoked the interest of many scholars, and then many studies about DTRS have been done in recent years [43–45]. Greco et al. [43] combined the decision-theoretic rough set with the dominance rough set. Yao and Herbert [44] combined DTRS with game-theoretic rough set. Liang [45] proposed a naive model of intuitionistic fuzzy decision-theoretic rough sets and applied it to single-period and multi-period decision making problems.

Three-way decision theory provides a more general methods, which combine the universal decision-making thoughts and the intuitive information processing patterns together. Recently, Yao [46] further presented a useful sequential three-way decision theory for solving practical decision-making problems when information is unavailable and is acquired on demands with associated cost. Liang and Wang [47,49] introduced the three-way decisions into multi-attribute group decision making and designed an intelligent Dempster-Shafer theory based group sorting method.

(3) Multigranulation decision-theoretic rough set model

With the development of the economy and society, more and more complex uncertain decision-making problems come into people's daily life and social administration. Many existing single-granulation rough set models including decision-theoretic rough set model can be used to solve some of the uncertain decision problems. When facing high-dimensional, large-group and inconsistent multi-attribute decision-making problems, these single-granulation models have difficulties in approximating concepts and deriving rules. In [48], Qian et al. proposed a multigranulation decision-theoretical rough set model, which combined with the thoughts of multigranulation rough set [33] and decision-theoretic rough set [42], to solve the complex decision-making problems by fusing multiple relations induced from the decision systems.

Three typical multigranulation decision-theoretic rough set models including the first two extreme models and the last moderate model, are defined for obtaining better decision results.

- Optimistic multigranulation decision-theoretic rough set model (OMGDTRS) In decision analysis, an "optimistic" decision maker is used to express the idea that the lower approximation of a concept only needs at leat one granular structure to satisfy with the inclusion condition. Then, the *min* operator is used to fuse the conditional possibilities of different knowledge granularities.

- Pessimistic multigranulation decision-theoretic rough set model (PMGDTRS)
  In decision analysis, a "pessimistic" decision maker often have the decision strategy, seeking common ground while eliminating differences, which is a conservative decision strategy. Then, the max operator is used to fuse the conditional possibilities of different knowledge granularities.
- Mean multigranulation decision-theoretic rough set model (MMGDTRS)
  In decision analysis, most of decision makers are with the "moderate" attitude. Then, the arithmetic $mean$ operator is used to fuse the conditional possibilities of different knowledge granularities for approximating the objective concept.

MGDTRS can derive many existing rough set models when the parameters satisfy special constraints. It provides a new perspective for rough decision analysis.

(4) Rough set based multi-attribute group decision-making model

Multi-attribute group decision making (MAGDM) is an important component of the modern decision science. The theories and methods of MAGDM have been widely applied in economic, management and many other fields. In many practical fields, the experts in group provide not only the decision matrixes, but also the primary sorting for alternatives by intuition. Many existing multi-attribute group decision-making methods focus on integrating the evaluations under multiple attributes and multiple experts, the intuition judgments of experts are, however, not considered. In Ref. [49], Liang and Wang proposed a rough set based multi-attribute group decision model (RS-MAGDM) to take a full consideration of the primary sorting.

A multi-attribute group decision system is denoted as $S^k = (U, C \cup \{d^k\}, f^k, V^k)$, where $U = \{x_1, x_2, ..., x_n\}$ is a set of alternatives, $C = \{c_1, c_2, ..., c_m\}$ is a set of attributes. A group of experts $E = \{e_1, e_2, ..., e_l\}$ evaluated the alternatives under the attribute set C and d. $f^k : U \times C \cup d^k \rightarrow V^k$ is an evaluation function.

The main processes of RS-MAGDM are shown as follows.

- Determination of weights of attributes
  The local weight of c_j^k is determined by the consistent degree that measures the dominance granulation structures between the attribute c_j^k and the primary decision attribute d^j, i.e., $w_j^k = sim(\succeq_{c_j^k}, \succeq_{d^k})$. The overall weight of c_j, w_j, is computed from the average of w_j^k.
- Determination of weights of experts
  The decision similarity degree $sim(e_k, e_{k'})$ between e_k and $e_{k'}$ is calculated by measuring the close degree between the structures of rough sets. The representative expert e^* is chosen for the most consistency with the other experts. Then, ρ_k, the weight of e_k, is determined by calculating the similarity degree between e_k and the representative expert e^*.
- Aggregation of decision information

The overall evaluation value of x is computed by integrating the values under multiple attributes and multiple experts. Then the corresponding decision tasks can be accomplished on the basis of overall values.

Moreover, based on dominance granular structures of DRS, Pang and Liang [50] designed six indices for evaluating the results of multi-attribute group decision making. In [50], three key evaluation indices C^k, T^k and U^k were proposed for measuring the decision consistency, closeness and uniformity of an expert in multi-attribute group decision making. Based on the three indices for individual experts, the other three indices C, T and K were designed for measuring the group consistency, group closeness, and group uniformity degrees.

RS-MAGDM introduces the intuitive judgements of experts into the process of the multi-attribute group decision making. The weights of attributes and experts are calculated from the dominance granular structures in decision matrix and the rough approximation structures of expert's primary judgements. The introduction of intuitive judgments and the natural computation of weights does brighten the model for solving the general multi-attribute group decision-making problems.

3   Concluding Remarks and Future Perspectives

The focal point of interest in the paper is to introduce four kinds of decision-closely-related rough set models. These models and methods have received much attention of the researchers from the fields of artificial intelligence, decision analysis, mathematics and others. They have also been successfully applied to solve many practical decision-making problems. Nevertheless, there still exist some challenging issues in the theoretical and practical researches of decision-oriented rough sets.

- Hybrid decision systems are common in practise. It is worth the effort to construct suitable decision-oriented rough set models for solving the hybrid decision-making problems.
- The fusion of rough set with other decision methods can open a door for the complex decision making.
- Many practical decision-making problems emerge with the characteristics of dynamic, large-sale of alternatives and high-dimensional attributes. It is another research hotspot to design effective and fast algorithms for this kind of problems.
- It is the objective of practitioners to apply diverse rough set models to real decision processes for obtaining reasonable and better decision results.

Acknowledgments. This work was supported by the National Natural Science Foundation of China (Nos. 61432011, U1435212), Research Project Supported by Shanxi Scholarship Council of China (No. 2013-101), the Key Problems in Science and Technology Project of Shanxi Province (No. 20110321027-01) and the Construction Project of the Science and Technology Basic Condition Platform of Shanxi Province (No. 2012091002-0101).

References

1. Zopounidis, C., Doumpos, M.: Multicriteria classification and sorting methods: a literature review. Eur. J. Oper. Res. **138**, 229–246 (2002)
2. Hwang, C.L., Yoon, K.: Multiple Attribute Decision Making-Methods and Applications: A State of the Art Survey. Lecture Notes in Economics and Mathematical Systems. Springer-Verlag, New York (1981)
3. Saaty, T.L.: The Analytic Hierarchy Process. McGraw-Hill, Now York (1980)
4. Benayoun, R., Roy, B., Sussman, N.: Manual de refrence du programme electre. Note de Synthese et Formation, No. 25. Paris: Direction Scientifique SEMA (1966)
5. Brans, J.P., Mareschal, B.: The promethee vi procedure: how to differentiate hard from soft multicriteria problems. J. Decis. Syst. **4**, 213–223 (1995)
6. Hwang, C.L., Lai, Y.J., Liu, T.Y.: A new approach for multiple objective decision making. Comput. Oper. Res. **20**, 889–899 (1993)
7. Pawlak, Z.: Rough sets. Int. J. Comput. Inf. Sci. **11**, 341–356 (1982)
8. Pawkak, Z.: Rough set approach to knowledge-based decision support. Eur. J. Oper. Res. **99**, 48–57 (1997)
9. Pawlak, Z.: Rough Sets: Theoretical Aspects of Reasoning about Data. Kluwer Academic Publishers, Boston (1991)
10. Greco, S., Matarazzo, B., Slowinski, R.: Rough approximation of a preference relation by dominance relations. Eur. J. Oper. Res. **117**, 63–83 (1999)
11. Skowron, A., Rauszer, C.: The Discernibility Matrices and Functions in Information Systems. In: Slowinski, R. (eds.) Intelligent Decision Support - Handbook of Applications and Advances of the Rough Sets Theory, vol. 11, pp. 331–362. Springer (1991)
12. Wang, G.Y., Yu, H., Yang, D.C.: Decision table reduction based on conditional information entropy. Chin. J. Comput. **25**, 759–766 (2002)
13. Liang, J.Y., Wang, F., Dang, C.Y., Qian, Y.H.: An efficient rough feature selection algorithm with a multi-granulation view. Int. J. Approx. Reason. **53**, 912–926 (2010)
14. Liang, J.Y., Wang, F., Dang, C.Y., Qian, Y.H.: A group incremental approach to feature selection applying rough set technique. IEEE Trans. Knowl. Data Eng. **26**, 294–308 (2014)
15. Slezak, D.: Approximate entropy reducts. Fund. Inform. **53**, 365–390 (2002)
16. Grzymala-Busse, J.W.: LERS: A system for learning from examples based on rough sets. In: Slowinski, R. (ed.) Intelligent Decision Support: Handbook of Applications and Advances of the Rough Set theory, vol. 11, pp. 3–18. Kluwer Academic Publishers, Dordrecht (1992)
17. Greco, S., Matarazzo, B., Slowinski, R.: The use of rough sets and fuzzy sets in MCDM. In: Gal, T., Hanne, T., Stewart, T. (eds.) Advances in Multiple Criteria decision Making. Kluwer Academic Publishers, Dordrecht (1999)
18. Grzymala-Busse, J.W., Stefanowski, J.: Three discretization methods for rule induction. Int. J. Intell. Syst. **26**, 29–38 (2001)
19. Leung, Y., Fischer, M.M., Wu, W.Z., Mi, J.S.: A rough set approach for the discovery of classification rules in interval-valued information systems. Int. J. Approx. Reason. **47**, 233–246 (2008)
20. Dubois, D., Prade, H.: Rough fuzzy sets and fuzzy rough sets. Int. J. Gen. Syst. **17**, 191–209 (1990)
21. Dubois, D., Prade, H.: Putting rough sets and fuzzy sets together. In: Slowinski, R. (ed.) Intelligent Decision Support: Handbook of Applications and Advances of the

Rough Sets Theory, vol. 11, pp. 203–232. Kluwer Academic Publishers, Dordrecht (1992)

22. Hu, Q.H., Xie, Z.X., Yu, D.R.: Hybrid attribute reduction based on a novel fuzzy rough model and information granulation. Pattern Recogn. **40**, 3509–3521 (2007)

23. Greco, S., Matarazzo, B., Slowinski, R.: Rough sets theory for multicriteria decision analysis. Eur. J. Oper. Res. **129**, 1–7 (2001)

24. Greco, S., Matarazzo, B., Slowinski, R.: Rough sets methodology for sorting problems in presence of multiple attributes and criteria. Eur. J. Oper. Res. **138**, 247–259 (2002)

25. Greco, S., Matarazzo, B., Slowinski, R., Zanakis, S.: Global investing risk: a case study of knowledge assessment via rough sets. Annal Oper. Res. **185**, 105–138 (2011)

26. Greco, S., Slowinski, R., Zielniewicz, P.: Putting dominance-based rough set approach and robust ordinal regression together. Dec. Support Syst. **54**, 891–903 (2013)

27. Wong, S.K.M., Ziarko, W.: Comparison of the probabilistic approximate classification and the fuzzy set model. Fuzzy Sets Syst. **21**, 357–362 (1987)

28. Pawlak, Z., Wong, S.K.M., Ziarko, W.: Rough sets: probabilistic versus deterministic approach. Int. J. Man-Mach. Stud. **29**, 81–95 (1988)

29. Ziarko, W.: Variable precision rough set model. J. Comput. Syst. Sci. **46**, 39–59 (1993)

30. Yao, Y.Y., Wong, S.K.M.: A decisoin theoretic framework for approximating concepts. Int. J. Man-Mach. Stud. **37**, 793–809 (1992)

31. Yao, Y.Y., Zhou, B.: Naive Bayesian Rough Sets. In: Yu, J., Greco, S., Lingras, P., Wang, G., Skowron, A. (eds.) RSKT 2010. LNCS (LNAI), vol. 6401, pp. 719–726. Springer, Heidelberg (2010)

32. Zhu, W., Wang, F.Y.: Reduction and axiomization of covering generalized rough sets. Inf. Sci. **152**, 217–230 (2003)

33. Qian, Y.H., Liang, J.Y., Yao, Y.Y., Dang, C.Y.: MGRS: A multi-granulation rough set. Inf. Sci. **180**, 949–970 (2010)

34. Yang, X.B., Song, X.N., Chen, Z.H., Yang, J.Y.: On multi-granulation rough sets in incomplete information system. Int. J. Mach. Learn. Cyber. **3**, 223–232 (2011)

35. Xu, W.H., Sun, W.X., Zhang, X.Y., Zhang, W.X.: Multiple granulation rough set approach to ordered information systems. Inter. J. General Syst. **41**, 475–501 (2012)

36. Lin, G.P., Liang, J.Y., Qian, Y.H.: Multigranulation rough sets: from partition to covering. Inf. Sci. **241**, 101–118 (2013)

37. Liou, J.J.H., Tzeng, G.H.: A dominance-based rough set approach to customer behavior in the airline market. Inf. Sci. **180**, 2230–2238 (2010)

38. Hu, Q.H., Yu, D.R., Guo, M.Z.: Fuzzy preference based rough sets. Inf. Sci. **180**, 2003–2022 (2010)

39. Szelag, M., Greco, S., Slowinski, R.: Variable consistency dominance-based rough set approach to preference learning in multicriteria ranking. Inf. Sci. **277**, 525–552 (2014)

40. Song, P., Liang, J.Y., Qian, Y.H.: A two-grade approach to ranking interval data. Knowl.-Based Syst. **27**, 234–244 (2012)

41. Yao, Y.Y.: An Outline of a Theory of Three-Way Decisions. In: Yao, J., Yang, Y., Słowiński, R., Greco, S., Li, H., Mitra, S., Polkowski, L. (eds.) RSCTC 2012. LNCS (LNAI), vol. 7413, pp. 1–17. Springer, Heidelberg (2012)

42. Yao, Y.Y., Wong, S.K.M., Lingras, P.: A decision-theoretic rough set model. In: Ras, Z.W., Zemankova, M., Emrich, M.L. (eds.) Methodologies for Intelligent Systems, vol. 5, pp. 17–25. North-Holland, New York (1990)

43. Greco, S., Słowiński, R., Yao, Y.Y.: Bayesian Decision Theory for Dominance-Based Rough Set Approach. In: Yao, J.T., Lingras, P., Wu, W.-Z., Szczuka, M.S., Cercone, N.J., Ślęzak, D. (eds.) RSKT 2007. LNCS (LNAI), vol. 4481, pp. 134–141. Springer, Heidelberg (2007)

44. Herbert, J.P., Yao, J.T.: Game-Theoretic Risk Analysis in Decision-Theoretic Rough Sets. In: Wang, G., Li, T., Grzymala-Busse, J.W., Miao, D., Skowron, A., Yao, Y. (eds.) RSKT 2008. LNCS (LNAI), vol. 5009, pp. 132–139. Springer, Heidelberg (2008)

45. Liang, D.C., Liu, D.: Deriving three-way decisions from intuitionistic fuzzy decision theoretic rough sets. Inf. Sci. **200**, 28–48 (2015)

46. Yao, Y.Y.: Granular Computing and Sequential Three-Way Decisions. In: Lingras, P., Wolski, M., Cornelis, C., Mitra, S., Wasilewski, P. (eds.) RSKT 2013. LNCS (LNAI), vol. 8171, pp. 16–27. Springer, Heidelberg (2013)

47. Wang, B.L., Liang, J.Y.: A Novel Intelligent Multi-attribute Three-Way Group Sorting Method Based on Dempster-Shafer Theory. In: Miao, D.Q., Pedrycz, W., Slezak, D., Peters, G., Hu, Q., Wang, R. (eds.) RSKT 2014. LNCS (LNAI), vol. 8818, pp. 789–800. Springer, Heidelberg (2014)

48. Qian, Y.H., Zhang, H., Sang, Y.L., Liang, J.Y.: Multi-granulation decision-theoretic rough sets. Int. J. Approx. Reason. **55**, 225–237 (2014)

49. Liang, J.Y., Wang, B.L.: Rough set based multi-attribute group decision making model. In: Jia, X.Y., Shang, L., Zhou X. Z. et al. Three-way Decision Theory and Applications, pp. 131–148. Nanjing University Press, Nanjing (2012)

50. Pang, J.F., Liang, J.Y.: Evaluation of the results of multi-attribute group decision-making with linguistic information. OMEGA **40**, 294–301 (2012)

On Generalized Decision Functions: Reducts, Networks and Ensembles

Dominik Ślęzak[1,2]([⊠])

[1] Institute of Mathematics, University of Warsaw, Ul. Banacha 2, 02-097
Warsaw, Poland
slezak@mimuw.edu.pl,slezak@infobright.com
http://www.dominikslezak.org
[2] Infobright Inc., Poland, Ul. Krzywickiego 34 Pok. 219, 02-078 Warsaw, Poland

Abstract. We summarize our observations on utilizing generalized decision functions to define dependencies between attributes in decision systems. We refer to well-known criteria for attribute selection and less-known results linking generalized decisions with the notions of multivalued dependency and conditional independence. We formulate the problem of finding the simplest ensembles of subsets of attributes which allow to retrieve original decision values of considered objects by intersecting the sets of possible decisions induced by particular attributes.

Keywords: Rough sets · Generalized decision functions · Decision reducts

1 Introduction

Generalized decision function is one of the fundamental notions of rough sets [1,2]. It is used to characterize decision reducts in inconsistent decision systems, to express uncertainty corresponding to rough set approximations of decision classes and so on. In this paper, we recall some properties and approaches related to this slightly forgotten but very important notion. We also present new results concerning decomposition and synthesis of decision systems which lead toward novel opportunities in the area of rough-set-based classifier ensembles.

Sections 2–4 gather definitions and facts which are already known or remain simple modifications of already published theorems. Section 5 introduces a new kind of approximate decision reducts based on generalized decisions. Sections 6–8 refer to our previous research on generalized-decision-based criteria for decomposing attribute sets in decision systems [3,4], now enriched by new specification of decomposition optimization problem, its complexity characteristics, Boolean representation and discussion on possible heuristic solutions. Finally, Sect. 9 outlines some ideas how to define and use generalized decisions for large data sets with complex non-categorical attributes and concludes the paper.

Partially supported by Polish National Science Centre grants DEC-2012/05/B/-ST6/03215 and DEC-2013/09/B/ST6/01568, and by Polish National Centre for Research and Development grants PBS2/B9/20/2013 and O ROB/0010/03/001.

© Springer International Publishing Switzerland 2015
Y. Yao et al. (Eds.): RSFDGrC 2015, LNAI 9437, pp. 13–23, 2015.
DOI: 10.1007/978-3-319-25783-9_2

2   Generalized Decision Functions

We assume that a data set is represented by a decision system $\mathbb{A} = (U, A \cup D)$, where U is a set of objects, and A and D are their conditional and decision attributes, respectively [1,5]. For $B \subseteq A \cup D$, we denote by $B(u)$ a vector of values of $u \in U$ over B. For simplicity we assume that values are categorical, so it is reasonable to describe data using equality-based conditions.

Definition 1. *Let decision system* $\mathbb{A} = (U, A \cup D)$ *be given. For an object* $u \in U$ *and an attribute subset* $B \subseteq A$, *generalized decision takes a form of a set* $\partial_{D/B}(u) = \{D(u') : u' \in [u]_B\}$, *where* $[u]_B = \{u' \in U : B(u') = B(u)\}$ *denotes indiscernibility class of* u *induced by* B.

Definition 2. *Let* $\mathbb{A} = (U, A \cup D)$ *be given. We say that* $B \subseteq A$ *is a* ∂-*decision superreduct, if and only if one of the following equivalent conditions holds:*

$$\forall_{u \in U} \; \partial_{D/B}(u) = \partial_{D/A}(u) \;\; or \;\; \forall_{u,u' \in U} \; \partial_{D/A}(u) \neq \partial_{D/A}(u') \Rightarrow B(u) \neq B(u')$$

We say that B *is a* ∂-*decision reduct, if and only if it is a* ∂-*decision superreduct and it has no proper subsets that are* ∂-*decision superreducts.*

The following relationship shows that generalized decisions allow to construct rough set approximations of decision classes and their set-theoretic sums.

Proposition 1. *[4] Let* $\mathbb{A} = (U, A \cup D)$ *be given. Consider an arbitrary subset* $X \subseteq U$ *which is definable by* D, *i.e., such that it is possible to represent it as a set-theoretic sum of some indiscernibility classes induced by* D. *Consider the following rough set approximations of* X *induced by a subset* $B \subseteq A$:

$$\underline{B}(X) = \{u \in U : [u]_B \subseteq X\} \quad \overline{B}(X) = \{u \in U : [u]_B \cap X \neq \emptyset\}$$

Then B *is a* ∂-*decision reduct, if and only if for each* X *definable by* D *we have* $\underline{B}(X) = \underline{A}(X)$ *and* $\overline{B}(X) = \overline{A}(X)$, *and for each proper subset of* B *at least one of those equalities does not hold for some subset* $X \subseteq U$ *definable by* D.

Generalized decisions are also related to other methods of expressing dependencies in data. For example, let us consider their correspondence to the notion of a multivalued dependency which is widely known in relational databases [6]. Let us reformulate this classical notion in terms of decision systems.

Definition 3. *Let* $\mathbb{A} = (U, A \cup D)$ *be given. For subsets* $C \subseteq B \subseteq A$, *we say that an embedded multivalued dependency* $C \twoheadrightarrow D|B$ *holds, if and only if for each* $u, u' \in U$ *such that* $C(u) = C(u')$, *there are objects* $x, x' \in U$ *such that* $B(x) = B(u)$, $B(x') = B(u')$, $D(x) = D(u')$ *and* $D(x') = D(u)$. *If* $B = A$, *we use simplified notation* $C \twoheadrightarrow D$ *and we call it a multivalued dependency.*

The following fact shows that occurrence of ∂-decision reducts in data is directly connected to normal forms studied in the theory of relational databases.

Proposition 2. *[7] Let* $\mathbb{A} = (U, A \cup D)$ *be given. Multivalued dependency* $B \twoheadrightarrow D$ *holds in* $\mathbb{A}$, *if and only if* B *is a* ∂-*decision superreduct in* $\mathbb{A}$.

3 Simplified Conditional Independence

Operating with generalized decision functions does not require a strict distinction between conditions and decisions. Below we assume that attributes in A and D can occur in different configurations. Therefore, for simplicity, in this section we use notation $\mathbb{A} = (U, A)$ instead of $\mathbb{A} = (U, A \cup D)$.

Definition 4. *Let* $\mathbb{A} = (U, A)$ *be given. Consider arbitrary pairwise disjoint subsets* $B_1, B_2, B_3 \subseteq A$. *We say that* B_1 *is* ∂-*independent from* B_3 *subject to* B_2, *denoted as* $B_1|B_2|B_3$, *if and only if the following holds:*

$$\forall_{u \in U} \; \partial_{B_1/B_2}(u) = \partial_{B_1/B_2 \cup B_3}(u)$$

One can treat the above as a kind of simplified independence statement which – unlikely in probabilistic calculus – focuses only on a possibility of occurrence of particular combinations of values. Below we recall some properties of ∂-independence, as reported in [4]. By analogy to Proposition 2, they are equivalent to classical properties of embedded multivalued dependencies [6].

Proposition 3. *[4] Let* $\mathbb{A} = (U, A)$ *be given. Consider arbitrary pairwise disjoint subsets* $B_1, B_2, B_3, B_4 \subseteq A$. *We have the following:*

$$B_1|B_2|B_3 \cup B_4 \Rightarrow B_1|B_2|B_3 \qquad B_1|B_2|B_3 \cup B_4 \Rightarrow B_1|B_2 \cup B_3|B_4$$
$$B_1|B_2|B_3 \Rightarrow B_3|B_2|B_1 \qquad B_1|B_2 \cup B_3|B_4 \wedge B_1|B_2|B_3 \Rightarrow B_1|B_2|B_3 \cup B_4$$

In probabilistic reasoning, analogous properties are called decomposition, weak union, symmetry and contraction [8]. Let us now consider ∂-related version of graphical representation of conditional independence statements.

Definition 5. *Let* $\mathbb{A} = (U, A)$ *be given. We say that a directed acyclic graph* $\mathbb{G} = (A, E)$ *is a* ∂-*map for* $\mathbb{A}$, *if and only if, for each* $B_1, B_2, B_3 \subseteq A$, *if* B_2 *d-separates* B_3 *from* B_1 – *i.e., each path between* B_1 *and* B_3 *is either covered by* B_2 *or contains a fragment* $\rightarrow a \leftarrow$, *where* a *is not in* B_2 *and has no directed path leading to any element of* B_2 – *then* $B_1|B_2|B_3$ *holds in* $\mathbb{A}$.

Efficiency of a ∂-map – i.e., the amount of ∂-independencies $B_1|B_2|B_3$ that it encodes graphically – grows if we manage to decrease the amount of its edges. This leads to the following optimization problem whose complexity can be proved using the same technique as described for Bayesian networks in [9].

Theorem 1. *The problem of finding, for an arbitrary input decision system* $\mathbb{A} = (U, A)$, *a* ∂-*map with minimum number of edges is NP-hard.*

The following fact – which is again analogous to Bayesian networks – shows that construction of a (sub-)optimal ∂-map can be based on heuristic algorithms searching for ∂-decision reducts along a predefined order over A.

Theorem 2. *[4] Let* $\mathbb{A} = (U, A)$ *be given. Consider an arbitrary linear order over* A *and, for* $a \in A$, *denote by* Π_a *a set of all attributes preceding* a *in that order. Consider a directed acyclic graph* $\mathbb{G} = (A, E)$ *and put* $\pi_a = \{b \in A : (b, a) \in E\}$. *If for each* $a \in A$ *there is inclusion* $\pi_a \subseteq \Pi_a$ *and for each* $u \in U$ *there is equality* $\partial_{\{a\}/\pi_a}(u) = \partial_{\{a\}/\Pi_a}(u)$, *then* $\mathbb{G}$ *is a* ∂-*map for* $\mathbb{A}$.

4   Generalized Decision Measures

From now on, we will assume a fixed set of decisions. We go back to notation $\mathbb{A} = (U, A \cup D)$. Moreover, for simplicity, we will write ∂_B instead of $\partial_{D/B}$. The following measures can be used to evaluate subsets of attributes.

Definition 6. *Let $\mathbb{A} = (U, A \cup D)$ be given. Functions $g_\partial, e_\partial : 2^A \to (0, 1]$ and $h_\partial : 2^A \to [0, +\infty)$ are defined as follows, for each $B \subseteq A$:*

$$g_\partial(B) = \frac{1}{|U|} \sum_{u \in U} \frac{1}{|\partial_B(u)|} \quad e_\partial(B) = \frac{1}{|U|} \sum_{u \in U} \frac{1}{2^{|\partial_B(u)|-1}} \quad h_\partial(B) = \sum_{u \in U} \frac{\log |\partial_B(u)|}{|U|}$$

For subsets $C \subseteq B \subseteq A$, there are always inequalities $g_\partial(C) \leq g_\partial(B)$, $e_\partial(C) \leq e_\partial(B)$ and $h_\partial(C) \geq h_\partial(B)$. Moreover, equalities $g_\partial(B) = 1$, $e_\partial(B) = 1$ and $h_\partial(B) = 0$ hold, if and only if $B \subseteq A$ determines D within U, i.e., all generalized decisions induced by B are singletons. Last but not least, $B \subseteq A$ is a ∂-decision reduct, if and only if $g_\partial(B) = g_\partial(A)$, $e_\partial(B) = e_\partial(A)$ and $h_\partial(B) = h_\partial(A)$, and there are no proper subsets of B which satisfy the same equalities.

g_∂ and h_∂ can be interpreted as related to gini index and information gain measures [10]. Moreover, e_∂ satisfies the following property which is interesting especially when we recall interpretation of lower and upper approximations as belief and plausibility functions in the theory of evidence [11].

Proposition 4. *[4] Let $\mathbb{A} = (U, A \cup D)$ be given. For every $B \subseteq A$ we have:*

$$e_\partial(B) = 1 - \frac{1}{|Def(D)|} \sum_{X \in Def(D)} \left(\frac{|\overline{B}(X)|}{|U|} - \frac{|\underline{B}(X)|}{|U|} \right)$$

where $Def(D)$ gathers all subsets $X \subseteq U$ that are definable by D.

All above measures can be further utilized to specify criteria for deriving minimal subsets of attributes which keep approximately the same level of information about decisions as the whole set of conditional attributes.

Definition 7. *Let $\mathbb{A} = (U, A \cup D)$ be given. Consider an approximation threshold $\varepsilon \in [0, 1)$. We say that $B \subseteq A$ is a $(g_\partial, \varepsilon)$-decision reduct, an $(e_\partial, \varepsilon)$-decision reduct and an $(h_\partial, \varepsilon)$-decision reduct, if and only if, respectively*

$$g_\partial(B) \geq (1 - \varepsilon) g_\partial(A) \quad e_\partial(B) \gtrless (1 - \varepsilon) e_\partial(A) \quad h_\partial(B) \leq h_\partial(A) + \log \frac{1}{1-\varepsilon}$$

and there are no proper subsets of B holding analogous inequalities.

The following result was proved for the case of g_∂ in [12]. The case of e_∂ can be shown in almost the same way. The case of h_∂ can be proved using exactly the same technique as described for approximate entropy reducts in [13].

Theorem 3. *Let $\varepsilon \in [0, 1)$ be given. The problems of finding a $(g_\partial, \varepsilon)$-decision reduct, an $(e_\partial, \varepsilon)$-decision reduct and an $(h_\partial, \varepsilon)$-decision reduct with minimum number of attributes for an arbitrary input decision system are NP-hard.*

The above characteristics can be further strengthened toward inapproximability theorems using mathematical apparatus introduced in [14]. There is also an ongoing research aimed at utilizing measures such as those discussed in this section to develop models of approximate conditional independence [7].

5   Embedded Decision Reducts

The notion of a ∂-decision reduct remains in the heart of rough-set-based methodology of data analysis [1]. The notions of approximate ∂-decision reducts discussed in the previous section make it more flexible with respect to inconsistencies and noises in data, allowing to use approximation thresholds to tune a balance between model generality and validity [4]. On the other hand, it is not always so obvious how to choose the level of $\varepsilon \in [0, 1)$. Moreover, as discussed in the next section, setting up an explicit threshold is not always necessary. Consequently, let us propose an alternative formulation of a subset of attributes that approximately maintains original ∂-based information.

Definition 8. *Let $\mathbb{A} = (U, A \cup D)$ be given. We say that subset $B \subseteq A$ is an embedded ∂-decision reduct, if and only if for every proper subset $C \subsetneq B$ there exists at least one $u \in U$ such that $\partial_C(u) \neq \partial_B(u)$.*

Below we outline basic properties of this new notion. Firstly, let us focus on its relationship to embedded multivalued dependencies.

Proposition 5. *Let $\mathbb{A} = (U, A \cup D)$ be given. Subset $B \subseteq A$ is an embedded ∂-decision reduct, if and only if there is no proper subset $C \subsetneq B$ such that embedded multivalued dependency $C \twoheadrightarrow D|B$ holds in $\mathbb{A}$.*

Proof. As we did in Sects. 2 and 3, we refer to the fact that, for a given $C \subsetneq B$, $C \twoheadrightarrow D|B$ is equivalent to $\forall_{u \in U} \, \partial_C(u) = \partial_B(u)$.

Another straightforward property shows a correspondence between embedded ∂-decision reducts and the formulations in Definition 7:

Proposition 6. *Let $\mathbb{A} = (U, A \cup D)$ be given. Subset $B \subseteq A$ is an embedded ∂-decision reduct, if and only if it is a $(g_\partial, \varepsilon)$-decision reduct, $(e_\partial, \varepsilon)$-decision reduct and $(h_\partial, \varepsilon)$-decision reduct for approximation thresholds $\varepsilon = 1 - \frac{g_\partial(B)}{g_\partial(A)}$, $\varepsilon = 1 - \frac{e_\partial(B)}{e_\partial(A)}$ and $\varepsilon = 1 - 2^{-(h_\partial(B) - h_\partial(A))}$, respectively.*

Proof. Let us consider g_∂ as an example. Each $B \subseteq A$ satisfies equality $g_\partial(B) = (1 - \varepsilon)g_\partial(A)$ for $\varepsilon = 1 - \frac{g_\partial(B)}{g_\partial(A)}$. Assume that there is $C \subsetneq B$ such that $C \twoheadrightarrow D|B$ holds, i.e., we have $\forall_{u \in U} \, \partial_C(u) = \partial_B(u)$. This would mean that $g_\partial(C) = g_\partial(B)$, so also $g_\partial(C) = (1 - \varepsilon)g_\partial(A)$. Thus, B would not be a $(g_\partial, \varepsilon)$-decision reduct. Oppositely, assume that there is no $C \subsetneq B$ such that $C \twoheadrightarrow D|B$ holds. This means that, for an arbitrary $C \subsetneq B$, there is $u \in U$ such that $\partial_C(u) \neq \partial_B(u)$. This leads to sharp inequality $g_\partial(C) < g_\partial(B)$ which means that $g_\partial(C) \geq (1 - \varepsilon)g_\partial(A)$ cannot be satisfied, i.e., B is a $(g_\partial, \varepsilon)$-decision reduct.

The above result can be also rephrased as follows, in order to emphasize that being an embedded ∂-decision reduct is something more generic and worth investigating regardless of fixed approximation thresholds.

Proposition 7. *Let* $\mathbb{A} = (U, A \cup D)$ *and threshold* $\varepsilon \in [0, 1)$ *be given. If subset* $B \subseteq A$ *is a* $(g_\partial, \varepsilon)$-*decision reduct,* $(e_\partial, \varepsilon)$-*decision reduct or* $(h_\partial, \varepsilon)$-*decision reduct, then it is also an embedded* ∂-*decision reduct.*

Proof. It is analogous to the proof of Proposition 6.

6   Ensembles of Complementary Reducts

For a decision system $\mathbb{A} = (U, A \cup D)$, cardinalities of generalized decisions reflect a kind of imprecision of describing D by particular subsets $B \subseteq A$. One can use the content of $\mathbb{A}$ to generate rules with antecedents based on values of B over objects $u \in U$ and consequents pointing at disjunctions of possible decisions gathered in sets $\partial_B(u)$. Such rules can be applied to classify objects outside U, i.e., to assign decisions based on their values observed over B. According to well-known principles of data-based induction, rules with less conditions (thus based on smaller attribute subsets) are likely to provide more efficient classification models, if only cardinalities of generalized decisions do not grow too much comparing to more specific rules generated using the whole A.

In [3, 4], it was observed that different subsets $B \subseteq A$ can help each other to build more precise classifications by intersecting their corresponding sets $\partial_B(u)$. For example, for rules $(a = v_a) \wedge (b = v_b) \Rightarrow (d = 1) \vee (d = 2)$ and $(b = v_b) \wedge (c = v_c) \Rightarrow (d = 1) \vee (d = 3)$, if a new object satisfies $(a = v_a) \wedge (b = v_b) \wedge (c = v_c)$ over $a, b, c \in A$, then we can label it with decision $(d = 1)$. This style of utilizing rules based on generalized decision functions can be to some extent interpreted within Gentzen systems [15]. Synthesis of such decision sets for new objects can be also further tuned using feedforward neural networks [16].

Definition 9. *Let* $\mathbb{A} = (U, A \cup D)$ *be given. We say that subsets* $B_1, ..., B_m \subseteq A$, $m \geq 0$, *are an ensemble of complementary embedded* ∂-*decision reducts, if and only if the following holds:*

$$\forall_{u \in U} \; \bigcap_{i=1}^{m} \partial_{B_i}(u) = \partial_A(u)$$

and it is impossible to replace any B_i, $i = 1, ..., m$, *with its proper subset without losing the above condition.*

Normally, one should expect inclusions $\bigcap_{i=1}^{m} \partial_{B_i}(u) \supseteq \partial_A(u)$. Requiring perfect equalities means that each subset of attributes can lose some ∂-related information – i.e., we may observe $\partial_{B_i}(u) \supsetneq \partial_A(u)$ – but the same ingredients of information cannot be lost by all $B_1, ..., B_m$ in the same time.

Thinking about $B_1, ..., B_m$ as embedded ∂-decision reducts follows the fact that any $B_i \subseteq A$ which does not satisfy conditions of Definition 8 could be replaced with its smaller subset inducing the same generalized decisions. On the other hand, even if all subsets $B_1, ..., B_m$ are indeed embedded ∂-decision reducts, then it might be still possible to replace some of them with smaller components – being embedded ∂-decision reducts too – without changing the overall outcome of intersection $\bigcap_{i=1}^{m} \partial_{B_i}(u)$ for every $u \in U$.

In summary, subsets $B_1, ..., B_m$ need to be minimal with respect to partial ability to describe decisions by each single B_i and joint ability to avoid the same inconsistencies by all components. Comparing to ensembles of classifiers based on approximate decision reducts [17,18], now we do not need to explicitly tune any thresholds. Actually, the most useful solutions may correspond to subsets of attributes that are $(g_\partial, \varepsilon_i)$-decision reducts, $(e_\partial, \varepsilon_i)$-decision reducts or $(h_\partial, \varepsilon_i)$-decision reducts for diverse values of $\varepsilon_i \in [0, 1)$, $i = 1, ..., m$.

7   Attribute Decomposition Problem

Let us now focus on three questions which are traditionally important for rough-set-based methods [2,5] – how to formalize criteria for extracting optimal ensembles of complementary embedded ∂-decision reducts from data, how to design heuristics aimed at searching for reasonable solutions, and whether there are any Boolean-reasoning-based representations that might help to better understand the nature of considered optimization problems.

Intuitively, the corresponding optimization problem should be stated as a task of finding possibly smallest subsets $B_1, ..., B_m$ satisfying conditions of Definition 9. Let us note that m can be arbitrarily large, if only we could decompose information within an input decision system onto a larger number of rules which are shorter, more general, maybe less precise individually but still jointly able to reconstruct valid decisions for objects in the training data.

We therefore propose to search through a space of all ensembles of complementary embedded ∂-decision reducts $B_1, ..., B_m$ for variable $m \geq 0$, paying special attention to cardinalities of their largest components along a kind of cardinality-based lexicographic order. This is because the largest subsets of attributes correspond to the largest collections of the longest rules, i.e., they affect complexity of the model more significantly than other subsets.

Definition 10. *Let* $\mathbb{A} = (U, A \cup D)$ *and two ensembles of complementary embedded ∂-decision reducts* $B_1, ..., B_m$ *and* $C_1, ..., C_n$, $m, n \geq 0$, *be given. Let us consider the following procedure:*

1. *If* $m > n$ *(m < n), add* $m - n$ *(n − m) empty sets to* $C_1, ..., C_n$ *(B$_1$, ..., B$_m$).*
2. *Sort sequences of cardinalities of attribute subsets in a descending order.*
3. *Find the first position for which sorted sequences differ from each other.*

We say that $B_1, ..., B_m$ *is simpler than* $C_1, ..., C_n$, *if and only if a value at the above-found position is lower for* $B_1, ..., B_m$ *than for* $C_1, ..., C_n$.

Let us note that the above procedure induces a linear order over ensembles of complementary embedded ∂-decision reducts for a given $\mathbb{A}$.

Theorem 4. *The problem of finding the simplest (i.e. the lowest according to the order introduced in Definition 10) ensemble of complementary embedded ∂-decision reducts for an arbitrary* $\mathbb{A} = (U, A \cup D)$ *is NP-hard.*

Proof. Let us show it by polynomial reduction of the minimum dominating set problem. Consider an undirected graph $\mathbb{G} = (V, E)$ and create binary decision system $\mathbb{A}_{\mathbb{G}} = (U_{\mathbb{G}} \cup \{u_*\}, A_{\mathbb{G}} \cup \{d\})$, where $a_v \in A_{\mathbb{G}}$ corresponding to $v \in V$ takes 1 on $u_{v'} \in U_{\mathbb{G}}$ corresponding to $v' \in V$, i.e. $a_v(u_{v'}) = 1$, if and only if $v = v'$ or $(v, v') \in E$, and where $a_v(u_*) = 0$, $d(u_{v'}) = 0$ and $d(u_*) = 1$ [5,19]. One can see that a subset $B \subseteq V$ is a dominating set in $\mathbb{G}$, if and only if it corresponds to a ∂-decision superreduct in $\mathbb{A}_{\mathbb{G}}$. Moreover, each ensemble of complementary embedded ∂-decision reducts for $\mathbb{A}_{\mathbb{G}}$ has to contain a classical ∂-decision reduct to reconstruct decision $d(u_*) = 1$. Consequently, the simplest ensemble for $\mathbb{G}$ takes a form of a single subset of attributes which is the smallest ∂-decision reduct in $\mathbb{A}_{\mathbb{G}}$, that is – the smallest dominating set in $\mathbb{G}$.

8   Heuristics and Boolean Representation

In practice it is difficult to choose upfront a number of elements for an ensemble. Thus, one can adapt top-down methods to decompose a set of attributes step by step. Let us follow an analogy to decision tree induction [10,19] and imagine a binary ∂-decomposition tree with its root representing the whole A, where each non-leaf node corresponding to $B \subseteq A$ is split onto two nodes corresponding to non-empty subsets $B_l, B_r \subseteq B$ such that $\partial_{B_l}(u) \cap \partial_{B_r}(u) = \partial_B(u)$ holds for each $u \in U$. Then, starting from a root-only tree, we can search for splits of consecutive nodes with a natural stopping criterion – a given node will remain a leaf, if and only if there are no further splits possible. One can show that the collection of all leaves of a tree created using this kind of criterion needs to correspond to attribute subsets meeting conditions of Definition 9.

In the above scenario, a solution of the problem formulated in Theorem 4 is heuristically replaced with a chain of solutions of a problem related to pairs of complementary embedded ∂-decision reducts. Such problem is NP-hard too and – referring again to [14] – one can prove its inapproximability. On the other hand, it is surely easier to design a heuristic algorithm for fixed $m = 2$ than for $m \geq 2$. In the rough set literature, a popular way to better understand complexity details and draft first solutions of an optimization problem is to encode it as a task of finding prime implicants for a data-related Boolean formula [2,19].

Proposition 8. *Let* $\mathbb{A} = (U, A \cup D)$ *and* $B \subseteq A$ *be given. Consider two sets of Boolean variables* $L_B = \{l_a : a \in B\}$ *and* $R_B = \{r_a : a \in B\}$, *where each* $a \in B$ *is assigned to* $l_a \in L_B$ *and* $r_a \in R_B$. *Define formula* τ_B^∂ *as follows:*

$$\bigwedge_{u, u_l, u_r \in U:\, D(u_l) \notin \partial_B(u) \wedge D(u_l) = D(u_r)} \left(\bigvee_{a \in B:\, a(u) \neq a(u_l)} l_a \vee \bigvee_{a \in B:\, a(u) \neq a(u_r)} r_a \right)$$

A formula α *is a prime implicant for* τ_B^∂, *if and only if it is a conjunction of some non-negated elements of* $L_B \cup R_B$ *and attribute subsets defined as* $B_l = \{a \in B : l_a \in \alpha\}$ *and* $B_r = \{a \in B : r_a \in \alpha\}$ *form a pair of minimal subsets such that the equality* $\partial_{B_l}(u) \cap \partial_{B_r}(u) = \partial_B(u)$ *holds for every* $u \in U$.

Proof. τ_B^∂ contains no negations, so its prime implicants correspond to minimal subsets of $L_B \cup R_B$ overlapping with all sets $\{l_a \in L_B : a(u) \neq a(u_l)\} \cup \{r_a \in R_B : a(u) \neq a(u_r)\}$, $u, u_l, u_r \in U$. Thus, it is enough to observe that, for any $B_l, B_r \subseteq B$ and $u \in U$, existence of $w \in \partial_{B_l}(u) \cap \partial_{B_r}(u) \setminus \partial_B(u)$ is equivalent to existence of $u_l \in [u]_{B_l}, u_r \in [u]_{B_r}$ such that $D(u_l) = D(u_r) = w \notin \partial_B(u)$.

One can treat such representation as a purely theoretical result although it does give us an insight how to heuristically derive sufficiently small pairs $B_l, B_r \subseteq B$ while constructing ∂-decomposition trees. For instance, one can start with $B_l = B_r = B$ and follow a randomly generated ordering over elements of $L_B \cup R_B$, each time attempting to remove some attribute from B_l or B_r (depending on the next element in the ordering) under the constraint that $\partial_{B_l}(u) \cap \partial_{B_r}(u) = \partial_B(u)$ still needs to hold for each $u \in U$. According to analogous studies in [18,20], repeating this procedure for a reasonable number of appropriately diversified orderings should enable to sufficiently explore a space of all possibilities.

9 Conclusions and Future Directions

We summarized basic ideas related to generalized decision functions [1,2]. We recalled their connections to other concepts of the theory of rough sets and to some other notions such as multivalued dependencies in relational databases or belief and plausibility functions in the theory of evidence [7,11]. We referred to our previous research on utilizing generalized decisions in the processes of approximate attribute reduction and attribute decomposition [3,4]. We also investigated a new optimization problem of searching for ensembles of attribute subsets which induce complementary generalized-decision-based information, including its complexity, Boolean characteristics and heuristic solutions. Efficient derivation of such ensembles from data may become a basis for new applications in the domains of data classification and knowledge representation.

Among challenges and opportunities in front of methods based on the notion of a generalized decision function, it is certainly worth mentioning a need of extending its meaning for complex non-categorical attributes. From this perspective, it is important to refer to rough-set-based approaches which replace classical indiscernibility relations with, e.g., rankings or similarities [21,22]. Although our initial analysis leads to conclusion that most of results reported in this paper will remain valid for most of non-equivalence relations considered in the rough set literature, a lot of research is still required in this area.

An emphasis should be put also on complex decision attributes. In this case, generalized decisions need to roughly describe subspaces of possible decision values rather than enumerate explicitly defined decision classes. Such rough descriptions can take a form of, e.g., intervals for numeric decisions or common prefixes for alphanumeric decisions. Ability to handle such extensions of classical generalized decisions can be useful, for instance, to accelerate data processing and data mining algorithms by letting them work with rough descriptions of bigger blocks of objects instead of precise values of particular objects [23,24].

References

1. Pawlak, Z., Skowron, A.: Rudiments of rough sets. Inf. Sci. **177**(1), 3–27 (2007)
2. Pawlak, Z., Skowron, A.: Rough sets and boolean reasoning. Inf. Sci. **177**(1), 41–73 (2007)
3. Ślęzak, D.: Decomposition and synthesis of decision tables with respect to generalized decision functions. In: Pal, S.K., Skowron, A. (eds.) Rough Fuzzy Hybridization - A New Trend in Decision Making, pp. 110–135. Springer, Singapore (1999)
4. Ślęzak, D.: Approximate Decision Reducts (in Polish). Ph.D. thesis under Supervision of A. Skowron. University of Warsaw, Poland (2002)
5. Skowron, A., Rauszer, C.: The discernibility matrices and functions in information systems. In: Słowiński, R. (ed.) Intelligent Decision Support - Handbook of Applications and Advances of the Rough Sets Theory. System Theory, Knowledge Engineering and Problem Solving, vol. 11, pp. 331–362. Kluwer, Dordrecht (1992)
6. Garcia-Molina, H., Ullman, J., Widom, J.: Database Systems: The Complete Book, 2nd edn. Prentice-Hall, Englewood Cliff (2008)
7. Ślęzak, D.: Degrees of conditional (in)dependence: a framework for approximate bayesian networks and examples related to the rough set-based feature selection. Inf. Sci. **179**(3), 197–209 (2009)
8. Pearl, J.: Probabilistic Reasoning in Intelligent Systems: Networks of Plausible Inference. Morgan Kaufmann, San Mate (1988)
9. Betliński, P., Ślęzak, D.: The problem of finding the sparsest bayesian network for an input data set is NP-hard. In: Chen, L., Felfernig, A., Liu, J., Raś, Z.W. (eds.) ISMIS 2012. LNCS, vol. 7661, pp. 21–30. Springer, Heidelberg (2012)
10. Rokach, L., Maimon, O.Z.: Data Mining with Decision Trees: Theory and Applications. World Scientific, Singapore (2008)
11. Skowron, A., Grzymała-Busse, J.W.: From rough set theory to evidence theory. In: Yager, R.R., Kacprzyk, J., Fedrizzi, M. (eds.) Advances in the Dempster-Shafer Theory of Evidence, pp. 193–236. Wiley, New York (1994)
12. Ślęzak, D.: Normalized decision functions and measures for inconsistent decision tables analysis. Fundamenta Informaticae **44**(3), 291–319 (2000)
13. Ślęzak, D.: Approximate entropy reducts. Fundamenta Informaticae **53**(3–4), 365–390 (2002)
14. Moshkov, M.J., Piliszczuk, M., Zielosko, B.: Partial Covers, Reducts and Decision Rules in Rough Sets - Theory and Applications. Studies in Computational Intelligence, vol. 145. Springer, Heidelberg (2008)
15. Kleene, S.C.: Mathematical Logic. Wiley, New York (1967)
16. Szczuka, M.S., Ślęzak, D.: Feedforward neural networks for compound signals. Theor. Comput. Sci. **412**(42), 5960–5973 (2011)
17. Widz, S., Ślęzak, D.: Rough set based decision support - models easy to interpret. In: Peters, G., Lingras, P., Ślęzak, D., Yao, Y. (eds.) Rough Sets: Selected Methods and Applications in Management & Engineering. Advanced Information and Knowledge Processing, pp. 95–112. Springer, London (2012)
18. Wróblewski, J.: Adaptive aspects of combining approximation spaces. In: Pal, S.K., Polkowski, L., Skowron, A. (eds.) Rough-Neural Computing - Techniques for Computing with Words. Cognitive Technologies, pp. 139–156. Springer, Heidelberg (2003)
19. Nguyen, H.S.: Approximate boolean reasoning: foundations and applications in data mining. In: Peters, J.F., Skowron, A. (eds.) Transactions on Rough Sets V. LNCS, vol. 4100, pp. 334–506. Springer, Heidelberg (2006)

20. Ślęzak, D.: Rough sets and functional dependencies in data: foundations of association reducts. In: Gavrilova, M.L., Tan, C.J.K., Wang, Y., Chan, K.C.C. (eds.) Transactions on Computational Science V. LNCS, vol. 5540, pp. 182–205. Springer, Heidelberg (2009)
21. Dembczyński, K., Greco, S., Kotłowski, W., Słowiński, R.: Optimized generalized decision in dominance-based rough set approach. In: Yao, J.T., Lingras, P., Wu, W.-Z., Szczuka, M.S., Cercone, N.J., Ślęzak, D. (eds.) RSKT 2007. LNCS (LNAI), vol. 4481, pp. 118–125. Springer, Heidelberg (2007)
22. Stefanowski, J., Tsoukiás, A.: Incomplete information tables and rough classification. Comput. Intell. **17**(3), 545–566 (2001)
23. Ślęzak, D., Synak, P., Wojna, A., Wróblewski, J.: Two database related interpretations of rough approximations: data organization and query execution. Fundamenta Informaticae **127**(1–4), 445–459 (2013)
24. Ganter, B., Meschke, C.: A formal concept analysis approach to rough data tables. In: Peters, J.F., Skowron, A., Sakai, H., Chakraborty, M.K., Slezak, D., Hassanien, A.E., Zhu, W. (eds.) Transactions on Rough Sets XIV. LNCS, vol. 6600, pp. 37–61. Springer, Heidelberg (2011)

Formalization of Medical Diagnostic Rules

Shusaku Tsumoto$^{(\boxtimes)}$ and Shoji Hirano

Faculty of Medicine, Department of Medical Informatics, Shimane University,
89-1 Enya-cho Izumo, Matsue 693-8501, Japan
{tsumoto,hirano}@med.shimane-u.ac.jp
http://www.med.shimane-u.ac.jp/med_info/tsumoto/

Abstract. This paper dicusses formalization of medical diagnostic rules which is closely related with rough set rule model. The important point is that medical diagnostic reasoning is characterized by focusing mechanism, composed of screening and differential diagnosis, which corresponds to upper approximation and lower approximation of a target concept. Furthermore, this paper focuses on detection of complications, which can be viewed as relations between rules of different diseases.

Keywords: Rough sets · Medical diagnostic rules · Focusing mechanism · Exclusive rules · Inclusive rules · Complications detection

1 Introduction

Classical medical diagnosis of a disease assumes that a disease is defined as a set of symptoms, in which the basic idea is *symptomatology*. Symptomatolgy had been a major diagnostic rules before laboratory and radiological examinations. Although the power of symptomatology for differential diagnosis is now lower, it is true that change of symptoms are very important to evaluate the status of chronic status. Even when laboratory examinations cannot detect the change of patient status, the set of symptoms may give important information to doctors.

Symptomatological diagnostic reasoning is conducted as follows. First, doctors make physical examations to a patient and collect the observed symptoms. If symptoms are observed enough, a set of symptoms give some confidence to diagnosis of a corresponding disease. Thus, correspondence between a set of manifestations and a disease will be useful for differential diagnosis. Moreover, similarity of diseases will be inferred by sets of symptoms.

The author has been discussed modeling of symptomatological diagnostic reasoning by using the core ideas of rough sets since [16]: selection of candidates (screening) and differential diagnosis are closely related with diagnostic rules obtained by upper and lower approximations of a given concept. Thus, this paper dicusses formalization of medical diagnostic rules which is closely related with rough set rule model. The important point is that medical diagnostic reasoning

---

This research is supported by Grant-in-Aid for Scientific Research (B) 15H2750 from Japan Society for the Promotion of Science(JSPS).

Y. Yao et al. (Eds.): RSFDGrC 2015, LNAI 9437, pp. 24–35, 2015.
DOI: 10.1007/978-3-319-25783-9_3

is characterized by focusing mechanism, composed of screening and differential diagnosis, which corresponds to upper approximation and lower approximation of a target concept. Furthremore, this paper focuses on detection of complications, which can be viewed as relations between rules of different diseases.

The paper is organized as follows. Section 2 shows characteristics of medical diagnostic process. Section 3 introduces rough sets and basic definition of probabilistic rules. Section 4 gives two style of formalization of medical diagnostic rules. The first one is a deterministic model, which corresponds to Pawlak's rough set model. And the other one gives an extension of the above ideas in probabilistic domain, which can be viewed as application of variable precision rough set model [17]. Section 5 proposes a new rule induction model, which includes formalization of rules for detection of complications. Section 6 discussed what has not been achieved yet. Finally, Sect. 7 concludes this chapter

2   Background: Medical Diagnostic Process

This section focuses on medical diagnostic process as rule-based reasoning. The fundamental discussion of medical diagnostic reasoning related with rough sets is given in [12].

2.1   RHINOS

RHINOS is an expert system which diagnoses clinical cases on headache or facial pain from manifestations. In this system, a diagnostic model proposed by Matsumura [1] is applied to the domain, which consists of the following three kinds of reasoning processes: exclusive reasoning, inclusive reasoning, and reasoning about complications.

First, exclusive reasoning excludes a disease from candidates when a patient does not have a symptom which is necessary to diagnose that disease. Secondly, inclusive reasoning suspects a disease in the output of the exclusive process when a patient has symptoms specific to a disease. Finally, reasoning about complications suspects complications of other diseases when some symptoms which cannot be explained by the diagnostic conclusion are obtained.

Each reasoning is rule-based and all the rules needed for diagnostic processes are acquired from medical experts in the following way.

Exclusive Rules. These rule correspond to exclusive reasoning. In other words, the premise of this rule is equivalent to the necessity condition of a diagnostic conclusion. From the discussion with medical experts, the following six basic attributes are selected which are minimally indispensable for 1defining the necessity condition: *1. Age, 2. Pain location, 3. Nature of the pain, 4. Severity of the pain, 5. History since onset, 6. Existence of jolt headache.* For example, the exclusive rule of common migraine is defined as:

```
In order to suspect common migraine,
the following symptoms are required:
```

```
pain location: not eyes,
nature :throbbing or persistent or radiating,
history: paroxysmal or sudden and
jolt headache: positive.
```

One of the reasons why the six attributes are selected is to solve an interface problem of expert systems: if all attributes are considered, all the symptoms should be input, including symptoms which are not needed for diagnosis. To make exclusive reasoning compact, we chose the minimal requirements only. It is notable that this kind of selection can be viewed as the ordering of given attributes, which is expected to be induced from databases. This issue is discussed later in Sect. 6.

Inclusive Rules. The premises of inclusive rules are composed of a set of manifestations specific to a disease to be included. If a patient satisfies one set, this disease should be suspected with some probability. This rule is derived by asking the medical experts about the following items for each disease: *1. a set of manifestations by which we strongly suspect a disease. 2. the probability that a patient has the disease with this set of manifestations:SI(Satisfactory Index) 3. the ratio of the patients who satisfy the set to all the patients of this disease:CI(Covering Index) 4. If the total sum of the derived CI(tCI) is equal to 1.0 then end. Otherwise, goto 5. 5. For the patients with this disease who do not satisfy all the collected set of manifestations, goto 1.* Therefore a positive rule is described by a set of manifestations, its satisfactory index (SI), which corresponds to *accuracy measure*, and its covering index (CI), which corresponds to *total positive rate*. Note that SI and CI are given empirically by medical experts.

For example, one of three positive rules for common migraine is given as follows.

```
If history: paroxysmal, jolt headache: yes,
nature: throbbing or persistent,
prodrome: no, intermittent symptom: no,
persistent time: more than 6 hours,
and location: not eye,
then common migraine is suspected with
accuracy 0.9 (SI=0.9) and this rule covers
60 percent of the total cases (CI=0.6).
```

Disease Image: Complcations Detection. This rule is used to detect complications of multiple diseases, acquired by all the possible manifestations of the disease. By the use of this rule, the manifestations which cannot be explained by the conclusions will be checked, which suggest complications of other diseases. For example, the disease image of common migraine is:

```
The following symptoms can be explained by
common migraine: pain location: any or
depressing: not or jolt headache: yes or ...
```

Therefore, when a patient who suffers from common migraine is depressing, it is suspected that he or she may also have other disease.

2.2 Focusing Mechanism

The most important process in medical differential diagnosis shown above is called a focusing mechanism [8,9]. Even in differential diagnosis of headache, medical experts should check possibilities of more than 100 candidates, though frequent diseases are 5 or 6. These candidates will be checked by past and present history, physical examinations, and laboratory examinations. In diagnostic procedures, a candidate is excluded one by one if symptoms necessary for diagnosis are not observed.

Focusing mechanism consists of the following two styles: exclusive reasoning and inclusive reasoning. Relations of this diagnostic model with another diagnostic model are discussed in [5,12], which is summarized in Fig. 1: First, exclusive reasoning excludes a disease from candidates when a patient does not have symptoms that is necessary to diagnose that disease. Second, inclusive reasoning suspects a disease in the output of the exclusive process when a patient has symptoms specific to a disease. Based on the discussion with medical experts, these reasoning processes are modeled as two kinds of rules, negative rules (or exclusive rules) and positive rules; the former corresponds to exclusive reasoning, the latter to inclusive reasoning [1].[1]

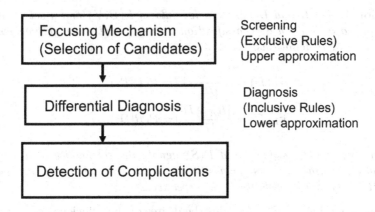

Fig. 1. Focusing mechanism.

---

[1] Implementation of detection of complications is not discussed here because it is derived after main two process, exclusive and inclusive reasoning. The way to deal with detection of complications is discussed in Sect. 5.

3   Basics of Rule Definitions

3.1   Rough Sets

In the following sections, we use the following notation introduced by Grzymala-Busse and Skowron [4], based on rough set theory [2]. Let U denote a nonempty finite set called the universe and A denote a nonempty, finite set of attributes, i.e., $a : U \to V_a$ for $a \in A$, where V_a is called the domain of a, respectively. Then a decision table is defined as an information system, $A = (U, A \cup \{d\})$. The atomic formulas over $B \subseteq A \cup \{d\}$ and V are expressions of the form $[a = v]$, called descriptors over B, where $a \in B$ and $v \in V_a$. The set $F(B, V)$ of formulas over B is the least set containing all atomic formulas over B and closed with respect to disjunction, conjunction, and negation.

For each $f \in F(B, V)$, f_A denotes the meaning of f in A, i.e., the set of all objects in U with property f, defined inductively as follows:

1. If f is of the form $[a = v]$, then $f_A = \{s \in U | a(s) = v\}$.
2. $(f \wedge g)_A = f_A \cap g_A$; $(f \vee g)_A = f_A \vee g_A$; $(\neg f)_A = U - f_a$.

3.2   Classification Accuracy and Coverage

Definition of Accuracy and Coverage. By use of the preceding framework, classification accuracy and coverage, or true positive rate are defined as follows.

Definition 1. *Let R and D denote a formula in $F(B, V)$ and a set of objects that belong to a decision d. Classification accuracy and coverage(true positive rate) for $R \to d$ is defined as:*

$$\alpha_R(D) = \frac{|R_A \cap D|}{|R_A|}(= P(D|R)), \tag{1}$$

$$\kappa_R(D) = \frac{|R_A \cap D|}{|D|}(= P(R|D)), \tag{2}$$

where $|S|$, $\alpha_R(D)$, $\kappa_R(D)$, and $P(S)$ denote the cardinality of a set S, a classification accuracy of R as to classification of D, and coverage (a true positive rate of R to D), and probability of S, respectively.

It is notable that $\alpha_R(D)$ measures the degree of the sufficiency of a proposition, $R \to D$, and that $\kappa_R(D)$ measures the degree of its necessity. For example, if $\alpha_R(D)$ is equal to 1.0, then $R \to D$ is true. On the other hand, if $\kappa_R(D)$ is equal to 1.0, then $D \to R$ is true. Thus, if both measures are 1.0, then $R \leftrightarrow D$.

3.3   Probabilistic Rules

By use of accuracy and coverage, a probabilistic rule is defined as:

$$R \xrightarrow{\alpha, \kappa} d \quad s.t. \ R = \wedge_j [a_j = v_k], \alpha_R(D) \ \delta_\alpha \text{ and } \kappa_R(D) \ \delta_\kappa, \tag{3}$$

where D denotes a set of samples that belong to a class d. If the thresholds for accuracy and coverage are set to high values, the meaning of the conditional part of probabilistic rules corresponds to the highly overlapped region. This rule is a kind of probabilistic proposition with two statistical measures, which is an extension of Ziarko's variable precision model (VPRS) [17].[2]

It is also notable that both a positive rule and a negative rule are defined as special cases of this rule, as shown in the next sections.

4   Formalization of Medical Diagnostic Rules

4.1   Deterministic Model

Positive Rules. A positive rule is defined as a rule supported by only positive examples. Thus, the accuracy of its conditional part to a disease is equal to 1.0. Each disease may have many positive rules. If we focus on the supporting set of a rule, it corresponds to a subset of the lower approximation of a target concept, which is introduced in rough sets [2]. Thus, a positive rule is defined as:

$$R \to d \quad s.t. \quad R = \wedge_j [a_j = v_k], \quad \alpha_R(D) = 1.0 \tag{4}$$

where D denotes a set of samples that belong to a class d.

This positive rule is often called a deterministic rule. However, we use the term, positive (deterministic) rules, because a deterministic rule supported only by negative examples, called a negative rule, is introduced below.

Negative Rules. The important point is that a negative rule can be represented as the contrapositive of an exclusive rule [8]. An exclusive rule is defined as a rule whose supporting set covers all the positive examples. That is, the coverage of the rule to a disease is equal to 1.0. That is, an exclusive rule represents the necessity condition of a decision. The supporting set of an exclusive rule corresponds to the upper approximation of a target concept, which is introduced in rough sets [2]. Thus, an exclusive rule is defined as:

$$R \to d \quad s.t. \quad R = \vee_j [a_j = v_k], \quad \kappa_R(D) = 1.0, \tag{5}$$

where D denotes a set of samples that belong to a class d.

Next, let us consider the corresponding negative rules in the following way. An exclusive rule should be described as:

$$d \to \vee_j [a_j = v_k],$$

because the condition of an exclusive rule corresponds to the necessity condition of conclusion d. Since a negative rule is equivalent to the contrapositive of an exclusive rule, it is obtained as:

$$\wedge_j \neg [a_j = v_k] \to \neg d,$$

---

[2] This probabilistic rule is also a kind of *rough modus ponens* [3].

which means that if a case does not satisfy any attribute value pairs in the condition of a negative rule, then we can exclude a decision d from candidates.

Thus, a negative rule is represented as:

$$\wedge_j \neg[a_j = v_k] \rightarrow \neg d \quad s.t. \quad \forall[a_j = v_k]\kappa_{[a_j=v_k]}(D) = 1.0, \qquad (6)$$

where D denotes a set of samples that belong to a class d.

Negative rules should also be included in a category of deterministic rules, because their coverage, a measure of negative concepts, is equal to 1.0. It is also notable that the set supporting a negative rule corresponds to a subset of negative region, which is introduced in rough sets [2].

In summary, positive and negative rules correspond to positive and negative regions defined in rough sets. Figure 2 shows the Venn diagram of those rules.

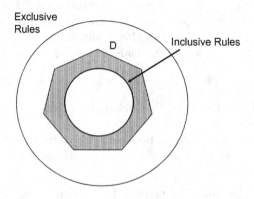

Fig. 2. Venn diagram of exclusive and positive rules.

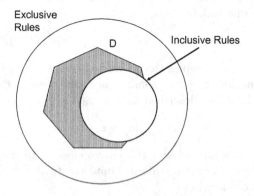

Fig. 3. Venn diagram of exclusive and inclusive rules.

4.2   Probabilistic Model

Although the above deterministic model exactly corresponds to original Pawlak rough set model, rules for differential diagnosis is strict for clinical setting, because clinical diagnosis may include elements of uncertainty.[3] Tsumoto [5] relaxes the condition of positive rules and defines an inclusive rules, which models the inclusive rules of RHINOS model. The definition is almost the same as probabilistic rules defined in Sect. 3, except for the constraints for accuracy: the threshold for accuracy is sufficiently high. Thus, the definitions of rules are summarized as follows.

Exclusive Rules

$$R \to d \quad s.t. \ R = \vee_j [a_j = v_k], \quad (s.t. \ \kappa_{[a_j = v_k]}(D) > \delta_\kappa) \quad \kappa_R(D) = 1.0. \quad (7)$$

Inclusive Rules

$$R \overset{\alpha,\kappa}{\to} d \quad s.t. \ R = \wedge_j [a_j = v_k], \quad \alpha_R(D) > \delta_\alpha \text{ and } \kappa_R(D) > \delta_\kappa. \quad (8)$$

In summary, positive and negative rules correspond to positive and negative regions defined in variable rough set model [17]. Figure 3 shows the Venn diagram of those rules.

Tsumoto introduces an algorithm for induction of exclusive and inclusive rules as PRIMEROSE-REX and conducted experimental validation and compared induced results with rules manually acquired from medical experts [5]. The results show that the rules do not include components of hierarchical diagnostic reasoning. Medical experts classify a set of diseases into groups of similar diseases and their diagnostic reasoning is multi-staged: first, different groups of diseases are checked, then final differential diagnosis is performed with the selected group of diseases. In order to extend the method into induction of hierarchical diagnostic rules, one of the authors proposes several approach to mining taxonomy from a dataset in [6,7,11].

5   New Rule Induction Model

The former rule induction models do not include reasoning about detection of complications, which is introduced as *disease image* as shown in Sect. 1. The core idea is that medical experts detect the symptoms which cannot be frequently occurred in the final diagnostic candidates. For example, let us assume that a patient suffering from muscle contraction headache, who usually complains of persistent pain, also complains of paroxysmal pain, say he/she feels a strong pain every one month. The situation is unusual and since paxosysmal pain is frequently observed by migraine, medical experts suspect that he/she suffers from

---

[3] However, determinic rule induction model is still powerful in knowledge discovery context as shown in [10].

muscle contraction headache and common migraine. Thus, a set of symptoms which are not useful for diagnosis of a disease may be important if they belong to the set of symptoms frequently manifested in other diseases. In other means, such the set of symptoms will be elements of detection of complications. Based on these observations, complications detection rules can be defined as follows:

Complications Detection Rules. Complications detection rule of a diseases are defined as a set of rules each of which is included into inclusive rules of other diseases.[4]

$$\{R \to d \quad s.t. \ R = [a_i = v_j], \ \alpha_R(D) > \delta_\alpha, \kappa_R(D) > \delta_\kappa\} \tag{9}$$

Figure 4 depicts the relations between exclusive, inclusive and complications detection rules.

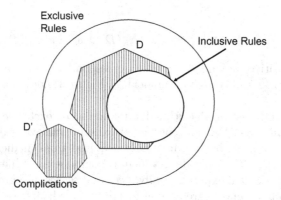

Fig. 4. Venn diagram of exclusive, inclusive and complications detection rules.

The relations between three types of rules can be visualized in a two dimensional plane, called (α, κ)-*plane*, as shown in Fig. 5. The vertical and horizontal axis denotes the values of accuracy and coverage, respectively. Then, each rule can be plotted in the plane with its accuracy and coverage values. The region for inclusive rules is shown in upper right, whereas the region for candidates of detection of complications is in lower left. When a rule of that region belongs to an inclusive rule of other disease, it is included into complications detection rule of the target diseases.

---

[4] The first term $R = [a_i = v_j]$ may not be needed theoretically. However, since deriving conjunction in an exhaustive way is sometimes computationally expensive, here this constraint is imposed for computational efficiency.

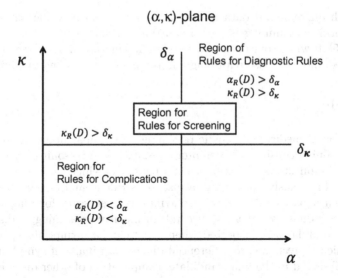

Fig. 5. Two dimensional plot: $(\alpha, \kappa)\text{-}plane$

6   Discussion: What Has Not Been Achieved?

In [12], one of the authors discusses the characteristics of differential diagnosis headache as follows: (a) Hierarchical classification is used. (b) A set of symptoms is used to describe each disease. (c) Description is based on specificity weighted over sensitivity, which shows that reasoning about frequency is implicitly included. (d) For coverage, exceptions are described. (e) Diagnostic criteria gives temporal information about episodes of headache. In the previous studies, automated extraction of knowledge with respect to (a), (b), (c) has been solved.

However, (d) and (e) still remains. Dealing with exceptions is related with complications detection, so partially (d) is solved. However, in some cases, exceptions are used for case-based reasoning by medical experts. Thus, combination of rule-based and case-based reasoning should be introduced.

Acquisition of temporal knowledge is important because medical experts use temporal reasoning in a flexible way. When one of the author interviewed the domain expert for RHINOS, he found that temporal reasoning is very important for complicated cases. For example, one patient suffers from both common migraine and tension headache. According to the diagnostic rules, RHINOS diagnoses the case as migraine. However, the main complaint came from tension headache. Since the onset of tension headache is persistent but the severity is mild, the patient focuses on the symptoms of migraine. If the system can focus on the differences in temporal natures of headaches, then it can detect the complications of migraine and tension headache. Thus, temporal reasoning is a key to diagnose complicated cases especially when all the symptoms may give a contradict interpretation.

Research on temporal data mining is ongoing, and now the authors show that temporal data mining is very important for risk management in several fields [13–15]. It will be our future work to develop methodologies for combination of rule-based and case-base reasoning and temporal rule mining in clinical data.

7   Conclusion

Formalization of medical diagnostic reasoning based on symptomatology is discussed. Reasoning consists of three processes, exclusive reasoning, inclusive reasoning and complications detection, the former two of which belongs to a focusing mechanism. In exclusive reasoning, a disease is ruled out from diagnostic candidates when a patient does not have symptoms necessary for diagnosis. The process corresponds to screening. Second, in inclusive reasoning, a disease out of selected candidates is suspected when a patient has symptoms specific to a disease, which corresponds to differential diagnosis. Finally, if symptoms which are rarely observed in the final candidate, complication of other diseases will be suspected.

Previous studies are surveyed: one of the author concentrate on the focusing mechanism. First, in a deterministic version, two steps are modeled as two kinds of rules obtained from representations of upper and lower approximation of a given disease. Then, he extends it into probabilistic rule induction, which can be viewed as an application of VPRS.

Then, the authors formalize complications detection rules in this paper. The core idea is that the rules are not simply formalized by the relations between a set of symptoms and a disease, but by those between a symptoms, a target disease and other diseases. The next step will be to introduce an efficient algorithm to generate complication detection rules from data.

Acknowledgments. The author would like to thank past Professor Pawlak for all the comments on my research and his encouragement. Without his influence, one of the authors would neither have received Ph.D on computer science, nor become a professor of medical informatics. The author also would like to thank Professor Jerzy Grzymala-Busse, Andrezj Skowron, Roman Slowinski, Yiyu Yao, Guoyin Wang, Wojciech Ziarko for their insightful comments.

References

1. Matsumura, Y., Matsunaga, T., Maeda, Y., Tsumoto, S., Matsumura, H., Kimura, M.: Consultation system for diagnosis of headache and facial pain: "rhinos". In: Wada, E. (ed.) LP 1985. LNCS, vol. 221, pp. 287–298. Springer, Heidelberg (1985)
2. Pawlak, Z.: Rough Sets. Kluwer Academic Publishers, Dordrecht (1991)
3. Pawlak, Z.: Rough modus ponens. In: Proceedings of International Conference on Information Processing and Management of Uncertainty in Knowledge-Based Systems 1998, Paris (1998)

4. Skowron, A., Grzymala-Busse, J.: From rough set theory to evidence theory. In: Yager, R., Fedrizzi, M., Kacprzyk, J. (eds.) Advances in the Dempster-Shafer Theory of Evidence, pp. 193–236. Wiley, New York (1994)
5. Tsumoto, S.: Automated induction of medical expert system rules from clinical databases based on rough set theory. Inf. Sci. **112**, 67–84 (1998)
6. Tsumoto, S.: Extraction of experts' decision rules from clinical databases using rough set model. Intell. Data Anal. **2**(3), 215–227 (1998)
7. Tsumoto, S.: Extraction of hierarchical decision rules from clinical databases using rough sets. Information Sciences (2003)
8. Tsumoto, S., Tanaka, H.: Automated discovery of medical expert system rules from clinical databases based on rough sets. In: Proceedings of the Second International Conference on Knowledge Discovery and Data Mining 1996, pp. 63–69. AAAI Press, Palo Alto (1996)
9. Tsumoto, S.: Modelling medical diagnostic rules based on rough sets. In: Polkowski, L., Skowron, A. (eds.) RSCTC 1998. LNCS (LNAI), vol. 1424, pp. 475–482. Springer, Heidelberg (1998)
10. Tsumoto, S.: Automated discovery of positive and negative knowledge in clinical databases based on rough set model. IEEE Eng. Med. Biol. Mag. **19**, 56–62 (2000)
11. Tsumoto, S.: Extraction of structure of medical diagnosis from clinical data. Fundam. Inform. **59**(2–3), 271–285 (2004)
12. Tsumoto, S.: Rough sets and medical differential diagnosis. In: Skowron, A., Suraj, Z. (eds.) Rough Sets and Intelligent Systems - Professor Zdzisław Pawlak in Memoriam. ISRL, vol. 42, pp. 605–621. Springer, Heidelberg (2013)
13. Tsumoto, S., Hirano, S.: Risk mining in medicine: application of data mining to medical risk management. Fundam. Inform. **98**(1), 107–121 (2010)
14. Tsumoto, S., Hirano, S.: Detection of risk factors using trajectory mining. J. Intell. Inf. Syst. **36**(3), 403–425 (2011)
15. Tsumoto, S., Hong, T.P.: Special issue on data mining for decision making and risk management. J. Intell. Inf. Syst. **36**(3), 249–251 (2011)
16. Tsumoto, S., Tanaka, H.: Induction of probabilistic rules based on rough set theory. In: Tomita, E., Kobayashi, S., Yokomori, T., Jantke, K.P. (eds.) ALT 1993. LNCS, vol. 744, pp. 410–423. Springer, Heidelberg (1993). http://dx.doi.org/10.1007/3-540-57370-4_64
17. Ziarko, W.: Variable precision rough set model. J. Comput. Syst. Sci. **46**, 39–59 (1993)

Multi-granularity Intelligent Information Processing

Guoyin Wang[1]([✉]), Ji Xu[2,3], Qinghua Zhang[1], and Yuchao Liu[4]

[1] Chongqing Key Laboratory of Computational Intelligence,
Chongqing University of Posts and Telecommunications, Chongqing 400065, China
wanggy@ieee.org
[2] School of Information Science and Technology,
Southwest Jiaotong University, Chengdu 610031, China
[3] Institute of Electronic Information Technology,
Chongqing Institute of Green and Intelligent Technology, CAS,
Chongqing 401122, China
xuji@cigit.ac.cn
[4] Chinese Institute of Command and Control, Beijing 100089, China

Abstract. Multi-granularity thinking, computation and problem solving are effective approaches for human being to deal with complex and difficult problems. Deep learning, as a successful example model of multi-granularity computation, has made significant progress in the fields of face recognition, image automatic labeling, speech recognition, and so on. Its idea can be generalized as a model of solving problems by joint computing on multi-granular information/knowledge representation (MGrIKR) in the perspective of granular computing (GrC). This paper introduces our research on constructing MGrIKR from original datasets and its application in big data processing. Firstly, we have a survey about the study of the multi-granular computing (MGrC), including the four major theoretical models (rough sets, fuzzy sets, quotient space,and cloud model) for MGrC. Then we introduce the five representative methods for constructing MGrIKR based on rough sets, computing with words(CW), fuzzy quotient space based on information entropy, adaptive Gaussian cloud transformation (A-GCT), and multi-granularity clustering based on density peaks, respectively. At last we present an MGrC based big data processing framework, in which MGrIKR is built and taken as the input of other machine learning and data mining algorithms.

Keywords: Multi-granularity · Fuzzy sets · Rough sets · Quotient space · Cloud model · Hierarchical clustering · Density peaks · Deep learning

1 Introduction

Deep learning has become one of the hot frontiers in the field of machine learning, because it has brought about breakthroughs in processing images, video,

© Springer International Publishing Switzerland 2015
Y. Yao et al. (Eds.): RSFDGrC 2015, LNAI 9437, pp. 36–48, 2015.
DOI: 10.1007/978-3-319-25783-9_4

speech and audio, as well as text [1]. The common features shared by all deep learning models are: they are composed of multiple processing layers and learn representations of data with multiple levels of abstraction, which are essentially in accordance with the spirit of MGrC. With the representation from a nine-layer deep neural network, DeepFace has reached the accuracy in face recognition up to 97.35%, almost equals to that of human (97.53 %)[2]. Vinyals et al succeeded in automatically adding caption to images by training a convolutional neural network (CNN) and a recurrent neural network (RNN)[3]. Graves A. et al. trained the deep Long Short-term Memory RNNs achieving a test set error of 17.7 % on the TIMIT phoneme recognition benchmark [4].

The success of deep learning attributes to that it solves a complex problem by dividing it into many layers, and a relatively simple task is fulfilled on each single layer. Some similar multi-layer artificial neural network models with such kind of deep structures have been developed in 1990's, such as ANFIS [5] an TMLNN [6].

For a long time, human beings have developed a multi-granular view of the world. Mankind perceives, measures, conceptualizes and reasons the objects from natural world and human society on multi-granular level [7]. This philosophy of granular computing (GrC) leads to quite a few benefits such as high efficiency, low energy consumption and robustness against some trivial incompleteness, uncertainty and even errors. Therefore, it has been recognized as one of the ongoing or underlying techniques and technologies to harness Big Data [8], and the research on granular computing has attracted the interests of many researchers and practitioners [9,10].

"GrC is a superset of the theory of fuzzy information granulation, rough set theory and interval computations, and is a subset of granular mathematics," Zadeh stated in 1997. As the objects of processing, granules are any subsets, classes, objects, clusters, and elements of a universe as they are drawn together by distinguishability, similarity, or functionality [10]. Skowron further generalized the concept of granules to include "functional elements" such as classifiers, agents and groups of agents [11]. Yao considers GrC to be a label of theories, methodologies, techniques, and tools that make use of granules in the process of problem solving [12].

Evolving from GrC, MGrC emphasizes on jointly utilizing multiple levels of information granules in problem solving [13], instead of choosing an optimal granular layer or switching between different granular layers. By "jointly utilizing", we mean that some problems cannot be solved on any single layer of their MGrIKR with the sense of finer result or coarser result, but need to integrate the solution components on each layer to form the right solution of the entire problem. Rough sets, fuzzy sets, quotient space, and cloud model can be taken as the major models to construct MGrIKR.

The fuzzy set theory developed by Zadeh in 1965 starts with definitions of membership function [14], with the more functions defined about an attribute, the attribute is granulated into the finer fuzzy information granules (IG). The reason for fuzzy IG is that crisp IG (e.g. an interval is partitioned by exact values) does not reflect the fact that the granules are fuzzy in almost all of human reasoning and concept formation [15, 16].

The rough set theory developed by Pawlak in 1982 is an effective model to acquire knowledge in information system with upper approximation and lower approximation as its core concepts [17], making decisions according to the definition of distinguishable relation and attribute reduct. Rough set can be used to granulate a set of objects into IGs. The grain size of an IG is determined by how many attributes and how many discrete values each attribute takes in the subset of the whole attribute set, which is selected to do the granulation. Generally, the more attributes and the more values each attribute takes, the finer the IGs will be generated.

The quotient space theory proposed by Zhang B and Zhang L is a model for problem solving with the basic idea of conceptualizing the world at different granularities and shifting the focus of thinking onto different abstract level [18, 19]. It has attracted the attention of researchers from the fields of information science, automatic control, and applied mathematics [20, 21]. Integrating the idea of fuzzy mathematics into quotient space theory, Zhang proposed fuzzy quotient space theory subsequently, which provides a powerful mathematical model and tool for GrC [22]. Fuzzy quotient space theory introduces fuzzy equivalence relation into the construction of quotient space, in which different threshold values of the membership function will lead to quotient spaces of different grain size. By setting different threshold values, an MGrIKR can be derived. Zhang Q further proposed to hierarchically construct normalized isosceles distance function between different quotient spaces, and to extend a fuzzy quotient space theory with arbitrary threshold [23–25].

The cloud model proposed by Li realizes the uncertain transformation between qualitative concepts and quantitative values and can be further used to realize the bidirectional cognition, i.e. from concept intension to extension and vice versa [26]. Liu improved cloud model and proposed A-GCT to adaptively construct a concept tree from given data [27], which is essentially a method to build MGrIKR.

GrC can be categorized into three basic models: granularity space optimization, granularity level switching, multi-granularity joint problem solving. In granularity space optimization, the most suitable granular level in MGrIKR for a specific domain is chosen, and the most efficient and satisfactory enough solution is computed on it [28–30]. Granularity level switching means the working granularity layer will be switched between higher layer and lower layer, to meet the requirements of solving a problem [31–33]. Multi-granularity joint problem solving takes a problem oriented MGrIKR as input, every layers of the MGrIKR are employed jointly to obtain a correct solution to the problem. The most flourishing and promising machine learning paradigm deep learning can be taken as an example of Multi-granularity joint computation [13]. Each of the three mechanisms has its particular type of problem to deal with. Multi-granularity joint problem solving should be the most important problem in future MGrC researches.

Table 1. A toy information system

Objects	Shape	Color	Size
x_1	Round	Green(lime, Li)	Big
x_2	Triangle	Green(ForestGreen, FG)	Big
x_3	Star	Green(LawnGreen, LG)	Big
x_4	Square	Red(Pink,P)	Big
x_5	Round	Red(Pink,P)	Small
x_6	Triangle	Red(Salmon,S)	Small
x_7	Star	Blue(Cyan,C)	Small
x_8	Square	Blue(Navy,N)	Small

The paper introduces the researches on the multi-granular intelligent information processing in recent years. The rest of the paper is organized as follows. Sections 2 and 3 discuss rough set theory and CW for multi-granularity computing, respectively. Section 4 introduces multi-granular fuzzy quotient space theory. Section 5 presents the A-GCT method to generate concept tree [27]. A hierarchical clustering method based on density peaks recently developed by us is briefly introduced in Sect. 6. In Sect. 7, we analyze the importance of building MGrIKR in big data processing, and draw a conclusion.

2 Multi-granularity Rough Set Theory

The classic rough set theory can be used to construct MGrIKR by selecting different attribute subsets and(or) altering the numbers of discrete values each selected attribute takes. This can be illustrated by Example 1.

Example 1. Consider the information system shown in Table 1.

We can get the MGrIKR by changing the condition attribute subsets as shown in Fig. 1a, or by changing the vaules of a particular attribute as shown in Fig. 1b.

A more detailed discussion on multi-granular rough set theory based on changing the attribute subsets can be found in [34], and that based on the attribute values' coarsening and refining can be found in [35].

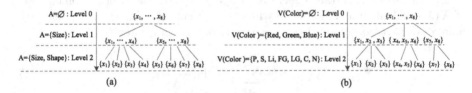

(a)   (b)

Fig. 1. (a)The MGrIKR constructed by varing the attribute subsets, and (b) by varing the values of attribute.

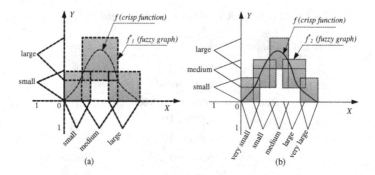

Fig. 2. Two fuzzy graphs approximates to function f at different granularities. (a) f is approximated to with 3 cartesian granules [36,37] (b) f is approximated to with 5 cartesian granules

3   Multi-granularity Computing with Words

Computing with words (CW) is a methodology in which the objects of computation are words and propositions drawn from a natural language rather than numbers [36]. CW is beneficial when the available information is too imprecise to justify the use of numbers, and when there is a tolerance for imprecision which can be exploited to achieve tractability, robustness, low solution cost, and better rapport with reality [37]. We can vary the number of concepts NC on attributes to build an MGrIKR. The larger NC is, the finer the granules in the form of concepts are.

We extend the example of fuzzy graph introduced in [36,37] to demonstrate how CW can be used to build an MGrIKR. See Fig. 2.

4   Multi-granularity Fuzzy Quotient Space Theory

Zhang Q. defined the information entropy sequence of hierarchical quotient space structure and analyzed the relationship between hierarchical quotient space structure and information entropy sequence in detail, based on the definition of fuzzy similarity relation and fuzzy equivalence relation [18], the precedent research on fuzzy quotient space [38,39] and measuring method of uncertainty [40].

Similar to the uncertainty measure proposed in [41], the information entropy of $X(\lambda_i)$ is defined as follows [25],where λ_i is a cut-off value, L_i is the clusters number of partition induced by λ_i, and X_{ik} is the kth cluster:

$$H_X(\lambda_i) = -\sum_{k=1}^{L_i} \frac{|X_{ik}|}{|X|} \ln(\frac{|X_{ik}|}{|X|}), \tag{1}$$

Then we get Theorem 1 as:

Theorem 1. *[25] Let R be a fuzzy equivalence relation on X, let $\pi_X(R) = \{X(\lambda_1), X(\lambda_2), ..., X(\lambda_t)\}$ be its corresponding hierarchical quotient space structure, then the entropy sequence $H(\pi_X(R)) = \{H_X(\lambda_1), H_X(\lambda_2), ..., H_X(\lambda_t)\}$ is a strictly monotonic increasing sequence.*

Theorem 1 shows that the information entropy sequence of a hierarchical quotient space structure is a monotonic increasing sequence with the partition (or quotient space) becoming finer.From the viewpoint of information entropy, if two hierarchical quotient space structures are isomorphic, then they have the same information entropies. This is given as Theorem 2.

Theorem 2. *[25] Let $\pi_X(R_1)$ and $\pi_X(R_2)$ be two hierarchical quotient space structures derived from fuzzy equivalence relations R_1 and R_2 on X, respectively. If $L(\pi_X(R_1)) = L(\pi_X(R_2))$ and $H(\pi_X(R_1)) = H(\pi_X(R_2))$, then $\pi_X(R_1) \simeq \pi_X(R_2)$.*

Theorem 2 reveals that the uncertainty of a hierarchical quotient space structure may be characterized by the information entropy sequence, the partition sequence, and the subblock sequence.

With the fuzzy similarity relation $\widetilde{R}$ on X, a fuzzy equivalence relation R is derived from the fuzzy similarity relation $\widetilde{R}$ by the transitive closure operation. Finally, a method is developed to construct hierarchical quotient space structure (HQSS). See Algorithm 1.

Algorithm 1. [25,42] constructing HQSS

Input Data set $X = \{x_i | i = 1, 2, ..., n\}$, and a fuzzy similarity relation matrix $M(\widetilde{R}) = (\widetilde{r}_{ij})_{n \times n}$

Output A hierarchical quotient space structure $\pi_X(R)$ (Where R is a fuzzy equivalence relation derived from $\widetilde{R}$)

Step1 By $M(\widetilde{R}) = (\widetilde{r}_{ij})_{n \times n}$, a similarity degree sequence $\lambda = \{\lambda_1, \lambda_2, ..., \lambda_t\}$ can be obtained.

Step2 For $k=1$ to t

      For $i=1$ to $n-1$

          For $j=i+1$ to n

          {if $\widetilde{r}_{ij} \geq \lambda_k$, x_i and x_j are in the same subblock, else x_i and x_j

          stay in different subblocks as before.}

          {if $X(\lambda_k) = \{X\}$, goto **Step3**}

          EndFor

      EndFor

   EndFor

Step3 An HQSS $\pi_X(R)$ is obtained.

The limitation of Algorithm 1 is that the clustering process directly uses the similarity between any pairs of objects, which may lead to too many clustering layers and too many clusters on the bottom layers if the number of objects is large and the similarity values vary from each other. And the time complexity of Algorithm 1 is $O(t \times n^2)$.

5   Multi-granularity Cloud Model

The process of GMM parameters estimated by the EM algorithm does not consider the concept cognition law, because many Gaussian distributions are overlapped. This causes concept confusion when GMM is used to express concepts. Therefore, we instead use Gaussian Cloud Model (GCM) to represent the extension of a concept and measure its confusion degree.

Definition 1. [26] *Let U be a universal set described by precise numbers, and C be the qualitative concept containing three numerical characters (Ex, En, He) related to U. If there is a number $x \in U$, which is a random realization of the concept C and satisfies $x = R_N(Ex, y)$, where $y = R_N(En, He)$ and the certainty degree of x on U is $\mu(x) = exp\left\{-\frac{(x-Ex)^2}{2y^2}\right\}$, then the distribution of x on U is a 2nd-order normal cloud or normal cloud, and each x is defined as a 2nd-order normal cloud drop.*

The contributions of cloud drops in different regions to the concept are different. 99.7 % of cloud drops to the concept C in the universal domain U lie in the domain $[Ex - 3En, En + 3En]$. As a result, we can neglect the contribution to the concept C by the cloud drops out of the domain $[Ex - 3En, Ex + 3En]$. This is the "$3En$" rule of the Gaussian cloud. In a Gaussian distribution $G(\mu_k, \sigma_k)$, if its weak peripheral elements region is separated with that in other Gaussian distribution, its relevant concept parameter is $En_k = \sigma_k, He_k = 0$. Otherwise, their standard variance is zoomed at the same scale α to guarantee their weak peripheral element region not overlap.

Gaussian cloud transformation is a data clustering method in nature. We propose two algorithms to extract concepts from cloud drops(data points).H-GCT can extract concepts with given number by prior knowledge, and A-GCT can automatically extract concepts on rational granularity without prior knowledge.

If two Gaussian distributions are close to each other, their standard variance is zoomed at the same scale to guarantee their weak peripheral element region not overlap. Then two scale parameter α_1, α_2 can be calculated by the formulas $\mu_{k-1} + 3 * \alpha_1 * \sigma_{k-1} = \mu_k - 3 * \alpha_1 * \sigma_k$ and $\mu_k + 3 * \alpha_2 * \sigma_k = \mu_{k+1} - 3 * \alpha_2 * \sigma_{k+1}$ The standard variance range of $G(\mu_k, \sigma_k)$ is $[\alpha * \sigma_k, \sigma_k]$, and $\alpha = min(\alpha_1, \alpha_2)$.

Algorithm 2. H-GCT [27]

Input Data set $X\{x_i | i = 1, 2...., N\}$,Concept quantity M

Output Gaussian clouds $C(Ex_k, En_k, He_k)|k = 1, ..., M$

Step1 Using GMM to transfer X to M Gaussian distributions: $G(\mu_k, \sigma_k)|k = 1, ..., M$.

Step2 for each $G(\mu_k, \sigma_k)$ compute α_k and parameters of Gaussian cloud: $Ex_k = \mu_k, En_k = (1 + \alpha_k)\sigma_k/2, He_k = (1 - \alpha_k)\sigma_k/6, He_k/En_k = (1 - \sigma_k)/(3(1 + \sigma_k))$.

Adaptive Gaussian Cloud Transformation (A-GCT) can transfer real data sample set to multiple concepts in different granularities automatically without the pre-specified number of concepts. In commonsense, relative to the low frequency data, high frequency data values have more contribution to the qualitative concept. So the wave number of data sample frequency distribution can be taken as the initial concept quantity M, then call the H_GCT to generate M Gaussian clouds, according to each concept clarity index He/En, making Gaussian cloud transform strategy. $He/En \leq 0.2116$ can guarantee a concepts key elements region is independent; $He/En \leq 0.1668$ can separate two concepts basic elements region; and so on.

Given a strategy $He/En \leq \beta$ as input, A-GCT forms a tree of concepts automatically.

Algorithm 3. A-GCT [27]

Input Data set $X\{x_i | i = 1, 2..., N\}$, Concept clarity β

Output Gaussian clouds $C(Ex_k, En_k, He_k) | k = 1, ..., m$

Step1 Count the wave number of data frequency distribution, as an initial concepts quantity m.

Step2 Using H-GCT to transfer data set X to Gaussian Clouds:

$C(Ex_k, En_k, He_k) | k = 1, ..., m$.

Step3 for each $C(Ex_k, En_k, He_k)$ if $He_k/En_k > \beta$, then $m = m - 1$.

Step4 Loop steps 2–3 until form m Gaussian clouds in which $He_k/En_k \leq \beta, k = 1, ..., m$.

We interpret the algorithm process of Gaussian cloud transformation by user age clustering in ArnetMiner (www.arnetminer.com). ArnetMiner is a platform for researchers society mining. At present, there are 988,645 users from 196 countries, and their age is between 25 and 85. Five concepts about age of ArnetMiner users are formed based on the definition of parameter concept clarity, as shown in Fig. 3.

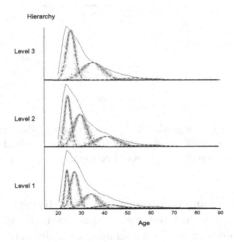

Fig. 3. Concept tree formed by A-GCT [27]

A-GCT forms a hierarchical structure of concepts in the process of computing the "clear" concepts, through iteratively calling H-GCT and taking the predetermined value β as threshold to decide when the process terminates. The time complexity of A-GCT is $O(m^2 \times N)$.

6   Multi-granularity Clustering Based on Density Peaks

Recently, Rodriguez and Laio proposed a clustering method based on fast searching and finding density peaks (DP_CLUS), which is efficient and accurate in clustering datasets of any shape [43]. It picks out centers by defining two simple measures: local density ρ and the distance to the nearest neighbor of higher density δ. The process of assigning noncenter data points to centers is very efficient(with the time complexity of $O(n)$, n is the number of data points). DP_CLUS is a flat clustering method returning one partition on the dataset per running. Besides, it includes interactive selection of the centers before assigning the objects to clusters, which enables the users to embrace human intuition. Meanwhile it brings inconvenience and even allows wrong choice of centers.

When applying DP_CLUS algorithm to cluster dataset, we notice that the centers take γ value (originally defined as $\gamma = \rho \times \delta$) on different levels when the dataset can be clustered hierarchically. Figure 4 illustrates the key observation of this study.

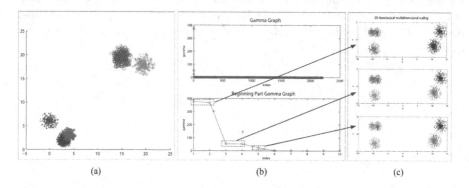

Fig. 4. (a) the dataset may be regarded as 5, 4, or 2 clusters. (b) γ curve of the dataset using DP_CLUS algorithm. (c) Corresponding to the three levels of values of potential centers, the dataset is clustered with 3 sets of centers.

To develop an efficient and robust hierarchical clustering method, we improve on DP_CLUS by leveraging the parameter γ . This leads to a hierarchical clustering method based on density peaks, which automatically detects all the possible centers and build a hierarchy presentation for the dataset (if it is hierarchically structured in nature). Subsequently, the relationship between the clusters on a higher layer and those on a lower layer is mathematically studied, which forms the solid foundation for a density peaks based efficient agglomerative hierarchical clustering(DenPEACH). DenPEACH builds the hierarchy based on the finest grained clustering result with an additional computational complexity of $O(m)$, where m is the number of clusters.

Using DenPEACH, we automatically construct the MGrIKR for an artificial datasets consists of 5 spirals, among which one is separated and the other four are arranged as 2 pairs.The experiments conducted on real datasets also show the superior efficiency and competitive accuracy against state-of-the-art hierarchical clustering methods.

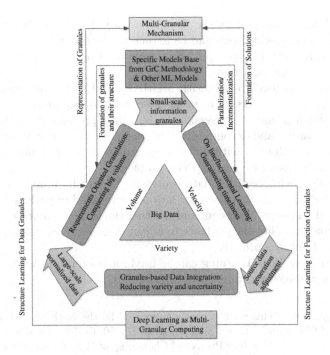

Fig. 5. MGrC based big data processing framework [44]

7 Conclusion

The researchers from the society of GrC and big data analytics have reached the consensus that GrC is an underlying methodology to harness big data. Intuitively, granular computing can reduce the data size into different level of granularity [8]. And the reduction of data size is also very helpful to meet the "timeliness" requirement in big data processing since smaller input size results in shorter running time for any algorithm. Besides, the granulation procedure in GrC actually performs simplification and abstraction on original data, so the "Variety" characteristic of big data is successfully coped with. Based on the analysis above, we proposed the uniform MGrC based big data processing framework, as shown in Fig. 5.

This framework describes the following points of big data processing based on MGrC:

(1) Use data integration and filtering techniques to extract, transform, and granulate the multi-sourced heterogeneous data into a normalized data table. In this way, the variety and uncertainty in the original data are reduced (or even removed).

(2) For specific problems in a given domain, the size of input data can be reduced to a suitable degree by granulating the input data into granules with clear semantics. In this process, the particular models (such as rough set, fuzzy set, quotient

space, cloud model, etc.) under the "umbrella" of GrC will be chosen to granulate the data on different layers and construct the hierarchy.

(3) The MGrIKR is taken as input to be computed by other machine learning and (or) data mining algorithms.

(4) The algorithms used to deal with MGrIKR need to be converted into distributed or incremental version to meet the timeliness requirement.

(5) The problem granular conversion in big data processing consists of two aspects: composition and decomposition of granules on different layers, and fast reconstruction of solutions with different granularity. And the inter-granularity mechanism is necessary for some special problems.

(6) The whole procedure of big data processing should be brought under close monitoring to judge whether the raw data are generated or gathered at proper frequency and on proper precision, and whether the features sampled are incomplete or redundant. Thus the guidance to adjust the data generation or gathering can be worked out accordingly.

(7) The thought of Deep Learning can be used for reference to learn the appropriate size of granules and number of layers in MGrIKR.

Acknowledgement. This work is partly supported by the National Natural Science Foundation of China under Grant numbers of 61272060, 61472056 and 61305055, and Natural Science Foundation Key Project of Chongqing of P. R. China under Grant No. CSTC2013jjB40003.

References

1. LeCun, Y., Bengio, Y., Hinton, G.: Deep learning. Nature **521**, 436–444 (2015)
2. Taigman, Y., Yang, M., Ranzato, M.A., et al.: Deepface: Closing the gap to human-level performance in face verification. In: IEEE Conference on Computer Vision & Pattern Recognition, CVPR 2014, pp. 1701–1708 (2014)
3. Vinyals, O., Toshev, A., Engio, S., Erhan, D.: Show and Tell: A Neural Image Caption Generator. In: IEEE Conference on Computer Vision & Pattern Recognition, CVPR 2015, pp. 3156–3164 (2015)
4. Graves, A., Mohamed, A., Hinton, G.: Speech recognition with deep recurrent neural networks. In: IEEE International Conference on Acoustics, Speech & Signal Processing, ICASSP 2013, pp. 6645–6649 (2013)
5. Jang, J.S.R.: ANFIS: adaptive-network-based fuzzy inference system. IEEE Trans SMC. **23**, 665–685 (1993)
6. Wang, G., Shi, H.: TMLNN: triple-valued or multiple-valued logic neural network. IEEE Trans. Neural Netw. **9**, 1099–1117 (1998)
7. Yager, R.R., Filev, D.: Operations for granular computing: mixing words with numbers. In: Proceeding 1998 IEEE International Conference Fuzz System, pp. 123–128 (1998)
8. Chen, C.L.P., Zhang, C.: Data-intensive applications, challenges, techniques and technologies: a survey on big data. Inf Sci. **275**, 314–347 (2014)
9. Miao, D., Wang, G., Liu, Q., Lin, T., Yao, Y.: Granular Computing: Past, Present and Prospects. Science Press, Beijing (2007)

10. Yao, J., Vasilakos, A., Pedrycz, W.: Granular computing: perspectives and challenges. IEEE Trans. Cybern. **43**, 1977–1989 (2013)
11. Skowron, A., Wasilewski, P.: Information systems in modeling interactive computations on granules. Theor. Comput. Sci. **412**, 5939–5959 (2011)
12. Yao, Y.: Granular Computing: basic issues and possible solutions. In: Proceeding of 5th Joint Conference on Information Science, vol. I, pp. 186–189. Atlantic (2000)
13. Wang, G., Xu, J.: Granular computing with multiple granular layers for brain big data processing. Brain Inform. **1**, 1–10 (2014)
14. Zadeh, L.: Fuzzy sets. Inf. Control **8**, 338–353 (1965)
15. Zadeh, L.: Toward a theory of fuzzy information granulation and its centrality in human reasoning and fuzzy logic. Fuzz Sets Syst. **90**, 111–127 (1997)
16. Zadeh, L.: Is there a need for fuzzy logic? Inf. Sci. **178**, 2751–2779 (2008)
17. Pawlak, Z.: Rough sets. Int. J. Comput. Inf. Sci. **11**, 341–356 (1982)
18. Zhang, B., Zhang, L.: Theory of Problem Solving and its Applications, 2nd edn. Tsinghua University Press, Beijing (2007). in Chinese
19. Zhang, L., Zhang, B.: The quotient space theory of problem solving. Fund inf. **59**, 287–298 (2004)
20. Wu, D., Ban, X., Oquendo, F.: An architecture model of distributed simulation system based on quotient space. Appl Math. **6**, 603S–609S (2012)
21. Dong, Q., et al.: Algebraic properties and topological properties of the quotient space of fuzzy numbers based on Mareˇ equivalence relation. Fuzz Sets Syst. **245**, 63–82 (2014)
22. Zhang, L., Zhang, B.: Fuzzy reasoning model under quotient space structure. Inf. Sci. **173**, 353–364 (2005)
23. Zhang, Q., Wang, G., Liu, X.: Hierarchical structure analysis of fuzzy quotient space. PR&AI. **21**, 627–634 (2008)
24. Zhang, Q.: Research on Hierarchy Granular Computing Theory and its Application[D]. Southwest Jiaotong University, Chengdu (2009)
25. Zhang, Q., Wang, G.: The uncertainty measure of hierarchical quotient space structure. Math. Prob. Eng. **6**, 505–515 (2011)
26. Li, D., Du, Y.: Artificial Intelligence with Uncertainty. Chapman & Hall/CRC Press, Boca Raton (2008)
27. Liu, Y., Li, D., He, W., Wang, G.: Granular computing based on gaussian cloud transformation. Fund. Inf. **127**, 385–398 (2013)
28. Nakatsuji, M., Fujiwara, Y.: Linked taxonomies to capture users' subjective assessments of items to facilitate accurate collaborative filtering. Artif. Intell. **207**, 52–68 (2014)
29. Pedrycz, W., Homenda, W.: Building the fundamentals of granular computing: a principle of justifiable granularity. Appl. Soft. Comput. **13**, 4209–4218 (2013)
30. Pedrycz, W.: Allocation of information granularity in optimization and decision-making models: towards building the foundations of granular computing. Eur. J. Oper. Res. **232**, 137–145 (2014)
31. McCalla, G., Greer, J., Barrie, B.: Granularity hierarchies. Comput. Math. App. **23**, 363–375 (1992)
32. Zhu, P., Hu, Q.: Adaptive neighborhood granularity selection and combination based on margin distribution optimization. Inf. Sci. **249**, 1–12 (2013)
33. Yao, Y.: A partition model of granular computing. In: Peters, J.F., Skowron, A., Grzymała-Busse, J.W., Kostek, B., Swiniarski, R.W., Szczuka, M.S. (eds.) Transactions on Rough Sets I, pp. 232–253. Springer, Heidelberg (2004)
34. Qian, Y., Liang, J., Yao, Y., Dang, C.: MGRS: a multi-granulation rough set. Inf. Sci. **180**, 949–970 (2010)

35. Chen, H., Li, T., Luo, C., Hong, S., Wang, G.: A rough set-based method for updating decision rules on attribute values' coarsening and refining. IEEE Trans. Knowl. Data Eng. **26**, 2886–2899 (2014)
36. Zadeh, L.A.: Fuzzy logic=computing with words. IEEE Trans. Fuzz Syst. **4**, 103–111 (1996)
37. Zadeh, L.A.: From computing with numbers to computing with words-from manipulation of measurements to manipulation of perceptions. IEEE Trans. Circ. Syst-I: Fund Theor. Appl. **45**, 105–119 (1999)
38. Tang, X.Q., Zhu, P., Cheng, J.X.: Cluster analysis based on fuzzy quotient space. J. Softw. **19**, 861–868 (2008)
39. Zhang, C.: Fuzzy sets and quotient spaces. In: Proceedings of the IEEE International Conference on Granular Computing, pp. 350–353 (2005)
40. Liang, J., Chin, K.S., Dang, C., Yam, R.C.M.: A new method for measuring uncertainty and fuzziness in rough set theory. Int. J. Gen. Syst. **31**, 331–342 (2002)
41. Wierman, M.J.: Measuring uncertainty in rough set theory. Int. J. Gen. Syst. **28**, 283–297 (1999)
42. Tang, X., Zhu, P., Cheng, J.: Cluster analysis based on fuzzy quotient space. J. Softw. **19**, 861–868 (2008)
43. Rodriguez, A., Laio, A.: Clustering by fast search and find of density peaks. Science **344**, 1492–1496 (2014)
44. Xu, J., Wang, G., Yu, H.: Review of big data processing based on granular computing. Chin. J. Comput. **38**, 1497–1517 (2015)

Granular Structures Induced by Interval Sets and Rough Sets

Ning Zhong[1,2](✉) and Jia-jin Huang[2]

[1] Deptartment of Life Science and Informatics, Maebashi Institute of Technology,
Maebashi City 371-0816, Japan
[2] International WIC Institute, Beijing University of Technology,
Beijing 100124, People's Republic of China
zhong@maebashi-it.ac.jp

Abstract. An interval set is a family of sets restricted by a upper bound and lower bound. Interval-set algebras are concrete models of granular computing. The triarchic theory of granular computing focuses on a multilevel and multi-view granular structure. This paper discusses granular structures of interval sets under inclusion relations between two interval sets from a measurement-theoretic perspective and set-theoretic perspective, respectively. From a measurement-theoretic perspective, this paper discusses preferences on two objects represented by interval sets under inclusion relations on interval sets. From a set-theoretic perspective, this paper uses different inclusion relations and operations on interval sets to construct multilevel and multi-view granular structures.

Keywords: Granular computing · Granular structure · Interval set

1 Introduction

To meet needs in the real world, the rough set theory [10] has been used in the feature selection [17], rule discovery [18,19], inductive logic programming [8], and so on. The interval set theory which is extended from the interval number theory is related to many studies of rough sets. The interval set theory, an appropriate mathematics tool to deal with vague and uncertain information systems, has attracted many attentions in the field of rough sets [12]. An interval set is a family of sets restricted by a upper bound and lower bound. The key problems about interval sets are representations, operations, and applications of interval sets [13].

Granular computing is a computing paradigm. The triarchic theory of granular computing consists of the structured thinking, structured problem solving and structured information processing with multilevel granular structures [14]. One of the central notions of the triarchic theory of granular computing is hierarchical multilevel granular structures defined by granules and levels [15]. By constructing a family of hierarchies, it is possible to obtain multiple different views. Granular structures are a family of complementary hierarchies working

© Springer International Publishing Switzerland 2015
Y. Yao et al. (Eds.): RSFDGrC 2015, LNAI 9437, pp. 49–60, 2015.
DOI: 10.1007/978-3-319-25783-9_5

together for a complete and comprehensive multi-view understanding and representation [15]. Based on the triarchic theory of granular computing, set-theoretic and graph-theoretic formulation, interpretation and representation of multilevel granular structures are proposed [2,3,15].

A crucial issue of granular computing is the search for an appropriate level of granularity by ignoring unimportant and irrelevant details. This issue can be discussed from a measurement-theoretic perspective and set-theoretic perspective, respectively [14]. As interval-set algebras are concrete models of granular computing [13], the granular structure of interval set can be also discussed from the two perspectives.

From a measurement-theoretic perspective, the comparison problem and the numerical representation problem deal with the qualitative construction and quantitative measurement of preference relations over each pair of alternatives [6]. A weak order, bi-weak order and interval order consist of an arrangement of objects from the best one to the worst one, respectively [4,6]. In the view of granular computing, a measure of granularity is a quantification of some intuitive, qualitative relationships between granularity [14]. This paper aims to use preference relations to compare objects represented by interval sets from the qualitative and quantitative views.

From a set-theoretic perspective, Yao et. al proposed a framework for set-theoretic approaches to granular computing [15]. The standard set-inclusion relation is used to construct multilevel granular structures [15]. Inspired by the framework for set-theoretic approaches to granular computing [14], this paper discusses the multilevel and multi-view granular structure of interval sets based on inclusion relations on interval sets.

The rest of the paper is organized as follows. Section 2 gives a brief introduction to interval sets. And then Sects. 3 and 4 gives the granular structure in interval sets from a measurement-theoretic and set-theoretic perspective, respectively. Section 5 discusses connections of rough sets and interval sets. The conclusion is presented in Sect. 6.

2   Interval Sets

2.1   Interval Sets and Interval-Set Algebras

An interval set is a family of sets restricted by a upper bound and a lower bound. The following definitions define the interval set and operations [12,13].

Definition 1. Let U be a finite set, called the universe or the reference set, and 2^U be its power set. A subset of 2^U of the form $\mathcal{A} = [A_l, A_u] = \{A \in 2^U | A_l \subseteq A \subseteq A_u\}$ is called an interval set, where it is assumed $A_l \subseteq A_u$. The set of all interval sets is denoted by $I(2^U) = \{[A_l, A_u] | A_l, A_u \subseteq U, A_l \subseteq A_u\}$.

Definition 2. Let $\cap$, $\cup$ and $-$ be the usual set intersection, union and difference defined on 2^U, respectively. A parallel definition of binary operations $\sqcap$, $\sqcup$, $\setminus$ two interval sets $\mathcal{A} = [A_l, A_u] \in I(2^U)$, $\mathcal{B} = [B_l, B_u] \in I(2^U)$ and a unitary operation $\neg$ on a single interval set $\mathcal{A}$ are defined as follows:

$$\mathcal{A} \sqcap \mathcal{B} = \{A \cap B | A \in \mathcal{A}, B \in \mathcal{B}\}$$
$$= [A_l \cap B_l, A_u \cap B_u]$$
$$\mathcal{A} \sqcup \mathcal{B} = \{A \cup B | A \in \mathcal{A}, B \in \mathcal{B}\}$$
$$= [A_l \cup B_l, A_u \cup B_u]$$
$$\mathcal{A} \backslash \mathcal{B} = \{A - B | A \in \mathcal{A}, B \in \mathcal{B}\}$$
$$= [A_l - B_l, A_u - B_u]$$
$$\neg \mathcal{A} = \{A^c | A \in \mathcal{A}\}$$
$$= [U, U] \backslash [A_l, A_u]$$

2.2   Inclusion Relations in Interval Sets

The inclusion relations among the lower and upper bounds could be defined the inclusion relation between two interval sets. There are four inclusion relations on interval sets [7].

Definition 3. Let $\mathcal{A} = [A_l, A_u] \in I(2^U)$, $\mathcal{B} = [B_l, B_u] \in I(2^U)$, $\mathcal{B}$ is small-weak subset to $\mathcal{A}$, written $\mathcal{A} \sqsubseteq_w \mathcal{B}$, if and only if $\forall B \in \mathcal{B}, \exists A \in \mathcal{A}$ satisfies $A \subseteq B$, and it is called that $\mathcal{A}$ and $\mathcal{B}$ have weak inclusion relation.

According to Definition 3, if an interval set $\mathcal{A}$ is a small-weak subset of interval set $\mathcal{B}$, then for any ordinary set B in $\mathcal{B}$, there exists an ordinary set A in $\mathcal{A}$ satisfying that A is the subset of B.

Definition 4. Let $\mathcal{A} = [A_l, A_u] \in I(2^U)$, $\mathcal{B} = [B_l, B_u] \in I(2^U)$, $\mathcal{B}$ is large-weak subset to $\mathcal{A}$, written $\mathcal{A} \sqsubseteq_{w+} \mathcal{B}$, if and only if $\forall A \in \mathcal{A}, \exists B \in \mathcal{B}$ satisfies $A \subseteq B$, and it is called that $\mathcal{A}$ and $\mathcal{B}$ have weak+ inclusion relation.

According to Definition 4, if an interval set $\mathcal{A}$ is a large-weak subset of interval set $\mathcal{B}$, then for any ordinary set A in $\mathcal{A}$, there exists an ordinary set B in $\mathcal{B}$ satisfying that A is the subset of B.

Definition 5. Let $\mathcal{A} = [A_l, A_u] \in I(2^U)$, $\mathcal{B} = [B_l, B_u] \in I(2^U)$, $\mathcal{B}$ is a possible subset of $\mathcal{A}$, written $\mathcal{A} \sqsubseteq_p \mathcal{B}$, if and only if $\forall B \in \mathcal{B}, \exists A \in \mathcal{A}$ satisfies $A \subseteq B$, and $\forall A \in \mathcal{A}, \exists B \in \mathcal{B}$ satisfies $A \subseteq B$, and it is called that $\mathcal{A}$ and $\mathcal{B}$ have possible inclusion relation.

According to Definition 5, if an interval set $\mathcal{A}$ is a possible subset of interval set $\mathcal{B}$, then for any ordinary set A in $\mathcal{A}$, there exists an ordinary set B in $\mathcal{B}$ satisfying that A is the subset of B. And for any ordinary set B in $\mathcal{B}$, there exists an ordinary set A in $\mathcal{A}$ satisfying that A is the subset of B.

Definition 6. Let $\mathcal{A} = [A_l, A_u] \in I(2^U)$, $\mathcal{B} = [B_l, B_u] \in I(2^U)$, $\mathcal{A}$ is a certain subset of $\mathcal{B}$, written $\mathcal{A} \sqsubseteq_c \mathcal{B}$, if and only if $\forall B \in \mathcal{B}, \forall A \in \mathcal{A}$ satisfies $A \subseteq B$, and it is called that $\mathcal{A}$ and $\mathcal{B}$ have certain inclusion relation.

According to Definition 6, if an interval set $\mathcal{A}$ is a certain subset of interval set $\mathcal{B}$, then for any ordinary set B in $\mathcal{B}$, there exists any ordinary set A in $\mathcal{A}$ satisfying that A is the subset of B.

When considering the lower bound and upper bound of the interval sets, interval set-inclusion relations can be presented using lower bound and upper bound by Eqs. (1)–(4) [7].

$$\mathcal{A} \sqsubseteq_w \mathcal{B} \Leftrightarrow A_l \subseteq B_l \tag{1}$$

$$\mathcal{A} \sqsubseteq_{w+} \mathcal{B} \Leftrightarrow A_u \subseteq B_u \tag{2}$$

$$\mathcal{A} \sqsubseteq_p \mathcal{B} \Leftrightarrow (A_l \subseteq B_l) \wedge (A_u \subseteq B_u) \tag{3}$$

$$\mathcal{A} \sqsubseteq_c \mathcal{B} \Leftrightarrow A_u \subseteq B_l \tag{4}$$

2.3   Granular Structure in Interval Sets

The standard set-inclusion relation $\subseteq$ have been used to arrange granules [14] to construct granular structures. These studies inspire us to arrange granules represented by interval sets through interval set-inclusion relation. Let $\sqsubseteq_*$ denote one of interval set-inclusion relations, we have

Definition 7. For $g, g' \in I(2^U)$, if there is an inclusion relation between g and g', denoted by $g \sqsubseteq_* g'$, we call g a sub-granule of g' and g' a super-granule of g. Suppose $G \subseteq I(2^U)$ is a nonempty family of subsets of $I(2^U)$. The poset $(G, \sqsubseteq_*)$ is called a granular structure. The $\sqsubseteq_*$ can be replaced by one of $\sqsubseteq_w, \sqsubseteq_{w+}, \sqsubseteq_p, \sqsubseteq_c$ in interval sets.

3   The Granular Structure Based on Order Relation in Interval Sets

3.1   Preference

Let X be a finite set, for any two elements $a, b \in X$,

$$a \prec b \Leftrightarrow a \text{ is less prefered than } b. \tag{5}$$

The relation is known as a strict preference relation. In the absence of strict preference, i.e., if both $\neg(a \prec b)$ and $\neg(b \prec a)$ hold, we say that a and b are indifferent. An indifference relation $\sim$ on X is defined as:

$$a \sim b \Leftrightarrow \neg(a \prec b) \wedge \neg(b \prec a). \tag{6}$$

Based on the strict preference and indifference, one can define a preference-indifference relation $\preceq$ on X:

$$a \preceq b \Leftrightarrow a \prec b \vee a \sim b. \tag{7}$$

If $a \preceq b$ holds, we say that b is not preferred to a, or b is at least as good as a. In general, the indifference relation $\sim$ is a symmetric relation and the preference-indifference $\preceq$ is a reflexive relation. Furthermore, any two pairs of elements are comparable under $\preceq$. In terms of the preference-indifference relation $\preceq$, the strict preference relation can be defined as $a \prec b \Leftrightarrow (a \preceq b) \wedge \neg(b \preceq a)$. In modelling preferences, we can use either the strict preference relation $\prec$ or the preference-indifference relation $\preceq$. If a preference relation $\prec$ satisfies the two properties:

Asymmetry: $a \prec b \Rightarrow \neg(a \prec b)$,
Negative transitivity: $\neg(a \prec b) \wedge \neg(b \prec a) \Rightarrow \neg(a \prec b)$,

it is called a weak order.

A preference relation can be used to model qualitative statements. Under certain conditions, we can translate such qualitative statements into quantitative statements. The following theorem, taken from measurement theory [5,11], shows that it is possible to quantitatively measure a preference relation $\prec$ if and only if the relation $\prec$ is a weak order.

Theorem 1. [5] A relation $\prec$ on a finite set A is a weak order if and only if there exists a real-valued function $f : X \to \Re$ satisfying:

$$\forall a, b \in X, a \prec b \Leftrightarrow f(a) < f(b) \tag{8}$$

The number $f(a), f(b), \dots$ as ordered by $<$ faithfully reflect the order of $a, b, \dots$ under $\prec$. The function f is referred to as an order-preserving utility function. If one prefers b to a, we can conclude that $f(a) < f(b)$. Conversely, if $f(a) < f(b)$, we can conclude that one prefers b to a.

For a strict partial order, $\sim$ is not necessarily an equivalence relation. An equivalence relation $\approx$ could be defined as $a \approx b \Leftrightarrow (a \sim c \Leftrightarrow b \sim c)$, for all $c \in X$ [5,14] .

Let a relation $\prec$ on a finite nonempty set X be a strict partial order. We have
(a) the relation $\approx$ is an equivalence relation;
(b) exactly one of $a \prec b$, $b \prec a$, $a \approx b$, $(a \sim b, \neg(a \approx b))$ holds for every $a, b \in X$.

The above principle explains structures of a strict partial order [5]. The following theorem shows one-way inferences for the family of strict partial orders [5,14].

Theorem 2. [5] Suppose X is a finite nonempty set and $\prec$ a binary relation on X. If $\prec$ is a strict partial order, then there exists a real valued function $f : X \to \Re$ such that for all $a, b \in X$, $a \prec b \Rightarrow f(a) < f(b)$, $a \approx b \Rightarrow f(a) = f(b)$.

For a strict partial order, a function f satisfying conditions in the Theorem 2 only partially truthfully preserves $a \prec b$, $b \prec a$, $a \approx b$, and one may have any one of $f(a) = f(b), f(a) < f(b)$ and $f(b) < f(a)$ for $(a \sim b, \neg(a \approx b))$ [14]. As an extension of the weak order, a bi-weak order is proposed as the intersection of two weak orders. The quantitative characterization of a bi-weak order is measured by the following theorem.

Theorem 3. [5] A relation $\prec$ on a finite set A is a bi-weak order if and only if there exist two real-valued function $f_1 : X \to \Re$ and $f_2 : X \to \Re$ satisfying

$$\forall a, b \in X, a \prec b \Leftrightarrow f_2(a) < f_2(b) \text{ and } f_1(a) < f_1(b), \tag{9}$$

$$\forall a \in X, f_1(a) \leq f_2(b). \tag{10}$$

A well known problem of weak orders is the indifference relation induced by weak orders is transitive [4,6]. However, the indifference is not necessarily transitive. An interval order is proposed to handle the case. The interval order shows a is less preferred than b if and only if the evaluation of a is smaller than the evaluation of b plus a threshold. The quantitative characterization of a interval order is measured by the following theorem.

Theorem 4. [5] A relation $\prec$ on a finite set X is a interval order if and only if there exist two real-valued function $f_1 : X \to \Re$ and $f_2 : X \to \Re$ on X satisfying

$$\forall a, b \in X, a \prec b \Leftrightarrow f_2(a) < f_1(b), \tag{11}$$

$$\forall a \in X, f_1(a) \leq f_2(a). \tag{12}$$

An interval number is an ordered pair of real numbers, $[a_1, a_2]$, with $a_1 \leq a_2$ [9,12]. It is also a set of real numbers defined by $[a_1, a_2] = \{x | a_1 \leq x \leq a_2\}$. According to the position relations among lower endpoints and upper endpoints of two interval numbers, Ozturk [6] discussed the interval value comparisons. As an extension of the interval number theory, the interval set theory provides mathematics tool to deal with the vague and uncertain information system from the set-theoretic perspective [12]. Inspired by the results in the theory of interval numbers, we can draw some similar conclusions. With respect to the interval number, for two interval sets $\mathcal{A} = [A_l, A_u]$ and $\mathcal{B} = [B_l, B_u]$, we have (1) $A_l \subseteq A_u \subseteq B_l \subseteq B_u$, (2) $A_l \subseteq B_l \subseteq A_u \subseteq B_u$, (3) $A_l \subseteq B_l \subseteq B_u \subseteq A_u$, (4)$B_l \subseteq A_l \subseteq A_u \subseteq B_u$, (5) $B_l \subseteq A_l \subseteq B_u \subseteq A_u$, (6) $B_l \subseteq B_u \subseteq A_l \subseteq A_u$.

As mentioned in [6], weak orders, bi-weak orders and interval orders provide useful structures for comparing two object. In the next section, we will apply these theorems to measuring the preferences in interval sets and how to compare two objects represented by interval sets.

3.2   Interval Set Comparisons

Let x be represented by $\mathcal{A} = [A_l, A_u]$ and y represented by $\mathcal{B} = [B_l, B_u]$. We first define a preference relation according to the inclusion relation between two interval sets. The preference relationship between x and y could be measured by $\mathcal{A}$ and $\mathcal{B}$, namely, $x \preceq_w y$ if and only if $\mathcal{A} \sqsubseteq_w \mathcal{B}$. According to Eq. (1), we can see that $\subseteq$ can be used to compare the lower bound of $\mathcal{A}$ and $\mathcal{B}$ if $\mathcal{A} \sqsubseteq_w \mathcal{B}$. As $\subseteq$ is irreflexive and transitive, we have a function $m : 2^U \to \Re$ satisfying $A_l \subseteq B_l \Rightarrow m(A_l) \leq m(B_l)$ [14] . Let $f(x) = m(A_l)$ and $f(y) = m(B_l)$, we have $f(x) \leq f(y)$ if $A_l \subseteq B_l$. According to the above analysis, there is a function

f satisfying the condition: $x \preceq_w y \Leftrightarrow f(x) \leq f(y)$ if and only if f satisfies $\mathcal{A} \sqsubseteq_w \mathcal{B} \Rightarrow f(x) \leq f(y)$. This relation could be represented by $(I(2^U), \sqsubseteq_w, \preceq_w)$.

In order to measure the preference relationship $x \preceq_{w+} y$ if and only if $\mathcal{A} \sqsubseteq_{w+} \mathcal{B}$, we can use $\subseteq$ to compare the up bound of $\mathcal{A}$ and $\mathcal{B}$ if $\mathcal{A} \sqsubseteq_{w+} \mathcal{B}$ according to Eq. (2). And then we have a function $m : 2^U \to \Re$ satisfying $m(A_u) \leq m(B_u)$ if $A_u \subseteq B_u$ [14]. For an interval set $\mathcal{A} = [A_l, A_u]$, let $f(x) = m(A_u)$. So we can measure $x \preceq_{w+} y$ by using $f(x) \leq f(y)$ if $A_u \subseteq B_u$. This relation could be represented by $(I(2^U), \sqsubseteq_{w+}, \preceq_{w+})$.

We need two functions to measure the preference relationship $x \preceq_p y$ if and only if $\mathcal{A} \sqsubseteq_p \mathcal{B}$. According to Eq. (3), we can use $\subseteq$ to compare the lower and up bound of $\mathcal{A}$ and $\mathcal{B}$ if $\mathcal{A} \sqsubseteq_p \mathcal{B}$. And then we have the function $m : 2^U \to \Re$ satisfying $m(A_l) \leq m(B_l)$ and $m(A_u) \leq m(B_u)$ if $(A_l \subseteq B_l) \wedge (A_u \subseteq B_u)$ [14]. For an interval set $\mathcal{A} = [A_l, A_u]$, let $f_1(x) = m(A_l)$ and $f_2(x) = m(A_u)$, and we have two function f_1, f_2 satisfying $x \preceq_p y \Leftrightarrow f_1(x) \leq f_1(y)$ and $f_2(x) \leq f_2(y)$. Obviously, $f_1(x) \leq f_2(x)$ as $A_l \subseteq A_u$. This relation could be represented by $(I(2^U), \sqsubseteq_p, \preceq_p)$

As $x \preceq_c y$ if and only if $\mathcal{A} \sqsubseteq_c \mathcal{B}$, we have $m(A_u) \leq m(B_l)$ if $A_u \subseteq B_l$. For an interval set $\mathcal{A} = [A_l, A_u]$, let $f_1(y) = m(B_l)$ and $f_2(x) = m(A_u)$. So we have $x \preceq_c y \Leftrightarrow f_2(x) \leq f_1(y)$ if $\mathcal{A} \sqsubseteq_c \mathcal{B}$. Obviously, $f_1(x) \leq f_2(x)$ as $A_l \subseteq A_u$. This relation could be represented by $(I(2^U), \sqsubseteq_c, \preceq_c)$.

The Table 1 summaries that we can obtain three different preference structures, namely, weak orders, bi-weak orders and interval orders from four inclusion relations between two interval sets.

There is obvious relation among the inclusion relations between two interval sets, that is, for two interval sets $\mathcal{A}$ and $\mathcal{B}$, $\mathcal{A} \sqsubseteq_c \mathcal{B} \Rightarrow \mathcal{A} \sqsubseteq_p \mathcal{B} \Rightarrow \mathcal{A} \sqsubseteq_w \mathcal{B}$, and $\mathcal{A} \sqsubseteq_c \mathcal{B} \Rightarrow \mathcal{A} \sqsubseteq_p \mathcal{B} \Rightarrow \mathcal{A} \sqsubseteq_{w+} \mathcal{B}$ [7]. Figure 1 shows the relation among the preference structures. The line with arrows in Fig. 1 shows a structure can be obtained by another structure.

4   Granular Structures in Interval Sets from Set-Theoretic Perspectives

A set-theoretic granular structure was proposed in [14]. In [14], the lattice operation ($\vee$ and $\wedge$) and the set-theoretic operation ($\cap$ and $\cup$) are used to construct models for granular structures. In this paper, we will show how to use interval set-theoretic operation $\sqcap$ and $\sqcup$ to construct granular structures from the perspective of interval sets.

Table 1. Preference Structure and Relation

Preference structure	Relation
Weak order	$(\sqsubseteq_w, \preceq_w)$ or $(\sqsubseteq_{w+}, \preceq_{w+})$
Bi-weak order	$(\sqsubseteq_p, \preceq_p)$
Interval order	$(\sqsubseteq_c, \preceq_c)$

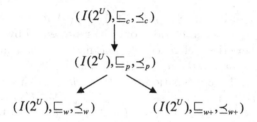

Fig. 1. Relationships among preference structures in interval sets

Definition 8. A granular structure $(G, \sqsubseteq_*)$ is a $\sqcap$-closed granular structure, denoted by $(G, \sqsubseteq_*, \sqcap)$, if $\forall \mathcal{A}, \mathcal{B}$, $\mathcal{A} \sqcap \mathcal{B}$ is in G. A granular structure $(G, \sqsubseteq_*)$ is a $\sqcup$-closed granular structure, denoted by $(G, \sqsubseteq_*, \sqcup)$, if $\forall \mathcal{A}, \mathcal{B} \in G$, $\mathcal{A} \sqcup \mathcal{B}$ is in G. A granular structure $(G, \sqsubseteq_*)$ is a $(\sqcap, \sqcup)$-closed granular structure, denoted by $(G, \sqcap, \sqcup)$, if $\forall \mathcal{A}, \mathcal{B} \in G$, both $\mathcal{A} \sqcap \mathcal{B}$ and $\mathcal{A} \sqcup \mathcal{B}$ are in G.

The above definition describes the granular structure based on $\sqcap$ and $\sqcup$ operations. How do we construct the granular structure by the use of granular structure based on the standard set? Fortunately, Li et al. proposed how to use the lower or up bound to represent inclusion relations on interval sets as shown in Eqs. (1)–(4) [7], which bridges the granular structure of standard sets and interval sets. The following sessions will discuss how to construct the granular structure by the use of lower or up bound of interval sets.

4.1   Granular Structure for $(I(2^U), \sqsubseteq_w)$

The weak inclusion relation $\sqsubseteq_w$ defines a kind of inclusion relation between two interval sets. According to Eq. (1), the inclusion relation between two interval sets could be measured by their lower bounds. It means that the inclusion relation between two interval sets could be judged through two sets. Based on the granular structure framework proposed by [14], the lattice operation ($\vee$ and $\wedge$) and the set-theoretic operation ($\cap$ and $\cup$) could be used to construct the granular structure. The following two definitions shows the lower bounds of two interval sets contributes the granular structure based on the weak inclusion.

Definition 9. Let L be the lower bound set of interval sets in G. A granular structure $(G, \sqsubseteq_w)$ is a meet-semilattice, denoted by $(G, \sqsubseteq_w, \wedge)$, if $\forall \mathcal{A} = [A_l, A_u], \mathcal{B} = [B_l, B_u] \in G$, $A_l \wedge B_l$ is in L. A granular structure $(G, \sqsubseteq_w)$ is a join-semilattice, denoted by $(G, \sqsubseteq_w, \vee)$, if $\forall \mathcal{A}, \mathcal{B} \in G$, $A_l \vee B_l$ is in L. A granular structure $(G, \sqsubseteq_w)$ is a lattice, denoted by $(G, \sqsubseteq_w, \wedge, \vee)$, if both $A_l \wedge B_l$ and $A_l \vee B_l$ are in L.

Definition 10. Let L be the lower bound set of interval sets in G. A granular structure $(G, \sqsubseteq_w)$ is a $\cap$-closed granular structure, denoted by $(G, \sqsubseteq_w, \cap)$, if $\forall \mathcal{A} = [A_l, A_u], \mathcal{B} = [B_l, B_u] \in G$, $A_l \cap B_l$ is in L; a granular structure $(G, \sqsubseteq_w)$ is a $\cup$-closed granular structure, denoted by $(G, \sqsubseteq_w, \cup)$, if $\forall \mathcal{A}, \mathcal{B} \in G$, $A_l \cup B_l$ is in L; a granular structure $(G, \sqsubseteq_w)$ is a $(\cap, \cup)$-closed granular structure, denoted by $(G, \sqsubseteq_w, \cap, \cup)$, if both $A_l \cap B_l$ and $A_l \cup B_l$ are in G.

4.2   Granular Structure for $(I(2^U), \sqsubseteq_{w+})$

According to Eq. (2), the following two definitions shows the up bounds of two interval sets contributes the granular structure based on the weak+ inclusion.

Definition 11. Let U be the up bound set of interval sets in G. A granular structure $(G, \sqsubseteq_{w+})$ is a meet-semilattice, denoted by $(G, \sqsubseteq_{w+}, \wedge)$, if $\forall \mathcal{A} = [A_l, A_u], \mathcal{B} = [B_l, B_u] \in G$, $A_u \wedge B_u$ is in U. A granular structure $(G, \sqsubseteq_{w+})$ is a join-semilattice, denoted by $(G, \sqsubseteq_{w+}, \vee)$, if $\forall \mathcal{A} = [A_l, A_u], \mathcal{B} = [B_l, B_u] \in G$, $A_u \vee B_u$ is in U. A granular structure $(G, \sqsubseteq_{w+})$ is a lattice, denoted by $(G, \sqsubseteq_{w+}, \wedge, \vee)$, if both $A_u \wedge B_u$ and $A_u \vee B_u$ are in U.

Definition 12. Let U be the up bound set of interval sets in G. A granular structure $(G, \sqsubseteq_{w+})$ is a $\cap$-closed granular structure, denoted by $(G, \sqsubseteq_{w+}, \cap)$, if $A_u \cap B_u$ is in U; a granular structure $(G, \sqsubseteq_{w+})$ is a $\cup$-closed granular structure, denoted by $(G, \sqsubseteq_{w+}, \cup)$, for $\forall \mathcal{A} = [A_l, A_u], \mathcal{B} = [B_l, B_u] \in G$, $A_u \cup B_u$ is in U; a granular structure $(G, \sqsubseteq_{w+})$ is a $(\cap, \cup)$-closed granular structure, denoted by $(G, \sqsubseteq_{w+}, \cap, \cup)$, if both $A_u \cap B_u$ and $A_u \cup B_u$ are in G.

4.3   Granular Structure for $(I(2^U), \sqsubseteq_p)$

The relationship between the lower and upper bounds and the possible inclusion on interval sets could be given in Eq. (3). We have the following definitions.

Definition 13. Let LU be the lower and up bound set of interval sets in G. A granular structure $(G, \sqsubseteq_p)$ is a meet-semilattice, denoted by $(G, \sqsubseteq_p, \wedge)$, if $\forall \mathcal{A} = [A_l, A_u], \mathcal{B} = [B_l, B_u] \in G$, both $A_l \wedge B_l$ and $A_u \wedge B_u$ are in LU. A granular structure $(G, \sqsubseteq_p)$ is a join-semilattice, denoted by $(G, \sqsubseteq_p, \vee)$, if $\forall \mathcal{A} = [A_l, A_u], \mathcal{B} = [B_l, B_u] \in G$, both $A_l \vee B_l$ and $A_u \vee B_u$ are in LU. A granular structure $(G, \sqsubseteq_p)$ is a lattice, denoted by $(G, \sqsubseteq_p, \wedge, \vee)$, if $A_l \vee B_l$, $A_u \vee B_u$, $A_l \wedge B_l$ and $A_u \wedge B_u$ are in LU.

Definition 14. Let LU be the lower and up bound set of interval sets in G. A granular structure $(G, \sqsubseteq_p)$ is a $\cap$-closed granular structure, denoted by $(G, \sqsubseteq_p, \cap)$, if $\forall \mathcal{A} = [A_l, A_u], \mathcal{B} = [B_l, B_u] \in G$, both $A_l \cap B_l$ and $A_u \cap B_u$ are in LU; a granular structure $(G, \sqsubseteq_p)$ is a $\cup$-closed granular structure, denoted by $(G, \sqsubseteq_p, \cup)$, if $\forall \mathcal{A} = [A_l, A_u], \mathcal{B} = [B_l, B_u] \in G$, $A_l \cup B_l$ and $A_u \cup B_u$ are in LU; a granular structure $(G, \sqsubseteq_p)$ is a $(\cap, \cup)$-closed granular structure, denoted by $(G, \sqsubseteq_p, \cap, \cup)$, if $A_l \cap B_l$, $A_u \cap B_u$, $A_l \cup B_l$ and $A_u \cup B_u$ are in LU.

4.4   Granular Structure for $(I(2^U), \sqsubseteq_c)$

The relationship between the bounds and the certain inclusion on interval sets could be given in Eq. (4), and then we have:

Definition 15. Let LU be the lower and up bound set of interval sets in G. A granular structure $(G, \sqsubseteq_c)$ is a meet-semilattice, denoted by $(G, \sqsubseteq_c, \wedge)$, if $\forall \mathcal{A} = [A_l, A_u], \mathcal{B} = [B_l, B_u] \in G$, $A_u \wedge B_l$ is in LU. A granular structure $(G, \sqsubseteq_c)$ is a join-semilattice, denoted by $(G, \sqsubseteq_c, \vee)$, if $\forall \mathcal{A} = [A_l, A_u], \mathcal{B} = [B_l, B_u] \in G$, $A_u \vee B_l$ is in LU. A granular structure $(G, \sqsubseteq_c)$ is a lattice, denoted by $(G, \sqsubseteq_c, \wedge, \vee)$, both $A_u \vee B_l$ and $A_u \wedge B_l$ are in LU.

Definition 16. Let LU be the lower and up bound set of interval sets in G. A granular structure $(G, \sqsubseteq_c)$ is a $\cap$-closed granular structure, denoted by $(G, \sqsubseteq_c, \cap)$, if $\forall \mathcal{A} = [A_l, A_u], \mathcal{B} = [B_l, B_u] \in G$, $A_u \cap B_l$ is in LU; a granular structure $(G, \sqsubseteq_c)$ is a $\cup$-closed granular structure, denoted by $(G, \sqsubseteq_c, \cup)$, if $\forall \mathcal{A} = [A_l, A_u], \mathcal{B} = [B_l, B_u] \in G$, $A_u \cup B_l$ is in LU; a granular structure $(G, \sqsubseteq_p)$ is a $(\cap, \cup)$-closed granular structure, denoted by $(G, \sqsubseteq_c, \cap, \cup)$, both $A_u \cap B_l$ and $A_u \cup B_l$ are in LU.

5   Connections of Rough Sets and Interval Sets

The rough set model provides a method to construct an interval set [12,16]. For a subset of objects $A \subseteq U$, it may be approximated by a pair of lower and upper approximations:

$$\underline{apr}(A) = \cup\{X \in \sigma(U/E) | X \subseteq A\}$$
$$\overline{apr}(A) = \cap\{X \in \sigma(U/E) | A \subseteq X\} \tag{13}$$

According to the property, $\underline{apr}(A) \subseteq \overline{apr}(A)$, a rough set $(\underline{apr}(A), \overline{apr}(A))$ forms an interval set $[\underline{apr}(A), \overline{apr}(A)]$, which means A lies in $[\underline{apr}(A), \overline{apr}(A)]$. Suppose $[\underline{apr}(A), \overline{apr}(A)]$ and $[\underline{apr}(B), \overline{apr}(B)]$ are the interval sets generated by the rough sets of A and B, the pair of approximations satisfies the following properties [10].

$$\underline{apr}(A \cap B) = \underline{apr}(A) \cap \underline{apr}(B)$$
$$\underline{apr}(A \cup B) \supseteq \underline{apr}(A) \cup \underline{apr}(B)$$
$$\overline{apr}(A \cup B) = \overline{apr}(A) \cup \overline{apr}(B)$$
$$\overline{apr}(A \cap B) \subseteq \overline{apr}(A) \cap \overline{apr}(B) \tag{14}$$

These properties conform to the definition of operations on interval sets. From these properties, it is evident that the rough set model provides a method for constructing an interval set and justifies the use of interval set operations [13]. So we can extend the method of constructing the granular structure based on the set theory into the interval set. Taking $(\cap, \cup)$-closed structure as an example, we have:

for $(G, \sqsubseteq_w, \cap, \cup)$,

$$\underline{apr}(A) = \cup\{X | X \in L, X \subseteq A\}$$
$$\overline{apr}(A) = \cap\{X | X \in L, A \subseteq X\} \tag{15}$$

for $(G, \sqsubseteq_{w+}, \cap, \cup)$,

$$\underline{apr}(A) = \cup\{X|X \in U, X \subseteq A\}$$
$$\overline{apr}(A) = \cap\{X|X \in U, A \subseteq X\} \qquad (16)$$

for $(G, \sqsubseteq_p, \cap, \cup)$,

$$\underline{apr}(A) = \cup\{X|X \in LU, X \subseteq A\}$$
$$\overline{apr}(A) = \cap\{X|X \in LU, A \subseteq X\} \qquad (17)$$

for $(G, \sqsubseteq_c, \cap, \cup)$,

$$\underline{apr}(A) = \cup\{X \cap Y|X \in L, Y \in U, X \cap Y \subseteq A\}$$
$$\overline{apr}(A) = \cap\{X \cup Y|X \in L, Y \in U, A \subseteq X \cup Y\} \qquad (18)$$

6   Conclusions

Based on the thriarchic theory of granular computing, this paper extends the results of granular structures in standard sets to interval sets from a measurement-theoretic perspective and set-theoretic perspective, respectively.

On the one hand, this paper discussed the order structures for objects represented by interval sets using inclusion relations on interval sets, which are extended from the results of representing preferences using interval numbers in [6]. This paper provides useful insights for ordering objects represented by interval sets from the measurement-theoretic perspective.

On the other hand, the granular structures of interval sets are discussed from the perspectives of interval set operations and set operations. Just like the results of granular structures in standard sets, the objects represented by interval sets can be organized in multi-level structures by the use of the proposed granular structures. In the future work, we will investigate real world applications.

Acknowledgements. Authors thanks Prof. Yiyu Yao from the University of Regina for his constructive suggestions on this paper.

References

1. Allen, J.F.: Maintaining knowledge about temporal intervals. Commun. ACM **26**(11), 832–843 (1983)
2. Chen, G., Zhong, N.: Granular structures in graphs. In: Yao, J.T., Ramanna, S., Wang, G., Suraj, Z. (eds.) RSKT 2011. LNCS, vol. 6954, pp. 649–658. Springer, Heidelberg (2011)
3. Chen, G., Zhong, N.: Three granular structure models in graphs. In: Li, T., Nguyen, H.S., Wang, G., Grzymala-Busse, J., Janicki, R., Hassanien, A.E., Yu, H. (eds.) RSKT 2012. LNCS, vol. 7414, pp. 351–358. Springer, Heidelberg (2012)
4. Fishburn, P.C.: Generalizations of semiorders: a review note. J. Math. Psychol. **41**(1997), 357–366 (1997)

5. Fishburn, P.C.: Utility Theory for Decision Making. Wiley, New York (1970)
6. Ozturk, M., Pirlot, M., Tsoukias, A.: Representing preferences using intervals. Artifi. Intell. **175**(2011), 1194–1222 (2011)
7. Li, H.X., Wang, M.H., Zhou, X.Z., Zhao, J.B.: An interval set model for learning rules from incomplete information table. Int. J. Approx. Reason. **53**(2012), 24–37 (2012)
8. Liu, C., Zhong, N.: Rough problem settings for ILP dealing with imperfect data. Comput. Intell. **17**(3), 446–459 (2001)
9. Moore, R.E.: Interval Analysis. Prentice-Hall, Englewood Cliffs (1966)
10. Pawlak, Z.: Rough sets. Int. J. Comput. Inf. Sci. **11**, 341–356 (1982)
11. Roberts, F.S.: Measurement Theory. Addeison-Wesley, Reading (1979)
12. Yao, Y.Y.: Interval-set algebra for qualitative knowledge representation. In: Proceedings of the 5th International Conference on Computing and Information, pp. 370–374 (1993)
13. Yao, Y.Y.: Interval sets and interval-set algebras. In: Proceedings of the 8th IEEE International Conference on Cognitive Informatics, pp. 307–314 (2009)
14. Yao, Y.Y., Zhao, L.Q.: A measurement theory view on the granularity of partitions. Inf. Sci. **213**, 1–13 (2012)
15. Yao, Y.Y., Zhang, N., Miao, D.Q., Xu, F.F.: Set-theoretic approaches to granular computing. Fundam. Informaticae **115**(2–3), 247–264 (2012)
16. Yao, Y., Hu, M.: A definition of structured rough set approximations. In: Kryszkiewicz, M., Cornelis, C., Ciucci, D., Medina-Moreno, J., Motoda, H., Raś, Z.W. (eds.) RSEISP 2014. LNCS, vol. 8537, pp. 111–122. Springer, Heidelberg (2014)
17. Zhong, N., Dong, J.Z., Ohsuga, S.: Using rough sets with heuristics to feature selection. J. Intell. Inf. Syst. **16**(3), 199–214 (2001)
18. Zhong, N., Dong, J.Z., Liu, C., Ohsuga, S.: A hybrid model for rule discovery in data. Knowl. Based Syst. **14**(7), 397–412 (2001)
19. Zhong, N., Dong, J.Z.: Meningitis data mining by cooperatively using GDT-RS and RSBR. Pattern Recogn. Lett. **24**(6), 887–894 (2003)

Generalized Rough Sets

Empirical Risk Minimization for Variable Consistency Dominance-Based Rough Set Approach

Jerzy Błaszczyński[1], Yoshifumi Kusunoki[2], Masahiro Inuiguchi[3], and Roman Słowiński[1,4(✉)]

[1] Institute of Computing Science, Poznań University of Technology,
Piotrowo 3a, 60-965 Poznań, Poland
{jerzy.blaszczynski,roman.slowinski}@cs.put.poznan.pl
[2] Graduate School of Engineering, Osaka University,
2-1, Yamadaoka Suita, Osaka 565-0871, Japan
kusunoki@eei.eng.osaka-u.ac.jp
[3] Graduate School of Engineering Science, Osaka University,
1-3, Machikaneyama, Toyonaka, Osaka 560-8531, Japan
inuiguti@sys.es.osaka-u.ac.jp
[4] Systems Research Institute, Polish Academy of Sciences, 01-447 Warsaw, Poland

Abstract. The paper concerns reasoning about partially inconsistent ordinal data. We reveal the relation between Variable Consistency Dominance-based Rough Set Approach (VC-DRSA) and the empirical risk minimization problem. VC-DRSA is an extension of DRSA that admits some degree of inconsistency with respect to dominance, which is controlled by thresholds on a consistency measure. To prove this relation, we first solve an optimization problem to find thresholds ensuring assignment of a maximum number of objects under disjoint and balanced setting of extended lower approximations of two complementary unions of ordered decision classes: "at least" class i, and "at most" class $i-1$, for a given $i \in \{2, \ldots, p\}$, where p is the total number of classes. For a given i, each object is supposed to be assigned to at most one of the two extended lower approximations. Moreover, the assignment is not influenced by unions' cardinalities. Second, we prune the set of objects not assigned to any extended lower approximation. Then, from a definable set, for a given i, we derive a classification function, which indicates assignment of an object to one of the two unions of decision classes. We define empirical risk associated with the classification function as a hinge loss function. We prove that the classification function minimizing the empirical risk function corresponds to the extended lower approximation in VC-DRSA involving thresholds obtained from the above optimization problem, followed by the pruning.

Keywords: Rough sets · Variable consistency dominance-based rough set approach · Empirical risk minimization

© Springer International Publishing Switzerland 2015
Y. Yao et al. (Eds.): RSFDGrC 2015, LNAI 9437, pp. 63–72, 2015.
DOI: 10.1007/978-3-319-25783-9_6

1 Introduction

Dominance-based Rough Set Approach (DRSA) [1], provides a methodology for ordinal data analysis under partial inconsistency with respect to dominance. An analyzed data set has the form of a decision table, which consists of objects described by condition attributes and classified into a finite number of decision classes. In DRSA, different values of condition attributes and different ranks of decision classes are totally ordered. Inconsistency with respect to dominance between objects u and u' from the decision table occurs when object u having evaluations on all condition attributes not worse than those of object u', has been assigned to a class worse than the class u'. Thus, reasoning about ordinal data must respect the monotonicity constraint that exists between the evaluation of objects on condition attributes and the ranks of decision classes to which these objects are assigned. To handle this monotonic relationship between conditions and decision, in DRSA, lower and upper approximations concern upward and downward unions of decision classes. There are two extensions of DRSA permitting inclusion of some acceptably inconsistent objects to lower approximations: Variable Consistency DRSA (VC-DRSA) [3] and Variable Precision DRSA (VP-DRSA) [4]. They define extended lower approximations based on different inconsistency measures.

Defining the extended lower approximations can be seen as a classification problem in statistical learning theory [5]. It consists in finding a classification function, which indicates assignment of an object to a decision class. The best function is selected from predefined category of functions by minimizing its empirical risk for a decision table with respect to a specific loss function. In [6], we have shown the relationship between definition of approximations in VP-DRSA and minimization of an empirical risk function. We demonstrated that minimization of the hinge loss function is equivalent to calculation of extended lower approximations using inconsistency measure defined for VP-DRSA. In this paper, we undertake an analogical study with respect to VC-DRSA. The value of such studies lies in interpretation of various versions of definitions of lower approximations in terms of statistical learning theory. They bridge the gap between rough set theory and statistical learning theory or, in general, machine learning.

First, given p ordered decision classes, we formulate an optimization problem, for two complementary unions of decision classes: "at least" class i, and "at most" class $i - 1$, for $i \in \{2, \ldots, p\}$. The problem aims at maximizing the number of objects assigned to extended lower approximations of the complementary unions under constraint that each object is assigned to at most one of the lower approximations. The same optimization problem ensures that thresholds on the inconsistency measure, which condition assignment of objects to extended lower approximations, are proportional to the cardinality of the complementary unions of classes. Second, since the solution of the above optimization problem leaves some inconsistent objects unassigned to any of the two extended lower approximations, we introduce a pruning procedure, which removes these objects from consideration, for a given $i \in \{2, \ldots, p\}$. We define classification functions corresponding to acceptably inconsistent sets of objects for a given i, and we define an empirical risk function for these classification functions. We prove that, for each $i \in \{2, \ldots, p\}$, the classification functions corresponding to the extended lower approximations in VC-DRSA minimize the empirical risk function over the pruned set of objects. This paper is organized as follows.

In Sect. 2, VC-DRSA is briefly introduced. ϵ-consistency measure and the extended lower and upper approximations are defined. In Sect. 3, we define the empirical risk function, and characterize the lower approximation from the viewpoint of risk minimization. Concluding remarks are given in Sect. 4.

2   Variable Consistency Dominance-Based Rough Set Approach

A decision table is a tuple $(U, AT = C \cup \{d\}, V)$, where $U = \{u_1, \ldots, u_n\}$ is a finite set of objects, $C = \{c_1, \ldots, c_m\}$ is a finite set of condition attributes, d is a decision attribute, and V is a set of all attribute values. For each object $u \in U$ and attribute $a \in AT$, $a(u) \in V$ is a value of attribute a for object u. We denote by $V_a \subseteq V$ a value set of attribute $a \in AT$. For $A \subseteq C$, let $V_A = \Pi_{a \in A} V_a$ be the Cartesian product of value sets of condition attributes from subset A.

The value set of the single decision attribute d is denoted by $V_d = \{1, 2, \ldots, p\}$. This set is totally ordered by relation $\geq_d$. Decision attribute d makes a partition of the set of objects U into p subsets $\{X_1, X_2 \ldots, X_p\}$, each of which is called a decision class, where $X_i = \{u \in U : d(u) = i\}$. By total order $\geq_d$, an upward union and a downward union of classes with respect to decision class X_i are defined, respectively, as:

$$X_i^{\geq} = \bigcup_{j \geq i} X_j, \ i = 2, \ldots, p, \tag{1}$$

$$X_i^{\leq} = \bigcup_{j \leq i} X_j, \ i = 1, \ldots, p-1. \tag{2}$$

The set C of condition attributes is divided into C_N and C_C. Set C_N is composed of nominal attributes with value sets which are neither ordered with respect to preference, nor monotonically related with the value set of decision attribute d, while set C_C is composed of attributes with value sets which are totally ordered with respect to preference and monotonically related with the value set of decision attribute d. The latter attributes are called criteria when the decision table includes examples of multiple criteria classification. For $a \in C_C$, the total order on value set V_a is denoted by $\geq_a$. We assume that all attributes from C_C are of the gain-type, i.e., the greater $a(u)$, the better decision class X_i object u may belong.

For $A \subseteq C$, a dominance relation D_A on U is defined by:

$$D_A = \{(u, u') \in U^2 : a(u) \geq_a a(u'), \forall a \in C_C \cap A \text{ and}$$
$$a(u) = a(u'), \forall a \in C_N \cap A\}. \tag{3}$$

D_A satisfies reflexively and transitivity. For any object $u \in U$, its dominating set and its dominated set are defined, respectively, as:

$$D_A^+(u) = \{u' \in U : (u', u) \in D_A\}, \qquad (4)$$

$$D_A^-(u) = \{u' \in U : (u, u') \in D_A\}. \qquad (5)$$

For any subset of condition attributes $A \subseteq C$, two complementary unions of decision classes $X_i^{\geq}$, $X_{i-1}^{\leq}$ ($X_i^{\geq} \cup X_{i-1}^{\leq} = U$, $i = 2, \ldots, p$), and each $u \in U$, we define ϵ-consistency measure of u with respect to $X_i^{\geq}$ and $X_{i-1}^{\leq}$, respectively, as:

$$\epsilon_A^{X_i^{\geq}}(u) = \frac{|D_A^+(u) \cap X_{i-1}^{\leq}|}{|X_{i-1}^{\leq}|}, \qquad (6)$$

$$\epsilon_A^{X_{i-1}^{\leq}}(u) = \frac{|D_A^-(u) \cap X_i^{\geq}|}{|X_i^{\geq}|}, \qquad (7)$$

values of ϵ-consistency measure can be interpreted as estimates of conditional probability $Pr(u' \in D_A^+(u) \,|\, u' \in X_{i-1}^{\leq})$ and $Pr(u' \in D_A^-(u) \,|\, u' \in X_i^{\geq})$, respectively. In other words, it is the number of objects in the dominance cone of object u that do not belong to the considered union of classes, divided by the number of all those objects that do not belong to the considered union of classes. This type of conditional probability is called catch-all likelihood [2]. ϵ-consistency measure satisfies desirable monotonicity properties defined for consistency measures in [3].

Using the ϵ-consistency measure, we can define parameterized lower approximations of the two complementary unions $X_i^{\geq}$ and $X_{i-1}^{\leq}$, in the following way. Let $\lambda_i^{\geq}, \lambda_{i-1}^{\leq} \in (0, 1]$ be some consistency thresholds, then lower approximations of unions $X_i^{\geq}$ and $X_{i-1}^{\leq}$ are defined, respectively, as:

$$\underline{X_i^{\geq}}(\lambda_i^{\geq}) = \{u \in U : \epsilon_A^{X_i^{\geq}}(u) < \lambda_i^{\geq} \wedge \epsilon_A^{X_{i-1}^{\leq}}(u) > 0\}, \qquad (8)$$

$$\underline{X_{i-1}^{\leq}}(\lambda_{i-1}^{\leq}) = \{u \in U : \epsilon_A^{X_{i-1}^{\leq}}(u) < \lambda_{i-1}^{\leq} \wedge \epsilon_A^{X_i^{\geq}}(u) > 0\}. \qquad (9)$$

Remark 1. We omit "A" in further formulas because in the rest of the paper we consider only the complete set of condition attributes C. Moreover, we are considering only one pair of complementary unions with respect to some X_i because the situation is the same for any $i \in \{2, \ldots, p\}$.

Observe that, according to definitions (8), (9), the two lower approximations are not necessarily disjoint. The property of disjoint lower approximations of complementary unions is, however, one of the basic properties postulated for DRSA. It also holds for variable-precision and variable-consistency dominance-based rough set approaches [3,4]. In order to assure that this property also holds for parameterized lower approximations defined by (8), (9), i.e., $\underline{X_i^{\geq}}(\lambda_i^{\geq}) \cap \underline{X_{i-1}^{\leq}}(\lambda_{i-1}^{\leq}) = \emptyset$, we have to find appropriate values of $\lambda_i^{\geq}, \lambda_{i-1}^{\leq}$.

For all objects inconsistent with respect to assignment to both complementary unions of classes, i.e., $u : \epsilon^{X_i^{\geq}}(u) > 0 \wedge \epsilon^{X_{i-1}^{\leq}}(u) > 0$, we formulate the following optimization problem where $\lambda_i^{\geq}, \lambda_{i-1}^{\leq}$ are decision variables:

$$\text{minimize} \quad \sum_{u \,:\, \epsilon^{X_i^{\geq}}(u)>0 \,\wedge\, \epsilon^{X_{i-1}^{\leq}}(u)>0} (v_u + v_u') \tag{10}$$

$$\text{subject to} \quad \epsilon^{X_i^{\geq}}(u) - v_u \mathbf{M} < \lambda_i^{\geq}, \tag{11}$$

$$\epsilon^{X_{i-1}^{\leq}}(u) - v_u' \mathbf{M} < \lambda_{i-1}^{\leq}, \tag{12}$$

$$\epsilon^{X_i^{\geq}}(u) + v_u'' \mathbf{M} \geq \lambda_i^{\geq}, \tag{13}$$

$$\epsilon^{X_{i-1}^{\leq}}(u) + v_u''' \mathbf{M} \geq \lambda_{i-1}^{\leq}, \tag{14}$$

$$v_u'' + v_u''' \leq 1, \tag{15}$$

$$v_u, v_u', v_u'', v_u''' \in \{0, 1\}, \tag{16}$$

$$\lambda_{i-1}^{\leq}|X_i^{\geq}| = \lambda_i^{\geq}|X_{i-1}^{\leq}|, \tag{17}$$

$$0 \leq \lambda_i^{\geq} \leq 0.5, \tag{18}$$

$$0 \leq \lambda_{i-1}^{\leq} \leq 0.5, \tag{19}$$

$$\mathbf{M} \gg 0, \tag{20}$$

where v_u, v_u', v_u'', v_u''' are auxiliary 0–1 variables associated with each inconsistent object u, i.e., such an object that $\epsilon^{X_i^{\geq}}(u) > 0 \wedge \epsilon^{X_{i-1}^{\leq}}(u) > 0$, and $\mathbf{M}$ is a big positive number. The auxiliary variables multiplied by $\mathbf{M}$ control the assignment of an inconsistent object to one of the two lower approximations. Precisely, when $v_u = 0$, then due to constraint (11) object u is assigned to $\underline{X}_i^{\geq}(\lambda_i^{\geq})$, otherwise, when $v_u = 1$, then object u is not assigned to this lower approximation. Analogously, due to constraint (12), when $v_u' = 0$, then object u is assigned to $\underline{X}_{i-1}^{\leq}(\lambda_{i-1}^{\leq})$, otherwise, when $v_u' = 1$, then object u is not assigned to this lower approximation. Further, when $v_u'' = 0$, then due to constraint (13) object u cannot be assigned to $\underline{X}_i^{\geq}(\lambda_i^{\geq})$, otherwise, when $v_u'' = 1$, then this assignment is possible, and, analogously, when $v_u''' = 0$, then due to constraint (14) object u cannot be assigned to $\underline{X}_{i-1}^{\leq}(\lambda_{i-1}^{\leq})$, otherwise, when $v_u''' = 1$, then this assignment is possible. To ensure that object u is assigned to at most one of the two lower approximations, the constraint (15) has to be satisfied. Thus, objective (10) tends to maximize the number of inconsistent objects assigned disjointly to the lower approximations of the complementary unions $X_i^{\geq}, X_{i-1}^{\leq}$. Finally, constraint (17) makes thresholds $\lambda_i^{\geq}, \lambda_{i-1}^{\leq}$ inversely proportional to the cardinalities of the complementary unions. Constraints (18)–(19), impose an arbitrary 50 % limit on the part of objects from the complementary union in the dominance cone based on u. Note that the optimal solution of (10)–(20) always exists (in extreme case, it is equivalent to DRSA definition of lower approximations), however, there may be also many equivalent optimal solutions.

For complementary unions $X_i^{\geq}$, $X_{i-1}^{\leq}$, we prune the universe U to the following set U_i':

$$U_i' = U \setminus \{u \in U : \epsilon^{X_i^{\geq}}(u) > 0, \ \epsilon^{X_{i-1}^{\leq}}(u) > 0, \ v_u + v_u' = 2\}. \tag{21}$$

U_i' is composed of only those objects which are consistent enough to be included in some lower approximation. Regarding the set U_i', we mention the next remark.

Remark 2. The following statements hold.

(a) $\underline{X_i^{\geq}}(\lambda_i^{\geq})$ and $\underline{X_{i-1}^{\leq}}(\lambda_{i-1}^{\leq})$ form a partition of U_i'.

(b) There is no object u in U_i' which satisfies both of $\epsilon^{X_i^{\geq}}(u) \geq \lambda_i^{\geq}$ and $\epsilon^{X_{i-1}^{\leq}}(u) \geq \lambda_{i-1}^{\leq}$.

(c) For each $u \in U_i'$, if both of $\epsilon^{X_i^{\geq}}(u) < \lambda_i^{\geq}$ and $\epsilon^{X_{i-1}^{\leq}}(u) < \lambda_{i-1}^{\leq}$ are fulfilled, then exactly one of $\epsilon^{X_i^{\geq}}(u) = 0$ and $\epsilon^{X_{i-1}^{\leq}}(u) = 0$ holds.

3   Empirical Risk Minimization

We show a relation between VC-DRSA and empirical risk minimization problem. We consider optimal discriminant sets for $X_i^{\geq}$ and $X_{i-1}^{\leq}$ with respect to a risk function. For such a pair of unions of classes, we have a family of sets defined as:

$$\mathcal{W}^+ = \left\{ W^+ \subseteq U : W^+ = \bigcup_{u \in W^+} D^+(u) \right\}, \tag{22}$$

$$\mathcal{W}^- = \left\{ W^- \subseteq U : W^- = \bigcup_{u \in W^-} D^-(u) \right\}. \tag{23}$$

Any $W^+ \in \mathcal{W}^+$ and $W^- \in \mathcal{W}^-$ is a definable set of objects. Thus, it can be interpreted as a simple classifier that assigns any object from one set to another set. In the following, we assume that $W^+ \cap W^- = \emptyset$. Let $\mathcal{W} = \{(W^+, W^-) \in \mathcal{W}^+ \times \mathcal{W}^- : W^+ \cap W^- = \emptyset\}$. An inference rule of a classifier $W \in \mathcal{W}$ is defined as:

$$u \text{ is assigned to } \begin{cases} X_i^{\geq} & \text{if } u \in W^+, \\ X_{i-1}^{\leq} & \text{if } u \in W^-. \end{cases} \tag{24}$$

In consequence, any $u \in U$ can be found in one of three situations:

- u is assigned to W^+,
- u is assigned to W^-,
- u is assigned neither to W^+ nor to W^-.

Similarly to [6], we use hinge loss function. To introduce this function we consider real-valued classifiers f_W, which are based on W:

$$f_W(u) = \begin{cases} |D^-(u) \cap W^+| & \text{if } u \in W^+, \\ -|D^+(u) \cap W^-| & \text{if } u \in W^-. \\ 0 & \text{if } u \notin W^+ \cup W^- \end{cases} \tag{25}$$

Positive value of $f_W(u)$ indicates the number of objects belonging to W^+ supporting $u \in W^+$. Negative value of $f_W(u)$ indicates the number of objects belonging to W^- supporting $u \in W^-$. Figure 1 shows an illustration of W^-, W^+, and f_W. In this example, there are 10 objects (5 circles and 5 squares). f_W of the top circle object u_i is 5 because there are 5 objects dominated by u_i and included in W^+. Similarly, f_W of the bottom square object u_j is -3.

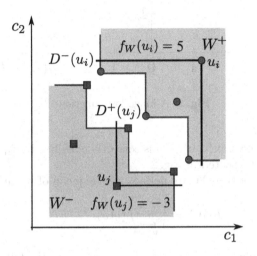

Fig. 1. W^-, W^+, and f_W

Analogically to logistic regression and support vector machine, the sign of f_W indicates classification:

$$u \text{ is classified to } \begin{cases} X_i^{\geq} & \text{if } f_W(u) > 0, \\ X_{i-1}^{\leq} & \text{if } f_W(u) < 0, \end{cases} \tag{26}$$

$f_W(u)$ can be seen as quantification of how far u is from the boundary between the two disjoint unions of classes.

Moreover, we introduce $y(u)$ function which assigns value according to labeling in the training set:

$$y(u) = \begin{cases} 1 & \text{if } u \in X_i^{\geq}, \\ -1 & \text{if } u \in X_{i-1}^{\leq}. \end{cases} \tag{27}$$

When $y(u)$ and $f_W(u)$ have the same sign, u is correctly classified by $f_W(u)$. On the other hand, when the signs of $y(u)$, and $f_W(u)$ differ, u is misclassified by $f_W(u)$.

For each $u \in U$, we define two hinge loss functions, in the following way:

$$L(y(u), f_W(u)) = [-y(u) \times f_W(u)]_+, \tag{28}$$

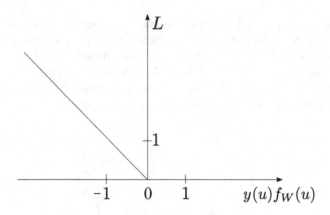

Fig. 2. A hinge loss function

where $[\diamond]_+$ stands for $\max\{\diamond, 0\}$. If u is correctly classified by $f_W(u)$, then $L = 0$. The shape of L is presented in Fig. 2.

Empirical risks of $W \in \mathcal{W}$ are defined by the mean of hinge loss over U, i.e.,

$$R(W) = \frac{1}{n} \sum_{u \in U} L(y(u), f_W(u)), \tag{29}$$

where n is the number of objects $u \in U$. We can prove the following theorems.

Theorem 1. *Let $\lambda_i^{\geq}$, $\lambda_{i-1}^{\leq}$ and U_i' are obtained from (10)–(20), and $W_* \in \mathcal{W}$. W_* minimizes the empirical risk function R under the constraint $W^+ \cup W^- = U_i'$, equivalent to pruning (21), if and only if*

$$
\begin{aligned}
W_*^+ &= \underline{X}_i^{\geq}(\lambda_i^{\geq}) \\
&= \{u \in U_i' : |D^+(u) \cap X_{i-1}^{\leq}| < \lambda_i^{\geq}|X_{i-1}^{\leq}| \wedge |D^-(u) \cap X_i^{\geq}| > 0\}, \tag{30} \\
W_*^- &= \underline{X}_{i-1}^{\leq}(\lambda_{i-1}^{\leq}) \\
&= \{u \in U_i' : |D^-(u) \cap X_i^{\geq}| < \lambda_{i-1}^{\leq}|X_i^{\geq}| \wedge |D^+(u) \cap X_{i-1}^{\leq}| > 0\}. \tag{31}
\end{aligned}
$$

Note that (30) corresponds directly to definition of parameterized lower approximation of union (8), while (31) corresponds directly to definition of parameterized lower approximation of union (9).

To prove Theorem 1, we first prove the following lemma.

Lemma 1. *We have,*

$$R(W) = \frac{1}{n}\left(\sum_{u' \in W^+} |D^+(u') \cap X_{i-1}^{\leq}| + \sum_{u' \in W^-} |D^-(u') \cap X_i^{\geq}| \right). \tag{32}$$

Proof. (Lemma 1)

$$nR(W) = \sum_{u \in U} L(y(u), f_W(u)) \tag{33}$$

$$= \sum_{u \in W^+ \cap X_{i-1}^\leq} |D^-(u) \cap W^+| + \sum_{u \in W^- \cap X_i^\geq} |D^+(u) \cap W^-| \tag{34}$$

$$= \left|(u, u') \in U^2 : u \in W^+ \cap X_{i-1}^\leq \wedge u' \in D^-(u) \cap W^+\right| \tag{35}$$

$$+ \left|(u, u') \in U^2 : u \in W^- \cap X_i^\geq \wedge u' \in D^+(u) \cap W^-\right| \tag{36}$$

$$= \left|(u, u') \in U^2 : u \in D^+(u') \cap W^+ \cap X_{i-1}^\leq \wedge u' \in W^+\right| \tag{37}$$

$$+ \left|(u, u') \in U^2 : u \in D^-(u') \cap W^- \cap X_i^\geq \wedge u' \in W^-\right| \tag{38}$$

$$= \sum_{u' \in W^+} |D^+(u') \cap X_{i-1}^\leq| + \sum_{u' \in W^-} |D^-(u') \cap X_i^\geq|. \tag{39}$$

This ends the proof of Lemma 1.

Proof. (Theorem 1) If we ignore $W_*^+ \in \mathcal{W}^+$ and $W_*^- \in \mathcal{W}^-$, the unique optimal solution W_* for minimizing R under $W_*^+ \cap W_*^- = \emptyset$ and $W_*^+ \cup W_*^- = U_i'$ is,

$$W_*^+ = \{u \in U_i' : |D^+(u) \cap X_{i-1}^\leq| < |D^-(u) \cap X_i^\geq|\},$$
$$W_*^- = \{u \in U_i' : |D^-(u) \cap X_i^\geq| < |D^+(u) \cap X_{i-1}^\leq|\}.$$

From Remark 2(b) and constraint (17), there is no object $u' \in U_i'$ such that $|D^+(u') \cap X_{i-1}^\leq| = \lambda_i^\geq |X_{i-1}^\leq| = \lambda_{i-1}^\leq |X_i^\geq| = |D^-(u') \cap X_i^\geq|$.
     Consider the lower approximations:

$$\underline{X}_i^\geq(\lambda_i^\geq) = \{u \in U : |D^+(u) \cap X_{i-1}^\leq| < \lambda_i^\geq |X_{i-1}^\leq| \wedge |D^-(u) \cap X_i^\geq| > 0\},$$

$$\underline{X}_{i-1}^\leq(\lambda_{i-1}^\leq) = \{u \in U : |D^-(u) \cap X_i^\geq| < \lambda_{i-1}^\leq |X_i^\geq| \wedge |D^+(u) \cap X_{i-1}^\leq| > 0\}.$$

Then, we show that $W_*^+ \cap \underline{X}_{i-1}^\leq(\lambda_{i-1}^\leq) = \{u \in U : |D^+(u) \cap X_{i-1}^\leq| < |D^-(u) \cap X_i^\geq| < \lambda_i^\geq |X_{i-1}^\leq| \wedge |D^+(u) \cap X_{i-1}^\leq| > 0\}$ is an empty set. From Remark 2(c) and constraint (17), $|D^-(u) \cap X_i^\geq| = 0$ for all $u \in W_*^+ \cap \underline{X}_{i-1}^\leq(\lambda_{i-1}^\leq)$, but it contradicts $|D^+(u) \cap X_{i-1}^\leq| < |D^-(u) \cap X_i^\geq|$. Hence, $W_*^+ \cap \underline{X}_{i-1}^\leq(\lambda_{i-1}^\leq) = \emptyset$. Similarly $W_*^- \cap \underline{X}_i^\geq(\lambda_i^\geq) = \emptyset$. Therefore, we have $W_*^+ = \underline{X}_i^\geq(\lambda_i^\geq)$ and $W_*^- = \underline{X}_{i-1}^\leq(\lambda_{i-1}^\leq)$, because $\underline{X}_i^\geq(\lambda_i^\geq)$ and $\underline{X}_{i-1}^\leq(\lambda_{i-1}^\leq)$ are also a partition of U_i'.
     Finally, we can easily check $\underline{X}_i^\geq(\lambda_i^\geq) \in \mathcal{W}^+$ and $\underline{X}_{i-1}^\leq(\lambda_{i-1}^\leq) \in \mathcal{W}^-$. Hence, $(\underline{X}_i^\geq(\lambda_i^\geq), \underline{X}_{i-1}^\leq(\lambda_{i-1}^\leq))$ is the unique optimal solution. $\square$

4   Concluding Remarks

In this paper, we have shown the connection between VC-DRSA and the empirical risk minimization. We have shown that minimization of the hinge loss function corresponds with calculation of extended lower approximations in VC-DRSA. To demonstrate this correspondence, for a given pair of complementary unions of ordered decision classes, we first introduced an optimization problem which aims at finding thresholds ensuring assignment of a maximum number of objects. In this problem, each object is supposed to be assigned to at most one extended lower approximation of the complementary unions of decision classes. Moreover, the assignment is not influenced by unions' cardinalities. Second, we pruned the set of objects by removing from consideration all objects which are not consistent enough to be assigned to any of the two lower approximations. Experimental study is one of our future works.

References

1. Greco, S., Matarazzo, B., Słowiński, R.: Rough sets theory for multicriteria decision analysis. Eur. J. Oper. Res. **129**, 1–47 (2001)
2. Fitelson, B.: Likelihoodism, Bayesianism, and relational confirmation. Synthese **156**, 473–489 (2007)
3. Błaszczyński, J., Greco, S., Słowiński, R., Szelag, M.: Monotonic variable consistency rough set approaches. Int. J. Approx. Reason. **50**, 979–999 (2009)
4. Inuiguchi, M., Yoshioka, Y., Kusunoki, Y.: Variable-precision dominance-based rough set approach and attribute reduction. Int. J. Approx. Reason. **50**, 1199–1214 (2009)
5. Hastie, T., Tibshirani, R., Friedman, J.: The Elements of Statistical Learning: Data Mining, Inference, and Prediction. Springer Series in Statistics. Second Edition, New York (2009)
6. Kusunoki, Y., Błaszczyński, J., Inuiguchi, M., Słowiński, R.: Empirical risk minimization for variable precision dominance-based rough set approach. In: Lingras, P., Wolski, M., Cornelis, C., Mitra, S., Wasilewski, P. (eds.) RSKT 2013. LNCS, vol. 8171, pp. 133–144. Springer, Heidelberg (2013)

Rough Set Approximations in Multi-scale Interval Information Systems

Shen-Ming Gu[1,2]([✉]), Ya-Hong Wan[1,2], Wei-Zhi Wu[1,2], and Tong-Jun Li[1,2]

[1] School of Mathematics, Physics and Information Science,
Zhejiang Ocean University, Zhoushan 316022, Zhejiang, People's Republic of China
{gsm,wuwz,litj}@zjou.edu.cn
[2] Key Laboratory of Oceanographic Big Data Mining & Application
of Zhejiang Province, Zhejiang Ocean University,
Zhoushan 316022, Zhejiang, People's Republic of China
1182179499@qq.com

Abstract. With the view point of granular computing, the notion of a granule may be interpreted as one of the numerous small particles forming a larger unit. There are different granules at different levels of scale in data sets having hierarchical structures. Human beings often observe objects or deal with data hierarchically structured at different levels of granulations. And in real-world applications, there may exist multiple types of data in interval information systems. Therefore, the concept of multi-scale interval information systems is first introduced in this paper. The lower and upper approximations in multi-scale interval information systems are then defined, and the accuracy and the roughness are also explored. Monotonic properties of these rough set approximations with different levels of granulations are analyzed with illustrative examples.

Keywords: Granular computing · Granules · Interval information systems · Multi-scale information systems · Rough sets

1 Introduction

Granular computing (GrC) is an approach for knowledge representation and data mining. The root of GrC comes from the concept of information granulation which was first introduced by Zadeh in 1979 [32,33]. The purpose of GrC is to seek for an approximation scheme which can effectively solve a complex problem at a certain level of granulation [25]. Ever since the introduction of the concept of "GrC", we have witnessed a rapid development and a fast growing interest in the topic [1,2,13,15,16,20,21,24,27–30].

With the view of GrC, a granule is a primitive notion which is a clump of objects drawn together by the criteria of indistinguishability, similarity or functionality [33]. The set of granules provides a representation of the unit with respect to a particular level of granularity [26]. An important and common used model for GrC is partition model proposed by Yao [29]. By employing the notion of labelled partition, Bittner and Smith [3] developed an ontologically motivated

© Springer International Publishing Switzerland 2015
Y. Yao et al. (Eds.): RSFDGrC 2015, LNAI 9437, pp. 73–81, 2015.
DOI: 10.1007/978-3-319-25783-9_7

formal theory of granular partitions which is relatively comprehensive and useful for granular levels. In order to represent hierarchical structure of data measured at different levels of granularities, Keet [12] explored a formal theory of granularity to build structure of the contents for different types of granularities. Recently, Wu and Leung [25] developed a new knowledge representation system, called multi-scale granular labelled partition structure, in which data are represented by different scales at different levels of granulations having a granular information transformation from a finer to a coarser labelled partition. Hence, the multi-scale information system is a new and interesting topic in the theory of GrC. More recently, Wu and Leung [26] explored optimal scale selection in multi-scale decision tables from the perspective of granular computation. Gu et al. [6–8] proposed some methods for knowledge acquisition in consistent or inconsistent multi-scale decision systems. In [9], multi-granulation rough sets were discussed in multi-scale information systems.

Various methods of GrC concentrating on concrete models in specific contexts have been proposed over the years. Rough set theory is perhaps one of the most advanced areas that popularize GrC [10,11,15,20,21,27–30]. It was originally proposed by Pawlak [18,19] as a formal tool for modelling and processing incomplete information. Rough set theory has been applied successfully in many areas, such as data mining, data analysis and decision making [4,5,23]. However, the classical methods of rough set are based on single valued information systems. Interval information systems are an important type of data tables, and generalized models of single valued information systems [22]. Several authors have studied about interval information systems and interval decision systems. For example, Yao et al. [31] presented a model for the interval set by using the lower and upper approximations in interval information systems. Leung et al. [14] investigated a rough set approach to discover classification rules through a process of knowledge induction which selects decision rules with a minimal set of features in interval information systems.

This paper mainly focuses on the study of rough set approximations in multi-scale interval information systems. The organization of this paper is as follows. In the next section, we first introduce some basic notions related to interval information systems. The concept of multi-scale interval information systems are explored in Sect. 3. In Sect. 4, we investigate the rough set approximations and the monotonic properties of these rough set approximations. We then conclude the paper with a summary and outlook for further research in Sect. 5.

2   Interval Information Systems

2.1   Information Systems

The notion of information systems provides a convenient tool for the representation of objects in terms of their attributes [17].

An information system is a pair (U, A), where $U = \{x_1, x_2, \ldots, x_n\}$ is a non-empty, finite set of objects called the universe of discourse and $A = \{a_1, a_2, \ldots, a_m\}$ is a non-empty, finite set of attributes, such that $a : U \to V_a$

for any $a \in A$, i.e., $a(x) \in V_a$, $x \in U$, where $V_a = \{a(x) : x \in U\}$ is called the domain of a [9].

Any non-empty attribute set $B \subseteq A$, it determines an indiscernibility relation on U as follows:

$$R_B = \{(x, y) \in U \times U : a(x) = a(y), \forall a \in B\}. \tag{1}$$

Since R_B is an equivalence relation on U, it partitions U into a family of disjoint subsets U/R_B of U:

$$U/R_B = \{[x]_B : x \in U\}, \tag{2}$$

where $[x]_B$ denotes the equivalence class determined by x with respect to (w.r.t.) B, i.e., $[x]_B = \{y \in U : (x, y) \in R_B\}$.

Any $X \subseteq U$ and $B \subseteq A$, the lower and upper approximations of X w.r.t. the equivalence relation R_B are defined as follows:

$$\underline{R_B}(X) = \{x \in U : [x]_B \subseteq X\}, \tag{3}$$

$$\overline{R_B}(X) = \{x \in U : [x]_B \cap X \neq \emptyset\}. \tag{4}$$

The lower approximation $\underline{R_B}(X)$ is the set of the objects that certainly belong to X w.r.t. B, while the upper approximation $\overline{R_B}(X)$ is the set of the objects that possibly belong to X w.r.t. B. The pair $\underline{R_B}(X), \overline{R_B}(X)$ is referred to as the Pawlak rough sets of X w.r.t. B.

Pawlak [18] proposed two numerical measures for evaluating the uncertainty of a rough set X: the accuracy and the roughness. The accuracy is simply defined by the ratio of the cardinalities of the lower and upper approximation sets of X. The roughness is calculated by subtracting the accuracy from one. The two measures are defined as follows:

$$\alpha_B(X) = \frac{|\underline{R_B}(X)|}{|\overline{R_B}(X)|}, \tag{5}$$

$$\beta_B(X) = 1 - \alpha_B(X). \tag{6}$$

where $|X|$ denotes the cardinality of the set X.

2.2   Interval Information Systems

An interval information system is still denoted without confusion by (U, A), where for each $x_i \in U$, and each $a_j \in A$, $a_j(x_i) = [l_{ij}, u_{ij}]$ denotes the interval data that object x_i holds on attribute a_j. The real numbers l_{ij} and u_{ij} are referred to as the left and right limits of $a_j(x_i)$ whereas $l_{ij} \leqslant u_{ij}$. In particular, $a_j(x_i)$ would degenerate into a real number if $l_{ij} = u_{ij}$. Under this consideration, we regard a single-valued information system as a special form of interval information systems.

For an interval information system (U, A) and $B \subseteq A$, a binary relation can be defined as:

$$S_B = \{(x, y) \in U \times U : a(x) \cap a(y) \neq \emptyset, \forall a \in B\}. \tag{7}$$

Obviously, S_B is reflexive and symmetric, while it is not transitive. So it is not an equivalence relation, it is a tolerance relation.

Given an interval information system (U, A) and $B \subseteq A$, the basic granules of knowledge induced by the tolerance relation S_B are the set of objects similar to x, called tolerance class, denoted as $S_B(x)$,

$$S_B(x) = \{y \in U : (x, y) \in S_B\}. \tag{8}$$

For any $X \subseteq U$ and $B \subseteq A$, the lower and upper approximations of X w.r.t. the tolerance relation S_B are defined as follows:

$$\underline{S_B}(X) = \{x \in U : S_B(x) \subseteq X\}, \tag{9}$$

$$\overline{S_B}(X) = \{x \in U : S_B(x) \cap X \neq \emptyset\}. \tag{10}$$

And the accuracy and the roughness are defined as follows:

$$\alpha_{S_B}(X) = \frac{|\underline{S_B}(X)|}{|\overline{S_B}(X)|}, \tag{11}$$

$$\beta_{S_B}(X) = 1 - \alpha_{S_B}(X). \tag{12}$$

3 Multi-scale Interval Information Systems

3.1 Multi-scale Information Systems

A multi-scale information system [25] is a tuple (U, A), where $U = \{x_1, x_2, \ldots, x_n\}$ is a non-empty, finite set of objects called the universe of discourse, $A = \{a_1, a_2, \ldots, a_m\}$ is a non-empty, finite set of attributes, and each $a_j \in A$ is a multi-scale attribute, i.e., for the same object in U, attribute a_j can take on different values at different scales. We assume that all the attributes have the same number I of levels of granulations. Hence a multi-scale information system can be represented as a system $(U, \{a_j^k : k = 1, 2, \ldots, I, j = 1, 2, \ldots, m\})$, where $a_j^k : U \to V_j^k$ is a surjective function and V_j^k is the domain of the k-th scale attribute a_j^k. For $1 \leqslant k \leqslant I - 1$, there exists a surjective function $g_j^{k,k+1} : V_j^k \to V_j^{k+1}$, such that $a_j^{k+1} = g_j^{k,k+1} \circ a_j^k$, i.e.

$$a_j^{k+1}(x) = g_j^{k,k+1}(a_j^k(x)), \tag{13}$$

where $g_j^{k,k+1}$ is called a granular information transformation function [25].

For $k \in \{1, 2, \ldots, I\}$, we denote $A^k = \{a_j^k : j = 1, 2, \ldots, m\}$. Then a multi-scale information system $(U, \{a_j^k : k = 1, 2, \ldots, I, j = 1, 2, \ldots, m\})$ can be decomposed into I information systems $(U, A^k), k = 1, 2, \ldots, I$.

Table 1. A multi-scale interval information system

U	a_1^1	a_2^1	a_1^2	a_2^2	a_1^3	a_2^3
x_1	[92, 98]	[90, 96]	[90, 100]	[88, 98]	[9, 10]	[8, 10]
x_2	[85, 91]	[88, 94]	[83, 93]	[86, 96]	[8, 10]	[8, 10]
x_3	[89, 95]	[83, 89]	[87, 97]	[81, 91]	[8, 10]	[8, 10]
x_4	[80, 86]	[79, 85]	[78, 88]	[77, 87]	[7, 9]	[7, 9]
x_5	[73, 79]	[81, 87]	[71, 81]	[79, 89]	[7, 9]	[7, 9]
x_6	[51, 57]	[62, 68]	[49, 59]	[60, 70]	[4, 6]	[6, 7]
x_7	[64, 70]	[73, 79]	[62, 72]	[71, 81]	[6, 8]	[7, 9]
x_8	[49, 55]	[53, 59]	[47, 57]	[51, 61]	[4, 6]	[5, 7]

3.2 Multi-scale Interval Information Systems

For a multi-scale information system $(U, \{a_j^k : k = 1, 2, \ldots, I, j = 1, 2, \ldots, m\})$, if (U, A^1) is an interval information system, given the granular information transformation functions $g_j^{1,2}, j = 1, 2, \ldots, m, \forall x, y \in U, a_j^1 \in A^1$, if $a_j^1(x) \cap a_j^1(y) \neq \emptyset$, then $g_j^{1,2}(a_j^1(x)) \cap g_j^{1,2}(a_j^1(y)) \neq \emptyset, j = 1, 2, \ldots, m$. Thus we can obtain an interval information system (U, A^2). Furthermore, the granular information transformation functions $g_j^{k,k+1}, j = 1, 2, \ldots, m, k = 1, 2, \ldots, I - 1, \forall x, y \in U, a_j^k \in A^k$, if $a_j^k(x) \cap a_j^k(y) \neq \emptyset$, then $g_j^{k,k+1}(a_j^k(x)) \cap g_j^{k,k+1}(a_j^k(y)) \neq \emptyset, j = 1, 2, \ldots, m$. Thus we can obtain interval information systems $(U, A^{k+1}), k = 1, 2, \ldots, I - 1$. Then we say $(U, \{a_j^k : k = 1, 2, \ldots, I, j = 1, 2, \ldots, m\})$ is a multi-scale interval information system.

Example 1. Table 1 depicts a multi-scale interval information system, where $U = \{x_1, x_2, \ldots, x_8\}$, $A^k = \{a_1^k, a_2^k\}, k = 1, 2, 3$. From the table we have $(U, A^k), k = 1, 2, 3$, this information system has three levels of granulations.

4 Rough Set Approximations

Given a multi-scale interval information system $(U, \{A^k : k = 1, 2, \ldots, I\})$, $1 \leqslant k \leqslant I$, (U, A^k) is an interval information system. For $B^k \subseteq A^k$ and $x \in U$, the tolerance class of x w.r.t. the tolerance relation S_{B^k} is defined as follows:

$$S_{B^k}(x) = \{y \in U : (x, y) \in S_{B^k}\}. \tag{14}$$

Proposition 1. *For a multi-scale interval information system* $(U, \{A^k : k = 1, 2, \ldots, I\})$, $1 \leqslant k \leqslant I - 1$, $x \in U$ *and* $B^k \subseteq A^k$, *then we have* $S_{B^k}(x) \subseteq S_{B^{k+1}}(x)$.

Proof. $\forall x \in U$,

$$
\begin{aligned}
S_{B^k}(x) &= \{y \in U : (x,y) \in S_{B^k}\} \\
&= \{y \in U : a_j^k(x) \cap a_j^k(y) \neq \emptyset, \forall a_j^k \in B^k\} \\
&\subseteq \{y \in U : g_j^{k,k+1}(a_j^k(x)) \cap g_j^{k,k+1}(a_j^k(y)) \neq \emptyset, \forall a_j^k \in B^k\} \\
&= \{y \in U : a_j^{k+1}(x) \cap a_j^{k+1}(y) \neq \emptyset, \forall a_j^{k+1} \in B^{k+1}\} \\
&= \{y \in U : (x,y) \in S_{B^{k+1}}\} \\
&= S_{B^{k+1}}(x).
\end{aligned}
$$

Therefore $S_{B^k}(x) \subseteq S_{B^{k+1}}(x)$.

Example 2. For the multi-scale interval information system $(U, \{a_j^k : k = 1,2,3, j = 1,2\})$ in Example 1. Let $B^k = \{a_1^k\}$, $k = 1,2,3$, then the tolerance classes can be obtained as follows:

$S_{B^1}(x_1) = \{x_1, x_3\}$, $S_{B^2}(x_1) = \{x_1, x_2, x_3\}$, $S_{B^3}(x_1) = \{x_1, x_2, x_3, x_4, x_5\}$;
$S_{B^1}(x_2) = \{x_2, x_3, x_4\}$, $S_{B^2}(x_2) = \{x_1, x_2, x_3, x_4\}$, $S_{B^3}(x_2) = \{x_1, x_2, x_3, x_4, x_5, x_7\}$;
$S_{B^1}(x_3) = \{x_1, x_2, x_3\}$, $S_{B^2}(x_3) = \{x_1, x_2, x_3, x_4\}$, $S_{B^3}(x_3) = \{x_1, x_2, x_3, x_4, x_5, x_7\}$;
$S_{B^1}(x_4) = \{x_2, x_4\}$, $S_{B^2}(x_4) = \{x_2, x_3, x_4, x_5\}$, $S_{B^3}(x_4) = \{x_1, x_2, x_3, x_4, x_5, x_7\}$;
$S_{B^1}(x_5) = \{x_5\}$, $S_{B^2}(x_5) = \{x_4, x_5, x_7\}$, $S_{B^3}(x_5) = \{x_1, x_2, x_3, x_4, x_5, x_7\}$;
$S_{B^1}(x_6) = \{x_6, x_8\}$, $S_{B^2}(x_6) = \{x_6, x_8\}$, $S_{B^3}(x_6) = \{x_6, x_7, x_8\}$;
$S_{B^1}(x_7) = \{x_7\}$, $S_{B^2}(x_7) = \{x_5, x_7\}$, $S_{B^3}(x_7) = \{x_2, x_3, x_4, x_5, x_6, x_7, x_8\}$;
$S_{B^1}(x_8) = \{x_6, x_8\}$, $S_{B^2}(x_8) = \{x_6, x_8\}$, $S_{B^3}(x_8) = \{x_6, x_7, x_8\}$.

Obviously, $S_{B^1}(x_i) \subseteq S_{B^2}(x_i) \subseteq S_{B^3}(x_i)$, $i = 1,2,\ldots,8$.

Given a multi-scale interval information system $(U, \{A^k : k = 1,2,\ldots,I\})$, $1 \leqslant k \leqslant I$, (U, A^k) is an interval information system. For $B^k \subseteq A^k$ and $X \subseteq U$, the lower and upper approximations of X w.r.t. the tolerance relation S_{B^k} are defined as follows:

$$
\underline{S_{B^k}}(X) = \{x \in U : S_{B^k}(x) \subseteq X\}, \tag{15}
$$

$$
\overline{S_{B^k}}(X) = \{x \in U : S_{B^k}(x) \cap X \neq \emptyset\}. \tag{16}
$$

And the accuracy and the roughness are defined as follows:

$$
\alpha_{S_{B^k}}(X) = \frac{|\underline{S_{B^k}}(X)|}{|\overline{S_{B^k}}(X)|}, \tag{17}
$$

$$
\beta_{S_{B^k}}(X) = 1 - \alpha_{S_{B^k}}(X). \tag{18}
$$

Proposition 2. *Given a multi-scale interval information system $(U, \{A^k : k = 1,2,\ldots,I\})$, $1 \leqslant k \leqslant I-1$, for $B^k \subseteq A^k$ and $X \subseteq U$, the lower and upper approximations of X w.r.t. the tolerance relation S_{B^k} satisfy the following properties:*

(1) $\underline{S_{B^{k+1}}}(X) \subseteq \underline{S_{B^k}}(X)$,
(2) $\overline{S_{B^{k+1}}}(X) \supseteq \overline{S_{B^k}}(X)$.

Proof. (1) $\forall x \in U$,

$$x \in \underline{S_{B^{k+1}}}(X) \Longleftrightarrow S_{B^{k+1}}(x) \subseteq X$$
$$\Longrightarrow S_{B^k}(x) \subseteq X$$
$$\Longleftrightarrow x \in \underline{S_{B^k}}(X).$$

Therefore $\underline{S_{B^{k+1}}}(X) \subseteq \underline{S_{B^k}}(X)$.

(2) $\forall x \in U$,

$$x \in \overline{S_{B^k}}(X) \Longleftrightarrow S_{B^k}(x) \cap X \neq \emptyset$$
$$\Longrightarrow S_{B^{k+1}}(x) \cap X \neq \emptyset$$
$$\Longleftrightarrow x \in \overline{S_{B^{k+1}}}(X).$$

Therefore $\overline{S_{B^{k+1}}}(X) \supseteq \overline{S_{B^k}}(X)$.

Proposition 3. *Given a multi-scale interval information system* $(U, \{A^k : k = 1, 2, \ldots, I\})$, $1 \leqslant k \leqslant I - 1$, *for* $B^k \subseteq A^k$ *and* $X \subseteq U$, *the accuracy and the roughness satisfy the following properties:*

(1) $\alpha_{S_{B^{k+1}}}(X) \leqslant \alpha_{S_{B^k}}(X)$,
(2) $\beta_{S_{B^{k+1}}}(X) \geqslant \beta_{S_{B^k}}(X)$.

Proof. (1) $\because \underline{S_{B^{k+1}}}(X) \subseteq \underline{S_{B^k}}(X)$, and $\overline{S_{B^{k+1}}}(X) \supseteq \overline{S_{B^k}}(X)$,
$\therefore |\underline{S_{B^{k+1}}}(X)| \leqslant |\underline{S_{B^k}}(X)|$, and $|\overline{S_{B^{k+1}}}(X)| \geqslant |\overline{S_{B^k}}(X)|$.

And $\alpha_{S_{B^{k+1}}}(X) = \dfrac{|\underline{S_{B^{k+1}}}(X)|}{|\overline{S_{B^{k+1}}}(X)|} \leqslant \dfrac{|\underline{S_{B^k}}(X)|}{|\overline{S_{B^k}}(X)|} = \alpha_{S_{B^k}}(X)$,

Therefore $\alpha_{S_{B^{k+1}}}(X) \leqslant \alpha_{S_{B^k}}(X)$.

(2) $\because \alpha_{S_{B^{k+1}}}(X) \leqslant \alpha_{S_{B^k}}(X)$,
$\therefore 1 - \alpha_{S_{B^{k+1}}}(X) \geqslant 1 - \alpha_{S_{B^k}}(X)$,
Therefore $\beta_{S_{B^{k+1}}}(X) \geqslant \beta_{S_{B^k}}(X)$.

Example 3. In Example 1, let $B^k = \{a_1^k\}$, $k = 1, 2, 3$, and $X = \{x_1, x_2, x_3, x_4, x_5\}$, then $\underline{S_{B^1}}(X) = \{x_1, x_2, x_3, x_4, x_5\}$, $\underline{S_{B^2}}(X) = \{x_1, x_2, x_3, x_4\}$, $\underline{S_{B^3}}(X) = \{x_1\}$;
$\overline{S_{B^1}}(X) = \{x_1, x_2, x_3, x_4, x_5\}$, $\overline{S_{B^2}}(X) = \{x_1, x_2, x_3, x_4, x_5, x_7\}$,
$\overline{S_{B^3}}(X) = \{x_1, x_2, x_3, x_4, x_5, x_7\}$.
We can obtain,

$$\underline{S_{B^1}}(X) \supseteq \underline{S_{B^2}}(X) \supseteq \underline{S_{B^3}}(X),$$

$$\overline{S_{B^1}}(X) \subseteq \overline{S_{B^2}}(X) \subseteq \overline{S_{B^3}}(X).$$

And furthermore,
$\alpha_{S_{B^1}}(X) = 1$, $\alpha_{S_{B^2}}(X) = \frac{2}{3}$, $\alpha_{S_{B^3}}(X) = \frac{1}{6}$;
$\beta_{S_{B^1}}(X) = 0$, $\beta_{S_{B^2}}(X) = \frac{1}{3}$, $\beta_{S_{B^3}}(X) = \frac{5}{6}$.
Obviously,

$$\alpha_{S_{B^1}}(X) \geqslant \alpha_{S_{B^2}}(X) \geqslant \alpha_{S_{B^3}}(X),$$

$$\beta_{S_{B^1}}(X) \leqslant \beta_{S_{B^2}}(X) \leqslant \beta_{S_{B^3}}(X).$$

5   Conclusion

In this paper, we have developed a new knowledge representation system called a multi-scale interval information system. A multi-scale interval information system can be used to represent interval data sets having hierarchical scale structures measured at different levels of granulations. By using tolerance relation to construct the tolerance classes, we also introduced rough set approximations in multi-scale interval information systems, and also examined the monotonic properties of these rough set approximations with illustrative examples. Our future work will focus on new approaches to attribute reduction and rules acquisition in multi-scale interval information systems.

Acknowledgments. This work is supported by grants from the National Natural Science Foundation of China (Nos. 61272021, 61202206 and 61173181), and the Zhejiang Provincial Natural Science Foundation of China (Nos. LZ12F03002, LY14F030001), and the Open Foundation from Marine Sciences in the Most Important Subjects of Zhejiang (No. 20130109).

References

1. Bargiela, A., Pedrycz, W.: Granular Computing: An Introduction. Kluwer Academic Publishers, Boston (2002)
2. Bargiela, A., Pedrycz, W.: Toward a theory of granular computing for human-centered information processing. IEEE Trans. Fuzzy Syst. **16**, 320–330 (2008)
3. Bittner, T., Smith, B.: A theory of granular partitions. In: Duckham, M., Goodchild, M.F., Worboys, M.F. (eds.) Foundations of Geographic Information Science, pp. 117–151. Taylor & Francis, London (2003)
4. Cornelis, C., Jensen, R., Hurtado, G., Slezak, D.: Attribute selection with fuzzy decision reducts. Inf. Sci. **180**, 209–224 (2010)
5. Dai, J., Tian, H., Wang, W., Liu, L.: Decision rule mining using classification consistency rate. Knowl. Based Syst. **43**, 95–102 (2013)
6. Gu, S.-M., Wu, W.-Z.: Knowledge acquisition in inconsistent multi-scale decision systems. In: Yao, J.T., Ramanna, S., Wang, G., Suraj, Z. (eds.) RSKT 2011. LNCS, vol. 6954, pp. 669–678. Springer, Heidelberg (2011)
7. Gu, S.M., Wu, W.Z.: On knowledge acquisition in multi-scale decision systems. Int. J. Mach. Learn. Cybern. **4**, 477–486 (2013)
8. Gu, S.M., Wu, W.Z., Zheng, Y.: Rule acquisition in consistent multi-scale decision systems. In: Proceedings of 8th International conference on Fuzzy System and Knowledge Discovery, pp. 390–393. IEEE Computer Society, Los Alamitos (2011)
9. Gu, S.M., Li, X., Wu, W.Z., Nian, H.: Multi-granulation rough sets in multi-scale information systems. In: Proceedings of the 2013 International Conference on Machine Learning and Cybernetics, pp. 108–113, Tianjin (2013)
10. Hu, Q.H., Liu, J.F., Yu, D.R.: Mixed feature selection based on granulation and approximation. Knowl. Based Syst. **21**, 294–304 (2008)
11. Inuiguchi, M., Hirano, S., Tsumoto, S.: Rough Set Theory and Granular Computing. Springer, Heidelberg (2002)
12. Keet, C.M.: A formal theory of granularity. Ph.D. thesis, KRDB Research Centre, Faculty of Computer Science, Free University of Bozen-Bolzano, Italy (2008)

13. Leung, Y., Zhang, J.S., Xu, Z.B.: Clustering by scale-space filtering. IEEE Trans. Pattern Anal. Mach. Intell. **22**, 1396–1410 (2000)
14. Leung, Y., Fischer, M., Wu, W., Mi, J.: A rough set approach for the discovery of classification rules in interval-valued information systems. Int. J. Approx. Reason. **47**, 233–246 (2008)
15. Lin, T.Y., Yao, Y.Y., Zadeh, L.A.: Data Mining, Rough Sets and Granular Computing. Physica- Verlag, Heidelberg (2002)
16. Ma, J.-M., Zhang, W., Wu, W.-Z., Li, T.-J.: Granular computing based on a generalized approximation space. In: Yao, J.T., Lingras, P., Wu, W.-Z., Szczuka, M.S., Cercone, N.J., Ślęzak, D. (eds.) RSKT 2007. LNCS (LNAI), vol. 4481, pp. 93–100. Springer, Heidelberg (2007)
17. Mi, J.S., Wu, W.Z., Zhang, W.X.: Approaches to knowledge reduction based on variable precision rough setsmodel. Inf. Sci. **159**, 255–272 (2004)
18. Pawlak, Z.: Rough sets. Int. J. Comput. Inf. Sci. **11**, 341–356 (1982)
19. Pawlak, Z.: Rough Sets: Theoretical Aspects of Reasoning about Data. Kluwer Academic Publishers, Boston (1991)
20. Qian, Y.H., Liang, J.Y., Dang, C.Y.: Knowledge structure, knowledge granulation and knowledge distance in a knowledge base. Int. J. Approx. Reason. **50**, 174–188 (2009)
21. Qian, Y.H., Liang, J.Y., Yao, Y.Y., Dang, C.Y.: MGRS: a multi-granulation rough set. Inf. Sci. **180**, 949–970 (2010)
22. Qian, Y.H., Liang, J.Y., Dang, C.Y.: Interval ordered information systems. Comput. Math. Appl. **56**, 1994–2009 (2008)
23. Tsumoto, S.: Automated extraction of hierarchical decision rules from clinical databases using rough set model. Expert Syst. Appl. **24**, 189–197 (2003)
24. Wu, W.-Z.: Rough set approximations based on granular labels. In: Sakai, H., Chakraborty, M.K., Hassanien, A.E., Ślęzak, D., Zhu, W. (eds.) RSFDGrC 2009. LNCS, vol. 5908, pp. 93–100. Springer, Heidelberg (2009)
25. Wu, W.Z., Leung, Y.: Theory and applications of granular labelled partitions in multi-scale decision tables. Inf. Sci. **181**, 3878–3897 (2011)
26. Wu, W.Z., Leung, Y.: Optimal scale selection for multi-scale decision tables. Int. J. Approx. Reason. **54**, 1107–1129 (2013)
27. Yao, Y.Y.: Stratified rough sets and granular computing. In: Dave, R.N., Sudkamp, T. (eds.) Proceedings of 18th International Conference of the North American Fuzzy Information Processing Society, pp. 800–804. IEEE Press, New York (1999)
28. Yao, Y.Y.: Information granulation and rough set approximation. Int. J. Intell. Syst. **16**, 87–104 (2001)
29. Yao, Y.: A partition model of granular computing. In: Peters, J.F., Skowron, A., Grzymała-Busse, J.W., Kostek, B., Swiniarski, R.W., Szczuka, M.S. (eds.) Transactions on Rough Sets I. LNCS, vol. 3100, pp. 232–253. Springer, Heidelberg (2004)
30. Yao, Y.Y.Y., Liau, C.-J., Zhong, N.: Granular computing based on rough sets, quotient space theory, and belief functions. In: Zhong, N., Raś, Z.W., Tsumoto, S., Suzuki, E. (eds.) ISMIS 2003. LNCS (LNAI), vol. 2871, pp. 152–159. Springer, Heidelberg (2003)
31. Yao, Y.Y., Li, X.: Comparison of rough-set and interval-set models for uncertain reasoning. Fundam. Inform. **27**, 289–298 (1996)
32. Zadeh, L.A.: Fuzzy sets and information granularity. In: Gupta, N., Ragade, R., Yager, R.R. (eds.) Advances in Fuzzy Set Theory and Applications, pp. 3–18. North-Holland, Amsterdam (1979)
33. Zadeh, L.A.: Towards a theory of fuzzy information granulation and its centrality in human reasoning and fuzzy logic. Fuzzy Sets Syst. **90**, 111–127 (1997)

A New Subsystem-Based Definition
of Generalized Rough Set Model

Caihui Liu[1,2](✉), Meizhi Wang[3], Yanfei Dai[1], and Yueli Luo[1]

[1] Department of Mathematics and Computer Science, Gannan Normal University,
Ganzhou 341000, Jiangxi Province, People's Republic of China
liu_caihui@163.com
[2] Department of Electrical and Computer Engineering, University of Alberta,
Edmonton, AB T6G 2G7, Canada
[3] Department of Physical Education, Gannan Normal University,
Ganzhou 341000, China

Abstract. The generalization of Pawlak rough set model always attracts the attentions of the researchers in the rough set society. In this paper, we propose a new subsystem-based definition of generalized rough set model and disclose the corresponding properties. We also discuss the interrelationships between our definition and the existing ones, the outputs show that our definition is effective and reasonable.

Keywords: Binary relation · Generalized rough set · Lower and upper approximations · σ-algebra

1   Introduction

Since Pawlak [1] introduced the rough set theory in early eighties, many proposals have been made for generalizing rough sets [2]. For example, Pawlak rough set model has been extended to a decision-theoretic rough set model [3,4], the arbitrary binary relations based rough set models [5,6], the covering-based rough sets models [7–9], the rough fuzzy sets and fuzzy rough sets models [10–12], etc. As we know, Pawlak rough approximations have many different but equivalent forms. Yao et al. [13] classified them into three categories, namely, the element-based definition, the granule-based definition and the subsystem-based definition, respectively. They think the three equivalent forms are the most commonly used ones in rough set theory and offer different interpretations of rough set approximations. Generally, if we use a general binary relation to replace an equivalence relations in the three forms, one can propose corresponding generalized rough approximations from the viewpoints of elements, granules, and subsystems, respectively. For example, Yao [14] proposed an element-based definition of generalized rough sets which is commonly used and induce many applications. Zhang et al. [15] introduced granule-based definitions of generalized rough sets. Recently, Liu et al. [16] proposed a subsystem-based definition of generalized rough sets and investigated the relationships of the element-based generalized

© Springer International Publishing Switzerland 2015
Y. Yao et al. (Eds.): RSFDGrC 2015, LNAI 9437, pp. 82–89, 2015.
DOI: 10.1007/978-3-319-25783-9_8

rough sets, the granule-based generalized rough sets, and the subsystem-based generalized rough sets. Based on the work of [16], the paper proposed a new subsystem-based definition of generalized rough sets by employing the concept of σ-algebra and discussed the relationships with the above mentioned definitions.

The remainder of the paper is organized as follows. Section 2 reviews three different but equivalent definitions of Pawlak rough sets and their three corresponding generalizations. Section 3 proposes the new subsystem-based definition of generalized rough sets by employing the concept of σ-algebra and discusses the interrelationships with the other generalized rough sets. Section 4 concludes the paper.

2   Preliminaries

This section will review three equivalent definitions of Pawlak rough sets as well as their corresponding generalizations based on binary relations.

Definition 1. Let U be a universe of discourse and R an equivalence relation on U. For any $X \subseteq U$, its corresponding element-based, granule-based, and subsystem-based definitions of Pawlak rough lower and upper approximations are defined as follows.

Element-based definition [17]:
$$\underline{Apr}(X) = \{x|\, [x]_R \subseteq X\}$$
$$\overline{Apr}(X) = \{x|\, [x]_R \cap X \neq \emptyset\}$$
Granule-based definition [17]:
$$\underline{Apr}(X) = \cup\{[x]_R|\, [x]_R \subseteq X\}$$
$$\overline{Apr}(X) = \cup\{[x]_R|\, [x]_R \cap X \neq \emptyset\}$$
Subsystem-based definition [17,24]:
$$\underline{Apr}(X) = \cup\{Y \mid Y \in B(U/R), Y \subseteq X\}$$
$$\overline{Apr}(X) = \cap\{Y \mid Y \in B(U/R)\,, X \subseteq Y\}$$
where $B(U/R) = \{\cup F|\, F \subseteq U/R\}$ is an atomic sub-Boolean algebra [24].

Because the three equivalent forms represent three research directions, many generalizations have been proposed from the viewpoint of elements, granules and subsystems by using arbitrary binary relations [2,3,6,16].

Yao et al. gave an element-based definition, Zhang et al. gave a granule-based definition and Yao et al. gave subsystem-based definition of generalized rough sets.

Definition 2. Let U be a universe of discourse and R a binary relation on U. For any $X \in U$, three types of generalized rough lower and upper approximations are defined as follows.

Element-based definition: [13,14]
$$\underline{R}(X) = \{x|\, r_R(x) \subseteq X\}$$
$$\overline{R}(X) = \{x|\, r_R(x) \cap X \neq \emptyset\}$$
where $r_R(x)$ is the successor neighborhood of x.

Granule-based definition: [15]
$$\underline{R_*}(X) = \cup\{r_R(x)\,|\,r_R(x) \subseteq X\}$$
$$\overline{R_*}(X) = \cup\{r_R(x)\,|\,r_R(x) \cap X \neq \emptyset\}$$
Subsystem-based definition: [2,13]
$$\underline{apr}(A) = \cup\{X\,|\,X \in \sigma(M), R_s(X) \subseteq A\}$$
$$\overline{apr}(A) = \cup\{X\,|\,X \in \sigma(M), \sim R_s(\sim X) \supseteq A\}$$
Where $M = \{m(A)\,|\,A \in F\}$, F is the family of focal sets, a subset with $m(A) \neq \emptyset$ is called a focal set; $R_s(X) = \bigcup_{x \in X} R_s(x)$, $R_s(x) = \{y \in U\,|\,xRy\}$.

Different to Yao et al.'s subsystem-based definition, Liu et al. [16] gave another kind of subsystem-based definition of generalized rough sets.

Definition 3. [16] Let U be a universe of discourse and R a binary relation on U. For any $X \in U$, its subsystem-based definition of generalized rough lower and upper approximations are defined as follows.

$$\underline{R_{**}}(X) = \cup\{Y\,|\,Y \in \theta_R, Y \subseteq X\}$$
$$\overline{R_{**}}(X) = \cap\{Y\,|\,Y \in \theta_R, X \subseteq Y\}$$
Where $\theta_R = \{X\,|\,\exists Y \subseteq U, X = \cup_{x \in Y} r_R(x)\} \cup \{\emptyset\}$.

3   A New Subsystem-Based Definition of Generalized Rough Set Model

The research on the subsystem-based generalized rough sets has attracted much attention [18–23]. As Yao et al. [18] pointed out that those subsystem-based definitions have been generalized by using different mathematical structures, such as topological spaces [19–21], closure systems [19,20], lattices [19,22], and posets [19,23]. Recently, Liu et al. [16] proposed a kind of subsystem-based generalized rough sets and disclosed some interesting relationships among the generalized rough lower and upper approximations. Although Liu et al.'s definition is meaningful and worth investigated, the θ_R in the definition is not closed under complement and thus θ_R is not a σ-algebra. In this section, we propose a new kind of subsystem-based generalized rough sets by employing the σ-algebra generated by quotient set $U/\equiv$, which is originated from the equivalence class containing x. Here, we mainly discuss the properties of new approximations and concentrate on the interrelationship between our definitions and the subsystem-based rough sets proposed by Yao et al. [13] and Liu et al [16]. We will provide further comparisons with other subsystem-based rough sets in our future work.

Definition 4. Let U be a universe of discourse and R a binary relation on U. For any $X \in U$, the new kind of subsystem-based definition of generalized rough lower and upper approximations are defined as follows.

$$\underline{R_{new}}(X) = \cup\{Y\,|\,Y \in \sigma(U/\equiv), Y \subseteq X\}$$
$$\overline{R_{new}}(X) = \cap\{Y\,|\,Y \in \sigma(U/\equiv), X \subseteq Y\}$$

where $x \equiv y \Leftrightarrow r_R(x) = r_R(y)$, $x, y \in U$ [14].

Example 1 is employed to show that our definitions are different from Definitions 2 and 3.

Example 1. Let U be a universe of discourse and R a binary relation on U, where $U = \{1, 2, 3, 4\}$, $R = \{(1, 1), (1, 2), (2, 2), (2, 4), (3, 1), (3, 2)\}$.
Obviously, we have that
$$R_s(1) = \{1, 2\}, R_s(2) = \{2, 4\}, R_s(3) = \{1, 2\}, R_s(4) = \emptyset.$$
then
$$U/\equiv = \{\{1, 3\}, \{2\}, \{4\}\}$$
Let $X_1 = \{1, 4\}$, $X_2 = \{4\}$, according to Definition 2, 3, 4, we can figure out the following.
$$\underline{apr}(X_1) = \{4\}, \overline{apr}(X_1) = \{1, 2, 3\},$$
$$\underline{apr}(X_2) = \{4\}, \overline{apr}(X_2) = \{2\};$$
$$\theta_R = \{X \mid \exists Y \subseteq U, X = \cup_{x \in Y} r_R(x)\} \cup \{\emptyset\} = \{\emptyset, \{1, 2\}, \{2, 4\}, \{1, 2, 4\}\}$$
$$\underline{R_{**}}(X_1) = \emptyset, \overline{R_{**}}(X_1) = \{1, 2, 4\},$$
$$\underline{R_{**}}(X_2) = \emptyset, \overline{R_{**}}(X_2) = \{2, 4\};$$

$$\sigma(U/\equiv) = \{\emptyset, \{1, 3\}, \{1, 2, 3\}, \{1, 3, 4\}, \{2\}, \{2, 4\}, \{4\}, U\},$$
$$\underline{R_{new}}(X_1) = \{4\}, \overline{R_{new}}(X_1) = \{1, 3, 4\},$$
$$\underline{R_{new}}(X_2) = \{4\}, \overline{R_{new}}(X_2) = \{4\}.$$

Therefore, $\underline{R_{new}}(X)$ is different from $\underline{apr}(X)$ and $\underline{R_{**}}(X)$, and at the same time, $\overline{R_{new}}(X)$ is different from $\overline{apr}(X)$ and $\overline{R_{**}}(X)$.

According to Example 1, we have the following remarks.

Remark 1. Let U be a universe of discourse and R a binary relation on U, for any $X \subseteq U$, $X \subseteq \overline{apr}(X)$ may not be satisfied.

Example 2 (Continued from Example 1). From Example 1, we have that $\overline{apr}(X_1) = \{1, 2, 3\}$ and $\overline{apr}(X_2) = \{2\}$.
Therefore,
$$X_1 = \{1, 4\} \not\subseteq \overline{apr}(X_1) = \{1, 2, 3\},$$
$$X_2 = \{4\} \not\subseteq \overline{apr}(X_2) = \{2\}.$$

Remark 2. Let U be a universe of discourse and R a binary relation on U, for any $X \subseteq U$, $\underline{apr}(X) \subseteq \overline{apr}(X)$ may not be satisfied.

Example 3 (Continued from Example 1). From Example 1, we have that $\underline{apr}(X_1) = \{4\}, \overline{apr}(X_1) = \{1, 2, 3\}$ and $\underline{apr}(X_2) = \{4\}, \overline{apr}(X_2) = \{2\}$.
Therefore,
$$\underline{apr}(X_1) = \{4\} \not\subseteq \overline{apr}(X_1) = \{1, 2, 3\},$$
$$\underline{apr}(X_2) = \{4\} \not\subseteq \overline{apr}(X_2) = \{2\}.$$

Proposition 1. Let U be a universe of discourse and R a binary relation on U, for any $X, Y \subseteq U$, we have

(1) $\underline{R_{new}}(\emptyset) = \emptyset$, $\overline{R_{new}}(U) = U$;

(2) $\underline{R_{new}}(X) \subseteq X \subseteq \overline{R_{new}}(X)$;

(3) $X \subseteq Y$ implies $\underline{R_{new}}(X) \subseteq \underline{R_{new}}(Y)$ and $\overline{R_{new}}(X) \subseteq \overline{R_{new}}(Y)$;

(4) $\underline{R_{new}}(X \cup Y) \supseteq \underline{R_{new}}(X) \cup \underline{R_{new}}(Y)$;

(5) $\overline{R_{new}}(X \cap Y) \subseteq \overline{R_{new}}(X) \cap \overline{R_{new}}(Y)$;

(6) $\underline{R_{new}}(\underline{R_{new}}(X)) = \underline{R_{new}}(X) \neq X$ and $\overline{R_{new}}(\overline{R_{new}}(X)) = \overline{R_{new}}(X) \neq X$;

(7) $\overline{R_{new}}(\underline{R_{new}}(X)) \neq \underline{R_{new}}(\overline{R_{new}}(X))$.

Proof. It can be proved easily according to Definition 4.

Proposition 2. Let U be a universe of discourse and R a binary relation on U, for any $X \subseteq U$, the following is satisfied.

$$\underline{R_{**}}(X) \subseteq \underline{apr}(X) \subseteq \underline{R_{new}}(X)$$

Proof. By [14], we have that $M = U/\equiv$, that is $\sigma(M) = \sigma(U/\equiv)$. According to [14], we know that $R_s(X) = \bigcup_{x \in X} R_s(x)$, if $R_s(X) \subseteq A$, then $R_s(X) \subseteq X$. Therefore, $\underline{apr}(X) \subseteq \underline{R_{new}}(X)$;

Note that $\theta_R = \{X \mid \exists Y \subseteq U, X = \cup_{x \in Y} r_R(x)\} \cup \{\emptyset\}$, one can easily prove that for any $X \in \theta_R$, there exists at least one $Y \in \sigma(U/\equiv)$, such that $X \subseteq Y$, therefore $\underline{R_{**}}(X) \subseteq \underline{apr}(X)$.

Corollary 1. Let U be a universe of discourse and R a binary relation on U, then the follows are satisfied.

(1) $\underline{R_{**}}(\emptyset) = \underline{apr}(\emptyset) = \underline{R_{new}}(\emptyset)$;

(2) $\underline{R_{**}}(U) = \underline{apr}(U) = \underline{R_{new}}(U)$;

(3) $\overline{R_{**}}(\emptyset) = \overline{apr}(\emptyset) = \overline{R_{new}}(\emptyset)$;

(4) $\overline{R_{**}}(U) = \overline{apr}(U) = \overline{R_{new}}(U)$.

Remark 3. $\overline{R_{**}}(X) \subseteq \overline{apr}(X) \subseteq \overline{R_{new}}(X)$ may not be satisfied.

The following example proves Remark 3.

Example 4 (Continued from Example 1). From Example 1, we have that $\overline{apr}(X_1) = \{1, 2, 3\}$, $\overline{R_{**}}(X_1) = \{1, 2, 4\}$, $\overline{R_{new}}(X_1) = \{1, 3, 4\}$ and $\overline{R_{**}}(X_2) = \{2, 4\}$, $\overline{apr}(X_2) = \{2\}$, $\overline{R_{new}}(X_2) = \{4\}$.
Therefore,

$\overline{apr}(X_1)$, $\overline{R_{**}}(X_1)$ and $\overline{R_{new}}(X_1)$ are not subset of each other;

but,

$\overline{apr}(X_2) = \{2\} \subset \overline{R_{**}}(X_2) = \{2, 4\}$, $\overline{R_{new}}(X_2) = \{4\} \subset \overline{R_{**}}(X_2) = \{2, 4\}$.

That is to say,

$\overline{R_{**}}(X) \subseteq \overline{apr}(X) \subseteq \overline{R_{new}}(X)$ may not be satisfied.

The relationships among the three kinds of subsystem-based approximations can be expressed by Fig. 1. In Fig. 1, each node denotes an approximation or a concept, and each line connects two approximations, where the lower node is a subset of the upper node.

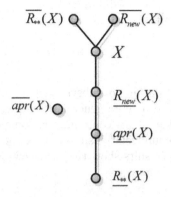

Fig. 1. Relationships of approximations

From Fig. 1, we can obtain the following meaningful conclusions.

First, from Fig. 1, we can see that among the three kinds of subsystem-based lower and upper approximations, the pair $(\underline{R_{new}}(X), \overline{R_{new}}(X))$ is the best to describe X, since in Fig. 1 the two approximations are closer to X than other approximations. And at the same time, the pair $(\underline{R_{**}}(X), \overline{R_{**}}(X))$ may be the worst to describe X, since in Fig. 1 the two approximations have the furthest distance with X with respect to other approximations.

Second, from Fig. 1, we can also find that $\overline{apr}(X)$ is an isolated point, that is, $\overline{apr}(X)$ is independent to any approximation or concept there.

Proposition 3. Let U be a universe of discourse and R a binary relation on U, for any $X \subseteq U$, $\underline{R_{**}}(X) = \underline{R_*}(X) \subseteq \underline{R_{new}}(X)$ is satisfied.

Proof. It straightforward according to Proposition 2 and Proposition 2.3 in [16].

Remark 4. $\overline{R_{**}}(X) = \overline{R_*}(X) \subseteq \overline{R_{new}}(X)$ may not be satisfied.

Proposition 4. Let U be a universe of discourse and R a binary relation on U, for any $X \subseteq U$, $\underline{R_{new}}(r_R(x)) = r_R(x)$ and $\overline{R_{new}}(r_R(x)) = r_R(x)$ are satisfied.

Proposition 4 illustrates that for any $x \in U$, the new defined approximations of its right neighbor are right neighbor itself.

Proposition 5. Let U be a universe of discourse and R a binary relation on U, for any $X \subseteq U$, $\underline{R_{new}}(X) = \sim \overline{R_{new}}(\sim X)$ and $\overline{R_{new}}(X) = \sim \underline{R_{new}}(\sim X)$ are satisfied.

Proof. According to Definition 4, we have

$$\underline{R_{new}}(X) = \cup\{Y \mid Y \in \sigma(U/\equiv), Y \subseteq X\}$$
$$= \sim \cap\{Y \mid Y \in \sigma(U/\equiv), X \subseteq \sim Y\}$$
$$= \sim \overline{R_{new}}(\sim X).$$
$$\overline{R_{new}}(X) = \cap\{Y \mid Y \in \sigma(U/\equiv), X \subseteq Y\}$$
$$= \sim \cup\{Y \mid Y \in \sigma(U/\equiv), Y \subseteq \sim X\}$$
$$= \sim \underline{R_{new}}(\sim X).$$

4   Conclusion

This paper has proposed a new kind of subsystem-based definition of generalized rough set by employing the σ-algebra generated by quotient set $U/\equiv$, which is originated from the equivalence class containing x. Some important properties of the approximations and the interrelationships among the new one with the existing ones are disclosed. The results show that the new definition is appropriate and reasonable.

Acknowledgements. This work was supported by the China National Natural Science Foundation of Youth Science Foundation under Grant No.: 61305052, 61403329, the State Scholarship Fund of China (File No. 201409865003), the Key Technology Research and Development Program of Education Bureau of Jiangxi Province of China under Grant No.: GJJ14660, the Key Technology Research and Development Program of Jiangxi Province of China under Grant No.: 20142BBF60010, 20151BBF60071.

References

1. Pawlak, Z.: Rough sets. Int. J. Comput. Inf. Sci. **11**, 341–356 (1982)
2. Yao, Y.Y.: Generalized rough set models. In: Polkowski, L., Skowron, A. (eds.) Rough Sets in Knowledge Discovery, pp. 286–318. Physica-Verlag, Heidelberg (1998)
3. Yao, Y.Y., Wong, S.K., Lingras, P.: A decision-theoretic rough setmodel. In: Ras, Z.W., Zemankova, M., Emrich, M.L. (eds.) Methodologies for Intelligent Systems, vol. 5, pp. 17–24. North-Holland, New York (1990)
4. Yao, Y.: Decision-theoretic rough set models. In: Yao, J.T., Lingras, P., Wu, W.-Z., Szczuka, M.S., Cercone, N.J., Ślęzak, D. (eds.) RSKT 2007. LNCS (LNAI), vol. 4481, pp. 1–12. Springer, Heidelberg (2007)
5. Pei, D.W., Xu, Z.B.: Transformation of rough set models. Knowl. Based Syst. **20**, 745–751 (2007)
6. Zhu, W.: Generalized rough sets based on relations. Inf. Sci. **177**(22), 4997–5011 (2007)
7. Zakowski, W.: Axiomatization in the space (U, π). Demonstratio Mathematica **XVI**, 761–769 (1983)
8. Bonikowski, Z., Bryniariski, E., Skardowska, U.W.: Extension and intentions in the rough set theory. Inf. Sci. **107**, 149–167 (1998)
9. Zhu, W., Wang, F.Y.: Reduction and axiomatization of covering generalized rough sets. Inf. Sci. **152**, 17–230 (2003)

10. Dubois, D., Prade, H.: Rough fuzzy sets and fuzzy rough sets. Int. J. Gen. Syst. **17**, 191–209 (1990)
11. He, Q., Wu, C.X., Chen, D.G.: Fuzzy rough set based attribute reduction for information systems with fuzzy decisions. Knowl. Based Syst. **24**(5), 689–696 (2011)
12. Huang, B.: Graded dominance interval-based fuzzy objective information systems. Knowl. Based Syst. **24**(7), 1004–1012 (2011)
13. Yao, Y.Y., Yao, B.X.: Covering based rough set approximations. Inf. Sci. **200**, 91–107 (2012)
14. Yao, Y.Y.: Constructive and algebraic methods of theory of rough sets. Inf. Sci. **109**, 21–47 (1998)
15. Zhang, W.X., Wu, W.Z., Liang, J.Y., Li, D.Y.: Theory and Method of Rough Sets. Science Press, Beijing (2001). (in Chinese)
16. Liu, G.L., Zhu, K.: The relationship among three types of rough approximation pairs. Knowl. Based Syst. **60**, 28–34 (2014)
17. Pawlak, Z.: Rough Sets: Theoretical Aspects of Reasoning About Data. Kluwer Academic Publishers, Boston (1991)
18. Yao, J.T., Ciucci, D., Zhang, Y.: Generalized rough sets. In: Kacprzyk, J., Pedrycz, W. (eds.) The Springer Handbook of Computational Intelligence, pp. 413–424. Springer, Heidelberg (2015)
19. Yao, Y.Y.: On generalizing Pawlak approximation operators. In: Polkowski, L., Skowron, A. (eds.) RSCTC 1998. LNCS (LNAI), vol. 1424, pp. 298–307. Springer, Heidelberg (1998)
20. Yao, Y.Y., Chen, Y.H.: Subsystem based generalizations of rough set approximations. In: Hacid, M.-S., Murray, N.V., Raś, Z.W., Tsumoto, S. (eds.) ISMIS 2005. LNCS (LNAI), vol. 3488, pp. 210–218. Springer, Heidelberg (2005)
21. Wiweger, A.: On topological rough sets. Bull. Pol. Acad. Sci. Math. **37**, 89–93 (1989)
22. Javinen, J.: On the structure of rough approximations. Fundam. Inf. **53**(2), 135–153 (2002)
23. Cattaneo, G.: Abstract approximation spaces for rough theories. Rough Sets Knowl. Discov. **1**, 59–98 (1998)
24. Yao, Y.Y.: The two sides of the theory of rough sets. Knowl. Based Syst. **80**, 67–77 (2015)

A Comparison of Two Types of Covering-Based Rough Sets Through the Complement of Coverings

Yanfang Liu[1][✉] and William Zhu[2][✉]

[1] Institute of Information Engineering, Longyan University, Longyan 364000, China
liuyanfang003@163.com
[2] Lab of Granular Computing, Minnan Normal University, Zhangzhou 363000, China
williamfengzhu@gmail.com

Abstract. In recent years, many types of covering-based rough set models were established and the study of their relationships was a hot research topic. Covering-based rough sets defined by Zhu and ones defined by Xu and Zhang were compared with each other through binary relations. And the relationships between these two types were further explored by introducing a concept of complementary neighborhood. However, the essential connections between these two types have not been revealed. In this paper, we consider these two types of covering-based rough sets by introducing a notion of the complement of coverings. In fact, these two types are expressed by each other through the complement of coverings. Based on the above results, we analyze a notion of the extension of a covering, which is introduced on the basis of the complement of the covering. Finally, we further explore the structure of these types of covering-based rough set models. This study suggests some internal connections between covering-based rough sets and demonstrates a new research tendency of them.

Keywords: Covering-based rough set · Approximation operators · Neighborhood · Complement · Extension

1 Introduction

Equivalence relation or partition is one of the key notions in Pawlak rough set theory [11,12]. The advantage of rough set theory is that it does not need any additional information about data, so it has been successfully applied to various fields, such as data mining, machine learning and pattern recognition. However, equivalence relation or partition is too restrictive for many practical applications. To address this issue, several interesting and meaningful extensions of rough set model [4,10,18,19] have been put forward.

Particularly, through extending a partition of a universe to a covering of the universe, Zakowski [25] first introduced covering-based rough sets. Pomykala [13] studied the second type of covering rough set model. The third type of covering upper approximation was first defined in [20]; Zhu and Wang defined the

© Springer International Publishing Switzerland 2015
Y. Yao et al. (Eds.): RSFDGrC 2015, LNAI 9437, pp. 90–101, 2015.
DOI: 10.1007/978-3-319-25783-9_9

fourth type of covering upper approximation in [29] and the fifth type of covering upper approximation in [26]. Other types of covering-based rough sets can be found in literatures [14,22]. Many scholars studied the properties of these covering-based rough set models to promote the development of these models. Zhu and Wang [27,28] first introduced a notion of reduct to study covering-based rough sets and detailedly investigated the properties of some types of covering approximation operators. Liu and Sai [7] studied the relationships between the covering-based rough set model defined by Zhu [26] and one defined by Xu and Zhang [22] through constructing a relation from a covering. Restrepo et al. [15–17] investigated the duality of lower and upper approximation operators and their partial order relation in covering-based rough sets and relation-based rough sets. Mani [9] applied a notion of complementary neighborhood [2] to study covering-based rough sets from a axiomatic granular perspective. Furthermore, Ma [8] used the concept of complementary neighborhood to consider these two types of covering-based rough sets. However, they do not give the essential relationships of these two covering-based rough set models. In this paper, we investigate these two types of covering-based rough sets through a notion of the complement of coverings and present their essential relationships.

We correct an error about the notion of complementary neighborhood in [8]. Based on this, we study a notion of the complement of a covering [21], and then proved that the covering approximations with respect to a covering defined by Xu and Zhang are equivalently expressed by the covering approximations with respect to the complement of the covering. From the above results, we study some properties of extension of a covering [21]. A relationship between the neighborhoods of a covering and ones of its extension is presented through a connection between the neighborhoods of the covering and the complementary neighborhoods of the covering. Moreover, the neighborhoods of extension of any covering of a universe are proved to be a partition of the universe. Finally, we further explore the matroidal structure of these two types of covering-based rough set models through the exact sets. The set of all exact sets is proved to satisfy the closed set axiom of matroids. Naturally, the set of all exact sets together with the set inclusion forms a lattice.

The remainder of this paper is organized as follows. In Sect. 2, some basic definitions about two types of covering-based rough sets are introduced. Section 3 first corrects an error in [8], and then presents an essential relationship between these two types of covering-based rough sets through a notion of the complement of a covering. In Sect. 4, we study some properties of extension of a covering based on the above results. Section 5 further investigates the matroidal structure of covering-based rough sets through the exact sets. Finally, Sect. 6 concludes this paper.

2   Preliminaries

In this section, we review some basic definitions and related results of two types of covering-based rough sets, which can be found in [14,22,24,26,27].

Definition 1 *(Covering). Let U be a universe of discourse and $\mathbf{C}$ a family of subsets of U. $\mathbf{C}$ is called a covering of U if none of subsets in $\mathbf{C}$ is empty and $\cup \mathbf{C} = U$. The ordered pair $(U, \mathbf{C})$ is called a covering approximation space if $\mathbf{C}$ is a covering of U.*

Definition 2 *(Neighborhood). Let $(U, \mathbf{C})$ be a covering approximation space and $x \in U$, $N_{\mathbf{C}}(x) = \cap \{K \in \mathbf{C} : x \in K\}$ is called the neighborhood of x with respect to $\mathbf{C}$.*

From the above definition, we see that the neighborhood of any element contains the element itself. Therefore, for a covering $\mathbf{C}$ of a universe U, $\{N_{\mathbf{C}}(x) : x \in U\}$ is also a covering of U.

Definition 3. *Let $(U, \mathbf{C})$ be a covering approximation space. We call $\{N_{\mathbf{C}}(x) : x \in U\}$ the induced covering of $\mathbf{C}$ and denote it as $Cov(\mathbf{C})$.*

For any $x \in U$, $N_{\mathbf{C}}(x)$ is the minimal set in $Cov(\mathbf{C})$ including x. Therefore, as an important concept of covering-based rough sets, the neighborhood has been widely applied to knowledge classification and feature selection. Moreover, several types of covering-based rough sets use the neighborhood to construct the covering lower and upper approximation operators.

Definition 4. *Let $(U, \mathbf{C})$ be a covering approximation space. For any $X \subseteq U$, the lower and upper approximations of X are defined respectively as*

$$FL_{\mathbf{C}}(X) = \{x \in U : N_{\mathbf{C}}(x) \subseteq X\};$$
$$FH_{\mathbf{C}}(X) = \{x \in U : N_{\mathbf{C}}(x) \cap X \neq \emptyset\}.$$

FL and *FH* are a pair of dual approximation operators defined by means of neighborhoods [14]. Zhu [26] has studied some properties of the upper approximation operator. Based on duality, we easily obtain properties of the lower approximation operator.

For a covering approximation space, Xu and Zhang [22] also introduced a pair of covering lower and upper approximations as follows.

Definition 5. *Let $(U, \mathbf{C})$ be a covering approximation space. For any $X \subseteq U$, the lower approximation $SL_{\mathbf{C}}(X)$ and the upper approximation $SH_{\mathbf{C}}(X)$ are defined as*

$$SL_{\mathbf{C}}(X) = \{x \in U : \forall u(x \in N_{\mathbf{C}}(u) \to u \in X)\};$$
$$SH_{\mathbf{C}}(X) = \cup \{N_{\mathbf{C}}(x) : x \in X\}.$$

The operators *SL* and *SH* are defined by neighborhoods and essentially dual. An interesting question is what connections exist between these two approximations? Liu and Sai [7] solved this issue through constructing a relation from a covering. Ma [8] used a notion of complementary neighborhood to consider these two types of covering-based rough sets. However, they do not give the essential relationships of these two approximations. Next we will further investigate these two approximations and obtain their essential relationships.

3   Relationships Between FL, FH and SL, SH Through the Complement of Coverings

Based on the existing results [7,8], we study FL, FH and SL, SH through a notion of the complement of a covering. We first correct an error which is related to complementary neighborhood in [8].

3.1   Complementary Neighborhood

Ma [8] used a notion of complementary neighborhood and presented some results of it. For any $X \subseteq U$, we denote $-X$ as the complement of X in U.

Definition 6 ([8]). *Let $(U, \mathbf{C})$ be a covering approximation space. For any $x \in U$, we call $M_{\mathbf{C}}(x) = \{y \in U : x \in N_{\mathbf{C}}(y)\}$ the complementary neighborhood of x.*

Proposition 1 ([8]). *Let $(U, \mathbf{C})$ be a covering approximation space and $x \in U$. Then,*
$$M_{\mathbf{C}}(x) = \cap\{-K : x \notin K \in \mathbf{C}\}.$$

However, the above proposition is wrong, and a counterexample is presented to illustrate it as follows.

Example 1. Suppose $U = \{1, 2, 3\}$ and $\mathbf{C} = \{\{1\}, \{1, 2\}, \{1, 3\}\}$. Since $N_{\mathbf{C}}(1) = \{1\}, N_{\mathbf{C}}(2) = \{1, 2\}, N_{\mathbf{C}}(3) = \{1, 3\}$, according to Definition 6, we have $M_{\mathbf{C}}(1) = \{1, 2, 3\}$. However, according to Proposition 1, $M_{\mathbf{C}}(1) = \emptyset$.

In the following proposition, we give an expression of the complementary neighborhood of any element.

Proposition 2. *Let $(U, \mathbf{C})$ be a covering approximation space and $x \in U$. Then,*
$$M_{\mathbf{C}}(x) = \begin{cases} U, & x \in \cap\mathbf{C}; \\ \cap\{-K : x \notin K \in \mathbf{C}\}, & else. \end{cases}$$

Proof. (1) If $x \in \cap\mathbf{C}$, i.e., $x \in K$ for all $K \in \mathbf{C}$, then for all $y \in U$, $x \in N_{\mathbf{C}}(y)$. Suppose there exists $y \in U$ such that $x \notin N_{\mathbf{C}}(y)$. Then there exists $K \in \mathbf{C}$ such that $y \in K$ and $x \notin K$ which is contradictory with $x \in \cap\mathbf{C}$. Therefore, according to Definition 6, $M_{\mathbf{C}}(x) = U$.

(2) If $x \notin \cap\mathbf{C}$, i.e., there exists at least one covering block $K \in \mathbf{C}$ such that $x \notin K$, then $x \in -K$. According to Definition 6,
$y \in M_{\mathbf{C}}(x) \Leftrightarrow x \in N_{\mathbf{C}}(y) \Leftrightarrow \forall K \in \mathbf{C}(y \in K \rightarrow x \in K) \Leftrightarrow \forall K \in \mathbf{C}(x \in -K \rightarrow y \in -K) \Leftrightarrow y \in \cap\{-K : x \notin K \in \mathbf{C}\}$.

For a covering approximation space, complementary neighborhood of an element is closely linked with the complement of any covering block which does not contain the element. In the following subsection, we will introduce a notion of the complement of a covering and study its relationship with the complementary neighborhood.

3.2 The Complement of a Covering and Relationships between the Two Types of Covering-Based Rough Sets

Wang and Hu [21] introduced the complement to get more knowledge from a covering, and then they defined a new notion of the complement of a covering.

Definition 7 *([21]). Let $(U, \mathbf{C})$ be a covering approximation space. We call $\mathbf{C}^{\sim} = \{-K : K \in \mathbf{C}\}$ the complement of $\mathbf{C}$.*

It is clear that the complement of a covering is not necessarily a covering. One of reasons is that $\emptyset$ maybe belong to the complement of a covering. Therefore, we present a remark about this situation.

Remark 1. Suppose $(U, \mathbf{C})$ is a covering approximation space. Then we denote $\mathbf{C}^{\sim} - \{\emptyset\}$ as $\mathbf{C}_{\emptyset}^{\sim}$. We call $\mathbf{C}_{\emptyset}^{\sim}$ the non-empty complement of $\mathbf{C}$.

In the following proposition, we present a relationship between the non-empty complement of a covering and the complementary neighborhoods of the covering.

Proposition 3. *Let $(U, \mathbf{C})$ be a covering approximation space and $x \in U$. Then,*

$$N_{\mathbf{C}_{\emptyset}^{\sim} \cup \{U\}}(x) = M_{\mathbf{C}}(x).$$

Proof. According to Proposition 2, we prove this problem under two cases: $x \in \cap \mathbf{C}$ and $x \notin \cap \mathbf{C}$.
(1) If $x \in \cap \mathbf{C}$, then $x \in K$ for all $K \in \mathbf{C}$. In other words, for all $C \in \mathbf{C}^{\sim}$, $x \notin C$. Therefore, $N_{\mathbf{C}_{\emptyset}^{\sim} \cup \{U\}}(x) = U$. Hence $N_{\mathbf{C}_{\emptyset}^{\sim} \cup \{U\}}(x) = M_{\mathbf{C}}(x)$ if $x \in \cap \mathbf{C}$.
(2) If $x \notin \cap \mathbf{C}$, then there exists $K \in \mathbf{C}$ such that $x \notin K$. In other words, $x \in -K$ where $-K \in \mathbf{C}_{\emptyset}^{\sim}$. Then $N_{\mathbf{C}_{\emptyset}^{\sim} \cup \{U\}}(x) = \cap\{-K \in \mathbf{C}_{\emptyset}^{\sim} : x \in -K\} \cap U = \cap\{-K \in \mathbf{C}_{\emptyset}^{\sim} : x \in -K\} = \cap\{-K : x \notin K \in \mathbf{C}\}$.
$y \in N_{\mathbf{C}_{\emptyset}^{\sim} \cup \{U\}}(x) \Leftrightarrow y \in \cap\{-K : x \notin K \in \mathbf{C}\} \Leftrightarrow \forall K \in \mathbf{C}(x \in -K \to y \in -K) \Leftrightarrow \forall K \in \mathbf{C}(y \in K \to x \in K) \Leftrightarrow x \in N_{\mathbf{C}}(y) \Leftrightarrow y \in M_{\mathbf{C}}(x)$.
Therefore, $N_{\mathbf{C}_{\emptyset}^{\sim} \cup \{U\}}(x) = M_{\mathbf{C}}(x)$ if $x \notin \cap \mathbf{C}$.

In order to study relationships between these two approximations in Definitions 4 and 5, we introduce another representation of the lower and upper approximations in Definition 5 through complementary neighborhoods.

Proposition 4. *Let $(U, \mathbf{C})$ be a covering approximation space. For any $X \subseteq U$,*

$$SL_{\mathbf{C}}(X) = \{x \in U : M_{\mathbf{C}}(x) \subseteq X\};$$
$$SH_{\mathbf{C}}(X) = \{x \in U : M_{\mathbf{C}}(x) \cap X \neq \emptyset\}.$$

Proof. According to Definitions 5 and 6, $SL_{\mathbf{C}}(X) = \{x \in U : \forall u(x \in N_{\mathbf{C}}(u) \to u \in X)\}$, $SH_{\mathbf{C}}(X) = \cup\{N_{\mathbf{C}}(x) : x \in X\}$ and $M_{\mathbf{C}}(x) = \{y \in U : x \in N_{\mathbf{C}}(y)\}$.
(1) $x \in SL_{\mathbf{C}}(X) \Leftrightarrow \forall u(x \in N_{\mathbf{C}}(u) \to u \in X) \Leftrightarrow (u \in M_{\mathbf{C}}(x) \to u \in X) \Leftrightarrow M_{\mathbf{C}}(x) \subseteq X \Leftrightarrow x \in \{x \in U : M_{\mathbf{C}}(x) \subseteq X\}$.
(2) $x \in SH_{\mathbf{C}}(X) \Leftrightarrow$ there exists $y \in X$ such that $x \in N_{\mathbf{C}}(y) \Leftrightarrow$ there exists $y \in X$ such that $y \in M_{\mathbf{C}}(x) \Leftrightarrow M_{\mathbf{C}}(x) \cap X \neq \emptyset \Leftrightarrow x \in \{x \in U : M_{\mathbf{C}}(x) \cap X \neq \emptyset\}$.

From the above results, it is straightforward to obtain the essential relationships between FL, FH and SL, SH.

Theorem 1. *Let $(U, \mathbf{C})$ be a covering approximation space. For any $X \subseteq U$,*
$$SL_{\mathbf{C}}(X) = FL_{\mathbf{C}_{\tilde{\emptyset}} \cup \{U\}}(X) \text{ and } SH_{\mathbf{C}}(X) = FH_{\mathbf{C}_{\tilde{\emptyset}} \cup \{U\}}(X).$$

Wang and Hu [21] have presented a necessary and sufficient condition of the complement of a covering to be a covering. Since a covering block of a covering including all the objects in a universe is useless for problem solving, they do not discuss coverings with this kind of covering block. Therefore, we will give a necessary and sufficient condition that the complement of any covering is a covering.

Proposition 5. *Let $(U, \mathbf{C})$ be a covering approximation space. Then $\mathbf{C}^{\sim}$ is a covering if and only if $\cap \mathbf{C} = \emptyset$ and $U \notin \mathbf{C}$.*

Proof. $\mathbf{C}^{\sim}$ is a covering if and only if $\underset{K \in \mathbf{C}^{\sim}}{\cup} K = U$ and $\emptyset \notin \mathbf{C}^{\sim}$ if and only if $\underset{-K \in \mathbf{C}}{\cup} K = U$ and $U \notin \mathbf{C}$ if and only if $\underset{-K \in \mathbf{C}}{\cap} - K = \emptyset$ and $U \notin \mathbf{C}$ if and only if $\cap \mathbf{C} = \emptyset$ and $U \notin \mathbf{C}$.

Example 2. Suppose $U = \{1, 2, 3\}$ and $\mathbf{C} = \{\{1, 2\}, \{1, 3\}, \{3\}\}$ is a covering of U. We see $\cap \mathbf{C} = \emptyset$ and $\{1, 2, 3\} \notin \mathbf{C}$. Then $\mathbf{C}^{\sim} = \{\{3\}, \{2\}, \{1, 2\}\}$ is a covering of U.

In the following proposition, the neighborhoods of a covering are compared with ones of a new covering which consists of the original covering and the universe.

Proposition 6. *Let $(U, \mathbf{C})$ be a covering approximation space. For any $x \in U$, $N_{\mathbf{C}}(x) = N_{\mathbf{C} \cup \{U\}}(x)$.*

Proof. $N_{\mathbf{C} \cup \{U\}}(x) = \cap \{K \in \mathbf{C} \cup \{U\} : x \in K\} = \cap \{K \in \mathbf{C} : x \in K\} \cap U = \cap \{K \in \mathbf{C} : x \in K\} = N_{\mathbf{C}}(x)$.

Corollary 1. *Let $(U, \mathbf{C})$ be a covering approximation space. For any $X \subseteq U$, $FL_{\mathbf{C}}(X) = FL_{\mathbf{C} \cup \{U\}}(X)$ and $FH_{\mathbf{C}}(X) = FH_{\mathbf{C} \cup \{U\}}(X)$.*

When the complement of a covering is still a covering, we can obtain the relationships between FL, FH and SL, SH as follows.

Proposition 7. *Let $(U, \mathbf{C})$ be a covering approximation space and $X \subseteq U$. If $\mathbf{C}^{\sim}$ is a covering of U, then $SL_{\mathbf{C}}(X) = FL_{\mathbf{C}^{\sim}}(X)$ and $FH_{\mathbf{C}}(X) = SH_{\mathbf{C}^{\sim}}(X)$.*

3.3 Conditions Under Which FH and SH Are Identical

The approximation ability of a covering can be improved by more exact approximations. Based on the duality, we present only a sufficient and necessary condition under which FH and SH are equal to each other. We first give an example to illustrate the relationships between neighborhoods and complementary neighborhoods.

Example 3. Let $U = \{1, 2, 3\}$ and $\mathbf{C} = \{\{1\}, \{1, 2\}, \{1, 2, 3\}\}$. We can obtain

$$M_{\mathbf{C}}(1) = \{1, 2, 3\} \quad N_{\mathbf{C}}(1) = \{1\}$$
$$M_{\mathbf{C}}(2) = \{2, 3\} \quad N_{\mathbf{C}}(2) = \{1, 2\}$$
$$M_{\mathbf{C}}(3) = \{3\} \qquad N_{\mathbf{C}}(3) = \{1, 2, 3\}$$

Proposition 8. *Let $(U, \mathbf{C})$ be a covering approximation space. Then $Cov(\mathbf{C})$ is a partition of U if and only if $N_{\mathbf{C}}(x) = M_{\mathbf{C}}(x)$ for any $x \in U$.*

Proof. $Cov(\mathbf{C})$ is a partition of $U \Leftrightarrow \forall x, y \in U(x \in N_{\mathbf{C}}(y) \leftrightarrow y \in N_{\mathbf{C}}(x)) \Leftrightarrow \{y \in U : y \in N_{\mathbf{C}}(x)\} = \{y \in U : x \in N_{\mathbf{C}}(y)\} \Leftrightarrow N_{\mathbf{C}}(x) = M_{\mathbf{C}}(x)$.

A sufficient and necessary condition, under which the upper approximation defined by Zhu is equal to the upper approximation defined by Xu and Zhang, is presented in the following proposition.

Proposition 9. *Let $(U, \mathbf{C})$ be a covering approximation space. Then for any $X \subseteq U$, $FH_{\mathbf{C}}(X) = SH_{\mathbf{C}}(X)$ if and only if $Cov(\mathbf{C})$ is a partition of U.*

Proof. According to Definition 4, Propositions 4 and 8, it is straightforward.

4   Extension of a Covering

Wang and Hu [21] introduced complement to get more knowledge from a covering, and then they defined a notion of the extension of a covering. Based on the lower and upper approximation operators FL and FH, they proved that the extension of a covering generates more exact approximations than itself generally. Therefore, the extension of a covering has been applied to attribute reduction. In this section, we mainly study the properties of the extension of a covering through the complement of the covering.

Definition 8 *([21]). Let $(U, \mathbf{C})$ be a covering approximation space. We call $\mathbf{C} \cup \mathbf{C}_{\emptyset}^{\sim}$ the extension of $\mathbf{C}$ and denote it as $\mathbf{C}^{\sharp}$.*

In the following proposition, we give an expression of the neighborhood of the extension of a covering through the covering itself.

Proposition 10. *Let $(U, \mathbf{C})$ be a covering approximation space. For any $x \in U$,*

$$N_{\mathbf{C}^{\sharp}}(x) = \{y \in U : N_{\mathbf{C}}(x) = N_{\mathbf{C}}(y)\}.$$

Proof. According to Definition 8 and Proposition 6, $N_{\mathbf{C}^{\sharp}}(x) = N_{\mathbf{C} \cup \mathbf{C}_{\emptyset}^{\sim}}(x) = N_{\mathbf{C} \cup \mathbf{C}_{\emptyset}^{\sim} \cup \{U\}}(x)$. According to Proposition 3, we have

$$\begin{aligned}
N_{\mathbf{C}^{\sharp}}(x) &= N_{\mathbf{C}}(x) \cap N_{\mathbf{C}_{\emptyset}^{\sim} \cup \{U\}}(x) \\
&= \{y \in U : y \in N_{\mathbf{C}}(x)\} \cap \{y \in U : x \in N_{\mathbf{C}}(y)\} \\
&= \{y \in U : y \in N_{\mathbf{C}}(x) \wedge x \in N_{\mathbf{C}}(y)\} \\
&= \{y \in U : N_{\mathbf{C}}(x) = N_{\mathbf{C}}(y)\}.
\end{aligned}$$

In order to further understand the above proposition, an example is given as follows.

Example 4 (Continued from Example 3). Since $\mathbf{C}^{\sharp} = \{\{1\}, \{1,2\}, \{1,2,3\}, \{2,3\}, \{3\}\}$ and $N_{\mathbf{C}}(1) = \{1\}, N_{\mathbf{C}}(2) = \{2\}, N_{\mathbf{C}}(3) = \{3\}$. Then,

$$N_{\mathbf{C}^{\sharp}}(1) = \{1\}, \quad \{y \in U : N_{\mathbf{C}}(1) = N_{\mathbf{C}}(y)\} = \{1\};$$
$$N_{\mathbf{C}^{\sharp}}(2) = \{2\}, \quad \{y \in U : N_{\mathbf{C}}(2) = N_{\mathbf{C}}(y)\} = \{2\};$$
$$N_{\mathbf{C}^{\sharp}}(3) = \{3\}, \quad \{y \in U : N_{\mathbf{C}}(3) = N_{\mathbf{C}}(y)\} = \{3\}.$$

Proposition 11. *Let $(U, \mathbf{C})$ be a covering approximation space. Then $Cov(\mathbf{C}^{\sharp})$ is a partition of U.*

Proof. According to Definition 3, $Cov(\mathbf{C}^{\sharp}) = \{N_{\mathbf{C}^{\sharp}}(x) : x \in U\}$. We see $\cup Cov(\mathbf{C}^{\sharp}) = U$. Therefore, we need to prove $N_{\mathbf{C}^{\sharp}}(x) \cap N_{\mathbf{C}^{\sharp}}(y) = \emptyset$ if $N_{\mathbf{C}^{\sharp}}(x) \neq N_{\mathbf{C}^{\sharp}}(y)$. According to Proposition 10, $N_{\mathbf{C}^{\sharp}}(x) = \{w \in U : N_{\mathbf{C}}(x) = N_{\mathbf{C}}(w)\}$ and $N_{\mathbf{C}^{\sharp}}(y) = \{z \in U : N_{\mathbf{C}}(y) = N_{\mathbf{C}}(z)\}$. Suppose $N_{\mathbf{C}^{\sharp}}(x) \cap N_{\mathbf{C}^{\sharp}}(y) \neq \emptyset$. Then there exists at least $z \in U$ such that $z \in N_{\mathbf{C}^{\sharp}}(x)$ and $z \in N_{\mathbf{C}^{\sharp}}(y)$, i.e., $N_{\mathbf{C}}(z) = N_{\mathbf{C}}(x)$ and $N_{\mathbf{C}}(z) = N_{\mathbf{C}}(y)$. Therefore, $N_{\mathbf{C}}(x) = N_{\mathbf{C}}(y)$, i.e., $N_{\mathbf{C}^{\sharp}}(x) = N_{\mathbf{C}^{\sharp}}(y)$ which is contradictory with $N_{\mathbf{C}^{\sharp}}(x) \neq N_{\mathbf{C}^{\sharp}}(y)$. Hence $N_{\mathbf{C}^{\sharp}}(x) \cap N_{\mathbf{C}^{\sharp}}(y) = \emptyset$ if $N_{\mathbf{C}^{\sharp}}(x) \neq N_{\mathbf{C}^{\sharp}}(y)$.

According to the above proposition, we see the lower and upper approximation operators FL and FH with respect to the induced covering of a covering degenerate to Pawlak lower and upper approximation operators respectively. Furthermore, the lower and upper approximations of the extension of a covering are actually Pawlak lower and upper approximations.

Proposition 12. *Let $(U, \mathbf{C})$ be a covering approximation space and $X \subseteq U$. Then, $FL_{\mathbf{C}^{\sharp}}(X) = FL_{Cov(\mathbf{C}^{\sharp})}(X)$ and $FH_{\mathbf{C}^{\sharp}}(X) = FH_{Cov(\mathbf{C}^{\sharp})}(X)$.*

Through the extension of coverings, Wang and Hu [21] got the less reduction attributes than the existing methods [1,6] in covering decision systems. In fact, Wang and Hu solved the attribute reduction problem of covering decision systems through degenerating covering-based rough sets to Pawlak rough sets.

5 Matroidal Approach and the Exact Sets

From the above sections, we see FL, FH and SL, SH are corresponding to each other. Therefore, we consider only FL and FH in this section.

Many authors [3,23] have studied the topologies induced by covering-based rough sets. FL is an interior operator and FH is a closure operator. Based on these results, some algebraic structures of all the exact sets are presented. Recently, the study of the combination of rough set and matroid is a hot research topic. In the following part, we investigate some properties of FL and FH by the matroidal approach. We first introduce the closed set axiom of matroids.

Proposition 13 *([5]). Let $\mathcal{L}$ be a collection of subsets of U. $\mathcal{L}$ is the set of all the closed sets of a matroid on U if and only if the following conditions hold:*
(L1) $U \in \mathcal{L}$;
(L2) If $L_1, L_2 \in \mathcal{L}$, then $L_1 \cap L_2 \in \mathcal{L}$;
(L3) If $L \in \mathcal{L}$ and $\{L_1, L_2, \cdots, L_m\}$ is the set of minimal members of $\mathcal{L}$ that properly contain L, then the sets $L_1 - L, L_2 - L, \cdots, L_m - L$ partition $U - L$.

In covering-based rough sets, when $FL_{\mathbf{C}}(X) = FH_{\mathbf{C}}(X)$, we say X is an exact set. This section investigates the exact sets for a covering approximation space $(U, \mathbf{C})$.

$$\mathcal{L} = \{X \subseteq U : FL_{\mathbf{C}}(X) = FH_{\mathbf{C}}(X)\}.$$

In the following proposition, we present some results about $\mathcal{L}$. The properties of FL and FH can be found in [27].

Proposition 14. *Let $(U, \mathbf{C})$ be a covering approximation space and $\mathcal{L} = \{X \subseteq U : FL_{\mathbf{C}}(X) = FH_{\mathbf{C}}(X)\}$. For any $X, Y \in \mathcal{L}$, we have*
(1) $-X \in \mathcal{L}$;
(2) $X \cap Y \in \mathcal{L}$;
(3) $X \cup Y \in \mathcal{L}$.

Proof. (1) If $X \in \mathcal{L}$, then $FL_{\mathbf{C}}(X) = FH_{\mathbf{C}}(X)$. Since $FL_{\mathbf{C}}(X) = -FH_{\mathbf{C}}(-X)$ and $FH_{\mathbf{C}}(X) = -FL_{\mathbf{C}}(-X)$, we have $FH_{\mathbf{C}}(-X) = FL_{\mathbf{C}}(-X)$. This implies $-X \in \mathcal{L}$.
(2) Since $X, Y \in \mathcal{L}$, then $FL_{\mathbf{C}}(X \cap Y) = FL_{\mathbf{C}}(X) \cap FL_{\mathbf{C}}(Y) = X \cap Y$. Since $X \cap Y \subseteq X$ and $X \cap Y \subseteq Y$, we have $FH_{\mathbf{C}}(X \cap Y) \subseteq FH_{\mathbf{C}}(X)$ and $FH_{\mathbf{C}}(X \cap Y) \subseteq FH_{\mathbf{C}}(Y)$, then $FH_{\mathbf{C}}(X \cap Y) \subseteq FH_{\mathbf{C}}(X) \cap FH_{\mathbf{C}}(Y)$, i.e., $FH_{\mathbf{C}}(X \cap Y) \subseteq X \cap Y$. Since $X \cap Y \subseteq FH_{\mathbf{C}}(X \cap Y)$, we have $FH_{\mathbf{C}}(X \cap Y) = X \cap Y$. Therefore, $X \cap Y \in \mathcal{L}$.
(3) Similar to the proof of (L2), we can prove $X \cup Y \in \mathcal{L}$.

Based on the above proposition, we prove that $\mathcal{L}$ is the set of all closed sets of a matroid on U.

Theorem 2. *Let $(U, \mathbf{C})$ be a covering approximation space and $\mathcal{L} = \{X \subseteq U : FL_{\mathbf{C}}(X) = FH_{\mathbf{C}}(X)\}$. Then $\mathcal{L}$ satisfies (L1), (L2) and (L3) of Proposition 13.*

Proof. (L1) Since $U = FL_{\mathbf{C}}(U) = FH_{\mathbf{C}}(U)$, we have $U \in \mathcal{L}$;
(L2) According to Proposition 14, we see if $X, Y \in \mathcal{L}$, then $X \cap Y \in \mathcal{L}$;
(L3) When $L \in \mathcal{L}$ and $\{L_1, L_2, \cdots, L_m\}$ is the set of minimal members of $\mathcal{L}$ that properly contain L, suppose the sets $L_1 - L, L_2 - L, \cdots, L_m - L$ do not partition $U - L$. Then there exist at least two members $L_i, L_j \in \{L_1, L_2, \cdots, L_m\}$ such that $(L_i - L) \cap (L_j - L) \neq \emptyset$. Let $N = (L_i - L) \cap (L_j - L)$. Then $L_i \cap L_j = N \cup L$. Since $L_i, L_j \in \mathcal{L}$, according to (L2), we have $L_i \cap L_j \in \mathcal{L}$, which is contradictory with the condition that L_i, L_j are the minimal members of $\mathcal{L}$ that properly contain L. Therefore, for any $L_i, L_j \in \{L_1, L_2, \cdots, L_m\}$, $(L_i - L) \cap (L_j - L) = \emptyset$. On the other hand, suppose $(L_1 - L) \cup (L_2 - L) \cup \cdots \cup (L_m - L) \neq U - L$. Then, there exists a non-empty set $Y \subseteq U$ such that $Y = U - (L_1 \cup L_2 \cup \cdots \cup L_m) = -L_1 \cap -L_2 \cap \cdots \cap -L_m$. According to Proposition 14, we see $-L_1, -L_2, \cdots, -L_m \in \mathcal{L}$.

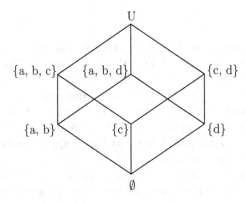

Fig. 1. $\mathcal{L}$

Hence $Y \in \mathcal{L}$. Since $L \in \mathcal{L}$, according to Proposition 14, $Y \cup L \in \mathcal{L}$. However, for any $L_i \in \{L_1, L_2, \cdots, L_m\}$, $L_i \not\subseteq Y \cup L$, which is contradictory with the condition that L_i, L_j are the minimal members of $\mathcal{L}$ that properly contain L. Therefore, $(L_1 - L) \cup (L_2 - L) \cup \cdots \cup (L_m - L) = U - L$.

In matroid theory, we see the set of all closed sets of a matroid together with the set inclusion is a lattice. Moreover, it is also a geometry lattice. In order to further understand $\mathcal{L}$, an example is given as follows.

Example 5. Let $U = \{a, b, c\}$ and $\mathbf{C} = \{\{a\}, \{a, b\}, \{c\}, \{d\}\}$ be a covering of U. Since $N_{\mathbf{C}}(a) = \{a\}, N_{\mathbf{C}}(b) = \{a, b\}, N_{\mathbf{C}}(c) = \{c\}$ and $N_{\mathbf{C}}(d) = \{d\}$, we have

$$
\begin{array}{ll}
FL_{\mathbf{C}}(\emptyset) = \emptyset & FH_{\mathbf{C}}(\emptyset) = \emptyset \\
FL_{\mathbf{C}}(\{a\}) = \{a\} & FH_{\mathbf{C}}(\{a\}) = \{a, b\} \\
FL_{\mathbf{C}}(\{b\}) = \emptyset & FH_{\mathbf{C}}(\{b\}) = \{b\} \\
FL_{\mathbf{C}}(\{c\}) = \{c\} & FH_{\mathbf{C}}(\{c\}) = \{c\} \\
FL_{\mathbf{C}}(\{d\}) = \{d\} & FH_{\mathbf{C}}(\{d\}) = \{d\} \\
FL_{\mathbf{C}}(\{a, b\}) = \{a, b\} & FH_{\mathbf{C}}(\{a, b\}) = \{a, b\} \\
FL_{\mathbf{C}}(\{a, c\}) = \{a, c\} & FH_{\mathbf{C}}(\{a, c\}) = \{a, b, c\} \\
FL_{\mathbf{C}}(\{a, d\}) = \{a, d\} & FH_{\mathbf{C}}(\{a, d\}) = \{a, b, d\} \\
FL_{\mathbf{C}}(\{b, c\}) = \{c\} & FH_{\mathbf{C}}(\{b, c\}) = \{b, c\} \\
FL_{\mathbf{C}}(\{b, d\}) = \{d\} & FH_{\mathbf{C}}(\{b, d\}) = \{b, d\} \\
FL_{\mathbf{C}}(\{c, d\}) = \{c, d\} & FH_{\mathbf{C}}(\{c, d\}) = \{c, d\} \\
FL_{\mathbf{C}}(\{a, b, c\}) = \{a, b, c\} & FH_{\mathbf{C}}(\{a, b, c\}) = \{a, b, c\} \\
FL_{\mathbf{C}}(\{a, b, d\}) = \{a, b, d\} & FH_{\mathbf{C}}(\{a, b, d\}) = \{a, b, d\} \\
FL_{\mathbf{C}}(\{a, c, d\}) = \{a, c, d\} & FH_{\mathbf{C}}(\{a, c, d\}) = \{a, b, c, d\} \\
FL_{\mathbf{C}}(\{b, c, d\}) = \{c, d\} & FH_{\mathbf{C}}(\{b, c, d\}) = \{b, c, d\} \\
FL_{\mathbf{C}}(U) = U & FH_{\mathbf{C}}(U) = U
\end{array}
$$

Therefore, $\mathcal{L} = \{\emptyset, \{a, b\}, \{c\}, \{d\}, \{a, b, c\}, \{a, b, d\}, \{c, d\}, U\}$. We use Fig. 1 to further illustrate $\mathcal{L}$.

6 Conclusions

In this paper, we applied a notion of the complement of coverings to study the relationships between covering-based rough sets defined by Zhu and ones defined by Xu and Zhang. We first corrected an existing result of complementary neighborhoods. Based on this result, we studied relationships between neighborhoods of a covering and ones of the complement of the covering. In fact, the lower and upper approximations with respect to a covering defined by Xu and Zhang were proved to be always equal to the lower and upper approximations with respect to the complement of the covering defined by Zhu. Finally, we presented some properties of extension of a covering based on the results of neighborhoods and complementary neighborhoods of the covering. This study suggests some internal connections between covering-based rough sets. Moreover, matroidal structures of covering-based rough sets are further investigated through the exact sets. Since matroids provide well-established platforms for greedy algorithms, the matroidal structure can be applied to design algorithms for attribute reduction.

Acknowledgments. This work is in part supported by the National Science Foundation of China under Grant Nos. 61170128, 61379049, 61379089 and 61440047.

References

1. Chen, D., Wang, C., Hu, Q.: A new approach to attribute reduction of consistent and inconsistent covering decision systems with covering rough sets. Inf. Sci. **177**, 3500–3518 (2007)
2. Inuiguchi, M.: Generalizations of rough sets and rule extraction. In: Peters, J.F., Skowron, A., Grzymała-Busse, J.W., Kostek, B., Świniarski, R.W., Szczuka, M.S. (eds.) Transactions on Rough Sets I. LNCS, vol. 3100, pp. 96–119. Springer, Heidelberg (2004)
3. Ge, X., Bai, X., Yun, Z.: Topological characterizations of covering for special covering-based upper approximation operators. Inf. Sci. **204**, 70–81 (2012)
4. Kondo, M.: On the structure of generalized rough sets. Inf. Sci. **176**, 589–600 (2005)
5. Lai, H.: Matroid Theory. Higher Education Press, Beijing (2001)
6. Li, F., Yin, Y.: Approaches to knowledge reduction of covering decision systems based on information theory. Inf. Sci. **179**, 1694–1704 (2009)
7. Liu, G., Sai, Y.: A comparison of two types of rough sets induced by coverings. Int. J. Approximate Reasoning **50**, 521–528 (2009)
8. Ma, L.: On some types of neighborhood-related covering rough sets. Int. J. Approximate Reasoning **53**, 901–911 (2012)
9. Mani, A.: Dialectics of counting and the mathematics of vagueness. In: Peters, J.F., Skowron, A. (eds.) Transactions on Rough Sets XV. LNCS, vol. 7255, pp. 122–180. Springer, Heidelberg (2012)
10. Meng, Z., Shi, Z.: A fast approach to attribute reduction in incomplete decision systems with tolerance relation-based rough sets. Inf. Sci. **179**, 2774–2793 (2009)
11. Pawlak, Z.: Rough Sets. Int. J. Comput. Inf. Sci. **11**, 341–356 (1982)
12. Pawlak, Z.: Rough Sets: Theoretical Aspects of Reasoning about Data. Kluwer Academic Publishers, Boston (1991)

13. Pomykala, J.A.: Approximation operations in approximation space. Bull. Pol. Acad. Sci. **35**, 653–662 (1987)
14. Qin, K., Gao, Y., Pei, Z.: On covering rough sets. In: Yao, J.T., Lingras, P., Wu, W.-Z., Szczuka, M.S., Cercone, N.J., Ślęzak, D. (eds.) RSKT 2007. LNCS (LNAI), vol. 4481, pp. 34–41. Springer, Heidelberg (2007)
15. Restrepo, M., Cornelis, C., Gmez, J.: Duality, conjugacy and adjointness of approximation operators in covering-based rough sets. Int. J. Approximate Reasoning **55**, 469–485 (2014)
16. Restrepo, M., Cornelis, C., Gmez, J.: Duality, partial order relation for approximation operators in covering based rough sets. Inf. Sci. **284**, 44–59 (2014)
17. Restrepo, M., Gómez, J.: Covering based rough sets and relation based rough sets. In: Kryszkiewicz, M., Cornelis, C., Ciucci, D., Medina-Moreno, J., Motoda, H., Raś, Z.W. (eds.) RSEISP 2014. LNCS, vol. 8537, pp. 143–152. Springer, Heidelberg (2014)
18. Skowron, A., Stepaniuk, J.: Tolerance approximation spaces. Fundamenta Informaticae **27**, 245–253 (1996)
19. Slowinski, R., Vanderpooten, D.: A generalized definition of rough approximations based on similarity. IEEE Trans. Knowl. Data Eng. **12**, 331–336 (2000)
20. Tsang, E., Cheng, D., Lee, J., Yeung, D.: On the upper approximations of covering generalized rough sets. In: Proceedings of 2004 International Conference on Machine Learning and Cybernetics, pp. 4200–4203. IEEE Press (2004)
21. Wang, G., Hu, J.: Attribute reduction using extension of covering approximation space. Fundamenta Informaticae **115**, 219–232 (2012)
22. Xu, W., Zhang, W.: Measuring roughness of generalized rough sets induced by a covering. Fuzzy Sets Syst. **158**, 2443–2455 (2007)
23. Yao, Y.Y.: Relational interpretations of neighborhood operators and rough set approximation operators. Inf. Sci. **111**, 239–259 (1998)
24. Yao, Y.Y., Yao, B.: Covering based rough set approximations. Inf. Sci. **200**, 91–107 (2012)
25. Zakowski, W.: Approximations in the space (u, π). Demonstratio Math. **16**, 761–769 (1983)
26. Zhu, W.: Topological approaches to covering rough sets. Inf. Sci. **177**, 1499–1508 (2007)
27. Zhu, W.: Relationship between generalized rough sets based on binary relation and covering. Inf. Sci. **179**, 210–225 (2009)
28. Zhu, W., Wang, F.: Reduction and axiomization of covering generalized rough sets. Inf. Sci. **152**, 217–230 (2003)
29. Zhu, W., Wang, F.: A new type of covering rough sets. In: 3rd International IEEE Conference on Intelligent Systems, pp. 444–449. IEEE Press, London (2006)

On the Nearness Measures of Near Sets

Keyun Qin[✉] and Bo Li

College of Mathematics, Southwest Jiaotong University,
Chengdu 610031, Sichuan, China
keyunqin@263.net, kingrayli@163.com

Abstract. In this paper, we focus our discussion on the nearness measures of near set approach. Some existing nearness measure is normalized with its basic properties being discussed. The connections between nearness relations and rough approximations are surveyed. The notions of strong nearness relations with respect to indiscernibility relation and weak indiscernibility relation are introduced. Some new nearness measures with respect to nearness relations and strong nearness relations are presented.

Keywords: Perceptual system · Near set · Rough set · Nearness measure · Strong nearness relation

1 Introduction

Near set approach was formally introduced by Peters in 2007 [1], motivated by image analysis and inspired by the study of perceptual nearness of physical objects. The near set approach focuses on discovering relations between samples relative to the degree of closeness of their specific descriptions. It provides a formal basis for observation, comparison, and classification of perceptual granules.

Recently, research on near set theory has been active, and great progress has been achieved. It has been shown that near sets can be used in perception-based approach to discover correspondences between images, such as face recognition systems, medical image analysis, image classification and so on [2–11]. Nearness measures play a central role in the applications of near set theory. It provided an approach to objects comparison. Henry [5] proposed the notion of nearness measure of near sets and presented some formulae to calculate the nearness measure. Based on tolerance space [12], some tolerance nearness measures were presented and applied to image correspondence [8,13]. In this paper, we conduct a further study of the nearness measures of near sets. Based on the analysis of existing nearness measures, we propose the notion of strong nearness relation and present some new nearness measures of near sets which are compatible with this nearness relation. Furthermore, the existing nearness measure is normalized in this paper with its basic properties being discussed.

© Springer International Publishing Switzerland 2015
Y. Yao et al. (Eds.): RSFDGrC 2015, LNAI 9437, pp. 102–111, 2015.
DOI: 10.1007/978-3-319-25783-9_10

2   Preliminaries

In near set theory, objects are known by their description. An object description is modeled as a vector function that represent object features. Object features are described by functions, e.g., color can be described by functions that measure intensity, hue and saturation. These functions are called probe functions. Formally, an object description $\varphi(x)$ associated with object x has the form:

$$\varphi(x) = (\varphi_1(x), \varphi_2(x), \cdots, \varphi_l(x)),$$

where φ_i is a probe function which assign a real value $\varphi_i(x)$ to x for every $i \leq l$.

Definition 1 *[11]. A perceptual system $< O, \mathbb{F} >$ is a real-valued information system where O is a non-empty set of perceptual objects, $\mathbb{F}$ is a set of probe functions.*

By this definition, a perceptual system $< O, \mathbb{F} >$ is a real valued information system with perceptual interpretation. For every $\mathfrak{B} \subseteq \mathbb{F}$, the indiscernibility relation $\sim_{\mathfrak{B}}$ induced by $\mathfrak{B}$ is defined as:

$$\sim_{\mathfrak{B}} = \{(x, y) \in O \times O; \forall \varphi \in \mathfrak{B}(\varphi(x) = \varphi(y))\}.$$

If $\mathfrak{B} = \{\varphi\}$ for some $\varphi \in \mathbb{F}$, instead of $\sim_{\{\varphi\}}$ we write $\sim_{\varphi}$. Clearly, $\sim_{\mathfrak{B}}$ is an equivalence relation. The equivalence class containing x is denoted by $x/\sim_{\mathfrak{B}}$, i.e., $x/\sim_{\mathfrak{B}} = \{y \in O; (x, y) \in \sim_{\mathfrak{B}}\}$. It follows that $(O, \sim_{\mathfrak{B}})$ is a Pawlak approximation space [14]. For every $X \subseteq O$, the lower approximation $\underline{\mathfrak{B}}(X)$ and upper approximation $\overline{\mathfrak{B}}(X)$ of X with respect to $(O, \sim_{\mathfrak{B}})$ are defined by [14]:

$$\underline{\mathfrak{B}}(X) = \{x \in O; x/\sim_{\mathfrak{B}} \subseteq X\} = \cup \{x/\sim_{\mathfrak{B}}; x/\sim_{\mathfrak{B}} \subseteq X\},$$
$$\overline{\mathfrak{B}}(X) = \{x \in O; x/\sim_{\mathfrak{B}} \cap X \neq \emptyset\} = \cup \{x/\sim_{\mathfrak{B}}; x/\sim_{\mathfrak{B}} \cap X \neq \emptyset\}.$$

The weak indiscernibility relation $\simeq_{\mathfrak{B}}$ induced by $\mathfrak{B}$ is defined as [11]:

$$\simeq_{\mathfrak{B}} = \{(x, y) \in O \times O; \exists \varphi \in \mathfrak{B}(\varphi(x) = \varphi(y))\}.$$

We observe that $\simeq_{\mathfrak{B}}$ is a reflexive and symmetric relation, but not necessarily transitive. In order to describe the closeness of objects, Peter et al. [11,13] introduced the notion of near sets and several kinds of nearness relations.

Definition 2 *[11]. Let $< O, \mathbb{F} >$ be a perceptual system and $X, Y \subseteq O$.*
(1) X is weakly near to Y within the perceptual system $< O, \mathbb{F} >$, denoted by $X \bowtie_{\mathbb{F}} Y$, if there exist $x \in X$, $y \in Y$ and $\mathfrak{B} \subseteq \mathbb{F}$ such that $x \sim_{\mathfrak{B}} y$.
(2) X is near to Y within the perceptual system $< O, \mathbb{F} >$, denoted by $X \bowtie_{\mathbb{F}} Y$, if there exist $\mathfrak{A}, \mathfrak{B} \subseteq \mathbb{F}$, $f \in \mathbb{F}$ and these are $A \in O/\sim_{\mathfrak{A}}$, $B \in O/\sim_{\mathfrak{B}}$, $C \in O/\sim_f$ such that $A \subseteq X$, $B \subseteq Y$, and $A, B \subseteq C$.

Definition 3 *[13]. Let $< O, \mathbb{F} >$ be a perceptual system, $\mathfrak{B} \subseteq \mathbb{F}$ and $X, Y \subseteq O$. $X \delta_{\mathfrak{B}} Y$ if and only if there exist $x \in X$ and $y \in Y$ such that $x \sim_{\mathfrak{B}} y$.*

Definition 4 *[11]. Let $< O, \mathbb{F} >$ be a perceptual system and $X \subseteq O$. X is said to be a perceptual near set if there exists $Y \subseteq O$ such that $X \bowtie_{\mathbb{F}} Y$.*

In what follows, the family of perceptual near sets of a perceptual system $< O, \mathbb{F} >$ is denoted by $Near_{\mathbb{F}}(O)$.

Theorem 1 *[11, 13]. Let $< O, \mathbb{F} >$ be a perceptual system, $X, Y, Z \subseteq O$, and $\mathfrak{B} \subseteq \mathbb{F}$.*

(1) $X \bowtie_{\mathbb{F}} Y$ if and only if there exist $x \in X$ and $y \in Y$ such that $x \simeq_{\mathbb{F}} y$.
(2) $X \in Near_{\mathbb{F}}(O)$ if and only if there is $A \in O/ \sim_{\mathbb{F}}$ such that $A \subseteq X$.
(3) $X \cap Y \neq \emptyset$ implies $X \bowtie_{\mathbb{F}} Y$ and $X \bowtie_{\mathbb{F}} Y$.
(4) $X \bowtie_{\mathbb{F}} Y$ and $Y \subseteq Z$ imply $X \bowtie_{\mathbb{F}} Z$.
(5) $X \bowtie_{\mathbb{F}} Y$ and $Y \subseteq Z$ imply $X \bowtie_{\mathbb{F}} Z$.
(6) $X \delta_{\mathfrak{B}} Y$ if and only if $\overline{\mathfrak{B}}(X) \cap \overline{\mathfrak{B}}(Y) \neq \emptyset$.

We notice that $\bowtie_{\mathbb{F}}$ and $\bowtie_{\mathbb{F}}$ are related to the set of all probe functions, whereas $\delta_{\mathfrak{B}}$ is based on a subset $\mathfrak{B}$ of $\mathbb{F}$. $\bowtie_{\mathbb{F}}$ and $\bowtie_{\mathbb{F}}$ can be easily extended to $\bowtie_{\mathfrak{B}}$ and $\bowtie_{\mathfrak{B}}$ respectively.

3   Nearness Measures

In order to measure the nearness degree of sets of perceptual objects, Henry et al. [5] proposed the following nearness measure.

Definition 5 *[5]. Let $< O, \mathbb{F} >$ be a perceptual system, $\mathfrak{B} \subseteq \mathbb{F}$ and $X, Y \subseteq O$. The nearness degree $NM_{\sim_{\mathfrak{B}}}(X, Y)$ between X and Y is defined by:*

$$NM_{\sim_{\mathfrak{B}}}(X, Y) = \frac{\sum_{x/\sim_{\mathfrak{B}} \in X/\sim_{\mathfrak{B}}} \sum_{y/\sim_{\mathfrak{B}} \in Y/\sim_{\mathfrak{B}}} \eta(x/\sim_{\mathfrak{B}}, y/\sim_{\mathfrak{B}})}{max(|X/\sim_{\mathfrak{B}}|, |Y/\sim_{\mathfrak{B}}|)} \quad (1)$$

where

$$\eta(x/\sim_{\mathfrak{B}}, y/\sim_{\mathfrak{B}}) = \begin{cases} min(|x/\sim_{\mathfrak{B}}|, |y/\sim_{\mathfrak{B}}|), & if \ (x, y) \in \sim_{\mathfrak{B}}, \\ 0, & otherwise. \end{cases} \quad (2)$$

In this definition, $X/\sim_{\mathfrak{B}} = \{X \cap Y; Y \in O/\sim_{\mathfrak{B}}\}$. By $x/\sim_{\mathfrak{B}} \in X/\sim_{\mathfrak{B}}$ we mean $x/\sim_{\mathfrak{B}} = X \cap x^O/\sim_{\mathfrak{B}}$, where $x^O/\sim_{\mathfrak{B}}$ is the equivalence class with respect to the set of all perceptual objects. The idea behind the $NM_{\sim_{\mathfrak{B}}}$ measure is that one reliable measure of nearness of sets in a perceptual system is based on the cardinality of equivalence classes that they share.

Theorem 2. *Let $< O, \mathbb{F} >$ be a perceptual system, $X, Y, Z \subseteq O$, and $\mathfrak{B} \subseteq \mathbb{F}$.*
(1) $NM_{\sim_{\mathfrak{B}}}(X, Y) > 0$ if and only if $X \delta_{\mathfrak{B}} Y$.
(2) $X \cap Y \neq \emptyset$ implies $NM_{\sim_{\mathfrak{B}}}(X, Y) > 0$.
(3) $NM_{\sim_{\mathfrak{B}}}(X, X) = \frac{|X|}{|X/\sim_{\mathfrak{B}}|}$.
(4) $X \subseteq Y \subseteq Z$ implies $NM_{\sim_{\mathfrak{B}}}(X, Z) \leq NM_{\sim_{\mathfrak{B}}}(X, Y)$ and $NM_{\sim_{\mathfrak{B}}}(X, Z) \leq NM_{\sim_{\mathfrak{B}}}(Y, Z)$.

Proof. (1) Suppose that $NM_{\sim_\mathfrak{B}}(X,Y) > 0$. Then there exist $x/\sim_\mathfrak{B} \in X/\sim_\mathfrak{B}$ and $y/\sim_\mathfrak{B} \in Y/\sim_\mathfrak{B}$ such that $\eta(x/\sim_\mathfrak{B}, y/\sim_\mathfrak{B}) > 0$. It follows that $x \in X$, $y \in Y$ and $(x,y) \in \sim_\mathfrak{B}$. Consequently, we have $X\delta_\mathfrak{B}Y$ as required.

Conversely, assume that $X\delta_\mathfrak{B}Y$. Then there exist $x \in X$ and $y \in Y$ such that $(x,y) \in \sim_\mathfrak{B}$. It follows that $x/\sim_\mathfrak{B} \in X/\sim_\mathfrak{B}$, $y/\sim_\mathfrak{B} \in Y/\sim_\mathfrak{B}$,

$$\eta(x/\sim_\mathfrak{B}, y/\sim_\mathfrak{B}) = min(|x/\sim_\mathfrak{B}|, |y/\sim_\mathfrak{B}|) > 0,$$

and hence $NM_{\sim_\mathfrak{B}}(X,Y) > 0$.

(2) follows from (1).

(3) Let $x/\sim_\mathfrak{B} \in X/\sim_\mathfrak{B}$, $y/\sim_\mathfrak{B} \in X/\sim_\mathfrak{B}$. If $x/\sim_\mathfrak{B} \neq y/\sim_\mathfrak{B}$, then we have $(x,y) \notin \sim_\mathfrak{B}$ and consequently $\eta(x/\sim_\mathfrak{B}, y/\sim_\mathfrak{B}) = 0$. It follows that

$$NM_{\sim_\mathfrak{B}}(X,X) = \frac{\sum_{x/\sim_\mathfrak{B} \in X/\sim_\mathfrak{B}} \sum_{y/\sim_\mathfrak{B} \in X/\sim_\mathfrak{B}} \eta(x/\sim_\mathfrak{B}, y/\sim_\mathfrak{B})}{max(|X/\sim_\mathfrak{B}|, |X/\sim_\mathfrak{B}|)}$$

$$= \frac{\sum_{x/\sim_\mathfrak{B} \in X/\sim_\mathfrak{B}} \eta(x/\sim_\mathfrak{B}, x/\sim_\mathfrak{B})}{|X/\sim_\mathfrak{B}|} = \frac{\sum_{x/\sim_\mathfrak{B} \in X/\sim_\mathfrak{B}} |x/\sim_\mathfrak{B}|}{|X/\sim_\mathfrak{B}|} = \frac{|X|}{|X/\sim_\mathfrak{B}|}.$$

(4) Assume that $Z/\sim_\mathfrak{B} = \{z_1/\sim_\mathfrak{B}, \cdots, z_m/\sim_\mathfrak{B}\}$. By $X \subseteq Y \subseteq Z$, we have

$$X/\sim_\mathfrak{B} = \{X \cap D; D \in Z/\sim_\mathfrak{B}, X \cap D \neq \emptyset\},$$
$$Y/\sim_\mathfrak{B} = \{Y \cap D; D \in Z/\sim_\mathfrak{B}, Y \cap D \neq \emptyset\}.$$

Without lose of generality, let

$$X/\sim_\mathfrak{B} = \{z_1/\sim_\mathfrak{B} \cap X, \cdots, z_l/\sim_\mathfrak{B} \cap X\},$$
$$Y/\sim_\mathfrak{B} = \{z_1/\sim_\mathfrak{B} \cap Y, \cdots, z_k/\sim_\mathfrak{B} \cap Y\},$$

where $l \leq k \leq m$. For each $i \in \{1, 2, \cdots, l\}$ and $j \in \{1, 2, \cdots, m\}$, we have $\eta(z_i/\sim_\mathfrak{B} \cap X, z_j/\sim_\mathfrak{B}) = 0$ if $i \neq j$, and

$$\eta(z_i/\sim_\mathfrak{B} \cap X, z_i/\sim_\mathfrak{B}) = min(|z_i/\sim_\mathfrak{B} \cap X|, |z_i/\sim_\mathfrak{B}|) = |z_i/\sim_\mathfrak{B} \cap X|,$$

$$NM_{\sim_\mathfrak{B}}(X,Z) = \frac{\sum_{x/\sim_\mathfrak{B} \in X/\sim_\mathfrak{B}} \sum_{z/\sim_\mathfrak{B} \in Z/\sim_\mathfrak{B}} \eta(x/\sim_\mathfrak{B}, z/\sim_\mathfrak{B})}{max(|X/\sim_\mathfrak{B}|, |Z/\sim_\mathfrak{B}|)}$$

$$= \frac{1}{m}(|z_1/\sim_\mathfrak{B} \cap X| + \cdots + |z_l/\sim_\mathfrak{B} \cap X|) = \frac{|X|}{m}.$$

Similarly, we have $NM_{\sim_\mathfrak{B}}(Y,Z) = \frac{|Y|}{m}$, $NM_{\sim_\mathfrak{B}}(X,Y) = \frac{|X|}{k}$, and consequently $NM_{\sim_\mathfrak{B}}(X,Z) \leq NM_{\sim_\mathfrak{B}}(X,Y)$ and $NM_{\sim_\mathfrak{B}}(X,Z) \leq NM_{\sim_\mathfrak{B}}(Y,Z)$.

Example 1. Let $< O, \mathbb{F} >$ be a perceptual system where $O = \{x_1, \cdots, x_9\}$, $\mathbb{F} = \{\varphi_1, \varphi_2\}$, and the values of probe functions are define in Table 1.

Table 1. Perceptual system $< O, \mathbb{F} >$

	x_1	x_2	x_3	x_4	x_5	x_6	x_7	x_8	x_9
φ_1	0	2	0	2	0	0	0	0	2
φ_2	1	5	1	5	1	3	3	3	3

The perceptual system $< O, \mathbb{F} >$ determines three indiscernibility relations $\sim_{\varphi_1}$, $\sim_{\varphi_2}$ and $\sim_{\mathbb{F}}$. The partitions determined by these relations are as follows:

$$O/\sim_{\varphi_1} = \{\{x_1, x_3, x_5, x_6, x_7, x_8\}, \{x_2, x_4, x_9\}\},$$
$$O/\sim_{\varphi_2} = \{\{x_1, x_3, x_5\}, \{x_2, x_4\}, \{x_6, x_7, x_8, x_9\}\},$$
$$O/\sim_{\mathbb{F}} = \{\{x_1, x_3, x_5\}, \{x_2, x_4\}, \{x_6, x_7, x_8\}, \{x_9\}\}.$$

Let $X = \{x_1, x_2, x_3, x_4, x_5\}$, $Y = \{x_1, \cdots, x_8\}$. It follows that $\eta(\{x_2, x_4\}, \{x_2, x_4\}) = 2$, $\eta(\{x_1, x_3, x_5\}, \{x_1, x_3, x_5\}) = 3$, $\eta(\{x_2, x_4\}, \{x_1, x_3, x_5\}) = \eta(\{x_2, x_4\}, \{x_6, x_7, x_8\}) = 0$, $\eta(\{x_1, x_3, x_5\}, \{x_2, x_4\}) = \eta(\{x_1, x_3, x_5\}, \{x_6, x_7, x_8\}) = 0$, and consequently, $NM_{\sim_{\mathbb{F}}}(X, Y) = \frac{5}{3}$.

By this Example and Theorem 2(3), $NM_{\sim_{\mathfrak{B}}}$ is not a normalized measure, i.e., $NM_{\sim_{\mathfrak{B}}}(X, Y) > 1$ is possible. We normalize this measure by the following definition.

Definition 6. *Let $< O, \mathbb{F} >$ be a perceptual system, $\mathfrak{B} \subseteq \mathbb{F}$ and $X, Y \subseteq O$. The nearness degree $NM^1_{\sim_{\mathfrak{B}}}(X, Y)$ between X and Y is defined by:*

$$NM^1_{\sim_{\mathfrak{B}}}(X, Y) = \frac{\sum_{x/\sim_{\mathfrak{B}} \in X/\sim_{\mathfrak{B}}} \sum_{y/\sim_{\mathfrak{B}} \in Y/\sim_{\mathfrak{B}}} \eta_1(x/\sim_{\mathfrak{B}}, y/\sim_{\mathfrak{B}})}{max(|X/\sim_{\mathfrak{B}}|, |Y/\sim_{\mathfrak{B}}|)} \quad (3)$$

where

$$\eta_1(x/\sim_{\mathfrak{B}}, y/\sim_{\mathfrak{B}}) = \begin{cases} \frac{min(|x/\sim_{\mathfrak{B}}|, |y/\sim_{\mathfrak{B}}|)}{max(|x/\sim_{\mathfrak{B}}|, |y/\sim_{\mathfrak{B}}|)}, & if \ (x, y) \in \sim_{\mathfrak{B}}, \\ 0, & otherwise. \end{cases} \quad (4)$$

We observe that $NM^1_{\sim_{\mathfrak{B}}}(X, Y) > 0$ if and only if $NM_{\sim_{\mathfrak{B}}}(X, Y) > 0$.

Theorem 3. *Let $< O, \mathbb{F} >$ be a perceptual system, $X, Y, Z \subseteq O$, and $\mathfrak{B} \subseteq \mathbb{F}$.*
(1) $0 \leq NM^1_{\sim_{\mathfrak{B}}}(X, Y) \leq 1$.
(2) $NM^1_{\sim_{\mathfrak{B}}}(X, Y) > 0$ if and only if $X\delta_{\mathfrak{B}}Y$.
(3) $X \cap Y \neq \emptyset$ implies $NM^1_{\sim_{\mathfrak{B}}}(X, Y) > 0$.
(4) $NM^1_{\sim_{\mathfrak{B}}}(X, Y) = 1$ if and only if for every $x/\sim_{\mathfrak{B}} \in X/\sim_{\mathfrak{B}}$, there exists $y/\sim_{\mathfrak{B}} \in Y/\sim_{\mathfrak{B}}$ such that $(x, y) \in \sim_{\mathfrak{B}}$, $|x/\sim_{\mathfrak{B}}| = |y/\sim_{\mathfrak{B}}|$ and for every $y/\sim_{\mathfrak{B}} \in Y/\sim_{\mathfrak{B}}$, there exists $x/\sim_{\mathfrak{B}} \in X/\sim_{\mathfrak{B}}$ such that $(y, x) \in \sim_{\mathfrak{B}}$, $|y/\sim_{\mathfrak{B}}| = |x/\sim_{\mathfrak{B}}|$.
(5) $NM^1_{\sim_{\mathfrak{B}}}(X, X) = 1$.
(6) $X \subseteq Y \subseteq Z$ implies $NM^1_{\sim_{\mathfrak{B}}}(X, Z) \leq NM^1_{\sim_{\mathfrak{B}}}(X, Y)$ and $NM^1_{\sim_{\mathfrak{B}}}(X, Z) \leq NM^1_{\sim_{\mathfrak{B}}}(Y, Z)$.

By Theorem 3(2), $NM^1_{\sim_\mathfrak{B}}$ is compatible with the nearness relation $\delta_\mathfrak{B}$, i.e., $NM^1_{\sim_\mathfrak{B}}(X,Y) > 0$ if and only if $X\delta_\mathfrak{B}Y$. The following example shows that $NM^1_{\sim_\mathfrak{B}}$ is not compatible with the nearness relations $\bowtie_\mathbb{F}$ and $\bowtie_\mathbb{F}$.

Example 2. We consider the perceptual system $< O, \mathbb{F} >$ as defined in Example 1. Let $X_1 = \{x_1, x_3, x_5\}$, $Y_1 = \{x_6, x_7, x_8\}$. It follows that $X_1\bowtie_\mathbb{F}Y_1$, $X_1 \bowtie_\mathbb{F} Y_1$ and $NM_{\sim_\mathbb{F}}(X_1, Y_1) = 0$. Let $X_2 = \{x_2, x_3\}$, $Y_2 = \{x_2, x_6\}$. By direct computation we know that $X_2 \bowtie_\mathbb{F} Y_2$ does not hold and $NM_{\sim_\mathbb{F}}(X_2, Y_2) = \frac{1}{2} \neq 0$.

Now, we present a nearness measure which is compatible with $\bowtie_\mathbb{F}$.

Definition 7. *Let $< O, \mathbb{F} >$ be a perceptual system and $X, Y \subseteq O$. The nearness degree $NM^2_{\sim_\mathbb{F}}(X,Y)$ between X and Y is defined by:*

$$NM^2_{\sim_\mathbb{F}}(X,Y) = \frac{\sum_{x/\sim_\mathbb{F}\in X/\sim_\mathbb{F}} \sum_{y/\sim_\mathbb{F}\in Y/\sim_\mathbb{F}} \eta_2(x/\sim_\mathbb{F}, y/\sim_\mathbb{F})}{|X/\sim_\mathbb{F}| \cdot |Y/\sim_\mathbb{F}|} \tag{5}$$

where

$$\eta_2(x/\sim_\mathbb{F}, y/\sim_\mathbb{F}) = \begin{cases} \frac{min(|x/\sim_\mathbb{F}|, |y/\sim_\mathbb{F}|)}{max(|x/\sim_\mathbb{F}|, |y/\sim_\mathbb{F}|)}, & if \ (x, y) \in\simeq_\mathbb{F}, \\ 0, & otherwise. \end{cases} \tag{6}$$

Theorem 4. *Let $< O, \mathbb{F} >$ be a perceptual system, $X, Y \subseteq O$.*
  (1) $0 \leq NM^2_{\sim_\mathbb{F}}(X,Y) \leq 1$.
  (2) $NM^2_{\sim_\mathbb{F}}(X,Y) > 0$ if and only if $X\bowtie_\mathbb{F}Y$.
  (3) $X \cap Y \neq \emptyset$ implies $NM^2_{\sim_\mathbb{F}}(X,Y) > 0$.
  (4) $NM^2_{\sim_\mathbb{F}}(X,Y) = 1$ if and only if $(x, y) \in\simeq_\mathbb{F}$ and $|x/\sim_\mathbb{F}| = |y/\sim_\mathbb{F}|$ for every $x \in X$, $y \in Y$.
  (5) $NM^2_{\sim_\mathbb{F}}(X,X) \geq \frac{1}{|X/\sim_\mathbb{F}|}$.

In order to present a nearness measure which is compatible with $\bowtie_\mathbb{F}$, we establish the connections between $\bowtie_\mathbb{F}$ and the rough approximations.

Theorem 5. *Let $< O, \mathbb{F} >$ be a conceptual system, $X, Y \subseteq O$. $X \bowtie_\mathbb{F} Y$ if and only if there exist $x \in \mathbb{F}(X)$, $y \in \mathbb{F}(Y)$ such that $x \simeq_\mathbb{F} y$.*

The following corollary follows directly from Theorem 1(1) and Theorem 5.

Corollary 1. *$X \bowtie_\mathbb{F} Y$ if and only if $\underline{\mathbb{F}}(X)\bowtie_\mathbb{F}\underline{\mathbb{F}}(Y)$.*

Based on Definition 7 and Corollary 1, we propose the following nearness measure.

Definition 8. *Let $< O, \mathbb{F} >$ be a perceptual system and $X, Y \subseteq O$. The nearness degree $NM^3_{\sim_\mathbb{F}}(X,Y)$ between X and Y is defined by:*

$$NM^3_{\sim_\mathbb{F}}(X,Y) = NM^2_{\sim_\mathbb{F}}(\underline{\mathbb{F}}(X), \underline{\mathbb{F}}(Y)). \tag{7}$$

It is clear that $NM^3_{\sim_\mathbb{F}}$ is compatible with $\bowtie_\mathbb{F}$.

4   Strong Nearness Relations

In this section, we propose a new kind of nearness relation, called strong nearness relation, between sets of perceptual objects, and in accordingly present some nearness measures. The motivation can be illustrated as follows.

We observe that the nearness relations $\bowtie_\mathbb{F}$, $\bowtie_\mathbb{F}$ and $\delta_\mathfrak{B}$ are defined by existence quantifier, i.e.,

(1) $X \bowtie_\mathbb{F} Y$ if and only if $\exists x \in X \exists y \in Y (x \simeq_\mathbb{F} y)$;

(2) $X \bowtie_\mathbb{F} Y$ if and only if $\exists x \in \mathbb{F}(X) \exists y \in \mathbb{F}(Y)(x \simeq_\mathbb{F} y)$;

(3) $X \delta_\mathfrak{B} Y$ if and only if $\exists x \in X \exists y \in Y (x \sim_\mathbb{F} y)$.

These nearness relations prescribes a minimum condition for nearness, namely, only one objet in one set has feature values that match those of at least one object in another set. It leads to the result that if $X \bowtie_\mathbb{F} Y$ (resp., $X \bowtie_\mathbb{F} Y$, or $X \delta_\mathfrak{B} Y$), then $X \bowtie_\mathbb{F} Z$ (resp., $X \bowtie_\mathbb{F} Z$, or $X \delta_\mathfrak{B} Z$) for every $Z \supseteq Y$. So, these nearness relations seem weak and need to be enhanced.

Peters et al. [1] proposed δ'−nearness relation by using existence and universal quantifier as follows: $X \delta'_\mathfrak{B} Y$ if and only if $\exists y \in Y \forall x \in X (x \sim_\mathfrak{B} y)$. It is pointed out in [1] that $\delta'_\mathfrak{B}$ is not symmetric, i.e., $X \delta'_\mathfrak{B} Y \Rightarrow Y \delta'_\mathfrak{B} X$ does not hold in general. Furthermore, $X \delta'_\mathfrak{B} Y$ implies that there exists $y \in Y$ such that $x \sim_\mathfrak{B} y$ for every $x \in X$, and consequently we have $x_1 \sim_\mathfrak{B} y \sim_\mathfrak{B} x_2$ for every $x_1, x_2 \in X$. Hence X is a subset of a $\mathfrak{B}$−equivalence class. That is to say, if X is not a subset of a $\mathfrak{B}$−equivalence class, then $X \delta'_\mathfrak{B} Y$ does not hold for every $Y \subseteq O$. So, the δ'−nearness relation seems narrow and can not be used conveniently. Based on the above observations, we introduce the notion of strong nearness relations.

Definition 9. *Let* $< O, \mathbb{F} >$ *be a conceptual system,* $X, Y \subseteq O$. X *is strongly near to* Y *with respect to* $\simeq_\mathbb{F}$ *within the perceptual system* $< O, \mathbb{F} >$, *denoted by* $X \bowtie_\mathbb{F} Y$, *if the following conditions hold:*

(1) $X \neq \emptyset$, $Y \neq \emptyset$;

(2) For each $x \in X$, *there exist* $y \in Y$ *such that* $x \simeq_\mathbb{F} y$;

(3) For each $y \in Y$, *there exist* $x \in X$ *such that* $y \simeq_\mathbb{F} x$.

It is noticed the relation δ' was firstly introduced in the rough set settings by Pawlak [14]. He called it rough top-equality and provided relationships with other types of rough equalities.

Theorem 6. *Let* $< O, \mathbb{F} >$ *be a conceptual system,* $X, Y, Z \subseteq O$.

(1) $X \bowtie_\mathbb{F} Y$ *implies* $Y \bowtie_\mathbb{F} X$.

(2) $X \bowtie_\mathbb{F} X$ *if and only if* $X \neq \emptyset$.

(3) If $X \bowtie_\mathbb{F} Z$, $X \subseteq Y \subseteq Z$, *then* $X \bowtie_\mathbb{F} Y$, $Y \bowtie_\mathbb{F} Z$.

(4) If $X \bowtie_\mathbb{F} M$, $Z \bowtie_\mathbb{F} M$, $X \subseteq Y \subseteq Z$, *then* $Y \bowtie_\mathbb{F} M$.

(5) $X \bowtie_\mathbb{F} Y$ *and* $X \bowtie_\mathbb{F} Z$ *imply* $X \bowtie_\mathbb{F} (Y \cup Z)$.

Definition 10. *Let* $< O, \mathbb{F} >$ *be a perceptual system,* $X, Y \subseteq O$. *The nearness degree* $NM^4_{\sim_\mathbb{F}}(X, Y)$ *between* X *and* Y *is defined by:*

$$NM^4_{\sim_\mathbb{F}}(X, Y) = \frac{min\{\varepsilon(X, Y), \varepsilon(Y, X)\}}{min\{|X/ \sim_\mathbb{F}|, |Y/ \sim_\mathbb{F}|\}} \tag{8}$$

where

$$\varepsilon(X,Y) = min_{x/\sim_\mathbb{F} \in X/\sim_\mathbb{F}} \sum_{y/\sim_\mathbb{F} \in Y/\sim_\mathbb{F}} \eta_2(x/\sim_\mathbb{F}, y/\sim_\mathbb{F}). \tag{9}$$

Theorem 7. *Let $< O, \mathbb{F} >$ be a perceptual system and $X, Y \subseteq O$.*
(1) $0 \leq NM_{\sim_\mathbb{F}}^4(X,Y) \leq 1$.
(2) $NM_{\sim_\mathbb{F}}^4(X,Y) = NM_{\sim_\mathbb{F}}^4(Y,X)$.
(3) $NM_{\sim_\mathbb{F}}^4(X,Y) > 0$ *if and only if* $X \bowtie_\mathbb{F} Y$.
(4) $NM_{\sim_\mathbb{F}}^4(X,X) \geq \frac{1}{|X/\sim_\mathbb{F}|}$.

We consider the strong nearness relation and nearness measure with respect to indiscernibility relation $\sim_\mathbb{F}$.

Definition 11. *Let $< O, \mathbb{F} >$ be a conceptual system, $X, Y \subseteq O$. X is strongly near to Y with respect to $\sim_\mathbb{F}$ within the perceptual system $< O, \mathbb{F} >$, denoted by $X \bowtie_\mathbb{F}' Y$, if the following conditions hold:*
(1) $X \neq \emptyset$, $Y \neq \emptyset$;
(2) For each $x \in X$, there exist $y \in Y$ such that $x \sim_\mathbb{F} y$;
(3) For each $y \in Y$, there exist $x \in X$ such that $y \sim_\mathbb{F} x$.

Theorem 8. *Let $< O, \mathbb{F} >$ be a conceptual system, $X, Y, Z \subseteq O$.*
(1) $X \bowtie_\mathbb{F}' Y$ *implies* $Y \bowtie_\mathbb{F}' X$.
(2) $X \bowtie_\mathbb{F}' X$ *if and only if* $X \neq \emptyset$.
(3) If $X \bowtie_\mathbb{F}' Z$, $X \subseteq Y \subseteq Z$, *then* $X \bowtie_\mathbb{F}' Y$, $Y \bowtie_\mathbb{F}' Z$.
(4) If $X \bowtie_\mathbb{F}' M$, $Z \bowtie_\mathbb{F}' M$, $X \subseteq Y \subseteq Z$, *then* $Y \bowtie_\mathbb{F}' M$.
(5) $X \bowtie_\mathbb{F}' Y$ *and* $X \bowtie_\mathbb{F}' Z$ *imply* $X \bowtie_\mathbb{F}' (Y \cup Z)$.

The following theorem shows the connections between $\bowtie_\mathbb{F}'$ and rough approximations.

Theorem 9. *Let $< O, \mathbb{F} >$ be a perceptual system, $X, Y \subseteq O$ and $X \neq \emptyset$, $Y \neq \emptyset$. $X \bowtie_\mathbb{F}' Y$ if and only if $\overline{\mathbb{F}}(X) = \overline{\mathbb{F}}(Y)$.*

Proof. Assume that $X \bowtie_\mathbb{F}' Y$ and $x \in \overline{\mathbb{F}}(X)$. It follows that $x/\sim_\mathbb{F} \cap X \neq \emptyset$. Suppose that $z \in x/\sim_\mathbb{F} \cap X$. We have $z \in X$ and $z/\sim_\mathbb{F} = x/\sim_\mathbb{F}$. Consequently, there exists $y \in Y$ such that $z \sim_\mathbb{F} y$ and hence $x/\sim_\mathbb{F} \cap Y = z/\sim_\mathbb{F} \cap Y \neq \emptyset$ by $y \in z/\sim_\mathbb{F} \cap Y$. Thus we have $x \in \overline{\mathbb{F}}(Y)$ and $\overline{\mathbb{F}}(X) \subseteq \overline{\mathbb{F}}(Y)$. $\overline{\mathbb{F}}(Y) \subseteq \overline{\mathbb{F}}(X)$ can be proved similarly.

Conversely, assume that $\overline{\mathbb{F}}(X) = \overline{\mathbb{F}}(Y)$. For every $x \in X$, we have $x \in \overline{\mathbb{F}}(X) = \overline{\mathbb{F}}(Y)$ and hence $x/\sim_\mathbb{F} \cap Y \neq \emptyset$. Let $y \in x/\sim_\mathbb{F} \cap Y$. It follows that $y \in Y$ and $x \sim_\mathbb{F} y$. Similarly, there exist $x \in X$ such that $y \sim_\mathbb{F} x$ for each $y \in Y$ and hence $X \bowtie_\mathbb{F}' Y$ as required.

Definition 12. *Let $< O, \mathbb{F} >$ be a perceptual system, $X, Y \subseteq O$. The nearness degree $NM_{\sim_\mathbb{F}}^5(X,Y)$ between X and Y is defined by:*

$$NM_{\sim_\mathbb{F}}^5(X,Y) = min\{\rho(X,Y), \rho(Y,X)\} \tag{10}$$

where

$$\rho(X,Y) = min_{x/\sim_\mathbb{F} \in X/\sim_\mathbb{F}} \sum_{y/\sim_\mathbb{F} \in Y/\sim_\mathbb{F}} \eta_1(x/\sim_\mathbb{F}, y/\sim_\mathbb{F}). \tag{11}$$

Theorem 10. *Let $< O, \mathbb{F} >$ be a perceptual system and $X, Y, Z \subseteq O$.*

(1) $0 \leq NM^5_{\sim_\mathbb{F}}(X, Y) \leq 1$.

(2) $NM^5_{\sim_\mathbb{F}}(X, Y) = NM^4_{\sim_\mathbb{F}}(Y, X)$.

(3) $NM^5_{\sim_\mathbb{F}}(X, Y) > 0$ if and only if $X \bowtie'_\mathbb{F} Y$.

(4) $NM^5_{\sim_\mathbb{F}}(X, X) = 1$.

(5) $X \subseteq Y \subseteq Z$ implies $NM^5_{\sim_\mathbb{F}}(X, Z) \leq NM^5_{\sim_\mathbb{F}}(X, Y)$.

(6) $X \bowtie'_\mathbb{F} Y$ implies $NM^5_{\sim_\mathbb{F}}(X, Y) = \rho(X, Y)$.

Now we point out the connections between various kinds of nearness relations. For a perceptual system $< O, \mathbb{F} >$ and $X, Y \subseteq O$, $X \neq \emptyset$, $Y \neq \emptyset$, we have:

(1) $X \bowtie_\mathbb{F} Y \Leftrightarrow \exists x \in X \exists y \in Y \exists \varphi \in \mathbb{F}(x \sim_\varphi y)$;

(2) $X \bowtie_\mathbb{F} Y \Leftrightarrow \exists x \in \underline{\mathbb{F}}(X) \exists y \in \underline{\mathbb{F}}(Y) \exists \varphi \in \mathbb{F}(x \sim_\varphi y)$;

(3) $X \delta_\mathbb{F} Y \Leftrightarrow \exists x \in X \exists y \in Y \forall \varphi \in \mathbb{F}(x \sim_\varphi y)$;

(4) $X \bowtie_\mathbb{F} Y \Leftrightarrow \forall x \in X \exists y \in Y \exists \varphi \in \mathbb{F}(x \sim_\varphi y) \wedge \forall y \in Y \exists x \in X \exists \varphi \in \mathbb{F}(x \sim_\varphi y)$;

(5) $X \bowtie'_\mathbb{F} Y \Leftrightarrow \forall x \in X \exists y \in Y \forall \varphi \in \mathbb{F}(x \sim_\varphi y) \wedge \forall y \in Y \exists x \in X \forall \varphi \in \mathbb{F}(x \sim_\varphi y)$.

Hence we have the following corollary:

Corollary 2. *Let $< O, \mathbb{F} >$ be a perceptual system and $X, Y \subseteq O$.*

(1) $X \bowtie'_\mathbb{F} Y \Rightarrow X \bowtie_\mathbb{F} Y \Rightarrow X \bowtie_\mathbb{F} Y$.

(2) $X \bowtie'_\mathbb{F} Y \Rightarrow X \delta_\mathbb{F} Y \Rightarrow X \bowtie_\mathbb{F} Y$.

(3) $X \bowtie_\mathbb{F} Y \Rightarrow X \bowtie_\mathbb{F} Y$.

5   Concluding Remarks

Nearness measures play a central role in the applications of near set approach. This paper is devoted to the further study of nearness measure of near sets. The existing nearness measure is normalized in this paper. The notion of strong nearness relation is proposed and new nearness measures between perceptual sets are presented. The connections between nearness relations and rough approximations are also mentioned. The new nearness measures presented in this paper may play an important role in some image analysis problems.

Acknowledgements. This work has been supported by the National Natural Science Foundation of China (Grant No. 61473239, 61175044), the Fundamental Research Funds for the Central Universities of China (Grant No. 2682014ZT28) and the Open Research Fund of Key Laboratory of Xihua University (szjj2014-052).

References

1. Peters, J.: Near sets, general theory about nearness of objects. Appl. Math. Sci. **53**, 2609–2629 (2007)
2. Gupta, S., Patnaik, K.: Enhancing performance of face recognition systems by using near set approach for selecting facial features. J. Theor. Appl. Inf. Technol. **4**(5), 433–441 (2008)

3. Hassanien, A., Abraham, A., Peters, J., Schaefer, G., Henry, C.: Rough sets and near sets in medical imaging: a review. IEEE Trans. Inf. Technol. Biomed. **13**(6), 955–968 (2009)

4. Henry, C., Peters, J.F.: Image pattern recognition using near sets. In: An, A., Stefanowski, J., Ramanna, S., Butz, C.J., Pedrycz, W., Wang, G. (eds.) RSFDGrC 2007. LNCS (LNAI), vol. 4482, pp. 475–482. Springer, Heidelberg (2007)

5. Henry C., Peters J.: Near set image segmentation quality index. In: GEOBIA 2008 Pixels, Objects, Intelligence. Geographic Object Based Image Analysis for the 21th Century, pp. 1–16. University of Calgary, Alberta (2008)

6. Henry, C.J., Ramanna, S.: Parallel computation in finding near neighbourhoods. In: Yao, J.T., Ramanna, S., Wang, G., Suraj, Z. (eds.) RSKT 2011. LNCS, vol. 6954, pp. 523–532. Springer, Heidelberg (2011)

7. Henry, C.J.: Perceptual indiscernibility, rough sets, descriptively near sets, and image analysis. In: Peters, J.F., Skowron, A. (eds.) Transactions on Rough Sets XV. LNCS, vol. 7255, pp. 41–121. Springer, Heidelberg (2012)

8. Henry, C., Ramanna, S.: Signature-based perceptual nearness: application of near sets to image retrieval. Math. Comput. Sci. **7**, 71–85 (2013)

9. EI-Monsef, M., Kozae, A.M., Marei, E.A.: Near set theory and its generalization. J. King Saud Univ. Sci. **23**, 41–45 (2011)

10. Peters, J.: Near sets, special theory about nearness of objects. Fundamenta Informaticae **75**(1–4), 407–433 (2007)

11. Peters, J., Wasilewski, P.: Foundations of near sets. Inf. Sci. **179**, 3091–3109 (2009)

12. Peters, J., Wasilewski, P.: Tolerance spaces: origins, theoretical aspects and applications. Inf. Sci. **195**, 211–225 (2012)

13. Peters, J., Skowron, A., Stepaniuk, J.: Nearness of objects: extension of approximation space model. Fundamenta Informaticae **79**, 497–512 (2007)

14. Pawlak, Z.: Rough sets. Int. J. Comput. Inf. Sci. **11**, 341–356 (1982)

Topological Properties for Approximation Operators in Covering Based Rough Sets

Mauricio Restrepo[1]([✉]) and Jonatan Gómez[2]

[1] Universidad Militar Nueva Granada, Bogotá, Colombia
mauricio.restrepo@unimilitar.edu.co
http://www.alife.unal.edu.co/
[2] Universidad Nacional de Colombia, Bogotá, Colombia
jgomezpe@unal.edu.co
http://www.alife.unal.edu.co/

Abstract. We investigate properties of approximation operators being closure and topological closure in a framework of sixteen pairs of dual approximation operators, for the study of covering based rough sets. We extended previous results about approximation operators related with closure operators.

Keywords: Covering rough sets · Approximation operators · Topological closure

1 Introduction

The main concept of rough set theory is the indiscernibility between objects given by an equivalence relation in a non-empty universe set U. In this paper, we give necessary conditions for covering-based upper approximation operators to be closure operators. Three different definitions of approximation operators were presented in a general framework for the study of covering based rough sets by Yao and Yao in [20], element based definition, granule based definition and system based definition. For the element based definition, Yao and Yao consider four different neighborhood operators. In the granule based definition, they consider six new coverings defined from a covering $\mathbb{C}$. The covering $\mathbb{C}$ and the six new coverings define fourteen pairs of approximation operators. For the system based definition two new coverings are defined: $\cap$-closure($\mathbb{C}$) and $\cup$-closure($\mathbb{C}$). From these neighborhoods operators, new coverings and systems, it is possible to obtain twenty pair of dual approximation operators. But, as Yao and Yao noted [20], there are other approximations out of this framework. For example, Yang and Li present in [18] a summary of seven non dual pairs of approximation operators used by Żakowski [21], Pomykala [9], Tsang [13], Zhu [28], Zhu and Wang [30] and Xu and Whang [16]. Restrepo et al. present a framework of sixteen pair of dual approximations, unifying the two above frameworks, from duality, conjugacy and adjointness [11].

© Springer International Publishing Switzerland 2015
Y. Yao et al. (Eds.): RSFDGrC 2015, LNAI 9437, pp. 112–123, 2015.
DOI: 10.1007/978-3-319-25783-9_11

Some topological connections with rough sets and generalized rough sets have been established. The relationships between topology and generalized rough sets induced by binary relations were studied in [1,5,8]. Q. Wu. proposes a study on rough sets which includes topological spaces, topological properties and homeomorphims [10]. L. Zhaowen investigates topological properties of compactness, separate and connectedness [23]. Finally W. Zhu [32] and G. Xun et al. [17] study topological characterization of some covering based approximation operators. Recently X. Bian et al. present a characterization of three types of approximation operators to be closure operators [2]. In this paper, we consider some previous characterizations of coverings and approximation operators to extend them to the framework proposed in [11].

The paper is organized as follows. Section 2 presents preliminary concepts about topology, rough sets, lower and upper approximations in covering based rough sets, the main neighborhood operators, and different coverings obtained from a covering $\mathbb{C}$. Section 3 presents topological characterization of coverings for some upper approximation operators. Section 4 presents necessary conditions for approximation operators to be closure operators. Finally, Sect. 5 presents some conclusions and future work.

2 Preliminaries

2.1 Pawlak's Rough Set Approximations

In Pawlak's rough set model an approximation space is an ordered pair $apr = (U, E)$, where E is an equivalence relation defined on a non-empty set U [7]. The equivalence relation E defines a partition of U, written as U/E. The set $[x]_E$ represents the equivalence class of x and $\mathscr{P}(U)$ represents the set of parts of U. According to Yao and Yao [19,20], there are three different, but equivalent ways to define lower and upper approximation operators: element based definition, granule based definition and subsystem based definition. According to the element based definition, for each $A \subseteq U$, the lower and upper approximations are defined by:

$$\underline{apr}(A) = \{x \in U : [x]_E \subseteq A\} = \bigcup\{[x]_E \in U/E : [x]_E \subseteq A\} \tag{1}$$

$$\overline{apr}(A) = \{x \in U : [x]_E \cap A \neq \emptyset\} = \bigcup\{[x]_E \in U/E : [x]_E \cap A \neq \emptyset\} \tag{2}$$

The first part of Eqs. (1) and (2) are called element based definition of approximation operators. The second part are called granule based definition.

Yao and Yao used the notion of a closure system over U, i.e., a family of subsets of U that contains U and is closed under set intersection [20]. Given a closure system $\mathbb{S}$ over U, it is possible to construct its dual system $\mathbb{S}'$, containing the complements of each K in $\mathbb{S}$, as follows:

$$\mathbb{S}' = \{\sim K : K \in \mathbb{S}\} \tag{3}$$

The system $\mathbb{S}'$ contains $\emptyset$ and it is closed under set union. Given $S = (\mathbb{S}', \mathbb{S})$, a dual pair of approximation operators can be defined as follows:

$$\underline{apr}_S(A) = \bigcup\{K \in \mathbb{S}' : K \subseteq A\} \qquad (4)$$

$$\overline{apr}_S(A) = \bigcap\{K \in \mathbb{S} : K \supseteq A\} \qquad (5)$$

2.2   Closures

The notion of closure operator usually is used on ordered sets and topological spaces. We present some concepts about ordered structures, according to Blyth [3].

A family $\mathscr{C}$ of subsets of U is called a closure system if it is closed under intersections.

Closure Operators

Definition 1. *A map* $c : \mathscr{P}(U) \to \mathscr{P}(U)$ *is a* **closure operator** *on* U *if it is such that, for all* $A, B \subseteq U$:

1. $c(A) = c[c(A)]$, *(idempotent).*
2. $A \subseteq B$ *implies* $c(A) \subseteq c(B)$, *(order preserving).*
3. $A \subseteq c(A)$, *(extensive).*

Definition 2. *A map* $c : \mathscr{P}(U) \to \mathscr{P}(U)$ *is a* **join morphims** *if it is such that* $c(A \cup B) = c(A) \cup c(B)$, *for all* $A, B \in \mathscr{P}(U)$.

It is easy to see that a join morphism is an order preserving: $A \subseteq B \Leftrightarrow A \cup B = B$, so $c(A) \cup c(B) = c(B) \Leftrightarrow c(A) \subseteq c(B)$.

Topological Closure

Definition 3. *A topology for* U *is a collection* τ *of subsets of* U *satisfying the following conditions:*

1. *The empty set and* U *belong to* τ.
2. *The union of the members of each sub-collection of* τ *is a member of* τ.
3. *The intersection of the members of each finite sub collection of* τ *is a member of* τ.

The pair (U, τ) is called a topological space. The elements in τ are called open sets. The complement of an open set is called a closed set.

A family $\mathscr{B}$ is called a base for (U, τ) if for every non-empty open subset O of U and each $x \in O$, there exists a set $B \in \mathscr{B}$ such that $x \in B$. Equivalently, a family $\mathscr{B}$ is called a base for (U, τ) if every non-empty open subset O of U can be represented as union of a subfamily of $\mathscr{B}$.

The closure of a subset A of a topological space U is the intersection of the members of the family of all closed sets containing A.

Definition 4. *A **topological closure** operator on U assigns to each subset A of U a subset $c(A)$ such that (Kuratowski axioms):*

1. *$c(\emptyset) = \emptyset$, (minimal element)*
2. *$c(A) = c[c(A)]$, (idempotent).*
3. *$A \subseteq c(A)$, (extensive).*
4. *$c(A \cup B) = c(A) \cup c(B)$, (join morphism).*

The interior of a subset A of a topological space U is the union of the members of the family of all open sets contained in A. The interior operator on U is an operator which assigns to each subset A of U a subset $A°$ such that the following statements are true.

1. $U° = U$
2. $(A°)° = A°$
3. $A° \subseteq A$
4. $(A \cap B)° = A° \cap B°$

Definition 5. *Let $f, g : \mathscr{P}(U) \to \mathscr{P}(U)$ be two self-maps. We say that g is the dual of f, if for all $A \in \mathscr{P}(U)$,*

$$g(\sim A) =\sim f(A),$$

where $\sim A$ represents the complement of $A \subseteq U$.

In a topological space an interior operator is the dual of a closure operator.

2.3 Covering Based Rough Sets

Covering based rough sets was proposed to extend the range of applications of rough set theory. In rough set theory the equivalence class of an element $x \in U$ can be considered as its neighborhood, but in covering based rough sets we need to consider the sets K in $\mathbb{C}$ such that $x \in K$.

Definition 6. [24] *Let $\mathbb{C} = \{K_i\}$ be a family of nonempty subsets of U. $\mathbb{C}$ is called a covering of U if $\bigcup K_i = U$. The ordered pair $(U, \mathbb{C})$ is called a covering approximation space.*

Definition 7. [4] *Let $(U, \mathbb{C})$ be a covering approximation space and $x \in U$. The set*

$$md(\mathbb{C}, x) = \{K \in \mathbb{C} : x \in K \wedge [\forall S \in \mathbb{C}(x \in S), (S \subseteq K \Rightarrow S = K)]\} \quad (6)$$

is called the minimal description of the object x.

Definition 8. [24] *A covering $\mathbb{C}$ is called unary if $|md(\mathbb{C}, x)| = 1$ for each $x \in U$.*

The notion of maximal description was introduced by W. Zhu and F. Wang in [31].

Definition 9. [31] *Let* $(U, \mathbb{C})$ *be a covering approximation space,* $K \in \mathbb{C}$. *If no other element of* $\mathbb{C}$ *contains* K, K *is called a maximal description in* $\mathbb{C}$. *All maximal descriptions for* $x \in U$ *in* $\mathbb{C}$ *are denoted as* $MD(\mathbb{C}, x)$.

Maximal description can be also be defined as:

$$MD(\mathbb{C}, x) = \{K \in \mathbb{C} : x \in K \wedge [\forall S \in \mathbb{C}(x \in S), (S \supseteq K \Rightarrow S = K)]\} \quad (7)$$

Definition 10. [20] *A mapping* $N : U \to \mathscr{P}(U)$, *such that* $x \in N(x)$ *is called a neighborhood operator.*

According to Eqs. (1) and (2), each neighborhood operator defines a pair of approximation operators, when we use the neighborhood $N(x)$ instead of the equivalence class $[x]_E$.

$$\underline{apr}_N(A) = \{x \in U : N(x) \subseteq A\} \quad (8)$$

$$\overline{apr}_N(A) = \{x \in U : N(x) \cap A \neq \emptyset\} \quad (9)$$

Element Based Definition. Equations (8) and (9) give the element based definition in covering based rough sets, analogous to Eqs. (1) and (2) in rough set theory.

From $md(\mathbb{C}, x)$ and $MD(\mathbb{C}, x)$, Yao and Yao define the following neighborhood operators:

1. $N_1^{\mathbb{C}}(x) = \bigcap \{K : K \in md(\mathbb{C}, x)\}$
2. $N_2^{\mathbb{C}}(x) = \bigcup \{K : K \in md(\mathbb{C}, x)\}$
3. $N_3^{\mathbb{C}}(x) = \bigcap \{K : K \in MD(\mathbb{C}, x)\}$
4. $N_4^{\mathbb{C}}(x) = \bigcup \{K : K \in MD(\mathbb{C}, x)\}$

The set $N_1^{\mathbb{C}}(x) = \bigcap md(\mathbb{C}, x)$ for each $x \in U$, is called the minimal neighborhood of x.

Granule Based Definition. Generalizing the granule based definitions given by the second parts of Eqs. (1) and (2), the following dual pairs of approximation operators based on a covering $\mathbb{C}$ were considered in [20]:

$$\underline{apr}'_{\mathbb{C}}(A) = \bigcup \{K \in \mathbb{C} : K \subseteq A\} \quad (10)$$

$$\overline{apr}'_{\mathbb{C}}(A) = \sim \underline{apr}'_{\mathbb{C}}(\sim A) \quad (11)$$

$$\underline{apr}''_{\mathbb{C}}(A) = \sim \overline{apr}''_{\mathbb{C}}(\sim A) \quad (12)$$

$$\overline{apr}''_{\mathbb{C}}(A) = \bigcup \{K \in \mathbb{C} : K \cap A \neq \emptyset\} \quad (13)$$

Generally a covering contains redundant information. For example, some definitions using all the sets of the covering and using the minimal sets only, are equivalent. Therefore, it is possible to consider only some particular elements of the covering.

From a covering $\mathbb{C}$ of U, we can define the coverings:

1. $\mathbb{C}_1 = \bigcup \{md(\mathbb{C}, x) : x \in U\}$
2. $\mathbb{C}_2 = \bigcup \{MD(\mathbb{C}, x) : x \in U\}$
3. $\mathbb{C}_3 = \{\bigcap (md(\mathbb{C}, x)) : x \in U\} = \{\bigcap (\mathscr{C}(\mathbb{C}, x)) : x \in U\}$
4. $\mathbb{C}_4 = \{\bigcup (MD(\mathbb{C}, x)) : x \in U\} = \{\bigcup (\mathscr{C}(\mathbb{C}, x)) : x \in U\}$
5. $\mathbb{C}_\cap = \mathbb{C} \setminus \{K \in \mathbb{C} : (\exists \mathbb{K} \subseteq \mathbb{C} \setminus \{K\}) \, (K = \bigcap \mathbb{K})\}$
6. $\mathbb{C}_\cup = \mathbb{C} \setminus \{K \in \mathbb{C} : (\exists \mathbb{K} \subseteq \mathbb{C} \setminus \{K\}) \, (K = \bigcup \mathbb{K})\}$

Coverings $\mathbb{C}_\cap$ and $\mathbb{C}_\cup$ are called the $\cap$-reduction and the $\cup$-reduction of $\mathbb{C}$, respectively. The idea is eliminate the sets that can be expressed as intersection or union of other sets in the covering.

Using Eqs. 10 to 13, each covering defines two pairs of approximation operators, therefore for each covering we have fourteen pairs of dual approximation operators.

Closure System Based Definition. As a particular example of a closure system, [20] considered the so-called intersection closure $S_{\cap,\mathbb{C}}$ of a covering $\mathbb{C}$, i.e., the minimal subset of $\mathscr{P}(U)$ that contains $\mathbb{C}$, $\emptyset$ and U, and is closed under set intersection. Similarly, the union closure of $\mathbb{C}$, denoted by $S_{\cup,\mathbb{C}}$, is the minimal subset of $\mathscr{P}(U)$ that contains $\mathbb{C}$, $\emptyset$ and U, and is closed under set union. It can be shown that the dual system $S'_{\cup,\mathbb{C}}$, defined by Eq. 3, forms a closure system. Both $S_\cap = ((S_{\cap,\mathbb{C}})', S_{\cap,\mathbb{C}})$ and $S_\cup = (S_{\cup,\mathbb{C}}, (S_{\cup,\mathbb{C}})')$ can be used to obtain two pairs of dual approximation operations, according to Eqs. (4) and (5).

According to the above three definitions Yao and Yao present twenty pairs of dual approximation operators, four from the element based definition, fourteen from granule based definition and two from the system based definition based on $\cap$-closure and $\cup$-closure.

The pairs of approximation operators $(\underline{apr}_N, \overline{apr}_N)$, $(\underline{apr}'_\mathbb{C}, \overline{apr}'_\mathbb{C})$ and $(\underline{apr}''_\mathbb{C}, \overline{apr}''_\mathbb{C})$ are dual pairs.

2.4   Other Framework of Lower and Upper Approximations

A summary of seven pairs of approximation operators for covering based rough sets was presented in [14,18]. In all cases only two lower approximations have been used. Żakowski first extended Pawlak's rough set theory from a partition to a covering in [21]. The second type of covering rough set model was presented by Pomykala in [9], Tsang [13] studied the third type. Zhu defined the fourth and the fifth types of covering-based approximation in [28,30]. Xu gave the definition of the sixth type in [16]. The seventh type of approximation operations can be found in [15].

Some of these approximation operators were already presented in Yao and Yao's framework, therefore we only present the different ones. The four upper approximations are listed as follows:

1. $H_1^\mathbb{C}(A) = L_1^\mathbb{C}(A) \cup (\bigcup \{md(\mathbb{C}, x) : x \in A - L_1^\mathbb{C}(A)\})$
2. $H_3^\mathbb{C}(A) = \bigcup \{N_2^\mathbb{C}(x) : x \in A\}$
3. $H_4^\mathbb{C}(A) = L_1^\mathbb{C}(A) \cup (\bigcup \{K : K \cap (A - L_1^\mathbb{C}(A)) \neq \emptyset\})$
4. $H_5^\mathbb{C}(A) = \bigcup \{N_1^\mathbb{C}(x) : x \in A\}$

where $L_1^{\mathbb{C}}(A)$ is the lower approximation defined as:

$$L_1^{\mathbb{C}}(A) = \bigcup\{K \in \mathbb{C} : K \subseteq A\} = \underline{apr}'_{\mathbb{C}}(A). \tag{14}$$

2.5   New Framework of Approximation Operators

Some equivalences and relationships among the operators in the two previous frameworks were studied and established by M. Restrepo et al. [11]. The Table 2 summarizes a framework of sixteen pairs of approximation operators, establishing equivalences among the two frameworks above, Yao and Yao's framework and Yang and Li's framework.

All the operators listed in Table 2 satisfy the relation $A \subseteq \overline{apr}(A)$ for all $A \subseteq U$, so they are upper approximations. Also, all they satisfy $\overline{apr}(\emptyset) = \emptyset$.

According to W. Zhu [24] operators 5, 6, 7 and 8 are not join morphisms and according to [25], operators 9, 10, 11 and 12 are not idempotent.

The following propositions show that $\underline{apr}_{N_3^{\mathbb{C}}}$ is an idempotent operator.

Proposition 1. *If $z \in N_3^{\mathbb{C}}(x)$ then $N_3^{\mathbb{C}}(z) \subseteq N_3^{\mathbb{C}}(x)$.*

Proof. If $z \in \underline{apr}_{N_3^{\mathbb{C}}}(x)$, we will show that $MD(\mathbb{C}, x) \subseteq MD(\mathbb{C}, z)$. In fact, if $K \in MD(\mathbb{C}, x)$, then $z \in K$, because $z \in N_3^{\mathbb{C}}(x)$. Now, if $S \supseteq K$ then $K = S$, because K is a maximal element. From $MD(\mathbb{C}, x) \subseteq MD(\mathbb{C}, z)$, we have that $\bigcap MD(\mathbb{C}, x) \supseteq \bigcap MD(\mathbb{C}, z)$ and therefore $N_3^{\mathbb{C}}(z) \subseteq N_3^{\mathbb{C}}(x)$.

Proposition 2. *The operator $\underline{apr}_{N_3^{\mathbb{C}}}$ is idempotent.*

Proof. Clearly $\underline{apr}_{N_3^{\mathbb{C}}}(\underline{apr}_{N_3^{\mathbb{C}}}(A)) \subseteq \underline{apr}_{N_3^{\mathbb{C}}}(A)$.

If $z \in \underline{apr}_{N_3^{\mathbb{C}}}(A)$, then $N_3^{\mathbb{C}}(z) \subseteq A$. We will show that $N_3^{\mathbb{C}}(z) \subseteq \underline{apr}_{N_3^{\mathbb{C}}}(A)$. In fact, if $w \in N_3^{\mathbb{C}}(z)$, by Proposition 1, we have that $N_3^{\mathbb{C}}(w) \subseteq N_3^{\mathbb{C}}(z) \subseteq A$, therefore $w \in \underline{apr}_{N_3^{\mathbb{C}}}(A)$. Since $N_3^{\mathbb{C}}(z) \subseteq \underline{apr}_{N_3^{\mathbb{C}}}(A)$ we have that $z \in \underline{apr}_{N_3^{\mathbb{C}}}(\underline{apr}_{N_3^{\mathbb{C}}}(A))$. So $\underline{apr}_{N_3^{\mathbb{C}}}(\underline{apr}_{N_3^{\mathbb{C}}}(A)) = \underline{apr}_{N_3^{\mathbb{C}}}(A)$ and $\underline{apr}_{N_3^{\mathbb{C}}}$ is an idempotent operator.

From duality, it is easy to establish the following corollary.

Corollary 1. *The operator $\overline{apr}_{N_3^{\mathbb{C}}}$ is idempotent.*

The following example shows that operators $\overline{apr}_{N_2^{\mathbb{C}}}$ and $\overline{apr}_{N_4^{\mathbb{C}}}$ are not idempotent operators.

Example 1. (Operators $\overline{apr}_{N_2^{\mathbb{C}}}$ and $\overline{apr}_{N_4^{\mathbb{C}}}$ are not idempotent).

Let us consider the covering $\mathbb{C} = \{\{1\}, \{1, 2\}, \{2, 3\}, \{4\}, \{1, 2, 3\}, \{1, 4\}\}$ of $U = \{1, 2, 3, 4\}$. The minimal description $md(\mathbb{C}, x)$, the maximal description $MD(\mathbb{C}, x)$ and the neighborhood operators are listed in Table 1.

$\overline{apr}_{N_2^{\mathbb{C}}}(\{1\}) = \{1, 2\}$, while $\overline{apr}_{N_2^{\mathbb{C}}}(\{1, 2\}) = \{1, 2, 3\}$, therefore $\overline{apr}_{N_2^{\mathbb{C}}}$ is not an idempotent operator. Similarly, $\overline{apr}_{N_4^{\mathbb{C}}}(\{2\}) = \{1, 2, 3\}$, while $\overline{apr}_{N_4^{\mathbb{C}}}(\{1, 2, 3\}) = \{1, 2, 3, 4\}$, therefore $\overline{apr}_{N_4^{\mathbb{C}}}$ is not an idempotent operator.

Table 1. Illustration of neighborhood operator for the covering $\mathbb{C}$.

x	$md(\mathbb{C},x)$	$MD(\mathbb{C},x)$	$N_1^{\mathbb{C}}(x)$	$N_2^{\mathbb{C}}(x)$	$N_3^{\mathbb{C}}(x)$	$N_4^{\mathbb{C}}(x)$
1	$\{\{1\}\}$	$\{\{1,2,3\},\{1,4\}\}$	$\{1\}$	$\{1\}$	$\{1\}$	$\{1,2,3,4\}$
2	$\{\{1,2\},\{2,3\}\}$	$\{\{1,2,3\}\}$	$\{2\}$	$\{1,2,3\}$	$\{1,2,3\}$	$\{1,2,3\}$
3	$\{\{2,3\}\}$	$\{\{1,2,3\}\}$	$\{2,3\}$	$\{2,3\}$	$\{1,2,3\}$	$\{1,2,3\}$
4	$\{\{4\}\}$	$\{\{1,4\}\}$	$\{4\}$	$\{4\}$	$\{1,4\}$	$\{1,4\}$

According to properties in Table 2, operators 2, 4, 9, 10, 11 do not satisfy idempotent property, operators from 5 to 8 are not join morphisms and operator 12 does not satisfy any property. Obviously each topological closure is a closure operator.

3   Topological Characterization of Upper Approximations

The following propositions are characterization of upper approximation operators presented in [17]. Similar results for lower approximations and their relation with interior operators, can be established from duality.

Table 2. Properties of upper approximations.

n	Upper approximation	Property		
		Idempotence	Order preserving	Join
1	$\overline{apr}_{N_1^{\mathbb{C}}} = \overline{apr}'_{\mathbb{C}_3} = H_6^{\mathbb{C}}$	Yes	Yes	Yes
2	$\overline{apr}_{N_2^{\mathbb{C}}}$	No	Yes	Yes
3	$\overline{apr}_{N_3^{\mathbb{C}}}$	Yes	Yes	Yes
4	$\overline{apr}_{N_4^{\mathbb{C}}} = \overline{apr}''_{\mathbb{C}} = \overline{apr}''_{\mathbb{C}_2} = \overline{apr}''_{\mathbb{C}_\cap} = H_2^{\mathbb{C}}$	No	Yes	Yes
5	$\overline{apr}'_{\mathbb{C}} = \overline{apr}'_{\mathbb{C}_1} = \overline{apr}'_{\mathbb{C}_\cup} = \overline{apr}_{S_\cup}$	Yes	Yes	No
6	$\overline{apr}'_{\mathbb{C}_2}$	Yes	Yes	No
7	$\overline{apr}'_{\mathbb{C}_4}$	Yes	Yes	No
8	$\overline{apr}'_{\mathbb{C}_\cap}$	Yes	Yes	No
9	$\overline{apr}''_{\mathbb{C}_1} = \overline{apr}''_{\mathbb{C}_\cup}$	No	Yes	Yes
10	$\overline{apr}''_{\mathbb{C}_3} = H_7^{\mathbb{C}}$	No	Yes	Yes
11	$\overline{apr}''_{\mathbb{C}_4}$	No	Yes	Yes
12	$\overline{apr}_{S_\cap}$	No	Yes	No
13	$H_1^{\mathbb{C}}$	Yes	No	No
14	$H_3^{\mathbb{C}}$	No	Yes	Yes
15	$H_4^{\mathbb{C}}$	Yes	No	No
16	$H_5^{\mathbb{C}}$	Yes	Yes	Yes

Proposition 3. [17] $H_1^{\mathbb{C}}$ is a topological closure if and only if $\mathbb{C}$ is unary.

Proposition 4. [17] $H_1^{\mathbb{C}}$ is a topological closure if and only if there exists a topology τ, on U such that $\mathbb{C}$ is a base for (U, τ).

Proposition 5. [17] $H_2^{\mathbb{C}}$ is a topological closure if and only if $\{N_4^{\mathbb{C}}(x) : x \in U\}$ forms a partition of U.

Proposition 6. [17] $H_3^{\mathbb{C}}$ is a topological closure if and only if exists a topology τ, on U such that, $\{N_2^{\mathbb{C}}(x) : x \in U\}$ is a base for (U, τ) and for all $x \in U$, $\{N_2\mathbb{C}(x)\}$ is a local base at x for (U, τ).

Proposition 7. [17] $H_4^{\mathbb{C}}$ is a topological closure if and only if $\mathbb{C}$ is a base for some topology τ on U and (U, τ) consists of two disjoints subspaces U_1 and U_2, satisfying:

1. For $K, K' \in \mathbb{C}$ with $K \neq K'$, we have: $K \cap U_2 = K' \cap U_2 = \emptyset$ or $K \cap U_2 \neq K' \cap U_2$ and $\{K \cap U_2 : K \in \mathbb{C}\}$ is a partition of U_2.
2. (U_1, τ_1) is a discrete space and (U_2, τ_2) is a pseudo-discrete space.

Propositions (3) to (7) are the characterization of operators 4, 13, 14 and 15 being topological closures, therefore closure operators. Also are closure operators 9, 10 and 11, because they are defined as $\overline{apr}_{\mathbb{C}}''$ for different coverings: $\mathbb{C}_1$, $\mathbb{C}_3$ and $\mathbb{C}_4$.

4   Algebraic and Topological Properties

From properties in Table 2, it is easy to establish the following propositions.

Proposition 8. Operators 1, 3 and 16 are topological closure operators.

Proposition 9. Operators 1, 3, 5, 6, 7, 8 and 16 are closure operators.

The following propositions show some properties of unary coverings.

Proposition 10. A covering $\mathbb{C}$ is unary if and only if there exists a topology τ, on U such that $\mathbb{C}$ is a base for (U, τ).

Proof. It is a consequence of Propositions 1 and 2.

Proposition 11. For any covering $\mathbb{C}$, the covering $\mathbb{C}_3$ is unary.

Proof. The elements in $\mathbb{C}_3$ are the neighborhoods $N_1^{\mathbb{C}}(x)$ for $x \in U$, so $\mathbb{C}_3 = \{N_1^{\mathbb{C}}(x)\}$ and it is unary, because the minimal description of each $x \in U$ has only an element, $|md(\mathbb{C}_3, x)| = 1$. Therefore $\mathbb{C}_3$ is unary.

Proposition 12. If $\mathbb{C}$ is unary, then $\mathbb{C}_1$ and $\mathbb{C}_{\cup}$ are unary.

Proof. The covering $\mathbb{C}_1$ is made of sets in minimal descriptions of $x \in U$. Since $\mathbb{C}$ is unary, clearly $\mathbb{C}_1$ is unary. $\mathbb{C}_{\cup}$ is unary, because $\mathbb{C}_1 = \mathbb{C}_{\cup}$, as was established in [12].

Corollary 2. *If* $\mathbb{C}$ *is unary,* $\overline{apr}'_{\mathbb{C}_1}$, $\overline{apr}'_{\mathbb{C}_3}$ *and* $\overline{apr}'_{\mathbb{C}_\cup}$ *are topological closures.*

Proposition 13. *If* $\mathbb{C}$ *is unary, then* $\overline{apr}_{N_1} = \overline{apr}_{N_2}$.

Proof. Using the number of elements in $md(\mathbb{C}, x)$, we can see that $\bigcup md(\mathbb{C}, x) = \bigcap md(\mathbb{C}, x)$, therefore, $N_1^{\mathbb{C}}(x) = N_2^{\mathbb{C}}(x)$ and $\overline{apr}_{N_1^{\mathbb{C}}} = \overline{apr}_{N_2^{\mathbb{C}}}$.

Corollary 3. *If* $\mathbb{C}$ *is unary,* $\overline{apr}_{N_2^{\mathbb{C}}}$ *is a topological closure.*

Corollary 4. *If* $\mathbb{C}$ *is unary,* $\overline{apr}_{N_2^{\mathbb{C}}}$ *is a closure operator.*

The following example shows that unary covering is not a condition of closure operator for $\overline{apr}_{N_4^{\mathbb{C}}}$.

Example 2. (An unary covering such that $\overline{apr}_{N_4^{\mathbb{C}}}$ is not an idempotent operator).

Let us consider the covering $\mathbb{C} = \{\{1\}, \{2,3\}, \{4\}, \{1,2,3\}, \{1,4\}\}$ of $U = \{1,2,3,4\}$. The minimal description $md(\mathbb{C}, x)$, the maximal description $MD(\mathbb{C}, x)$ and the neighborhood operators are listed in Table 3.

Table 3. Illustration of minimal and maximal description.

x	$md(\mathbb{C}, x)$	$MD(\mathbb{C}, x)$	$N_1^{\mathbb{C}}(x)$	$N_2^{\mathbb{C}}(x)$	$N_3^{\mathbb{C}}(x)$	$N_4^{\mathbb{C}}(x)$
1	$\{\{1\}\}$	$\{\{1,2,3\},\{1,4\}\}$	$\{1\}$	$\{1\}$	$\{1\}$	$\{1,2,3,4\}$
2	$\{\{2,3\}\}$	$\{\{1,2,3\}\}$	$\{2,3\}$	$\{2,3\}$	$\{1,2,3\}$	$\{1,2,3\}$
3	$\{\{2,3\}\}$	$\{\{1,2,3\}\}$	$\{2,3\}$	$\{2,3\}$	$\{1,2,3\}$	$\{1,2,3\}$
4	$\{\{4\}\}$	$\{\{1,4\}\}$	$\{4\}$	$\{4\}$	$\{1,4\}$	$\{1,4\}$

According to Table 3, $\mathbb{C}$ is unary.

$\overline{apr}_{N_4^{\mathbb{C}}}(\{2\}) = \{1,2,3\}$, while $\overline{apr}_{N_4^{\mathbb{C}}}(\{1,2,3\}) = \{1,2,3,4\}$, therefore $\overline{apr}_{N_4^{\mathbb{C}}}$ is not a idempotent operator.

Proposition 14. *If* $\mathbb{C}$ *is a covering of* U *and* $\{N_4^{\mathbb{C}}(x) : x \in U\}$ *forms a partition of* U, *then* $\overline{apr}''_{\mathbb{C}}$ *is a topological closure.*

Proof. It is a consequence of Proposition 5.

5   Conclusions

This paper studies the properties of upper approximation operators for the study of coverings based rough sets, extending the results presented in [32] to the framework presented in [11]. We show that $\overline{apr}_{N_3^{\mathbb{C}}}$ is an idempotent operator. We give necessary conditions for topological closures and closure operators. For an unary covering $\mathbb{C}$, we have that operators 1, 2, 5, 8 and 16 are topological closures and therefore closure operators. Necessary condition for closure operators in the new framework are given. As future work we will establish sufficient conditions for topological closure and closure operators.

References

1. Abo, E.A.: Rough sets and topological spaces based on similarity. Int. J. Mach. Learn. Cybern. **4**, 451–458 (2013)
2. Bian, X., Wang, P., Yu, Z., Bai, X., Chen, B.: Characterization of coverings for upper approximation operators being closure operators. Inf. Sci. **314**, 41–54 (2015)
3. Blyth, T.S.: Lattices and Ordered Algebraic Structures. Springer Universitext, London (2005)
4. Bonikowski, Z., Brynarski, E.: Extensions and Intensions in rough set theory. Inform. Sci. **107**, 149–167 (1998)
5. Hai, Y., Wan-rong, Z.: On the topological properties of generalized rough sets. Inform. Sci. **263**(1), 141–152 (2014)
6. Järvinen, J.: Lattice theory for rough sets. In: Peters, J.F., Skowron, A., Düntsch, I., Grzymała-Busse, J.W., Orłowska, E., Polkowski, L. (eds.) Transactions on Rough Sets VI. Lncs, vol. 4374, pp. 400–498. Springer, Heidelberg (2007)
7. Pawlak, Z.: Rough sets. Int. J. Comput. Inform. Sci. **11**(5), 341–356 (1982)
8. Pei, Z., Pei, D., Zheng, L.: Topology vs generalized rough sets. Int. J. Approximate Reasoning **52**, 231–239 (2011)
9. Pomykala, J.A.: Approximation operations in approximation space. Bull. Acad. Pol. Sci. **35**(9–10), 653–662 (1987)
10. Wu, Q., Wang, T., Huan, Y., Li, J.: Topology theory on rough sets. IEEE Trans. Syst. Man Cybern. **38**, 68–77 (2008)
11. Restrepo, M., Cornelis, C., Gómez, J.: Duality, conjugacy and adjointness of approximation operators in covering-based rough sets. Int. J. Approximate Reasoning **55**, 469–485 (2014)
12. Restrepo, M., Cornelis, C., Gómez, J.: Partial order relation for approximation operators in covering-based rough sets. Inf. Sci. **284**, 44–59 (2014)
13. Tsang, E., Chen, D., Lee J., Yeung, D.S.: On the upper approximations of covering generalized rough sets. In: Proceedings of the 3rd International Conference on Machine Learning and Cybernetics, pp. 4200–4203 (2004)
14. Wang, L., Yang, X., Yang, J., Wu, C.: Relationships among generalized rough sets in six coverings and pure reflexive neighborhood system. Inf. Sci. **207**, 66–78 (2012)
15. Xu, W., Zhang, W.: Measuring roughness of generalized rough sets induced by a covering. Fuzzy Sets Syst. **158**, 2443–2455 (2007)
16. Xu, Z., Wang, Q.: On the properties of covering rough sets model. J. Henan Normal Univ. Nat. Sci. **33**(1), 130–132 (2005)
17. Xun, G., Xiaole, B., Yun, Z.: Topological characterization of coverings for special covering based upper approximation operators. Inf. Sci. **204**, 70–81 (2012)
18. Yang, T., Li, Q.: Reduction about approximation spaces of covering generalized rough sets. Int. J. Approximate Reasoning **51**, 335–345 (2010)
19. Yao, Y.Y.: Constructive and algebraic methods of the theory of rough sets. Inf. Sci. **109**, 21–47 (1998)
20. Yao, Y.Y., Yao, B.: Covering based rough sets approximations. Inf. Sci. **200**, 91–107 (2012)
21. Zakowski, W.: Approximations in the space (u, π). Demonstratio Math. **16**, 761–769 (1983)
22. Zhang, Y., Li, J., Wu, W.: On axiomatic characterizations of three pairs of covering based approximation operators. Inf. Sci. **180**(2), 274–287 (2010)
23. Zhaowen, L.: Topological structure of generalized rough sets. IEEE Trans. Knowl. Data Eng. **19**(8), 1131–1144 (2012)

24. Zhu, W.: Properties of the first type of covering-based rough sets. In: Proceedings of Sixth IEEE International Conference on Data Mining - Workshops, pp. 407–411 (2006)

25. Zhu, W.: Properties of the second type of covering-based rough sets. In: Proceedings of the IEEE/WIC/ACM International Conference on Web Intelligence and Intelligent Agent Technology, pp. 494–497 (2006)

26. Zhu, W.: Basic concepts in covering-based rough sets. In: Proceedings of Third International Conference on Natural Computation, pp. 283–286 (2007)

27. Zhu, W.: Properties of the third type of covering-based rough sets. In: Proceedings of International Conference on Machine Learning and Cybernetics, pp. 3746–2751 (2007)

28. Zhu, W.: Relationship between generalized rough sets based on binary relation and covering. Inf. Sci. **179**, 210–225 (2009)

29. Zhu, W., Wang, F.: Reduction and axiomatization of covering generalized rough sets. Inf. Sci. **152**, 217–230 (2003)

30. Zhu, W., Wang, F.: A new type of covering rough set. In: Proceedings of Third International IEEE Conference on Intelligence Systems, pp. 444–449 (2006)

31. Zhu, W., Wang, F.: On three types of covering based rough sets. IEEE Trans. Knowl. Data Eng. **19**(8), 1131–1144 (2007)

32. Zhu, W.: Topological approach to covering rough sets. Inf. Sci. **177**, 1499–1508 (2007)

Rough Sets and Graphs

Preclusivity and Simple Graphs

Giampiero Chiaselotti[2], Davide Ciucci[1]([✉]), Tommaso Gentile[2],
and Federico Infusino[2]

[1] DISCo, University of Milano – Bicocca, Viale Sarca 336 – U14, 20126 Milan, Italy
ciucci@disco.unimib.it

[2] Department of Mathematics and Informatics, University of Calabria,
Via Pietro Bucci, Cubo 30B, 87036 Arcavacata di Rende, CS, Italy
giampiero.chiaselotti@unical.it, {gentile,f.infusino}@mat.unical.it

Abstract. The adjacency relation of a simple undirected graph is a preclusive (irreflexive and symmetric) relation. Hence, it originates a preclusive space enabling us to define the lower and upper preclusive approximations of graphs and two orthogonality graphs. Further, the possibility of defining the similarity lower and upper approximations and the sufficiency operator on graphs will be investigated, with particular attention to complete and bipartite graphs. All these mappings will be put in relation with Formal Concept Analysis and the theory of opposition.

Keywords: Undirected graphs · Preclusivity relation · Sufficiency operator · Formal concept analysis · Theory of opposition

1 Introduction

In exploring the possibilities to apply rough set theory to graphs, several directions are possible. In [6] we considered the adjacency matrix of a graph as a Boolean Information Table and studied the corresponding of classical rough set instruments to graphs. Here, we are focusing on the adjacency relation among vertices in a simple (i.e. with no loops) undirected graph. It can be seen that it is a preclusive relation (irreflexive and symmetric), hence it defines a preclusive space introduced by Cattaneo in [1]. Thus, the preclusive lower and upper approximations are discussed in the case of simple undirected graphs. Of course, the negation of a preclusive relation is a similarity relation, so standard similarity approximations can also be defined. A further connection can be established with Formal Concept Analysis (FCA). Indeed, in this framework, the so-called preclusive complement turns out to coincide with the standard extension operator in FCA. Finally, all these operators will be discussed under the light of the theory of opposition and a cube of opposition will be defined starting from the similarity relation [9,13]. Apart from the already mentioned approximations and preclusive complement, in this cube we can also accommodate the sufficiency operator [12] and study its corresponding on graphs. In this paper particular attention will

© Springer International Publishing Switzerland 2015
Y. Yao et al. (Eds.): RSFDGrC 2015, LNAI 9437, pp. 127–137, 2015.
DOI: 10.1007/978-3-319-25783-9_12

be given to complete and bipartite graphs, whereas n-cycles and n-paths will be analyzed in detail in a second part of this work [7].

In Sect. 2, a refresher of the basic notions of Graph Theory, Formal Concept Analysis and Preclusivity Spaces will be given. Then, the link between Graphs and Preclusivity Spaces will be discussed in Sect. 3. Finally, in Sect. 4 we will analyze the connection with the cube of opposition.

The main contribution of this work is to highlight the possibility to investigate graphs by preclusivity and similarity relations and to frame in the same picture graphs, similarity/preclusivity rough sets, formal concept analysis and the theory of opposition.

2   Preliminary Notions

The basic notions of Graph Theory [11], Preclusivity spaces [1] and Formal Concept Analysis [15] are now re-called.

2.1   Graphs

We denote by $G = (V(G), E(G))$ a finite simple (i.e. no loops and no multiple edges are allowed) undirected graph, with vertex set $V(G) = \{v_1, \ldots, v_n\}$ and edge set $E(G)$. If $v, v' \in V(G)$, we will write $v \sim v'$ if $\{v, v'\} \in E(G)$ and $v \nsim v'$ otherwise. We denote by $Adj(G)$ the adjacency matrix of G. We recall that $Adj(G)$ is the $n \times n$ matrix (a_{ij}) such that $a_{ij} := 1$ if $v_i \sim v_j$ and $a_{ij} := 0$ otherwise. If $v \in V(G)$ we set

$$N_G(v) := \{w \in V(G) : \{v, w\} \in E(G)\} \tag{1}$$

and $N_G(v)$ is usually called *neighborhood* of v in G. Graphs of particular interest for our discussion will be complete and bipartite ones.

Definition 2.1. *The* complete graph *on n vertices, denoted by K_n, is the graph with vertex set $\{v_1, \ldots, v_n\}$ and such that $\{v_i, v_j\}$ is an edge, for each pair of indexes $i \neq j$.*

Definition 2.2. *A graph $B = (V(B), E(B))$ is said* bipartite *if there exist two non-empty subsets B_1 and B_2 of $V(B)$ such that $B_1 \cap B_2 = \emptyset$, $B_1 \cup B_2 = V(B)$ and $E(B) \subseteq \{\{x, y\} : x \in B_1, y \in B_2\}$. In this case the pair (B_1, B_2) is called a* bipartition *of B and we write $B = (B_1|B_2)$. It is said that $B = (B_1|B_2)$ is a* complete bipartite graph *if $E(B) = \{\{x, y\} : x \in B_1, y \in B_2\}$. If p and q are two positive integers and $B_1 = \{x_1, \ldots, x_p\}$, $B_2 = \{y_1, \ldots, y_q\}$, we denote by $K_{p,q}$ the complete bipartite graph having bipartition (B_1, B_2).*

Complete and bipartite graphs play an important role in graph theory and they are used in several applications. For instance, complete graphs can model the method of pairwise comparison [10], complete bipartite sub-graphs coincide with concepts in formal concept analysis [5,16] and (general) bipartite graphs are used to model actor/item interactions in personalized recommender systems [14].

2.2   Preclusivity Spaces

We start by recalling some basic notions concerning the preclusivity spaces [1]. First of all let us consider the well-known generalization of rough sets based on a similarity (or tolerance) relation S, that is a reflexive and symmetric relation [19]. In this generalized framework the lower and upper approximations of a set H can be defined as $L_S(H) := \{x \in U : S(x) \subseteq H\}$ and $U_S(H) := \{x \in U : S(x) \cap H \neq \emptyset\}$, where $S(x) = \{y \in U : xSy\}$ is the neighborhood or similarity granule of x. Clearly, similarity relations give more flexibility than equivalence ones and they naturally arise in practical situations: just think to a distance measure or to the possibility to consider similar two objects which share 90 % of attributes instead of 100 % required by the equivalence relation approach.

In standard rough sets, we can consider also the negation of the equivalence relation, that is the so-called discernibility relation, which is used to defined the discernibility matrix and hence to compute reducts and rules. In this generalized context, the negation of a similarity negation is a preclusive, i.e., irreflexive and symmetric relation: $* : U \mapsto U$ defined as $x * y$ iff $\neg(xSy)$. It defines a preclusive space and a complementation (usually different from standard set complementation) as follows.

Definition 2.3. *A preclusivity space is a pair* $\mathfrak{P} = (U, *)$, *where U is a non empty set and $*$ is a binary relation on U which is irreflexive: if $u * u'$, then $u \neq u'$, and symmetric: if $u * u'$ then $u' * u$. If H and K are two subsets of U, they are called* mutually preclusive, *in symbols $H(*)K$, if $h * k$ for all $h \in H$ and $k \in K$. The subset*

$$H^* := \{u \in U : h * u \ \forall h \in H\} \qquad (2)$$

is called the preclusive orthocomplement *of H.*

Further, if $H \subseteq U$, we set $H^\flat := H^{c*c}$. Then, it results (see [1]) that $H^* \subseteq H^c \subseteq H^\flat$.

The importance of the preclusive complementation, with respect to data analysis, is that it can define a pair of lower and upper approximations closer, with respect to the similarity ones, to the set under approximation. Formally, for any subset $H \subseteq U$, the $*$-lower and $*$-upper approximations of H are $\mathbb{I}(H) = H^{\flat\flat} = H^{c**c} \subseteq H$ and $H \subseteq \mathbb{C}(H) = H^{**}$. It can be shown that $\mathbb{I}$ and $\mathbb{C}$ are a pair of dual interior and closure operators (see [1]). A set H is called $*$-closed if $H^{**} = H$. The family of all $*$-closed subsets is denoted by $CL(\mathfrak{P})$. Analogously, H is called $*$-open if H^c is $*$-closed. The family of all $*$-open subsets is denoted by $OP(\mathfrak{P})$. Moreover, due to the duality, we have

$$OP(\mathfrak{P}) = \{H^c : H \in CL(\mathfrak{P})\} \qquad (3)$$

It can be shown that $\mathbb{I}$ and $\mathbb{C}$ can be expressed in terms of open and closed elements as

$$\mathbb{I}(H) = \bigcup \{K : K \in OP(\mathfrak{P}) \wedge K \subseteq H\} \qquad (4)$$

$$\mathbb{C}(H) = \bigcap \{K : K \in CL(\mathfrak{P}) \wedge H \subseteq K\} \qquad (5)$$

At this point we use the following terminology.

Definition 2.4. *If $\mathfrak{P} = (U, *)$ is a preclusivity space and $H \subseteq U$, we say that the ordered pair $(\mathbb{I}(H), \mathbb{C}(H))$ is the* preclusive approximation *of H. We also say that H is* preclusive-exact, *or also* closed-open, *if $\mathbb{I}(H) = H = \mathbb{C}(H)$. We denote by $CO(\mathfrak{P})$ the family of all preclusive-exact subsets of U.*

By the above definitions, we have therefore the following equivalent characterization for the preclusive-exact subsets:

$$CO(\mathfrak{P}) = CL(\mathfrak{P}) \cap OP(\mathfrak{P}) \tag{6}$$

As we have anticipated before, preclusive approximations are closer to the set under approximation then the corresponding similarity ones. Formally, it turns out that $L_S(H) = H^{c*}$ and $U_S(H) = H^{*c}$ and the following chain holds [2]:

$$L_S(H) \subseteq \mathbb{I}(H) \subseteq H \subseteq \mathbb{C}(H) \subseteq U_S(H).$$

Let us also notice that by the above definitions and results, we easily get that the exterior region of the similarity approximation, that is $e(H) := U_S^c(H)$ (representing the objects surely not belonging to H) coincides with the preclusive complement: $e(H) = H^*$.

Finally, in [1] two interesting orthogonality properties $\perp_*$ and $\perp_\flat$ have been defined, respectively between *-closed and *-open subsets of a given preclusivity space. We recall these definitions.

Definition 2.5. *Let $\mathfrak{P} = (U, *)$ be a preclusivity space.*
(i) If $A, B \in OP(\mathfrak{P})$, it is said that A and B are $\flat$-orthogonal, denoted by the symbol $A \perp_\flat B$, if $A \subseteq B^\flat$.
*(i) If $A, B \in CL(\mathfrak{P})$, it is said that A and B are *-orthogonal, denoted by the symbol $A \perp_* B$, if $A \subseteq B^*$.*

Remark 2.1. It results that both the binary relations, $\perp_\flat$ on $OP(\mathfrak{P})$ and $\perp_*$ on $CL(\mathfrak{P})$, are irreflexive and symmetric (see [1]).

As a final remark, let us notice that preclusive spaces have their theoretical roots in the algebraic context of Brouwer-Zadeh posets, firstly defined in the area of quantum computing [3].

2.3   Formal Concept Analysis

We relate now Preclusive Space notions to the basic concepts of Formal Concept Analysis (FCA) [15]. A *formal context* is a triple $\mathbb{K} = (Z, M, \mathcal{R})$, where Z and M are sets and $\mathcal{R} \subseteq Z \times M$ is a binary relation involving them. The elements of Z and M are called objects and attributes (or properties) respectively. Usually one writes $g\mathcal{R}m$ instead of $(g, m) \in \mathcal{R}$ with the meaning that objects g has property m.

Hence, a preclusivity space $(U, *)$ is a particular formal context, where $Z = M = U$ and $\mathcal{R} = *$. In particular, the preclusive complement coincides with the so-called *intension* of a set of objects and also, in this particular case where objects are equal to attributes, to the *extension* of a set of attributes. Moreover, in standard FCA, we can derive a closure operator by a double application of the intension/extension operator. Of course, this is nothing else than our closure operator $\mathbb{C}$.

3   Simple Graphs as Preclusivity Spaces

We interpret now any simple undirected graph as a preclusivity space.

Definition 3.1. *Let $G = (V(G), E(G))$ be a simple undirected graph (not necessarily finite). Then, the binary adjacency relation $\sim$ is exactly the preclusivity relation $*$ on $V(G)$ since in G there are no loops. We set therefore $\mathfrak{P}[G] := (V(G), \sim)$ and we call it the* preclusivity space *of G. We denote respectively by $\mathbb{I}_G$ and $\mathbb{C}_G$ the lower and upper approximation maps for the preclusivity space $\mathfrak{P}[G]$, as defined in (4) and (5). We also set $CL(G) := CL(\mathfrak{P}[G])$, $OP(G) := OP(\mathfrak{P}[G])$ and $CO(G) := CO(\mathfrak{P}[G])$.*

Remark 3.1. For the preclusivity space $\mathfrak{P}[G]$ we use both the notations $\sim$ and $*$ in order to denote the preclusivity binary relation on this space.

Given this definition, it is immediate to give a geometrical interpretation to the preclusive complement. Indeed, if $H \subseteq V(G)$ then by (1) and (2):

$$H^* = \bigcap \{N_G(v) : v \in H\} \tag{7}$$

Moreover, the similarity granule $S(v)$ defined by the negation of the preclusive relation $*$, is the set complement of the neighborhood: $S(v) = (N_G(v))^c$.

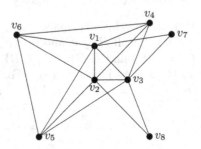

Example 3.1. If G is the graph below and $H = \{v_1, v_2, v_3\}$ then $H^* = \{v_4, v_5\}$, $\mathbb{I}(H) = \emptyset$ and $\mathbb{C}(H) = \{v_1, v_2, v_3, v_6\}$.

If we consider the similarity relation arising from the graph, we have $S(v_1) = \{v_1, v_8\}$, $S(v_2) = \{v_2, v_7\}$ and $S(v_3) = \{v_3, v_6\}$. So, $L_S(H) = \emptyset$ and $U_S(H) = \{v_1, v_2, v_3, v_6, v_7, v_8\}$.

In the next definition we adapt the orthogonality relations to the graph case.

Definition 3.2. *Let $G = (V(G), E(G))$ be a simple undirected graph. Then,*

- *We call* $\flat$*-orthogonality graph of G, denoted by $\mathbb{O}^\flat(G)$, the simple undirected graph having vertex set $OP(G)$ and edge set $\{\{A, B\} : A, B \in OP(G) \wedge A \perp_\flat B\}$.*
- *We call* $*$*-orthogonality graph of G, denoted by $\mathbb{O}^*(G)$, the simple undirected graph having vertex set $CL(G)$ and edge set $\{\{A, B\} : A, B \in CL(G) \wedge A \perp_* B\}$.*

We will now see how these definitions apply to complete and complete bipartite graphs.

3.1 Two Basic Cases

The first case that we investigate is the complete graph.

Proposition 3.1. *Let $G = K_n$ and $V = V(G)$. Then, $\mathbb{I}_G(H) = \mathbb{C}_G(H) = L_S(H) = U_S(H) = H$ for all vertex subsets $H \subseteq V(G)$. Hence, $CL(G) = OP(G) = CO(G) = \mathcal{P}(V)$.*

Proof. If $H \subseteq V(G)$ we have that $H^* = H^c$ by definition of K_n, therefore $\mathbb{I}_G(H) := H^{\flat\flat} = H^{c**c} = H$, $\mathbb{C}(H) := H^{**} = H$, $L_S(H) := H^{c*} = H$ and $U_S(H) := H^{*c} = H$. $\qquad\square$

So, every set of vertices is exact with respect to both pairs of approximations.

When G is the complete graph we have the following characterization of the orthogonality graphs.

Proposition 3.2. *Let $V = V(K_n)$. Then*

$$\mathbb{O}^*(K_n) = \mathbb{O}^\flat(K_n) = (\mathcal{P}(V), \{\{A, B\} : A \cap B = \emptyset\}) \qquad (8)$$

Proof. Let $G = \mathbb{O}^\flat(K_n)$. In the preclusivity space $\mathfrak{P}[K_n]$ we have that $H^\flat := H^{c*c} = H^c$ (because $H^* = H^c$) for all $H \subseteq V$ and $OP(K_n) = \mathcal{P}(V)$ by Proposition 3.1. Therefore $V(G) = \mathcal{P}(V)$ and $E(G) := \{\{A, B\} : A, B \in \mathcal{P}(V) \wedge A \subseteq B^c\}$, that is equivalent to the second identity of (8). For the first identity the proof is the same because $CL(K_n) = OP(K_n)$ and $H^* = H^\flat = H^c$. $\qquad\square$

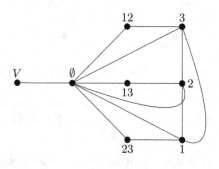

Example 3.2. Below we draw the graph $\mathbb{O}^b(K_3) = \mathbb{O}^*(K_3)$ (we represent the vertex subsets without braces):

Our next case of study is the complete bipartite graph.

Proposition 3.3. *Let* $G = K_{p,q} = (B_1|B_2)$, $V = V(G)$ *and* $H \subseteq V$ *such that* $H \neq \emptyset$ *and* $H \neq V$. *Then:*

$$\mathbb{I}_G(H) = \begin{cases} \emptyset & \text{if } H^c \cap B_1 \neq \emptyset \wedge H^c \cap B_2 \neq \emptyset \\ B_2 & \text{if } B_2 \subseteq H \\ B_1 & \text{if } B_1 \subseteq H \end{cases} \tag{9}$$

and

$$\mathbb{C}_G(H) = \begin{cases} V & \text{if } H \cap B_1 \neq \emptyset \wedge H \cap B_2 \neq \emptyset \\ B_1 & \text{if } H \subseteq B_1 \\ B_2 & \text{if } H \subseteq B_2 \end{cases} \tag{10}$$

Moreover

$$CL(G) = OP(G) = CO(G) = \{\emptyset, B_1, B_2, V\} \tag{11}$$

Proof. Eq. (9) has been proved in the FCA setting in [5]. From this we obtain Eq. (10) by duality. Then, by (9) we immediately deduce that $\mathbb{I}(H) = H$ if and only if $H = \emptyset$, $H = B_1$ or $H = B_2$. Hence $OP(G) = \{\emptyset, B_1, B_2, V\}$, because $\emptyset$ and V are always $*$-open sets (this is true in any preclusivity space). In a similar way, by (10) we also obtain $CL(G) = \{\emptyset, B_1, B_2, V\}$. $\square$

By definitions of L_S and U_S, we easily get the following result.

Proposition 3.4. *Let* $G = K_{p,q} = (B_1|B_2)$ *and* $H \subseteq V$. *Then:* $L_S(H) = \cup\{B_i : B_i \subseteq H\}$, $U_S(H) = \cup\{B_i : B_i \cap H \neq \emptyset\}$.

So, we have that also in this case the similarity lower and upper approximations coincide with the preclusive ones: $\mathbb{I}_G(H) = L_S(H)$ and $\mathbb{C}_G(H) = U_S(H)$.

The next result describes $\mathbb{O}^b(K_{p,q})$ and $\mathbb{O}^*(K_{p,q})$.

Proposition 3.5. *Let* $K_{p,q} = (B_1|B_2)$ *and* $V = V(K_{p,q})$. *Then* $\mathbb{O}^b(K_{p,q}) = \mathbb{O}^*(K_{p,q})$ *and it is the following graph:*

Proof. Let $G = \mathbb{O}^b(K_{p,q})$. By Proposition 3.3, we have that $V(G) = OP(G) = \{\emptyset, B_1, B_2, V\}$. Moreover, $\emptyset^b = V$ and $B_1^b := B_1^{c*c} = B_2^{*c} = B_1^c = B_2$. Hence, the unique non-trivial orthogonality relation is $B_1 \perp_b B_2$, and then the thesis follows. If we take $G = \mathbb{O}^*(K_{p,q})$, we obtain the same result since $B_1^* = B_2$ and $B_2^* = B_1$. $\square$

4   The Cube of Opposition Generated by the Preclusive Relation

The cube of opposition is a geometrical representation of the dependencies arising among four logical statements and their negations [13]. It is a generalization of the Aristotelian square of opposition and it can be defined in a standard way from a binary relation $R \subseteq X \times Y$ [9]. In our particular case, we consider $X = Y$ and as basic relation we take a similarity relation S. In this way, we are able to accommodate in the same cube the similarity approximations and the preclusive complement (intension/extension operator in FCA). In this framework the preclusive relation corresponding to S will be denoted as $\overline{S}$ ($x\overline{S}y$ if and only if $\neg(xSy)$).

An important consequence of S being a similarity relation is that it is also serial, a necessary requirement in order to respect all the dependencies among the different corners of the cube. Once denoted $\overline{H} := H^c$, the four corners originating from S are (the names derive from the classical ones taken from Aristotelian tradition):

(I) $S(H) = \{x \in X | \exists s \in H, xSs\} = \{x \in X | H \cap S(x) \neq \emptyset\}$
(O) $S(\overline{H}) = \{x \in X | \exists s \in \overline{H}, xSs\}$
(E) $\overline{S(H)} = \{x \in X | \forall s \in H, (x * s)\}$
(A) $\overline{S(\overline{H})} = \{x \in X | \forall s \in \overline{H}, (x * s)\} = \{x \in X | S(x) \subseteq H\}$

We remark that E and A are the complement of I and O, respectively, and that A is a subset of I and E a subset of O. So, we can see that (A) is the similarity lower approximation and (I) is the similarity upper approximation. Moreover, (E) is the preclusive complement, which coincides also with the exterior region $(U_S)^c$ and the standard operator of FCA. Finally, (O) is the upper approximation of H^c.

The remaining four corners are obtained using the preclusivity relation $* := \overline{S}$ in place of the similarity one:

(o) $\overline{S}(H) = \{x \in X | \exists s \in H, (x * s)\}$
(i) $\overline{S}(\overline{H}) = \{x \in X | \exists s \in \overline{H}, (x * s)\} = \{x \in X | H \cup S(x) \neq Y\}$
(a) $\overline{\overline{S}(H)} = \{x \in X | \forall s \in H, xSs\} = \{x \in X | H \subseteq S(x)\}$
(e) $\overline{\overline{S}(\overline{H})} = \{x \in X | \forall s \in \overline{H}, xSs\}$

The corners (a) and (i) are, respectively, the *sufficiency* operator, usually denoted as $[[\cdot]]$, and its dual [12]. In the case of a graph $G = (V, E)$, the sufficiency operator reads as:

$$[[H]]_G = \{v \in V | N_G(v) \subseteq H^c\}$$

That is, it represents the set of vertices not connected to all the elements in H. So, in case of a complete graph $G = K_n$, we always have $[[H]]_G = \emptyset$ for H not empty and $[[\emptyset]]_G = V$. In case of bipartite graphs $G = K_{p,q}$, we have:

$$[[H]]_G = \begin{cases} B_1 & H \subset B_2 \\ B_2 & H \subset B_1 \\ V & H = \emptyset \\ \emptyset & \text{otherwise} \end{cases}$$

The corner (o) applied to a graph G represents the set of vertices connected with at least one vertex in H: $\overline{\mathcal{S}}(H) = \{v \in V | N_G(v) \cap H \neq \emptyset\}$. It can be expressed in terms of neighborhood as $\overline{\mathcal{S}}(H) = \cup_{v \in H} N_G(v)$. So, in case of complete and complete bipartite graphs it is equal respectively to

$$\begin{cases} \emptyset & H = \emptyset \\ V \setminus \{v_i\} & H = \{v_i\} \\ V & \text{otherwise} \end{cases} \quad \text{and} \quad \begin{cases} \emptyset & H = \emptyset \\ B_1 \,(\text{resp.}\ B_2) & H \subseteq B_2 \,(\text{resp.}\ B_1) \\ V & \text{otherwise} \end{cases}$$

The corner (e) is the complement of (o) and it can be interpreted as the set of vertices connected with no vertex outside H: $\overline{\mathcal{S}}(\overline{H}) = \{v \in V | N_G(v) \subseteq H\}$.

Let us remark that the framework discussed here is different from the one presented in [5]. Indeed, in that case, we had two different cubes: one for the FCA context and another one for the rough set context. The FCA one was based on the relation corresponding to our $*$. So, it was the mirror of the present one, with $R = *$ and $\overline{R} = \mathcal{S}$ and where, however, the link with the similarity approximations and the sufficiency operator was not outlined. The other cube was based on the equivalence relation that can be defined in the standard rough set way on the adjacency matrix. Thus, the approximations in that case where the standard Pawlak ones and the sufficiency operator was always trivially the empty set or the set under approximation.

Finally, we notice that the preclusive and similarity relation can be also put in another structure of opposition, by considering the properties they satisfy instead of looking at the operators they generate. This particular structure was discussed in [8] and other generalizations are given in [18].

5   Conclusion

By interpreting the adjacency relation of a simple undirected graph as a preclusive relation, we were able to put in correspondence several fields of investigation: preclusive spaces, generalized rough set theory, formal concept analysis and the theory of opposition. All the instruments of these theories can then be applied to the starting graph in order to analyze it in the light of granular computing. Here, we focused on the general setting and on the case of complete and complete bipartite graphs, in [7] we will discuss the n-cycle and n-path graphs. Of course, the picture is far from being complete. From one side, other classes of graphs can be taken into account. For example, in generalized rough sets [19,20], approximation functions based on general binary relation are studied. So, it would be interesting to investigate the correspondence between classes of graphs and different binary relations. For instance, in our graph context, a preclusive relation that is also *serial* corresponds to a simple graph without isolated points, whereas a *Euclidean* preclusive relation corresponds to a simple graphs whose connected components are complete graphs.

Further, other instruments, for instance the discernibility matrix and reducts arising from the similarity relation are worth studying [17]. Apart from a new theoretical approach to graph theory, we plan to use this transfer of results from one field to the other to give new insights in graph-based applications. With respect to this issue, let us also mention that the preclusive relation arising from graph is at the basis of [4], where rough sets are used to test bipartiteness of simple undirected graphs.

References

1. Cattaneo, G.: Generalized rough sets (preclusivity fuzzy-intuitionistic (BZ) lattices). Stud. Logica **01**, 47–77 (1997)
2. Cattaneo, G.: Abstract approximation spaces for rough theories. In: Polkowski, L., Skowron, A. (eds.) Rough Sets in Knowledge Discovery 1: Methodology and Applications. Studies in Fuzziness and Soft Computing, pp. 59–98. Physica, Heidelberg (1998)
3. Cattaneo, G., Nisticò, G.: Brouwer-Zadeh posets and three valued Łukasiewicz posets. Fuzzy Sets Syst. **33**, 165–190 (1989)
4. Chen, J., Li, J.: An application of rough sets to graph theory. Inf. Sci. **201**, 114–127 (2012)
5. Chiaselotti, G., Ciucci, D., Gentile, T.: Simple undirected graphs as formal contexts. In: Baixeries, J., Sacarea, C., Ojeda-Aciego, M. (eds.) ICFCA 2015. LNCS, vol. 9113, pp. 287–302. Springer, Heidelberg (2015)
6. Chiaselotti, G., Ciucci, D., Gentile, T., Infusino, F.: Rough set theory applied to simple undirected graphs. In: Ciucci, D., Wang, G., Mitra, S., Wu, W. (eds.) RSKT 2015. LNCS, vol. 9436. Springer, Heidelberg (2015)
7. Chiaselotti, G., Ciucci, D., Gentile, T., Infusino, F.: Preclusivity and simple graphs: the n-cycle and n-path cases. In: Yao, Y., Hu, Q., Yu, H., Grzymala-Busse, J. (eds.) RSFDGrC 2015. LNCS, vol. 9437, pp. 138–148. Springer, Heidelberg (2015)
8. Ciucci, D., Dubois, D., Prade, H.: Opposition in rough set theory. In: Li, T., Nguyen, H.S., Wang, G., Grzymala-Busse, J., Janicki, R., Hassanien, A.E., Yu, H. (eds.) RSKT 2012. LNCS, vol. 7414, pp. 504–513. Springer, Heidelberg (2012)
9. Ciucci, D., Dubois, D., Prade, H.: The structure of opposition in rough set theory and formal concept analysis - toward a new bridge between the two settings. In: Beierle, C., Meghini, C. (eds.) FoIKS 2014. LNCS, vol. 8367, pp. 154–173. Springer, Heidelberg (2014)
10. David, H.A.: The Method of Paired Comparisons. Oxford University Press, New York (1988)
11. Diestel, R.: Graph Theory. Graduate Text in Mathematics, 4th edn. Springer, Heidelberg (2010)
12. Düntsch, I., Orlowska, E.: Beyond modalities: sufficiency and mixed algebras. In: Orlowska, E., Szalas, A. (eds.) Relational Methods for Computer Science Applications, pp. 263–285. Physica-Verlag, Springer, Heidelberg (2001)

13. Dubois, D., Prade, H.: From Blanché's hexagonal organization of concepts to formal concept analysis and possibility theory. Logica Universalis **6**, 149–169 (2012)
14. Epasto, A., Feldman, J., Lattanzi, S., Leonardi, S., Mirrokni, V., Reduce and aggregate: similarity ranking in multi-categorical bipartite graphs. In: Proceedings of World wide web (WWW 2014), pp. 349–360. ACM
15. Ganter, B., Wille, R.: Formal Concept Analysis. Mathematical Foundations. Springer, Heidelberg (1999)
16. Kuznetsov, S.O.: Mathematical aspects of concept analysis. J. Math. Sci. **80**, 1654–1698 (1996)
17. Skowron, A., Stepaniuk, J.: Tolerance approximation spaces. Fundam. Inform. **27**, 245–253 (1996)
18. Yao, Y.Y.: Duality in rough set theory based on the square of opposition. Fundam. Informaticae **127**, 49–64 (2013)
19. Yao, J.T., Ciucci, D., Zhang, Y.: Generalized rough sets. In: Kacprzyk, P. (ed.) Handbook of Computational Intelligence, Chapter 25, pp. 413–424. Springer, Heidelberg (2015)
20. Yao, Y.Y., Lin, T.Y.: Generalization of rough sets using modal logics. Intell. Autom. Soft Comput. **2**, 103–120 (1996)

Preclusivity and Simple Graphs: The n–cycle and n–path Cases

Giampiero Chiaselotti[2], Davide Ciucci[1(✉)], Tommaso Gentile[2], and Federico Infusino[2]

[1] DISCo, Università di Milano – Bicocca, Viale Sarca 336 – U14, 20126 Milan, Italy
ciucci@disco.unimib.it
[2] Department of Mathematics and Informatics, University of Milano–Bicocca, Via Pietro Bucci, Cubo 30B, 87036 Arcavacata di Rende, CS, Italy
giampiero.chiaselotti@unical.it, {gentile,f.infusino}@mat.unical.it

Abstract. Two classes of graphs, the n–cycles and n–paths, are interpreted as preclusivity spaces. In this way, it is possible to define two pairs of approximations on them: one based on a preclusive relation and another one based on a similarity relation. Further, two relations can be defined among the set of vertices and they define two different graphs, which are here studied.

Keywords: Undirected graphs · Preclusivity relation · Rough approximations

1 Introduction

In [4], a general framework to study undirected simple graphs under the light of preclusivity spaces has been introduced. Here, we continue this line of research by analyzing two basic and fundamental classes of graphs: n-paths and n-cycles.

A preclusivity relation is an irreflexive and symmetric relation, that enables to define a preclusive complement on subsets of the universe (in general different from the usual set complement) and also a lower and upper approximation, that are, respectively, an interior and a closure operator. As we will recall in the next section, the preclusive complement can be defined, in the case of graphs, in terms of the neighborhood of a vertex.

Given a preclusivity relation, its negation is a similarity one, so it is also possible to define a standard rough approximation based on this similarity relation [2]. Moreover, the preclusive complement coincides with the extension operator of formal concept analysis, once interpreted the graph as a formal context [3]. We will not develop here the links among preclusive spaces and formal concept analysis, nor the possibility to organize all these operators in a cube of opposition [5]. For the general approach, we refer to [4].

In order to keep the paper self-consistent, in Sect. 2, the basic notions of graph theory and preclusive spaces will be recalled. Sections 3 and 4 deal, respectively, with n-cycle and n-path graphs. Preclusive and similarity approximations are

Y. Yao et al. (Eds.): RSFDGrC 2015, LNAI 9437, pp. 138–148, 2015.
DOI: 10.1007/978-3-319-25783-9_13

discussed and the graphs originated by $\flat$-orthogonal and $*$-orthogonal preclusive relations studied.

2   Preliminary Notions

For the sake of completeness, we give here the notions that will be useful in the following. We also recall the basic results of the connection between preclusivity spaces and graphs. For further details, see [1,4,6,7].

2.1   Graphs

We denote by $G = (V(G), E(G))$ a finite simple (i.e. no loops and no multiple edges are allowed) undirected graph, with vertex set $V(G) = \{v_1, \ldots, v_n\}$ and edge set $E(G)$. We denote by $Adj(G)$ the adjacency matrix of G, that is the $n \times n$ matrix (a_{ij}) such that $a_{ij} := 1$ if $\{v, v'\} \in E(G)$ and $a_{ij} := 0$ otherwise. When two vertices are adjacent $(\{v, v'\} \in E(G))$, we will write $v \sim v'$ and $v \nsim v'$ otherwise. Given a vertex $v \in V(G)$, the *neighborhood* of v is defined as $N_G(v) := \{w \in V(G) : \{v, w\} \in E(G)\}$. In the following, we will be interested to two particular classes of graphs: n-cycles and n-paths.

Definition 2.1. *The n-cycle on n vertices, denoted by C_n, is the graph with vertex set $V(C_n) = \{v_1, \ldots, v_n\}$ and edge set $E(C_n) = \{\{v_1, v_2\}, \{v_2, v_3\}, \ldots, \{v_{n-1}, v_n\}, \{v_n, v_1\}\}$.*

Definition 2.2. *If n is a positive integer we denote by P_n the graph having n vertices $v_1, \ldots, v_n$ and such that $E(P_n) = \{\{v_i, v_{i+1}\} : i = 1, \ldots, n-1\}$. In literature P_n is sometimes called* path *on n vertices.*

Finally, we need to define the notion of isomorphism between graphs.

Definition 2.3. *If $G_1 = (V(G_1), E(G_1))$ and $G_2 = (V(G_2), E(G_2))$ are two undirected graphs, we say that G_1 is* isomorphic to *G_2, denoted by $G_1 \cong G_2$, if there exists a bijective map $\phi : V(G_1) \to V(G_2)$ such that for all $v, w \in V(G_1)$ $\{v, w\} \in E(G_1)$ iff $\{\phi(v), \phi(w)\} \in E(G_2)$.*

2.2   Preclusivity Spaces

A *preclusivity space* is a pair $\mathfrak{P} = (U, *)$, where U is a nonempty set and $*$ is an irreflexive and symmetric binary relation on U. The *preclusive orthocomplement* of a subset H is defined as $H^* := \{u \in U : h * u \ \forall h \in H\}$. By the interaction of the preclusive and standard complements we can also define another complement as $H^\flat := H^{c*c}$.

Thanks to these operators, it is possible to define the *preclusive approximation* of a subset H, made by the preclusive lower and upper approximations respectively defined as $\mathbb{I}(H) := H^{\flat\flat} = H^{c**c}$ and $\mathbb{C}(H) = H^{**}$. They are a pair of dual interior and closure operators, such that $\mathbb{I}(H) \subseteq H \subseteq \mathbb{C}(H)$. The family of all $*$-closed subsets is $CL(\mathfrak{P}) := \{H \subseteq U : H = \mathbb{C}(H)\}$, whereas the open sets are $OP(\mathfrak{P}) : \{H \subseteq U : H = \mathbb{I}(H)\}$. Hence, a set H is *preclusive-exact*, or also *closed-open*, if $\mathbb{I}(H) = H = \mathbb{C}(H)$ and the family of all preclusive-exact subsets will be denoted as $CO(\mathfrak{P}) = CL(\mathfrak{P}) \cap OP(\mathfrak{P})$.

In any preclusive space $(U, *)$, it is possible to define two preclusive relations among subsets of the universe U. Two open sets $A, B \in OP(\mathfrak{P})$ are $\flat$-orthogonal, denoted by the symbol $A \perp_\flat B$, if $A \subseteq B^\flat$. Dually, two closed sets $A, B \in CL(\mathfrak{P})$ are $*$-orthogonal, denoted by the symbol $A \perp_* B$, if $A \subseteq B^*$.

Finally, in a preclusive space it is also possible to define a pair of similarity lower and upper approximations. Indeed, the complement of a preclusive relation xSy iff $\neg(x * y)$ is a similarity relation. Hence, the similarity lower and upper approximations are defined as usual as $L_S(H) := \{x \in U : S(x) \subseteq H\}$ and $U_S(H) := \{x \in U : S(x) \cap H \neq \emptyset\}$, where $S(x) = \{y \in U : xSy\}$ is the neighborhood or similarity granule of x [8]. Moreover, we have that $L_S(H) = H^{c*}$ and $U_S(H) = H^{*c}$ and the following chain holds:

$$L_S(H) \subseteq \mathbb{I}(H) \subseteq H \subseteq \mathbb{C}(H) \subseteq U_S(H). \tag{1}$$

2.3   Simple Graphs as Preclusivity Spaces

Given a simple undirected graph $G = (V(G), E(G))$, the binary adjacency relation $\sim$ is a preclusivity relation. Therefore, we define the *preclusivity space* of G as $\mathfrak{P}[G] := (V(G), \sim)$. We also denote by $\mathbb{I}_G$ and $\mathbb{C}_G$ the lower and upper approximation maps for the preclusivity space $\mathfrak{P}[G]$ and we set $CL(G) := CL(\mathfrak{P}[G])$, $OP(G) := OP(\mathfrak{P}[G])$ and $CO(G) := CO(\mathfrak{P}[G])$. As an immediate consequence, the preclusive complement can be defined in terms of the neighborhood as

$$H^* = \bigcap \{N_G(v) : v \in H\} \tag{2}$$

In the next definition we adapt the orthogonality relations to the graph case.

Definition 2.4. *Let* $G = (V(G), E(G))$ *be a simple undirected graph.* (i) *We call* $\flat$-orthogonality graph *of* G, *denoted by* $\mathbb{O}^\flat(G)$, *the simple undirected graph having vertex set* $OP(G)$ *and edge set* $\{\{A, B\} : A, B \in OP(G) \wedge A \perp_\flat B\}$.

We call $*$-orthogonality graph *of* G, *denoted by* $\mathbb{O}^*(G)$, *the simple undirected graph having vertex set* $CL(G)$ *and edge set* $\{\{A, B\} : A, B \in CL(G) \wedge A \perp_* B\}$.

3   The Case of C_n

Let us study the case of the cycle graph C_n. Since $C_3 \cong K_3$, where K_3 is the complete graph with three vertices, and $C_4 \cong K_{2,2}$, where $K_{2,2}$ is a complete bipartite graph with four vertices, we can assume $n \geq 5$, and use the results in [4] for $n = 3, 4$. In the next proposition, we describe the $*$-closure function and the $*$-interior function for the simple graph C_n.

Proposition 3.1. *Let $n \geq 5$, $G = C_n$ and $V = V(G)$. Let $A \subseteq V(G)$ such that $A \neq \emptyset$ and $A \neq V$. Then:*

$$\mathbb{C}_G(A) = \begin{cases} A \text{ if } A = \{v_i\} \text{ or } A = \{v_i, v_{i+2}\} \\ V \text{ otherwise} \end{cases} \tag{3}$$

and

$$\mathbb{I}_G(A) = \begin{cases} A \text{ if } A = \{v_i\}^c \text{ or } A = \{v_i, v_{i+2}\}^c \\ \emptyset \text{ otherwise} \end{cases} \tag{4}$$

where $i = 1, \ldots, n$ and the above index sums $i + 2$ are taken $\mathrm{mod}(n)$.

Proof. At first, we observe that if $A = \{v_i\}$ is a singleton, then

$$A^* = \begin{cases} \{v_{i-1}, v_{i+1}\} \text{ if } i \neq 1, n \\ \{v_2, v_n\} & \text{if } i = 1 \\ \{v_{n-1}, v_1\} & \text{if } i = n \end{cases} \tag{5}$$

Now, it is clear that the only vertex in V connected with the two points in A^* is v_i itself, so $A^{**} = A$. The case $A = \{v_i, v_j\}$ with $j \equiv i + 2 \pmod{n}$ and $i = 1, \ldots, n$, is similar. Let A be a proper non-empty subset of $V(G)$ containing at least three vertices or only two vertices v_i, v_j such that $j \not\equiv i + 2 \pmod{n}$. By definition of cycle, a vertex not contained in A, say v_k, cannot be connected with all points of A, so it follows $A^{**} = V$. This shows (3). For the second part, if $A = \{v_i\}^c$, we have: $\mathbb{I}_G(A) = A^{bb} = A^{c**c} = \{v_i\}^c$. We get a similar computations in the case $A = \{v_i, v_j\}^c$ where $j \equiv i + 2 \pmod{n}$. Moreover, if A is a proper non-empty subset of $V(G)$ different from the previous, because of the relation $\mathbb{I}_G(A) = A^{bb} = A^{c**c}$ and the first part, we have that in this case $\mathbb{I}_G(A) = \emptyset$. This completes the proof of (4). □

Corollary 3.1. *Let $V = V(C_n) = \{v_1, \ldots, v_n\}$, with $n \geq 5$. Then $CL(C_n) = \{\emptyset, \{v_i\}, \{v_i, v_{i+2}\}, V\}$, $OP(C_n) = \{\emptyset, \{v_i\}^c, \{v_i, v_{i+2}\}^c, V\}$, where the index sums $i + 2$ are taken* $\mathrm{mod}(n)$, *and $CO(C_n) = \{\emptyset, V\}$.*

Proof. It follows immediately from the fact that $\emptyset$ and C_n are always $*-$open and $*-$closed sets and from Proposition 3.1. □

For the similarity upper and lower approximations, we have the following.

Proposition 3.2. *Let $n \geq 5$, $G = C_n$ and $V = V(G)$. Let $A \subseteq V(G)$ such that $A \neq \emptyset$ and $A \neq V$. Then:*

$$U_S(A) = \begin{cases} V \setminus \{v_{i-1}, v_{i+1}\} & \text{if } A = \{v_i\} \\ V \setminus \{v_{i+1}\} & \text{if } A = \{v_i, v_{i+2}\} \\ V & \text{otherwise} \end{cases} \tag{6}$$

$$L_S(A) = \begin{cases} \{v_{i-1}, v_{i+1}\} & \text{if } A = \{v_i\}^c \\ \{v_{i+1}\} & \text{if } A = \{v_i, v_{i+2}\}^c \\ \emptyset & \text{otherwise} \end{cases} \tag{7}$$

where $i = 1, \ldots, n$ and the above index sums $i + 2$ are taken $\mathrm{mod}(n)$.

Proof. First of all let us consider that due to property (1), the similarity lower approximation is empty whenever the preclusive one is empty and the similarity upper approximation is V whenever the preclusive one is V. Thus, given the above Proposition 3.1, the only cases left to determine are $\{v_i\}$, $\{v_i, v_{i+2}\}$ for the upper approximation and their set complement for the lower approximation. Now, the similarity granule for a given vertex is $S(v_i) = \{v_{i-1}, v_{i-1}\}$. So, the remaining cases follow by definition of the similarity lower and upper approximations.   □

In order to better describe the graph structure of $\mathbb{O}^*(C_n)$ when $n \geq 5$, we introduce a new graph family.

Definition 3.1. *If $n \geq 3$ we call* closed n-chandelier, *denoted by* Ch_n, *the simple graph having vertex set* $V(Ch_n) := \{\alpha_1, \ldots, \alpha_n, \beta_1, \ldots, \beta_n, \gamma, \delta\}$ *and the following edges:* $\{\delta, \alpha_i\}$, $\{\delta, \beta_i\}$, $\{\delta, \gamma\}$, $\{\alpha_i, \beta_i\}$ *and* $\{\alpha_i, \alpha_{i+1}\}$ *for all* $i = 1, \ldots, n$, *where $i + 1$ is taken* mod (n).

Example 3.1. Below we provide a graphical representation for the graph Ch_5:

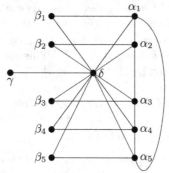

In the next result we determine the graph $\mathbb{O}^*(C_n)$ when $n \geq 5$.

Proposition 3.3. *If $n \geq 5$ then $\mathbb{O}^*(C_n) \cong Ch_n$.*

Proof. At first, observe that $V^* = \emptyset$ and $\emptyset^* = V$; if $A = \{v_i\}$, with $i = 1, \ldots, n$ fixed, then

$$A^* = \begin{cases} \{v_{i-1}, v_{i+1}\} & \text{if } i \neq 1, n \\ \{v_2, v_n\} & \text{if } i = 1 \\ \{v_{n-1}, v_1\} & \text{if } i = n \end{cases} \tag{8}$$

where the index sums are taken mod(n). If $A = \{v_i, v_{i+2}\}$, then $A^* = \{v_{i+1}\}$ where, once again, the index sum are taken mod(n). We may conclude that the $*$−orthogonal relations are:

$$\{v_i\} \perp_* B, \qquad \text{where } B = \begin{cases} \emptyset \\ \{v_{i-1}\} \\ \{v_{i+1}\} \\ \{v_{i-1}, v_{i+1}\} \end{cases} \tag{9}$$

and

$$\{v_i, v_{i+2}\} \perp_* B, \qquad \text{where } B = \begin{cases} \emptyset \\ \{v_{i+1}\} \end{cases} \tag{10}$$

where the index sums are all taken $\mod(n)$.

Now, let $G_1 = Ch_n$ and $G_2 = \mathbb{O}^*(C_n)$. Define $\phi : G_1 \mapsto G_2$ as $\alpha_i \mapsto \{v_i\}$, $\beta_i \mapsto \{v_i, v_{i+2}\}$, $\gamma \mapsto V$ and $\delta \mapsto \emptyset$. The map ϕ is clearly bijective and, using the $*-$orthogonality relations just found we have that all the incidence relations are preserved. For example, if $\{\delta, \alpha_i\}$ is an edge of G_1, then $\{\emptyset, \{v_i\}\}$ is an edge of G_2 because $\{v_i\}$ and $\emptyset$ are $*-$orthogonal. The way of proving that the other incidence relations are preserved is similar.   □

As in the case of $\mathbb{O}^*(C_n)$, we will introduce a new graph family to better describe the graph structure of $\mathbb{O}^\flat(C_n)$.

Definition 3.2. *If $n \geq 3$ we call n-weathercock, denoted by W_n, the simple graph having vertex set $V(W_n) := \{\alpha_1, \dots, \alpha_n, \beta_1, \dots, \beta_n, \gamma, \delta\}$ and the following edges: $\{\delta, \alpha_i\}$, $\{\delta, \beta_i\}$, $\{\delta, \gamma\}$, $\{\alpha_i, \beta_i\}$ for all $i = 1, \dots, n$.*

Example 3.2. Below we provide a graphical representation for the graph W_5:

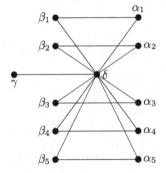

In the next result we determine the graph $\mathbb{O}^\flat(C_n)$ when $n \geq 5$.

Proposition 3.4. *If $n \geq 5$ then $\mathbb{O}^\flat(C_n) \cong W_n$.*

Proof. At first, observe that $V^\flat = \emptyset$ and $\emptyset^\flat = V$. We recall $OP(G) = \{\emptyset, \{v_i\}^c, \{v_i, v_{i+2}\}^c, V\}$, where the index sums are taken $\mod(n)$. Then:

$$A^\flat = \begin{cases} \{v_i, v_{i+2}, v_{i+3}\} & \text{if } A = \{v_i\}^c \\ \{v_{i+1}\}^c & \text{if } A = \{v_i, v_{i+2}\}^c \end{cases} \tag{11}$$

where the index sums are all taken $\mod(n)$. We may conclude that the $\flat-$orthogonal relations are: $\{v_i\}^c \perp_\flat B$, where $B = \emptyset$ or $B = \{v_i, v_{i+2}, v_{i+3}\}$ and $\{v_i, v_{i+2}\}^c \perp_\flat B$ where $B = \emptyset$ or $B = \{v_{i+1}\}^c$ (once again, the index sums are all taken $\mod(n)$).

Now, let $G_1 = W_n$ and $G_2 = \mathbb{O}^\flat(C_n)$. Define $\phi : G_1 \mapsto G_2$ as $\alpha_i \mapsto \{v_i\}^c$, $\beta_i \mapsto \{v_i, v_{i+2}\}^c$, $v \mapsto V$ and $w \mapsto \emptyset$. The map ϕ is clearly bijective and, using the $\flat$-orthogonality relations just found we have that all the incidence relations are preserved. For example, if $\{\delta, \alpha_i\}$ is an edge of G_1, then $\{\emptyset, \{v_i\}^c\}$ is an edge of G_2 because $\{v_i\}^c$ and $\emptyset$ are $\flat$-orthogonal. The way of proving that the other incidence relations are preserved is similar. $\square$

4   The Case of P_n

In this section we will focus our attention to the n-path P_n. At first, the $*$-interior and $*$-closure approximations are given.

Proposition 4.1. *Let* $n \geq 3$, $G = P_n$ *and* $V = V(G)$. *Let* $A \subseteq V(G)$ *such that* $A \neq \emptyset$ *and* $A \neq V$. *Then:*

$$\mathbb{C}_G(A) = \begin{cases} A & \text{if } A = \{v_i\} \vee A = \{v_{i-1}, v_{i+1}\}, \text{ where } i = 2, \ldots, n-1 \\ \{v_1, v_3\} & \text{if } A = \{v_1\} \\ \{v_{n-2}, v_n\} & \text{if } A = \{v_n\} \\ V & \text{otherwise} \end{cases}$$

(12)

and

$$\mathbb{I}_G(A) = \begin{cases} A & \text{if } A = \{v_i\}^c \vee A = \{v_{i-1}, v_{i+1}\}^c, \text{ where } i = 2, \ldots, n-1 \\ \emptyset & \text{otherwise.} \end{cases}$$

(13)

Proof. At first, we observe that if $A = \{v_i\}$, then

$$A^* = \begin{cases} \{v_{i-1}, v_{i+1}\} & \text{if } i \neq 1, n \\ \{v_2\} & \text{if } i = 1 \\ \{v_{n-1}\} & \text{if } i = n \end{cases}$$

(14)

Now, it's clear that if A is the singleton of the first case, the only vertex in V connected with the two points in A^* is v_i itself, so $A^{**} = A$. On the contrary, if $A = \{v_1\}$, then $A^{**} = \{v_2\}^* = \{v_1, v_3\}$. The case $A = \{v_n\}$ is similar and it gives as result $A^{**} = \{v_{n-2}, v_n\}$. Let $A = \{v_{i-1}, v_{i+1}\}$ with $i = 2, \ldots, n-1$. Then the only vertex linked to these points is v_i and, for the previous computations, we have $\{v_i\}^* = \{v_{i-1}, v_{i+1}\}$. This means $A^{**} = A$ even in this case. Let A be a proper non-empty subset of $V(G)$ containing at least two vertices v_i, v_j such that $i < j$ and $j \neq i+2$. In this case, by (2), $A^* \subseteq N_G(v_i) \cap N_G(v_j) = \emptyset$. This proves (12).

For the second part, if $A = \{v_i\}^c$, where $i = 2, \ldots, n-1$, we have: $\mathbb{I}_G(A) = A^{\flat\flat} = A^{c**c} = \{v_i\}^c$. We get a similar computations in the case $A = \{v_{i-1}, v_{i+1}\}^c$ where $i = 2, \ldots, n-1$. Moreover, if A is a proper non-empty subset of $V(G)$ different from the previous one, due to the relation $\mathbb{I}_G(A) = A^{\flat\flat} = A^{c**c}$ and the first part, we have that in this case $\mathbb{I}_G(A) = \emptyset$. This complete the proof of (13). $\square$

Corollary 4.1. *Let* $V = V(P_n)$. *Then* $CL(P_n) = \{\emptyset, \{v_i\}, \{v_{i-1}, v_{i+1}\}, V\}$, *where* $i = 2, \ldots, n - 1$. *Moreover* $OP(P_n) = \{\emptyset, \{v_i\}^c, \{v_{i-1}, v_{i+1}\}^c, V\}$, *where* $i = 2, \ldots, n - 1$.

Proof. It follows immediately from the fact that $\emptyset$ and C_n are always $*$−open and $*$−closed sets and from Proposition 4.1. □

We now give the expression of the similarity lower and upper approximations in case of the path graphs.

Proposition 4.2. *Let* $n \geq 3$, $G = P_n$ *and* $V = V(G)$. *Let* $A \subseteq V(G)$ *such that* $A \neq \emptyset$ *and* $A \neq V$. *Then:*

$$
U_S(A) = \begin{cases} V \setminus \{v_{i-1}, v_{i+1}\} & A = \{v_i\} \quad i \neq 1, n \\ V \setminus \{v_{i+1}\} & A = \{v_1\} \text{ or } A = \{v_i, v_{i+2}\} \quad i \leq (n-2) \\ V \setminus \{v_{n-1}\} & A = \{v_n\} \\ V & \text{otherwise} \end{cases} \tag{15}
$$

and

$$
L_S(A) = \begin{cases} \{v_{i-1}, v_{i+1}\} & \text{if } A = \{v_i\}^c \quad i \neq 1, n \\ \{v_2\} & \text{if } A = \{v_1\}^c \\ \{v_{n-1}\} & \text{if } A = \{v_n\}^c \\ \{v_{i+1}\} & \text{if } A = \{v_i, v_{i+2}\}^c \quad i \leq n - 2 \\ \emptyset & \text{otherwise} \end{cases} \tag{16}
$$

where $i = 1, \ldots, n$ *and the above index sums* $i + 2$ *are taken* $\mathrm{mod}(n)$.

Proof. Similar to Proposition 3.2. □

In order to better describe the graph structure of $\mathbb{O}^*(P_n)$ when $n \geq 5$, we introduce a new graph family.

Definition 4.1. *If* $n \geq 3$ *we call open* n-*chandelier, denoted by* OCh_n, *the simple graph having vertex set* $V(OCh_n) := \{\alpha_1, \ldots, \alpha_n, \beta_1, \ldots, \beta_n, \gamma, \delta\}$ *and the following edges:* $\{\delta, \alpha_i\}$, $\{\delta, \beta_i\}$, $\{\delta, \gamma\}$, $\{\alpha_i, \beta_i\}$ *and* $\{\alpha_i, \alpha_{i+1}\}$ *for all* $i = 1, \ldots, n - 1$.

Example 4.1. Below we provide a graphical representation for the graph OCh_5:

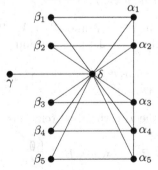

In the next result we determine the graph $\mathbb{O}^*(P_n)$.

Proposition 4.3. *If $n \geq 5$ then $\mathbb{O}^*(P_n) \cong OCh_{n-2}$.*

Proof. At first, observe that: (i) $V^* = \emptyset$ and $\emptyset^* = V$; (ii) if $A = \{v_i\}$, where $i = 2, \ldots, n-1$ then $A^* = \{v_{i-1}, v_{i+1}\}$; (iii) if $A = \{v_i, v_{i+2}\}$, for $i = 1, \ldots, n-2$, then $A^* = \{v_{i+1}\}$. We may conclude that the $*$−orthogonal relations are:

$$\{v_i\} \perp_* B, \quad \text{where } B = \begin{cases} \emptyset \\ \{v_{i-1}\} \\ \{v_{i+1}\} \\ \{v_{i-1}, v_{i+1}\} \end{cases} \tag{17}$$

if $i = 2, \ldots, n-2$;

$$\{v_2\} \perp_* B, \quad \text{where } B = \begin{cases} \emptyset \\ \{v_3\} \\ \{v_1, v_3\} \end{cases} \tag{18}$$

$$\{v_{n-1}\} \perp_* B, \quad \text{where } B = \begin{cases} \emptyset \\ \{v_{n-2}\} \\ \{v_{n-2}, v_n\} \end{cases} \tag{19}$$

and

$$\{v_{i-1}, v_{i+1}\} \perp_* B, \quad \text{where } B = \begin{cases} \emptyset \\ \{v_i\} \end{cases} \tag{20}$$

where $i = 1, \ldots, n-1$.

Now, let $G_1 = OCh_{n-2}$ and $G_2 = \mathbb{O}^*(P_n)$. Define $\phi : G_1 \mapsto G_2$ as $\alpha_i \mapsto \{v_{i+1}\}$, $\beta_j \mapsto \{v_j, v_{j+2}\}$, $\gamma \mapsto V$ and $\delta \mapsto \emptyset$ for $i, j = 1, \ldots, n-2$. The map ϕ is clearly bijective and, using the $*$−orthogonality relations just found we have that all the incidence relations are preserved. For example, if $\{\delta, \alpha_i\}$ is an edge of G_1, then $\{\emptyset, \{v_{i+1}\}\}$ is an edge of G_2 because $\{v_{i+1}\}$ and $\emptyset$ are $*$−orthogonal. The way of proving that the other incidence relations are preserved is similar. $\square$

In the next result we determine the graph $\mathbb{O}^\flat(P_n)$ when $n \geq 5$.

Proposition 4.4. *If $n \geq 5$ then $\mathbb{O}^\flat(P_n) \cong W_{n-2}$.*

Proof. At first, observe that $V^\flat = \emptyset$ and $\emptyset^\flat = V$. We recall that $OP(G) = \{\emptyset, \{v_i\}^c, \{v_j, v_{j+2}\}^c, V\}$, where $i = 2, \ldots, n-1$ and $j = 1, \ldots, n-2$. Then:

$$A^\flat = \begin{cases} \{v_{i-1}, v_{i+1}\} & \text{if } A = \{v_i\}^c \\ \{v_{j+1}\}^c & \text{if } A = \{v_j, v_{j+2}\}^c \end{cases} \tag{21}$$

We may conclude that the $\flat$−orthogonal relations are:

$$\{v_i\}^c \perp_\flat B, \quad \text{where } B = \begin{cases} \emptyset \\ \{v_{i-1}, v_{i+1}\} \end{cases} \tag{22}$$

and

$$\{v_j, v_{j+2}\}^c \perp_\flat B, \qquad \text{where } B = \begin{cases} \emptyset \\ \{v_{j+1}\}^c \end{cases} \tag{23}$$

Now, let $G_1 = W_{n-2}$ and $G_2 = \mathbb{O}^\flat(P_n)$. Define $\phi : G_1 \mapsto G_2$ as $\alpha_i \mapsto \{v_{i+1}\}^c$, $\beta_j \mapsto \{v_j, v_{j+2}\}^c$, $v \mapsto V$, $w \mapsto \emptyset$.

The map ϕ is clearly bijective and, using the $\flat$−orthogonality relations just found we have that all the incidence relations are preserved. For example, if $\{\delta, \alpha_i\}$ is an edge of G_1, then $\{\emptyset, \{v_{i+1}\}^c\}$ is an edge of G_2 because $\{v_{i+1}\}^c$ and $\emptyset$ are $\flat$−orthogonal. The way of proving that the other incidence relations are preserved is similar. □

5   Conclusion

The general framework outlined in [4] has been applied to n-cycle and n-path graphs. The importance of this work is twofold. From one side, it is an exemplification of how to apply some granular computing concepts, mainly those linked to preclusivity spaces, to graphs. From the other side, the two classes of graphs studied here play a basic and important role in graph theory. In particular, n-paths are simple trees and they are at the basis of the path decomposition problem [6], which has applications in VLSI design, computational linguistic and compiler design. So, natural directions to explore are applying our framework to trees and investigating possible applications connected to the path decomposition problem.

References

1. Cattaneo, G.: Generalized rough sets (Preclusivity fuzzy-intuitionistic (BZ) lattices). Studia Logica **01**, 47–77 (1997)
2. Cattaneo, G.: Abstract approximation spaces for rough theories. In: Polkowski, L., Skowron, A. (eds.) Rough Sets in Knowledge Discovery 1: Methodology and Applications. Studies in Fuzziness and Soft Computing, pp. 59–98. Physica Verlag, Heidelberg (1998)
3. Chiaselotti, G., Ciucci, D., Gentile, T.: Simple undirected graphs as formal contexts. In: Baixeries, J., Sacarea, C., Ojeda-Aciego, M. (eds.) ICFCA 2015. LNCS, vol. 9113, pp. 287–302. Springer, Heidelberg (2015)
4. Chiaselotti, G., Ciucci, D., Gentile, T., Infusino, F.: Preclusivity and simple graphs. In: Yao, Y., Hu, Q., Yu, H. Grzymala-Busse, J. (eds.) RSFDGrC 2015. LNCS, vol. 9437, pp. 127–137. Springer, Heidelberg (2015)
5. Ciucci, D., Dubois, D., Prade, H.: The structure of oppositions in rough set theory and formal concept analysis - toward a new bridge between the two settings. In: Beierle, C., Meghini, C. (eds.) FoIKS 2014. LNCS, vol. 8367, pp. 154–173. Springer, Heidelberg (2014)
6. Diestel, R.: Graph Theory. Graduate Text in Mathematics, 4th edn. Springer, Heidelberg (2010)

7. Ganter, B., Wille, R.: Formal Concept Analysis. Mathematical Foundations. Springer, Heidelberg (1999)
8. Yao, J.T., Ciucci, D., Zhang, Y.: Generalized rough sets. In: Kacprzyk, J., Pedrycz, W. (eds.) Springer Handbook of Computational Intelligence, Chapter 25, pp. 413–424. Springer, Heidelberg (2015)

Connectedness of Graph and Matroid
by Covering-Based Rough Sets

Hui Li$^{(\boxtimes)}$ and William Zhu

Lab of Granular Computing, Minnan Normal University, Zhangzhou, China
lihui_grc@126.com

Abstract. Covering-based rough sets provide an efficient theory to process information in data mining. Matroid theory is a generalization of both linear algebra and graph theory, and has a variety of applications in many fields, such as covering-based rough sets. In this paper, we study the connectedness of graphs and matroids through covering-based rough sets. First, we present an approach to induce a covering by a graph. Then we use the covering upper approximation operator and the rank of matrix representation of the covering to study the connectedness of the graph. Moreover, we give the expression of the number of the connected components of a graph. Second, we establish a matroid based on the covering induced by a graph and study the connectedness of this matroid.

Keywords: Covering-based rough set · Matroid · Graph · Connectedness · Granular computing

1 Introduction

Rough set theory, introduced by Pawlak [1,2], has proved to be very useful in dealing with granularity in information systems. It also has already been applied to environmental science, chemistry psychology, granular computing and other fields [3]. Traditional rough set theory is based on partitions generated from equivalence relations. However, in many real world applications, equivalence relation is too restrictive to use. Hence, various generalizations are proposed to extend equivalence relation based rough set model, such as relation based rough set model [4,5] and covering based rough set model [6,7].

Graph theory is highly utilized in computer science, especially in data mining [8], clustering [9]. Graph theory has also combined with other theories, for example, covering-based rough sets [10].

As a generalization of both linear algebra and graph theory, matroid theory [11,12] was originally proposed by Whitney. It has been found that matroids are effective to simplify various ideas in graph theory Matroids have powerful axiomatic systems which provide a well-platform to connect with other theories, such as classical rough sets [13], relation-based rough sets [14] and covering-based rough sets [15].

© Springer International Publishing Switzerland 2015
Y. Yao et al. (Eds.): RSFDGrC 2015, LNAI 9437, pp. 149–160, 2015.
DOI: 10.1007/978-3-319-25783-9_14

In this paper, we connect the graph theory with matroids from the viewpoint of covering-based rough sets and study whether graphs and matroids are connected or not by means of covering-based rough sets. First, we define a family of sets from an undirected graph. Then we prove the family of sets are a covering if and only if the graph has no isolated vertices. Based on the covering, we study the connectedness of the graph from the viewpoint of covering upper approximation operator. We get the number of the connected components of a graph. Second, based on the covering, we can establish a matroid from the viewpoint of covering upper approximation operator and the matroid is proved to be a partition-circuit matroid. The connectedness of the matroid is also studied in terms of covering-based rough sets. Third, we investigate the connectedness of a graph and the matroid induced by the graph. Especially, we get the conclusion that they have the same connectedness.

The rest of this paper is organized as follows: Sect. 2 recalls some fundamental definitions and properties of covering-based rough sets, matroids and graphs. In Sect. 3, a covering is induced by a graph without isolated vertices and the connectedness of the graph is studied. In Sect. 4, we establish a matroid by covering upper approximation operator and study the connectedness of the matroid. Finally, we present concluding remarks of this paper in Sect. 5.

2   Basic Definitions

In this section, we recall some basic concepts and important conclusions of covering-based rough sets, matroids and graphs.

2.1   Rough Set

Classical rough set theory is based on partitions on a universe. Covering-based rough set theory is obtained through extending a partition to a covering. To begin with, the concept of covering is introduced.

Definition 1 (*Covering [6]). Let E be a finite universe of discourse and C a family of subsets of E. If none of subsets in C is empty and $\bigcup\{C : C \in C\} = E$, then C is called a covering of E. The ordered pair (E, C) is called a covering approximation space.*

The following definition presents a widely used pair of approximation operators. We also give some properties of the covering approximation operator.

Definition 2 (*Approximations operators [16]). Let (E, C) be a covering approximation space and $X \subseteq E$. The covering upper and lower approximations of X, denoted by $\overline{C}(X)$ and $\underline{C}(X)$, respectively, are defined as:*

$$\overline{C}(X) = \bigcup\{K \in C : K \cap X \neq \emptyset\};$$
$$\underline{C}(X) = \bigcup\{K \in C : K \subseteq X\}.$$

Proposition 1 *[17]. Let C be a covering. The operators $\underline{C}$ and $\overline{C}$ have the following properties:*

(1) $\underline{C}(E) = E$; $\overline{C}(E) = E$;   (2) $\underline{C}(\emptyset) = \emptyset$; $\overline{C}(\emptyset) = \emptyset$;
(3) $\underline{C}(X) \subseteq X$; $X \subseteq \overline{C}(X)$;   (4) $\overline{C}(X \cup Y) = \overline{C}(X) \cup \overline{C}(Y)$;
(5) $X \subseteq Y \Rightarrow \underline{C}(X) \subseteq \underline{C}(Y)$;   (6) $X \subseteq Y \Rightarrow \overline{C}(X) \subseteq \overline{C}(Y)$.

The following definition shows how to induce a matrix from a covering.

Definition 3 *[18]. Let C be a covering of $E = \{x_1, x_2, \cdots, x_n\}$. Suppose $C = \{C_1, C_2, \cdots, C_m\}$. We define a matrix $A(C) = (a_{ij})_{m \times n}$ as*

$$a_{ij} = \begin{cases} 1, & x_j \in C_i \\ 0, & x_j \notin C_i \end{cases}$$

This matrix is called a matrix representation of covering.

2.2   Matroid

In the following, a matroid is defined from the viewpoint of independent sets.

Definition 4 *(Matroid [11]). A matroid is an ordered pair $M = (E, \mathcal{I})$ consisting of a finite set E, and a collection $\mathcal{I}$ (called independent sets) of subsets of E with the following three properties:*

(1) $\emptyset \in \mathcal{I}$;
(2) If $A \in \mathcal{I}$ and $B \subseteq A$, then $B \in \mathcal{I}$;
(3) If $A, B \in \mathcal{I}$ and $|A| < |B|$, then there exists $e \in B - A$ such that $A \cup \{e\} \in \mathcal{I}$, where $|A|$ denotes the cardinality of A.

Example 1. Let $E = \{a, b, c, d\}$, $\mathcal{I} = \{\emptyset, \{a\}, \{b\}, \{c\}, \{d\}, \{a, b\}, \{a, c\}, \{b, c\}, \{b, d\}, \{a, d\}, \{c, d\}, \{a, b, c\}, \{a, b, d\}, \{b, c, d\}\}$. Then $M = (E, \mathcal{I})$ is a matroid.

For convenience sake, we introduce some symbols as follows.

Definition 5 *[12]. Let E be a finite and nonempty set and $\mathcal{A} \subseteq 2^E$ a family of subsets of E. Then*
$Min(\mathcal{A}) = \{X \in \mathcal{A} : \forall Y \in \mathcal{A}, Y \subseteq X \Rightarrow X = Y\}$;
$Opp(\mathcal{A}) = \{X \subseteq E : X \notin \mathcal{A}\}$.

The complement of an independent set in the power set of the universe is a dependent set. A minimal set of the dependent sets is called a circuit of the matroid.

Definition 6 *(Circuit [12]). Let $M = (E, \mathcal{I})$ be a matroid. If the subset $X \in Opp(\mathcal{I}) = 2^E - \mathcal{I}$, then we call X is a dependent set of M. Any minimal dependent set in M is called a circuit of M, and we denote the family of all circuits of M by $\mathcal{C}(M)$, that is, $\mathcal{C}(M) = Min(Opp(\mathcal{I}))$.*

A matroid uniquely determines its circuits, and vice versa as shown in the following proposition.

Proposition 2 (*Circuit axiom [12]*). *Let C be a family of subsets of U. Then there exists $M = (E, \mathcal{I})$ such that $C = C(M)$ if and only if C satisfies the following three conditions:*

(C1) $\emptyset \notin C$;
(C2) *If* $C_1, C_2 \in C$ *and* $C_1 \subseteq C_2$, *then* $C_1 = C_2$;
(C3) *If* $C_1, C_2 \in C$, $C_1 \nsubseteq C_2$ *and* $e \in C_1 \cap C_2$, *then there exists* $C_3 \in C$ *such that* $C_3 \subseteq C_1 \cup C_2 - \{e\}$.

2.3   Graph

Theoretically, a graph is an ordered pair consisting of vertices and edges that connect these vertices. We assume the graphs studied in this paper are undirected.

Definition 7 (*Undirected graph [19]*). *A graph $G = (V, E)$ consists of a nonempty set of vertices V and a set of edges E. Each edge has either one or two vertices associated with it, called its endpoints. Generally, we write $e = uv$ or $e = vu$ for an edge e with endpoints u and v. If any edge of the graph is undirected, we say the graph is an undirected graph.*

In a graph G, if vertex v is an endpoint of edge e, then v and e are incident. When vertex u and vertex v are the endpoints of an edge, they are adjacent. We also say u is adjacent to v. When edge e_1 and edge e_2 have a common vertex, they are adjacent. We say e_1 is adjacent to e_2. An isolated vertex is a vertex not adjacent to any other vertex. A loop is an edge whose endpoints are equal. Multiple edges are edges having the same pair of endpoints. A simple graph is a graph having no loops or multiple edges. A graph, each of whose vertices belongs to V and each of whose edges belongs to E, is called a subgraph of G. The subgraph induced by a subset of vertices $K \subseteq V$ is called a vertex induced subgraph of G, and denoted by G_k. This subgraph has vertex set K, and its edge set $E' \subseteq E$ consists of those edges from E that have both their ends in K.

A walk of a graph G is a list $v_0, e_1, v_1, \cdots, e_k, v_k$ of vertices such that, for all $1 \leq i \leq k$, the edge e_i has endpoints v_{i-1} and v_i. A u, v-walk has first vertex u and last vertex v.

Definition 8 (*Connected graph [19]*). *A graph G is connected if each pair of vertices (u, v) in G have a u, v-walk.*

Every finite graph G can be partitioned into nonempty subgraphs $G(V_1, E_1), G(V_2, E_2), \cdots, G(V_s, E_s)$ such that two vertices u and v are connected if and only if both u and v belong to a subgraph $G(V_i, E_i)$. We call these subgraphs the components of G, and denote the number of components of G by $\omega(G)$. G is connected if and only if G has only one component.

3   Covering Induced by Graph

In [20], Wang, et al. construct a covering from the viewpoint of the vertices of a simple graph without isolated vertices. In this paper, we define a family of subsets of edges of a graph. Then we prove the family of subsets of edges is a covering if and only the graph has no isolated vertices.

Definition 9. *Let* $G = (V, E)$ *be a graph where* $|V| = p$, $|E| = n$. *We define a family of sets* $C(G) = \{K_1, K_2, \cdots, K_p\}$ *with respect to* G *where* $K_i (i = 1, 2, \cdots, p)$ *contains all the edges that* v_i *incidents.*

Based on the definition, we can see every edge of G appears in the $\mathbf{C}(G)$ two times when the graph has no loops. The following example illustrates the $\mathbf{C}(G)$ of graph G.

Example 2. Let $G = (V, E)$ be a graph as shown in Fig. 1. Then $\mathbf{C}(G) = \{\{e_1, e_2, e_3, \}, \{e_2, e_3, e_4, \}, \{e_4, e_5\}, \{e_1, e_5\}\}$.

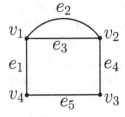

Fig. 1. Graph G

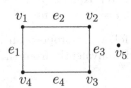

Fig. 2. Graph G

In the above example, $\mathbf{C}(G)$ is a covering of E. But, in general, $\mathbf{C}(G)$ is not necessary a covering. The following example can illustrate $\mathbf{C}(G)$ is not a covering of the edges of a graph.

Example 3. Let $G = (V, E)$ be a graph as shown in Fig. 2. Then $\mathbf{C}(G) = \{\{e_1, e_2\}, \{e_2, e_3\}, \{e_3, e_4\}, \{e_1, e_4\}, \emptyset\}$. We can see $\mathbf{C}(G)$ is not a covering of E.

$\mathbf{C}(G)$ is not necessary a covering, but, if the graph without isolated vertices, $\mathbf{C}(G)$ is a covering. In fact, we have the following proposition.

Proposition 3. *Let* $G = (V, E)$ *be a graph.* G *has no isolated vertices if and only if the family* $\mathbf{C}(G)$ *is a covering of* E.

Proof. ($\Rightarrow$) From the definition of $\mathbf{C}(G)$, every block K_i of $\mathbf{C}(G)$ is all the edges that v_i incidents. If G has not isolated vertices, then $K_i \neq \emptyset$. For every edge $e_i \in E$ of G, there exists $K_i \in C(G)$ such that $e_i \in K_i$. So $\bigcup K_i = E$, namely, $C(G)$ is a covering of E. ($\Leftarrow$) If $C(G)$ is a covering of E, then $\emptyset \notin C(G)$. So there exists at least one edge that adjacent to each vertex. Hence G has not isolated vertices.

According to the above proposition, an undirected graph without isolated vertices can be characterized by a covering of its edges. In the rest of this paper, we assume that the graphs of discourse are finite and without isolated vertices unless otherwise stated.

In [10], the connectedness of a graph is characterized by a family of subsets of vertices of the graph, which is a covering. In our discussions, we also use a covering to study the connectedness of the graph. But every subset in our covering consists of the edges of the graph, not vertices of the graph as in [10]. First, the connectedness of any pair of distinct edges of a graph is defined.

Definition 10. *Let $G = (V, E)$ be a graph and e_1, e_2 be two distinct edges of E. The edge e_1 is connected to e_2 if and only if $\exists K_i \in \mathbf{C}(G)$ such that $\{e_1, e_2\} \subseteq K_i \in C(G)$ where K_i is all the edges that v_i incidents($i = 1, 2, \cdots, p$) or there exist $v_1, v_2, \cdots, v_m$ such that e_1 incidents with v_1, e_2 incidents with v_m, namely there exist $K_1, K_2, \cdots, K_m \in \mathbf{C}(G)$ such that $e_1 \in K_1$, $e_2 \in K_m$ and $K_i \cap K_{i+1} \neq \emptyset$ for any $i = 1, 2, \cdots, m - 1$.*

The following proposition illustrates that a connected graph can be characterized equivalently from the viewpoint of the covering upper approximation operator.

Proposition 4. *Let $G = (V, E)$ be a graph. The graph is connected if and only if for any $\emptyset \neq X \subset E$, $\overline{\mathbf{C}(G)}(X) \neq X$.*

Proof. ($\Rightarrow$) Since $\overline{\mathbf{C}}(\emptyset) = \emptyset$, we need to prove that: $\forall \emptyset \neq X \subseteq E$, if $\overline{\mathbf{C}}(X) = X$, then $X = E$. Let $e_i \in X$. For all $e_j \in E - \{e_i\}$, if $\{e_i, e_j\} \subseteq \mathbf{C}(G)$, then $e_j \in \{e_i, e_j\} \subseteq \overline{\mathbf{C}(G)}(X) = X$ which implies $E - \{e_i\} \subseteq X$. For all $e_j \in E - \{e_i\}$, if e_i and e_j are not adjacent, then there exist $K_1, K_2, \cdots, K_m \in \mathbf{C}(G)$ such that $e_i \in K_1$, $e_j \in K_m$ and $K_i \cap K_{i+1} \neq \emptyset$ for any $i = 1, 2, \cdots, m - 1$. Since $e_i \in K_1 \cap X$, $K_1 \subseteq \overline{\mathbf{C}(G)}(X) = X$. Because $K_1 \cap K_2 \neq \emptyset$, $K_2 \subseteq \overline{\mathbf{C}(G)}(K_1)$. Also $K_2 \subseteq \overline{\mathbf{C}(G)}(K_1) \subseteq \overline{\mathbf{C}(G)}(X) = X$. In the same way, we can obtain $e_j \in K_m \subseteq X$, then $E - \{e_i\} \subseteq X$. Since $e_i \in X$, $E \subseteq X$. Because $X \subseteq E$, so $X = E$. ($\Leftarrow$) For all $e_i \in E$, let $P_{e_i} = \{e_j \in E : e_i \text{ is adjacent to } e_j\}$. Then $P_{e_i} \neq \emptyset$ because $e_i \in P_{e_i}$. We need to prove $\overline{\mathbf{C}(G)}(P_{e_i}) = P_{e_i}$. For all $e_j \in \overline{\mathbf{C}(G)}(P_{e_i})$, there exists $K \in \mathbf{C}(G)$ such that $e_j \in K$ and $K \cap P_{e_i} \neq \emptyset$. Suppose $e_k \in K \cap P_{e_i}$, then e_j is adjacent to e_k and e_k is adjacent to e_i, thus e_j is adjacent to e_i. Namely $e_j \in P_{e_i}$. Thus $\overline{\mathbf{C}(G)}(P_{e_i}) \subseteq P_{e_i}$. Because $P_{e_i} \subseteq \overline{\mathbf{C}(G)}(P_{e_i})$, then $\overline{\mathbf{C}(G)}(P_{e_i}) = P_{e_i}$. By assumption, we know $P_{e_i} = E$. Therefore G is connected.

In the following, we study the number of the connected components of a graph based on covering-based rough sets.

Proposition 5. *Let $G = (V, E)$ be a graph. The number of the connected components of G is $\omega_G = |Min\{X \subseteq E : \overline{C(G)}(X) = X \wedge X \neq \emptyset\}|$.*

Proof. A graph is connected if and only if for any $\emptyset \neq X \subset E$, $\overline{C(G)}(X) \neq X$. Then a graph is disconnected if and only if $\exists X \subset E, \overline{C(G)}(X) = X$. Let $\mathcal{A} = Min\{X \subseteq E : \overline{C(G)}(X) = X \wedge X \neq \emptyset\}$. We know $\forall X_1, X_2 \in \mathcal{A}$, $X_1 \cap X_2 = \emptyset$. Namely, $\overline{C(G)}(X_1) = X_1$ and $\overline{C(G)}(X_2) = X_2$. If $X_1 \cap X_2 \neq \emptyset$, let $X = X_1 \cap X_2$, then $\overline{C(G)}(X) = \overline{C(G)}(X_1 \cap X_2) \subseteq \overline{C(G)}(X_1) \cap \overline{C(G)}(X_2) = X_1 \cap X_2$. Since the elements of $\mathcal{A}$ are minimal, so $X = \emptyset$. Thus, $\forall X_1, X_2 \in \mathcal{A}$, $X_1 \cap X_2 = \emptyset$. We can get $\cup X_i = E$ for all $X_i \in \mathcal{A}$. If $\cup X_i \neq E$, then $\exists Y = E - \cup X_i$ and $Y \notin \mathcal{A}$ such that $\overline{C(G)}(Y) \neq Y$. Then G is connected. It is contradictory to the graph is disconnected. So $\cup X_i = E$. $X_i \in \mathcal{A}$ is the minimal subset of E, then the number of the connected components is $\omega_G = |Min\{X \subseteq E : \overline{C(G)}(X) = X \wedge X \neq \emptyset\}|$.

In order to further understand the characteristic of the graph, we give the following example to illustrate the number of the connected components.

Example 4. Let $G = (V, E)$ be a graph as shown in Fig. 3. Then the family of sets $\mathbf{C}(G) = \{\{e_1, e_2, e_3\}, \{e_4, e_5\}\}$. We can get $Min\{X \subseteq E : \overline{C(G)}(X) = X \wedge X \neq \emptyset\} = \{\{e_1, e_2, e_3\}, \{e_4, e_5\}\}$. Namely $\omega_G = 2$. So the number of the connected components of G is 2.

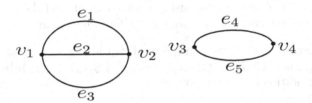

Fig. 3. Graph G

In the following, we use another method to study the connectedness of a graph. Firstly, we use complete incidence matrix of graph to express the relationship between vertices and edges. The definition of complete incidence matrix is shown as follows.

Definition 11 (*Complete incidence matrix [21]*). *Let $G = (V, E)$ be a graph with $|V| = p$, $|E| = q$. Let*

$$m_{ij} = \begin{cases} 1, & \text{if edge } e_j \text{ incidents to vertice } v_i \\ 0, & \text{otherwise} \end{cases}$$

Then the $p \times q$ matrix, denoted by M_e, which consists of elements m_{ij} ($i = 1, 2, \cdots, p, j = 1, 2, \cdots, q$), is called complete incidence matrix.

In graph theory, there is a lemma to judge whether a loopless graph is connected or not by the rank of complete incidence matrix. The operations of the above matrix are all conducted on the filed $\mathcal{F} = \{0, 1\}$.

Lemma 1 *[19]. Let $G = (V, E)$ be a loopless graph with $|V| = p$, $|E| = q$. Then the graph is connected if and only if the rank of complete incidence matrix for G is $p - 1$.*

Based on the above lemma and the definition of covering, we can use the rank of matrix representation of the covering to judge a connected graph.

Proposition 6. *Let $G = (V, E)$ be a loopless graph with $|V| = p$, $|E| = q$. G is connected if and only if $r(A(\mathbf{C}(G))) = p - 1$.*

Proof. Comparing the definition of $A(\mathbf{C}(G))$ and complete incidence matrix, we can get every row of $A(C(G))$ represents the edges that vertex v_i incidents. It has the same meaning of complete incidence matrix. Based the above lemma, the graph is connected if and only if $r(A(\mathbf{C}(G))) = p - 1$.

According to the above proposition, we have the following corollary.

Corollary 1. *Let $G = (V, E)$ be a loopless graph with $|V| = p$, $|E| = q$. The graph has k connected components if and only if $r(A(\mathbf{C}(G))) = p - k$.*

In order to illustrate the relationship between graph and the rank of matrix representation of the covering, we present the following example.

Example 5. Let $G = (V, E)$ be a graph as shown in Fig. 4. We can get $|V| = 7$ and $|E| = 6$. Then $\mathbf{C}(G) = \{\{e_1, e_2\}, \{e_1, e_3\}, \{e_2, e_3\}, \{e_4, e_5\}, \{e_6\}\}$. So the matrix representation of the covering is

$$
A(\mathbf{C}(G)) = \begin{array}{c} \\ C_1 \\ C_2 \\ C_3 \\ C_4 \\ C_5 \end{array} \begin{pmatrix} e_1 & e_2 & e_3 & e_4 & e_5 & e_6 \\ 1 & 1 & 0 & 0 & 0 & 0 \\ 1 & 0 & 1 & 0 & 0 & 0 \\ 0 & 1 & 1 & 0 & 0 & 0 \\ 0 & 0 & 0 & 1 & 1 & 0 \\ 0 & 0 & 0 & 0 & 0 & 1 \end{pmatrix} \tag{1}
$$

On the filed $\mathcal{F} = \{0, 1\}$, the rank of $A(\mathbf{C}(G))$ is 4.

4   The Connectedness of Matroid Induced by Covering

In the above section, we have defined a covering from a graph. Now, we define a matroid from a covering and study its connectedness. Firstly, a new family of sets are defined by the covering approximation operator. Then we prove the family of sets to satisfy the circuit axioms of matroids. So we can get a type of matroids. Next, we discuss the connectedness of the matroid. We start from the definition of the new family of sets.

Definition 12. *Let G be a graph and $C(G)$ a covering induced by G. We define a family of sets as follows*

$$C'(G) = Min\{X \subseteq E : \overline{C(G)}(X) = X \wedge X \neq \emptyset\}.$$

Next, we will prove $C'(G)$ to satisfy the circuit axioms of matroids.

Proposition 7. *Let $G = (V, E)$ be a graph. $C'(G)$ satisfies circuit axioms (C1), (C2) and (C3) of Proposition 2.*

Proof. (1) From the definition of $C'(G)$, it is clear that $\emptyset \notin C'(G)$.

(2) Let $C_1, C_2 \in C'(G)$ and $C_1 \subseteq C_2$. Then $\overline{C(G)}(C_1) = C_1$, $\overline{C(G)}(C_2) = C_2$. Because C_1 and C_2 are two minimal elements of $C'(G)$ and $C_1 \subseteq C_2$, so $C_1 = C_2$.

(3) Let $C_1, C_2 \in C'(G)$ and $C_1 \neq C_2$. Then $\overline{C(G)}(C_1) = C_1$, $\overline{C(G)}(C_2) = C_2$. On one hand, $\overline{C(G)}(C_1 \cap C_2) \subseteq \overline{C(G)}(C_1) \cap \overline{C(G)}(C_2) = C_1 \cap C_2$. On the other hand, $C_1 \cap C_2 \subseteq \overline{C(G)}(C_1 \cap C_2)$. Thus, $\overline{C(G)}(C_1 \cap C_2) = C_1 \cap C_2$. Because the elements of $C'(G)$ are minimal, so $C_1 \cap C_2 = \emptyset$. Therefore, for all $C_1, C_2 \in C'(G)$, and $C_1 \neq C_2$, we can get $C_1 \cap C_2 = \emptyset$. i.e., (C3) of Proposition 2 holds. To sum up, we can get $C'(G)$ satisfies circuit axioms.

$C'(G)$ satisfies circuit axioms, so it can generate a matroid.

Definition 13. *Let $G = (V, E)$ be a graph and $C(G)$ a covering induced by G. The matroid with $C'(G)$ as its circuit family is denoted by $M(C) = (E, \mathcal{I}(C))$, where $\mathcal{I}(C) = Opp(Upp(C'(G)))$. We call $M(C)$ the matroid induced by $C(G)$.*

The matroid induced by $C(G)$ can be illustrated by the following example.

Example 6. Let $G = (V, E)$ be a graph as shown in Fig. 5. Then $C(G) = \{\{e_1\}, \{e_1, e_2\}, \{e_2, e_3\}, \{e_3\}, \{e_4, e_5, e_6\}\}$. We can get $C'(G) = \{\{e_1, e_2, e_3\}, \{e_4, e_5, e_6\}\}$. Therefore, $M(C) = (E, \mathcal{I}(C))$ with $C'(G)$ as its circuit family.

Next, we give the definition of partition-circuit matroid in the following.

Definition 14 (*Partition-circuit matroid [13]*). *Let P be a partition on U. The matroid, whose family of all circuits is $C_P = P$, is denoted by $M_P = (E, \mathcal{I}_P)$ and called partition-circuit matroid.*

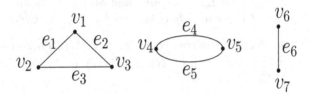

Fig. 4. Graph G

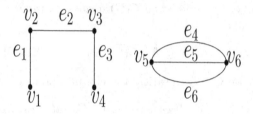

Fig. 5. Graph G

The following proposition shows that $M(\mathbf{C})$ is also a partition-circuit matroid.

Proposition 8. *Let G be a graph and $\mathbf{C}(G)$ a covering induced by G. $M(\mathbf{C})$ is the matroid induced by $\mathbf{C}(G)$, then matroid $M(\mathbf{C})$ is a partition-circuit matroid.*

Proof. Based on the definition of partition-circuit matroid, we need to prove that $\mathbf{C}'(G)$ is a partition. According to Proposition 7, $\forall C_i, C_j \in \mathbf{C}'(G)$, $C_i \cap C_j = \emptyset$. For all $C_i \in \mathbf{C}'(G)$, $\overline{\mathbf{C}(G)}(C_i) = C_i$. $\overline{\mathbf{C}(G)}(\cup C_i) = \cup \overline{\mathbf{C}(G)}(C_i) = \cup C_i$. The number of the connected components of G is $w_G = |Min\{X \subseteq E : \overline{\mathbf{C}(G)}(X) = X \wedge X \neq \emptyset\}|$. Namely every C_i is a connected component. So $\cup C_i = E$. Then $\mathbf{C}'(G)$ is a partition of E. Therefore, the matroid $M(\mathbf{C})$ is a partition-circuit matroid.

The connectedness is an important property not only in graph theory, but also in matroid theory. Then we give the definition of connected matroid as follows.

Definition 15 (*Connected matroid [12]*). *Let M be a matroid on E. For any two elements $e_1, e_2 \in E$, we define $e_1 \sim e_2$ if and only if $e_1 = e_2$ or $\exists C \in \mathcal{C}(M)$ such that $e_1, e_2 \in C$. In [11], the relation $\sim$ is an equivalence relation. Let $E_1, \cdots, E_c$ be equivalence classes on $E(M)$. $\forall i (1 \leq i \leq c)$, $M|E_i$ is called a connected component of M. If $E(M)$ has only one equivalence class, we call M is connected. Namely the matroid M is connected if and only if, for every pair of distinct elements of $E(M)$, there is a circuit containing both.*

The following proposition shows that whether $M(\mathbf{C})$ is connected or not.

Proposition 9. *Let $G = (V, E)$ be a graph and $M(\mathbf{C})$ the matroid induced by $\mathbf{C}(G)$. $M(\mathbf{C})$ is connected if and only if for any $\emptyset \neq X \subset E$, $\overline{\mathbf{C}(G)}(X) \neq X$.*

Proof. Based on the definition of connected matroid, $M(\mathbf{C})$ is connected if and only if there is only one element in $\mathbf{C}'(G)$, if and only if $\mathbf{C}'(G) = \{\{E\}\}$, if and only if for $\forall X \subset E$, $\overline{\mathbf{C}(G)}(X) \neq X$.

The following proposition shows how to compute the number of the connected components of a matroid.

Proposition 10. *Let $G = (V, E)$ be a graph and $M(\mathbf{C})$ the matroid induced by $C(G)$. The number of the connected components of $M(\mathbf{C})$ is $\omega_{M(\mathbf{C})} = |Min\{X \subseteq E : \overline{C(G)}(X) = X \wedge X \neq \emptyset\}|$.*

Proof. According to Proposition 8 and the proof of Proposition 7, the circuits $\mathbf{C}'(G)$ of $M(\mathbf{C})$ is a partition of E. Then the number of the connected components of $M(\mathbf{C})$ is equal to the cardinality of $\mathbf{C}'(G)$. Therefore, $\omega_{M(\mathbf{C})} = |Min\{X \subseteq E : \overline{C(G)}(X) = X \wedge X \neq \emptyset\}|$.

We give the connectedness between the graph and the matroid induced by $\mathbf{C}(G)$.

Proposition 11. *Let $G = (V, E)$ be a graph. The graph G is connected if and only if the matroid $M(\mathbf{C})$ is connected.*

Proof. The graph is connected if and only if for any $\emptyset \neq X \subset E$, $\overline{C(G)}(X) \neq X$ if and only if $M(\mathbf{C})$ is connected.

5 Conclusions

In this paper, we studied the connectedness between a graph and the matroid induced by the graph. Firstly, we defined a covering by an undirected graph without isolated vertices and studied the connectedness of the graph by covering approximation operators. Moreover, the rank of the matrix representation of the covering was also introduced to study the connectedness of the graph. Then, based on the covering induced by a graph, we established a matroid. We studied the connectedness of this matroid. Finally, we investigated the connection between the graph and the matroid induced by the graph and got that they had the same connectedness.

Acknowledgments. This work is in part supported by The National Nature Science Foundation of China under Grant Nos. 61170128, 61379049 and 61379089, the Key Project of Education Department of Fujian Province under Grant No. JA13192, and the Science and Technology Key Project of Fujian Province, China Grant No. 2012H0043.

References

1. Pawlak, Z.: Rough sets. Int. J. Comput. Inf. Sci. **11**, 341–356 (1982)
2. Pawlak, Z.: Rough classification. Int. J. Man-Mach. Stud. **20**, 469–483 (1984)
3. Wu, W., Zhang, W.: Rough set approximations vs. measurable spaces (in chinese). In: Granular Computing (GrC), 2006 IEEE International Conference on Granular Computing, pp. 329–332. IEEE Press, China (2006)
4. Diker, M.: Textural approach to generalized rough sets based on relations. Inf. Sci. **180**, 1418–1433 (2010)
5. Liu, G., Zhu, W.: The algebraic structures of generalized rough set theory. Inf. Sci. **178**, 4105–4113 (2008)

6. Zhu, W., Wang, F.: Reduction and axiomization of covering generalized rough sets. Inf. Sci. **152**, 217–230 (2003)
7. Wang, S., Zhu, W., Zhu, Q., Min, F.: Covering base. J. Inf. Comput. Sci. **9**, 1343–1355 (2012)
8. Nettleton, D.F.: Data mining of social networks represented as graphs. Comput. Sci. Revi. **7**, 1–34 (2013)
9. Fukami, T., Takahashi, N.: New classes of clustering coefficient locally maximizing graphs. Discrete Appl. Math. **162**, 202–213 (2014)
10. Huang, A., Zhu, W.: Connectedness of graphs and its application to connected matroids through covering-based rough sets. arXiv preprint arXiv:1312.4234 (2013)
11. Oxley, J.G.: Matroid Theory. Oxford University Press, New York (1993)
12. Lai, H.: Matroid Theory. Higher Education Press, Beijing (2001)
13. Liu, Y., Zhu, W.: Characteristic of partition-circuit matroid through approximation number. In: Granular Computing (GrC), 2012 IEEE International Conference on Granular Computing, pp. 314–319. IEEE Press, Hangzhou (2012)
14. Zhu, W., Wang, S.: Matroidal approaches to generalized rough sets based on relations. Int. J. Mach. Learn. Cybern. **2**, 273–279 (2011)
15. Liu, C., Miao, D.: Covering rough set model based on multi-granulations. In: Kuznetsov, S.O., Ślęzak, D., Hepting, D.H., Mirkin, B.G. (eds.) RSFDGrC 2011. LNCS, vol. 6743, pp. 87–90. Springer, Heidelberg (2011)
16. Pomykala, J.A.: Approximation operations in approximation space. Bul. Pol. Acad. Sci. **35**, 653–662 (1987)
17. Chen, J., Li, J., Lin, Y.: On the structure of definable sets in covering approximation spaces. Int. J. Mach. Learn. Cybern. **4**, 195–206 (2013)
18. Wang, S., Zhu, W., Zhu, Q., Min, F.: Characteristic matrix of covering and its application to boolean matrix decomposition. Inf. Sci. **263**, 186–197 (2014)
19. West, D.B., et al.: Introduction to Graph Theory. Prentice Hall, Upper Saddle River (2001)
20. Wang, S., Zhu, Q., Zhu, W., Min, F.: Equivalent characterizations of some graph problems by covering-based rough sets. J. Appl. Math. (2013)
21. Wang, Z.: Graph Theory. Beijing Institute of Technology Press, Beijing (2002)

Controllability in Directed Complex Networks: Granular Computing Perspective

Yunyun Yang, Gang Xie$^{(\boxtimes)}$, and Zehua Chen

College of Information Engineering, Taiyuan University of Technology,
Taiyuan 030024, Shanxi, China
chinayunyunyang@qq.com, xiegang@tyut.edu.cn

Abstract. Controlling complex networks to a desired state has been a widespread sense in contemporary science. Usually, we seek a maximum matching of complex networks by matching theory and control those unmatched nodes to achieve the purpose of controlling complex networks. However, for complex networks with high dimensions, it is hard to find its maximum matching or there are copious unmatched nodes that need to be controlled. Therefore, controlling complex networks is extremely strenuous. Motivated by the idea of granular computing (GrC), we take a fine graining preprocessing to the whole complex networks and obtain several different granules. Then find the maximum matching in every granule and control those unmatched nodes to procure the goal of controlling the entire network. At last, the related key problems in GrC-based controllability of complex networks processing framework are discussed.

1 Introduction

The quick development of information technology has pushed on the process of the whole world informationization and networkization. It has taken mankind into a new information age and there are a large number of networks emerge every day [1–3]. These networks are derived from a wealth of sources and increase quickly, mainly including social networks, Internet of Things, communications networks, etc. For such a large complex networks, people in various fields realize that it is an enormous challenge to control it [4,5]. Myriad researchers in the fields of controllability of complex networks seek solutions.

Complex networks have gained increasing attention in the past decade [6–9]. Within complex networks, the controllability of complex networks plays an important role. Currently, the controllability of complex networks mainly focused on two small branches: (1) the theory of pinning control by using stability theory of dynamical systems, (2) the theory evolved from classical theory of controllability. There are massive scholars in this in-depth study of the two

G. Xie–This research is supported by the National Natural Science Foundation of China (Grant No. 60975032 and Grant No. 61402319), and Research Project Supported by Shanxi Scholarship Council of China (Grant No. 2013-031).

© Springer International Publishing Switzerland 2015
Y. Yao et al. (Eds.): RSFDGrC 2015, LNAI 9437, pp. 161–169, 2015.
DOI: 10.1007/978-3-319-25783-9_15

branches. Among them, the theory of pinning control of complex networks proposed by Wang, Li and Chen is a hot academia widespread concern [10]. It has substantial applications in pinning synchronizability of complex networks and other aspects. By using stability theory of dynamical systems, pinning control makes all nodes in networks tend to balance state (node) through controlling a small part of nodes. Soon afterwards the ideas and methods of pinning control are widely used in the study of the synchronization of complex networks [11]. In addition, due to the large-scale and highly complex structures of complex networks system, it is no longer viable to find the optimal input position through the traverse method. In view of this, in 2011 Liu et al. published on structural controllability theory of complex networks in Nature [12]. By introducing matching theory, Liu gave the condition for structural controllability of complex networks. Thereafter, Nepusz and Vicsek [13] studies the structure controllability of network system with edges dynamics by nodes and edges interchangeable way. Wang et al. introduced a method of adding some edges to achieve the purpose of controlling the entire network through controlling only one node [14]. Currently, most of the research of the controllability of complex networks was still based on structural controllability theory proposed by Liu et al. But the scope of structural controllability theory is limited. Because it is necessary and requirement to determine maximum matching in directed networks in structural control theory for controlling real world complex networks. The popular classic method through the Hopcroft-Karp algorithm and other proposed algorithms requires the determination of the bipartite equivalent graph (i.e. network), which belongs to the NP-complete class of problems. Therefore, it is needed to develop new theories and methods to compensate for structural controllability theorys shortcomings.

Granular computing [15] as a methodology, in the process of solving the issue we choose a size suitable granule as the processing target. Under the premise of guaranteeing to obtain a satisfactory solution, improve the efficiency of resolving the issue. Since 1979 Zadeh [16] published his first paper on information granularity, numerous researchers have in-depth studied on granular computing theories and models. Moreover, through combining with artificial intelligence and machine learning technology, these theories and models made numerous achievements. Suitable granularity is often caused by a problem in itself and the background of the problem. Suitable granularity is very prominent for designing the data processing framework based on GrC. GrC simulates human thinking and can be applied to the current world facing big data and big networks. Recently, how to apply GrC to big data processing has been international and domestic researchers attention. In the paper, we take a fine graining preprocessing to the whole complex networks and obtain several different granules. Then find the maximum matching in every granule. This method can be efficiently and used to find maximum matching in complex networks.

The rest of the paper is organized as follows. In the remainder of this introductory section, the current situation of the development of controllability of complex networks is summarized. In Sect. 2, we review basic ideas of GrC that are relevant for our approach and we provide basic terminologies and notations used throughout this paper. In Sect. 3, the techniques to enhance controllability of directed complex networks based on GrC. Section 4, we present experiments to evaluate the techniques we propose. Section 5 concludes this paper.

2 Definitions

For convenience, some notations and definitions are given below.

2.1 Controllability of Complex Networks Based on the Linear System

In this section, we introduce the notions of a digraph and maximum matchings of a digraph.

Let $G_D = (V, E)$ be a digraph. A digraph $G_D = (V, E)$ consists of a non-empty finite set $V(G_D)$ of elements called nodes, and a finite set $E(G_D)$ of ordered pairs of distinct nodes called edges. We write $G_D = (V, E)$ which means that V and E are the node set and edge set of G_D, respectively. The size of G_D is the number of nodes in G_D, and is denoted by $|G_D|$. Here consider a linear time-invariant (LTI) dynamics Eq. 1 on a directed network $G_D(A)$

$$\dot{\mathbf{x}}(t) = \mathbf{A}\mathbf{x}(t) + \mathbf{Q}\mathbf{u}(t), \tag{1}$$

where $\mathbf{A}$ is the transpose of the (weighted) adjacency matrix of the network, $\mathbf{x}(t)$ is a time dependent vector of the state variables of the nodes, $\mathbf{u}(t)$ is the vector of input signals, and $\mathbf{Q}$ is the so called input matrix, which defines how the input signals are connected to the nodes of the network.

Based on the linear system control theory [17], a complex network is controllable if and only if it satisfies Kalmans controllability rank condition [18]. It can be controlled from any initial state to any desired state in finite time, if and only if the $N \times NM$ controllability matrix $\mathbf{C}$ has full rank. It can be formalized as follows

$$\text{rank}[\mathbf{Q}, \mathbf{A}\mathbf{Q}, \mathbf{A^2Q}, \cdots, \mathbf{A^{n-1}Q}] = N.$$

Where $\mathbf{x} = (x_1, x_2, \cdots, x_N)^T \in \Re^N$ is called the state vector, $\mathbf{A} = (a_{ij})_{N \times N}$ is the state matrix, and a_{ij} denotes the weight of a directed edge from node j to i. $Q \in \Re^{N \times M}$ is the input matrix, and $\mathbf{u} = (u_1, u_2, \cdots, u_M)^T \in \Re^M$ is the input or control vector.

Recently, Liu et al. [12] developed a minimum input theory to efficiently characterize the structural controllability of directed networks. According to the literature [12], the structural controllability of directed networks can be mapped into the problem of the maximum matching, where external control is necessary for every unmatched node. The maximum matching is defined and the method for finding a maximum matching respectively as follows:

Definition 1. Let $G_D = (V, E)$ be a digraph, a matching M_D in a digraph G_D is a set of edges with no common end-nodes. That is to say, no two edges in M_D share a common starting node or a common ending node. We also require that no element of M_D is a loop. If M_D is a matching, then we say that the edges of M_D are independent. A node is matched if it is an ending node of an edge in the matching. Otherwise, it is unmatched.

For the case of bipartite digraphs, a simple algorithm based on bipartite undigraphs is described. A maximum matching of a digraph $G_D = (V, E)$ can be easily found in its bipartite representation, denoted as $F(V)$ (see Fig. 1). The bipartite graph is defined as $F(V) = (V^+ \cup V^-, \Gamma)$. Here, $V^+ = \{x_1^+, x_2^+, \cdots, x_N^+\}$ and $V^- = \{x_1^-, x_2^-, \cdots, x_N^-\}$ are the sets of nodes corresponding to the N columns and rows of the state matrix $\mathbf{A}$, respectively. Edge set $\Gamma = \{(x_j^+, x_i^-) | a_{ij} \neq 0\}$. In the case of bipartite digraphs, a simple algorithm based on bipartite undigraphs is described. For a thorough review on the bipartite matching problem, we refer to the book by Burkard et al. [19]. For a general bipartite graph, its maximum matching can be found efficiently using the well-known Hopcroft-Karp algorithm, which runs in $O(\sqrt{V}E)$ time [20].

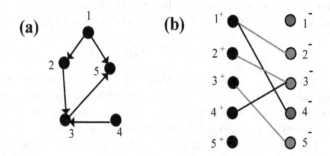

Fig. 1. A simple network (digraph), its bipartite representation and its maximum matching. (a), A simple network (digraph) with 5 nodes and 5 edges. (b), The bipartite representation of the network and its maximum matching. Matched nodes are shown in green and matching edges are shown in green (Color figure online).

It is easy to see a maximum matching in $G_D = (V, E)$ which covers the maximum possible number of nodes in Γ. In addition, a digraph has a cycle factor if and only if its bipartite representation $F(V)$ contains a perfect matching.

2.2 Granular Computing

Human tend to think of granular, abstract levels rather than on the level of detailed and precise data that are characteristic for computers. For example, recent studies confirm that too much information may lead to an information overload that reduces the quality of human decisions and actions. To address these challenges, GrC was proposed as a multidisciplinary field of research [15, 16].

One of the key concepts in Granular Computing is the concept of granulation. Granulation is an operation or a process of forming granules, with a granule being a collection of objects (points) that are drawn together by some constraints, such as indistinguishability, similarity or functionality. The process of constructing information granules is called information granulation. It granulates a universe of discourse into a family of disjoint or overlapping granules. In a broad sense, granulation involves decomposition of whole into parts. Another fundamental property of granules and levels is their granularity. The granularity of granules and levels enables us to construct a hierarchical structure called a hierarchy. The term hierarchy is utilized to denote a family of partially ordered levels, in which each level consists of a family of interacting and interrelated granules. Thus one of the main directions in the study of GrC is to deal with the construction, interpretation, representation of granules, and relations among granules which are represented by granular IF-THEN rules with granular variables and their relevant granular values.

Recently, C. L. Philip Chen [21] discussed several ongoing or underlying techniques and technologies to harness big data, and the first mentioned is GrC. GrC is a burgeoning conceptual and computing paradigm of knowledge discovery. In some degree, it will be motivated by the urgent need for efficient processing of complex networks. Theoretical foundations of granular computing are exceptionally sound and are a highly comprehensive treatment of this paradigm.

3   The Techniques to Enhance Controllability of Directed Complex Networks Based on GrC

When we talk about complex networks, the first property of it is its size. As GrC is a general computation theory for effectively using granules such as classes, subsets and groups to build an efficient computational model for complex applications with huge amounts of data (networks), information and knowledge, therefore it is very natural to employ granular computing techniques to explore complex networks. Intuitively, granular computing can reduce complex networks size into different level of granularity. The philosophy and general principles of GrC is of fundamental value to effective and efficient solving some problems of the huge complex networks by GrC is an interesting research area with great potential.

For complex networks with high dimensions, it is hard to find the maximum matching. The complex networks are composed of infinite sub-networks. There is a certain correlation between the sub-networks. Therefore, for complex networks with high dimensions, we can granulation them into numerous granules by using the method of granular computing, and then conducts the research on each granule.

Here let U be a non-empty classical set (may be finite or infinite), which will be the main universe of discourse throughout this paper. The collection of all subsets of U will be denoted as the set of all objects in the complex network. Complex networks model is expressed by the triplet (U, L, f), where

$U = \{v_1, v_2, \cdots, v_n\}$, L is the tagset, and $f : L \to 2^U$ is tag mapping. Obtain k categories by dividing U, namely $C = \{w_1, w_2, \cdots, w_n\}, 1 \le k \le n$, satisfying

$$w_i \ne \phi, \ w_i \cap w_j = \phi, \ \Sigma_{i=1}^k w_i = U, \ i, j = 1, 2, \cdots, k \ (i \ne j), \ k \le n$$

For a given directed complex network $G_D = (V, E), V = U$, the algorithm on enhancing controllability of directed complex networks based on GrC of the steps in the following:

Step 1: Granular the entire network U and attain different granules $C = \{w_1, w_2, \cdots, w_k\}, 1 \le k \le n$.

The principle of the granulation is each node in the every granule as much as possible to reach the perfect matching or near-perfect matching.

Step 2: Transform every granule into a bipartite graph $F_m(V) = (V_m^+ \cup V_m^-, \Gamma_m)$.

Step 3: According the algorithm [22], in every granule judge each node is matched. If it is not matched node, we need to control it.

Step 4: Through controlling these non matched nodes in each granule, achieve the purpose of controlling the whole network.

Specifically, in Step 3 we uses a heuristic for finding maximum matching in digraphs directly from its unipartite form [22], i.e., without obtaining the bipartite equivalent. The main advantage of the approach over the previously existing ones is that it determines the number of driver nodes. The steps given below, are to be followed for determining maximum matching in every granule:

Step 3.1: Determine the node in every granule of the network under consideration, which possesses the maximum out degree. Let this node be denoted by N_a. The node N_a obtained in this step is considered as the driver node.

Step 3.2: Determine the node connected to N_a (driver node) having the maximum out degree. Let the node be denoted as N_b.

Step 3.3: Consider the edge from node N_a to N_b, and discard all other edges connected to N_a, thus setting its degree to 1.

Step 3.4: Assign N_b to N_a and repeat step 3.2 and step 3.3 till the degree of N_a becomes 1. After completion of step 3.4, the augmenting path having N_a (from step 3.1) as its driver node is obtained.

Step 3.5: Repeat steps 3.1 to 3.4 till the granule has been considered.

4   Examples

Take for an example, a directed network $G_D = (V, E)$ with 10 nodes and 21 edges shown in Fig. 2.

We granular the entire network G_D and obtain two different granules $C = \{w_1, w_2\}$, where $w_1 = \{v_1, v_2, v_3, v_4, v_5\}$, and $w_2 = \{v_6, v_7, v_8, v_9, v_{10}\}$. The bipartite graph of the network w_1 is defined as $F_1(V) = (V_1^+ \cup V_1^-, \Gamma_1)$. Here, $V_1^+ = \{x_1^+, x_2^+, \cdots, x_5^+\}$ and $V_1^- = \{x_1^-, x_2^-, \cdots, x_5^-\}$ are the sets of nodes. Then according to Hopcroft-Karp algorithm its maximum matching can be found efficiently and we find that in the w_1 granule the set of the matched nodes is the

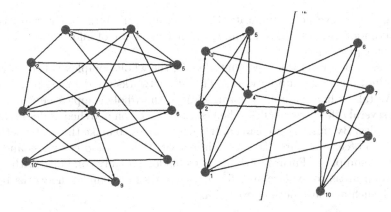

Fig. 2. The topological structure of a network and the topology of the network after granulating

nodes $\{v_1, v_2, v_3, v_4, v_5\}$, and the matching edges are $\{e_{1,5}, e_{5,2}, e_{2,3}, e_{3,4}, e_{4,1}\}$. These matching edges are just well formed a directed chain, according to the definition of the complex network controllability, and then as long as we control the starting node v_1 or the ending node v_5 to achieve the purpose of control the whole w_1 granule.

Likewise, the bipartite graph of the network w_2 is defined as $F_2(V) = (V_2^+ \cup V_2^-, \Gamma_2)$. Here, $V_2^+ = \{x_6^+, x_7^+, \cdots, x_{10}^+\}$ and $V_2^- = \{x_6^-, x_7^-, \cdots, x_{10}^-\}$ are the sets of nodes. Further, in the w_2 granule the set of the matched nodes is the nodes $\{v_6, v_7, v_8, v_9, v_{10}\}$, and the matching edges are $\{e_{6,8}, e_{8,7}, e_{7,10}, e_{10,9}\}$. These matching edges are also just well formed a directed chain, by the same token, and we control the starting node v_6 or the ending node v_9 to achieve the purpose of controlling the whole w_2 granule.

The complexity of maximum matching algorithm in a directed network is at most $O(N^{1/2}L)$ for a network with N nodes and L edges. The algorithm complexity of Step 1 is at most $O(k)$. The whole complexity of the algorithm is $O(N^{1/2}Lk)$.

5   Conclusions and Future Work

As we have entered an era of big data and networks, a new wave of scientific revolution is about to begin. As we know, the controllability of complex networks is a hot topic of current research. When analyzing the problem of the controllability of complex networks, we tend to omit some details, prominent features, and then express as a mathematical object of the appropriate analytical method. In other words, we tend to choose the main reason, namely the thought of coarse graining. Those existing techniques and tools to control complex networks are very limited to solving the real complex networks problems completely.

GrC, as a new and rapidly growing paradigm of information processing, may be viewed as a particular class of Ways to Think. It emphasizes on the effective use of multiple levels of granularity. GrC is an umbrella term to cover any theories, methodologies, techniques, and tools that make use of information granules in complex problem solving.

In this paper, from the viewpoint of GrC, the whole complex networks granulation into numerous sub-networks. The sub-networks are defined on the granular. We take a fine graining preprocessing to the whole complex networks and obtain several different granules. Then control those non matched nodes in different granules. By this way, we can control fewer nodes to make the whole network to get into a certain state. To a certain extent, this method costs less and less energy consumption. Future work needs to be done mainly combined with other application areas to explore specific characteristics of complex networks based on GrC, such as construction, evolution, etc.

References

1. Milo, R., Shen-Orr, S., Itzkovitz, S., Kashtan, N., Chklovskii, D., Alon, U.: Network motifs: simple building blocks of complex networks. Science **298**(5594), 824–827 (2002)
2. Barabsi, A.L., Albert, R.: Emergence of scaling in random networks. Science **286**(5439), 509–512 (1999)
3. Dorogovtsev, S.N., Mendes, J.F.F.: Evolution of networks. Adv. Phys. **51**(4), 1079–1187 (2002)
4. Yuan, Z.Z., Zhao, C., Di, Z.R., Wang, W.X., Lai, Y.C.: Exact controllability of complex networks. Nat. Commun. **4** (2013)
5. Liu, Y.Y., Slotine, J.J., Barabsi, A.L.: Control centrality and hierarchical structure in complex networks. Plos One **7**(9), e44459 (2012)
6. Albert, R., Barabsi, A.L.: Statistical mechanics of complex networks. Rev. Mod. Phys. **74**(1), 47 (2002)
7. Boccalettia, S., Latorab, V., Morenod, Y., Chavezf, M., Hwanga, D.U.: Complex networks: structure and dynamics. Phys. Rep. **424**(4), 175–308 (2006)
8. Strogatz, S.H.: Exploring complex networks. Nature **410**(6825), 268–276 (2001)
9. Newman, M.E.J.: The structure and function of complex networks. SIAM Rev. **45**(2), 167–256 (2003)
10. Li, X., Wang, X.F., Chen, G.R.: Pinning a complex dynamical network to its equilibrium. IEEE Trans. Circ. Syst. I Regul. Pap. **51**(10), 2074–2087 (2004)
11. Yu, W.W., Chen, G.R., Lü, J.H.: On pinning synchronization of complex dynamical networks. Automatica **45**(2), 429–435 (2009)
12. Liu, Y.Y., Jean, J.S., Albert, L.B.: Controllability of complex networks. Nature **473**(7346), 167–173 (2011)
13. Nepusz, T., Vicsek, T.: Controlling edge dynamics in complex networks. Nat. Phys. **8**(7), 568–573 (2012)
14. Wang, W.X., Ni, X., Lai, Y.C., Grebogi, C.: Optimizing controllability of complex networks by minimum structural perturbations. Phys. Rev. E **85**(2), 026115 (2012)
15. Yao, Y.Y.: Granular computing: basic issues and possible solutions. In: Proceedings of the 5th Joint Conference on Information Sciences, vol. 1, pp. 186–189 (2000)
16. Zadeh, L.A.: A theory of approximate reasoning. Mach. Intell. **9**, 149–194 (1979)
17. Garcia, C.E., Prett, D.M., Morari, M.: Model predictive control: theory and practicea survey. Automatica **25**(3), 335–348 (1989)

18. Sontag, E.D.: Kalmans controllability rank condition: from linear to nonlinear. In: Antoulas, A.C. (ed.) Mathematical System Theory, pp. 453–462. Springer, Heidelberg (1991)
19. Burkard, R., DellAmico, M., Martello, S.: Assignment Problems. Revised Reprint. Siam, Philadelphia (2009)
20. Hopcroft, J.E., Karp, R.M.: An n5/2 algorithm for maximum matchings in bipartite graphs. SIAM J. Comput. **2**(4), 225–231 (1973)
21. Chen, C.L.P., Zhang, C.Y.: Data-intensive applications, challenges, techniques and technologies: a survey on Big Data. Inf. Sci. **275**, 314–347 (2014)
22. Chatterjee, A., Das, D., Naskar, M.K., Pal, N., Mukherjee, A.: Heuristic for maximum matching in directed complex networks. In: Proceedings of the 2th IEEE International Conference on the Advances in Computing, Communications and Informatics, pp. 1146–1151 (2013)

Rough and Fuzzy Hybridization

Dynamic Maintenance of Rough Fuzzy Approximations with the Variation of Objects and Attributes

Yanyong Huang[1,2], Tianrui Li[1(✉)], and Shi-jinn Horng[1]

[1] School of Information Science and Technology,
Southwest Jiaotong University, Chengdu 610031, China
trli@swjtu.edu.cn, horngsj@yahoo.com.tw
[2] Tianfu Institute of Mathematics,
Tianfu College of Southwestern University of Finance and Economics,
Chengdu 610045, China
yyhswjtu@163.com

Abstract. In many fields including medical research, e-business and road transportation, data may vary over time, *i.e.*, new objects and new attributes are added. In this paper, we present a method for dynamically updating approximations based on rough fuzzy sets under the variation of objects and attributes simultaneously in fuzzy decision systems. Firstly, a matrix-based approach is proposed to construct the rough fuzzy approximations on the basis of relation matrix. Then the method for incrementally computing approximations is presented, which involves the partition of the relation matrix and partly changes its element values based the prior matrices' information. Finally, an illustrative example is employed to validate the effectiveness of the proposed method.

Keywords: Fuzzy decision system · Rough fuzzy set · Incremental learning · Matrix

1 Introduction

Rough set theory (RST), proposed by Pawlak [16,17], has been successfully applied in many fields, such as artificial intelligence, data mining, intelligent information processing and so on [7,15,18]. However, the Pawlak's RST could not efficiently tackle when decision attribute values are fuzzy and each object with different probability belongs to different decision classes. Dubois et al. [6] firstly presented rough fuzzy set to solve this problem. Rough fuzzy set has been investigated in many aspects until now. For example, Yao et al. compared the difference between fuzzy sets and rough sets [20]. Luo et al. discussed the property of rough fuzzy ideal in algebraic systems [14]. Xu et al. presented a novel granular computing approach based formal concept analysis in fuzzy datasets [19].

In many applications, information systems may vary over time, which means that the objects, attributes and attribute values may change [3,8,11,12]. For

© Springer International Publishing Switzerland 2015
Y. Yao et al. (Eds.): RSFDGrC 2015, LNAI 9437, pp. 173–184, 2015.
DOI: 10.1007/978-3-319-25783-9_16

example, new patients may be added to a medical diagnostic information system, which may have additional attributes due to the use of new diagnostic instruments. The patient's record may also be revised over time. How to utilize the prior knowledge and infuse the new data to update the knowledge becomes a crucial problem. The incremental learning technique is an effective way to solve this problem [1,4]. For example, Zhang et al. proposed an incremental method for updating approximations based on matrix computation when the object set varies with time [22]. Li et al. presented an incremental method for updating approximations based on the characteristic relation under the attribute generalization [9]. Luo et al. proposed two incremental algorithms based on dominance-based RST with the addition/removal of attribute values [13]. Chen et al. presented a matrix-based incremental method for updating approximations under the decision-theoretic rough sets when the attributes and objects are added simultaneously [2]. In rough fuzzy sets, Cheng et al. proposed an incremental updating approximations method when the attribute set evolves over time [5]. Zeng et al. presented an incremental updating rough fuzzy approximations method when the object set evolves over time [21]. However, they did not consider Fuzzy Decision System (FDS) may be changed over time, *i.e.*, the attributes and objects simultaneously vary. In this paper, we present an incremental approach based matrix for updating approximations in FDS under the variation of attributes and objects simultaneously.

The remainder of this paper is organized as follows. Section 2 introduces the basic concepts of rough fuzzy set in FDS. Section 3 proposes the matrix representation of the lower and upper rough fuzzy approximations. Section 4 presents an incremental updating method for rough fuzzy approximations when the attribute set and the object set vary simultaneously. Section 5 illustrates an example to validate the effectiveness of the proposed method. Section 6 concludes the paper and discusses our future work.

2   Preliminaries

In this section, the basic concepts of FDS and rough fuzzy sets are briefly reviewed [6].

Definition 1. *[6] A FDS is 4-tuple $S = \langle U, C \cup D, V, f \rangle$, where $U = \{x_i | i \in \{1, 2, \ldots, n\}\}$ is a non-empty finite set of objects, called the universe; C is a non-empty finite set of condition attributes and D is a non-empty finite set of fuzzy decision attributes, $C \cap D = \emptyset$; $V = V_C \cup V_D$, where V is the domain of all attributes, V_C is the domain of all condition attributes and V_D is the domain of decision attributes; f is an information function from $U \times (C \cup D)$ to V such that $f : U \times C \rightarrow V_C$, $f : U \times D \rightarrow [0, 1]$.*

Since the classical RST could not deal with the fuzzy concepts in a crisp approximation space, Dubois and Prade introduced the rough fuzzy set model.

Definition 2. *[6] Let $S = \langle U, C \cup D, V, f \rangle$ be a FDS and $A \subseteq C$. $\tilde{d}$ is a fuzzy subset on D, where $\tilde{d}(x)$ ($x \in U$) denotes the degree of membership of x in $\tilde{d}$. The lower and upper approximations of $\tilde{d}$ are a pair of fuzzy sets on D with respect to the equivalence relation R_A, and their membership functions are defined as follows:*

$$\underline{R_A}\tilde{d}(x) = min\{\tilde{d}(y) | y \in [x]_{R_A}\}$$
$$\overline{R_A}\tilde{d}(x) = max\{\tilde{d}(y) | y \in [x]_{R_A}\}$$

(1)

where $R_A = \{(x, y) \in U \times U | f(x, a) = f(y, a), \forall a \in A\}$, $[x]_{R_A}$ denotes the equivalence class of an element x under R_A and $[x]_{R_A} = \{y \in U | x R_A y\}$.

Example 1. Table 1 illustrates a medical diagnosis FDS, $S = \langle U, C \cup D, V, f \rangle$, where $U = \{x_i | i \in \{1, 2, \ldots, 10\}\}$ denotes the set of patients, the set of condition attributes is $C = \{headache, musclepain, sorethroat, temperature\} = \{c_1, c_2, c_3, c_4\}$, the set of fuzzy decision attribute is $D = \{Flu\} = \{d\}$. The domain $V_{c_1} = \{no, moderate, heavy\} = \{0, 1, 2\}$, $V_{c_2} = V_{c_3} = V_{c_4} = \{no, yes\} = \{0, 1\}$.

Let $A = \{c_1, c_2\} \subset C$. $U/R_A = \{\{x_1, x_3, x_4\}, \{x_2, x_5, x_7, x_9, x_{10}\}, \{x_6, x_8\}\}$. The fuzzy decision attribute Flu $\tilde{d} = \{\frac{0.8}{x_1}, \frac{0.3}{x_2}, \frac{1}{x_3}, \frac{0.7}{x_4}, \frac{0.1}{x_5}, \frac{0.3}{x_6}, \frac{0.2}{x_7}, \frac{0}{x_8}, \frac{0.4}{x_9}, \frac{0.2}{x_{10}}\}$.

According to Definition 2, the degrees of membership can be obtained as follows:

$$\underline{R_A}\tilde{d}(x_1) = \underline{R_A}\tilde{d}(x_3) = \underline{R_A}\tilde{d}(x_4) = 0.8 \wedge 1 \wedge 0.7 = 0.7;$$

$$\underline{R_A}\tilde{d}(x_2) = \underline{R_A}\tilde{d}(x_5) = \underline{R_A}\tilde{d}(x_7) = \underline{R_A}\tilde{d}(x_9) = \underline{R_A}\tilde{d}(x_{10}) = 0.1;$$

$$\underline{R_A}\tilde{d}(x_6) = \underline{R_A}\tilde{d}(x_8) = 0;$$

$$\overline{R_A}\tilde{d}(x_1) = \overline{R_A}\tilde{d}(x_3) = \overline{R_A}\tilde{d}(x_4) = 0.8 \vee 1 \vee 0.7 = 1;$$

$$\overline{R_A}\tilde{d}(x_2) = \overline{R_A}\tilde{d}(x_5) = \overline{R_A}\tilde{d}(x_7) = \overline{R_A}\tilde{d}(x_9) = \overline{R_A}\tilde{d}(x_{10}) = 0.4;$$

$$\overline{R_A}\tilde{d}(x_6) = \underline{R_A}\tilde{d}(x_8) = 0.3.$$

Then the lower and upper approximations of $\tilde{d}$ are as follows:

$$\underline{R_A}\tilde{d} = \{\frac{0.7}{x_1}, \frac{0.1}{x_2}, \frac{0.7}{x_3}, \frac{0.7}{x_4}, \frac{0.1}{x_5}, \frac{0}{x_6}, \frac{0.1}{x_7}, \frac{0}{x_8}, \frac{0.1}{x_9}, \frac{0.1}{x_{10}}\};$$

$$\overline{R_A}\tilde{d} = \{\frac{1}{x_1}, \frac{0.4}{x_2}, \frac{1}{x_3}, \frac{1}{x_4}, \frac{0.4}{x_5}, \frac{0.3}{x_6}, \frac{0.4}{x_7}, \frac{0.3}{x_8}, \frac{0.4}{x_9}, \frac{0.4}{x_{10}}\}.$$

3   Matrix Representation of the Lower and Upper Approximations in the FDS

In this section, we use the matrix-based method for representing the lower and upper approximations in the FDS. Firstly, the equivalent matrix is proposed in the FDS. Then two matrix operations are defined. Finally, it is shown that the lower and upper approximations in the FDS can be effectively computed by the matrix-based method.

Table 1. A decision table with a fuzzy decision attribute.

U	c_1	c_2	c_3	c_4	d
x_1	2	1	0	1	0.8
x_2	1	0	0	1	0.3
x_3	2	1	0	1	1
x_4	2	1	0	1	0.7
x_5	1	0	0	1	0.1
x_6	0	1	1	0	0.3
x_7	1	0	0	1	0.2
x_8	0	1	1	0	0
x_9	1	0	1	0	0.4
x_{10}	1	0	1	0	0.2

Definition 3. *[10] Let* $S = \langle U, C \cup D, V, f \rangle$ *be a FDS, where* $U = \{x_i | i \in \{1, 2, \ldots, n\}\}$, $A \subseteq C$. *The relation matrix is defined as* $M^A = (m_{ij}^A)_{n \times n}$, *where*

$$m_{ij}^A = \begin{cases} 1, & x_i \in [x_j]_{R_A} \\ 0, & other \end{cases} \tag{2}$$

Proposition 1. *[10]* $M^A = (m_{ij}^A)_{n \times n}$ *is a symmetric matrix, and* $m_{ii}^A = 1 (i = 1, \ldots, n)$.

Example 2. (Continuation of Example 1) Let $A = \{c_1, c_2\}$, $B = \{c_3, c_4\}$. Then the relation matrices M^A and M^B can be calculated as follows.

$$M^A = \begin{pmatrix} 1 & 0 & 1 & 1 & 0 & 0 & 0 & 0 & 0 & 0 \\ 0 & 1 & 0 & 0 & 1 & 0 & 1 & 0 & 1 & 1 \\ 1 & 0 & 1 & 1 & 0 & 0 & 0 & 0 & 0 & 0 \\ 1 & 0 & 1 & 1 & 0 & 0 & 0 & 0 & 0 & 0 \\ 0 & 1 & 0 & 0 & 1 & 0 & 1 & 0 & 1 & 1 \\ 0 & 0 & 0 & 0 & 0 & 1 & 0 & 1 & 0 & 0 \\ 0 & 1 & 0 & 0 & 1 & 0 & 1 & 0 & 1 & 1 \\ 0 & 0 & 0 & 0 & 0 & 1 & 0 & 1 & 0 & 0 \\ 0 & 1 & 0 & 0 & 1 & 0 & 1 & 0 & 1 & 1 \\ 0 & 1 & 0 & 0 & 1 & 0 & 1 & 0 & 1 & 1 \end{pmatrix} \qquad M^B = \begin{pmatrix} 1 & 1 & 1 & 1 & 1 & 0 & 1 & 0 & 0 & 0 \\ 1 & 1 & 1 & 1 & 1 & 0 & 1 & 0 & 0 & 0 \\ 1 & 1 & 1 & 1 & 1 & 0 & 1 & 0 & 0 & 0 \\ 1 & 1 & 1 & 1 & 1 & 0 & 1 & 0 & 0 & 0 \\ 1 & 1 & 1 & 1 & 1 & 0 & 1 & 0 & 0 & 0 \\ 0 & 0 & 0 & 0 & 0 & 1 & 0 & 1 & 1 & 1 \\ 1 & 1 & 1 & 1 & 1 & 0 & 1 & 0 & 0 & 0 \\ 0 & 0 & 0 & 0 & 0 & 1 & 0 & 1 & 1 & 1 \\ 0 & 0 & 0 & 0 & 0 & 1 & 0 & 1 & 1 & 1 \\ 0 & 0 & 0 & 0 & 0 & 1 & 0 & 1 & 1 & 1 \end{pmatrix}$$

From Example 2, it can be easily found that $m_{ii}^A = 1$, $m_{ii}^B = 1$, and $M^A = (M^A)^T$, $M^B = (M^B)^T$.

Definition 4. *Let* $S = \langle U, C \cup D, V, f \rangle$ *be a FDS. The corresponding relation matrices of* $A, B \subseteq C$ *are* $M^A = (m_{ij}^A)_{n \times n}$ *and* $M^B = (m_{ij}^B)_{n \times n}$, *respectively. Then the dot operation between* M^A *and* M^B *is defined as follows.*

$$M^A \bullet M^B = (m_{ij}^A \cdot m_{ij}^B)_{n \times n}, \tag{3}$$

where • *is the dot product of two matrices.*

Proposition 2. *Let $S = \langle U, C \cup D, V, f \rangle$ be a FDS. The corresponding relation matrices of $A, B \subseteq C$ are M^A and M^B, respectively. Then the relation matrix $M^{A \cup B}$ of $A \cup B$ is $M^{A \cup B} = M^A \bullet M^B$.*

Proof. If $m_{ij}^{A \cup B} = 1$, according to Definition 3, it follows $x_i \in [x_j]_{R_{A \cup B}}$. Then $x_i \in [x_j]_{R_A}$ and $x_i \in [x_j]_{R_B}$. We have $m_{ij}^A = 1$ and $m_{ij}^B = 1$, that is, $m_{ij}^{A \cup B} = m_{ij}^A \cdot m_{ij}^B$. If $m_{ij}^{A \cup B} = 0$, then $x_i \notin [x_j]_{R_{A \cup B}}$, that is, $x_i \notin [x_j]_{R_A}$ or $x_i \notin [x_j]_{R_B}$. Then $m_{ij}^A = 0$ or $m_{ij}^B = 0$. Hence $m_{ij}^A \cdot m_{ij}^B = m_{ij}^{A \cup B}$. The reverse is also true.

Example 3. (Continuation of Example 2) According to Definition 4, we can compute $M^A \bullet M^B$ as follows.

$$M^A \bullet M^B = \begin{pmatrix} 1011000000 \\ 0100101011 \\ 1011000000 \\ 1011000000 \\ 0100101011 \\ 0000010100 \\ 0100101011 \\ 0000010100 \\ 0100101011 \\ 0100101011 \end{pmatrix} \bullet \begin{pmatrix} 1111101000 \\ 1111101000 \\ 1111101000 \\ 1111101000 \\ 1111101000 \\ 0000010111 \\ 1111101000 \\ 0000010111 \\ 0000010111 \\ 0000010111 \end{pmatrix} = \begin{pmatrix} 1011000000 \\ 0100101000 \\ 1011000000 \\ 1011000000 \\ 0100101000 \\ 0000010100 \\ 0100101000 \\ 0000010100 \\ 0000000011 \\ 0000000011 \end{pmatrix}$$

It is easy to verify to the relation matrix $M^{A \cup B} = M^A \bullet M^B$.

Definition 5. *Let $S = \langle U, C \cup D, V, f \rangle$ be a FDS, where $U = \{x_i | i \in \{1, 2, \ldots, n\}\}$. $\tilde{d}$ is a fuzzy subset on D, $\tilde{d}(x)$ $(x \in U)$ is the degree of membership of x in $\tilde{d}$ and $M^A = (m_{ij}^A)_{n \times n}$ is the relation matrix of $A \subseteq C$. We define the $\otimes_{max}$ operation as follows.*

$$(M^A \otimes_{max} \tilde{d})(i) = max(m_{i1}^A \cdot \tilde{d}(x_1), m_{i2}^A \cdot \tilde{d}(x_2), \ldots, m_{in}^A \cdot \tilde{d}(x_n)) \quad (i = 1, 2, \ldots, n) \tag{4}$$

where max operation takes the maximum value among the n numbers.

Theorem 1. *Let $S = \langle U, C \cup D, V, f \rangle$ be a FDS, where $U = \{x_i | i \in \{1, 2, \ldots, n\}\}$. $\tilde{d}$ is a fuzzy subset on D and M^A is the relation matrix of $A \subseteq C$. The upper and lower approximations of $\tilde{d}$ are calculated as follows.*

$$\overline{R_A}\tilde{d} = M^A \otimes_{max} \tilde{d}; \tag{5}$$

$$\underline{R_A}\tilde{d} = 1 - M^A \otimes_{max} \tilde{d}^c \tag{6}$$

where 1 is the column vector that all elements are one and $\tilde{d}^c$ is the complement of $\tilde{d}$.

Proof. According to Definition 5, $(M^A \otimes_{max} \tilde{d})(i) = max(m_{i1}^A \cdot \tilde{d}(x_1), m_{i2}^A \cdot \tilde{d}(x_2), \ldots, m_{in}^A \cdot \tilde{d}(x_n))$. If $m_{ij}^A = 1$, it means that $x_j \in [x_i]_{R_A}$, then $m_{ij}^A \cdot \tilde{d}(x_j) = \tilde{d}(x_j)$, otherwise $m_{ij}^A \cdot \tilde{d}(x_j) = 0$. Therefore, according to Definition 2, $(M^A \otimes_{max} \tilde{d})(i) = \overline{R_A}\tilde{d}(x_i)$. In addition, according to $\underline{R_A}\tilde{d} = \sim \overline{R_A}\tilde{d}^c$ and Eq. 5, it is clearly that $\underline{R_A}\tilde{d}(x_i) = (1 - M^A \otimes_{max} \tilde{d}^c)(i)$.

Example 4. Given a FDS $S=\langle U, C \cup D, V, f\rangle$ as shown in Table 1. Let $A = \{c_1, c_2\}$, $\tilde{d} = \{\frac{0.8}{x_1}, \frac{0.3}{x_2}, \frac{1}{x_3}, \frac{0.7}{x_4}, \frac{0.1}{x_5}, \frac{0.3}{x_6}, \frac{0.2}{x_7}, \frac{0}{x_8}, \frac{0.4}{x_9}, \frac{0.2}{x_{10}}\}$. From the results of Example 2, then

$$\overline{R_A}\tilde{d} = M^A \otimes_{max} \tilde{d} = \begin{pmatrix} 1\,0\,1\,1\,0\,0\,0\,0\,0\,0 \\ 0\,1\,0\,0\,1\,0\,1\,0\,1\,1 \\ 1\,0\,1\,1\,0\,0\,0\,0\,0\,0 \\ 1\,0\,1\,1\,0\,0\,0\,0\,0\,0 \\ 0\,1\,0\,0\,1\,0\,1\,0\,1\,1 \\ 0\,0\,0\,0\,0\,1\,0\,1\,0\,0 \\ 0\,1\,0\,0\,1\,0\,1\,0\,1\,1 \\ 0\,0\,0\,0\,0\,1\,0\,1\,0\,0 \\ 0\,1\,0\,0\,1\,0\,1\,0\,1\,1 \\ 0\,1\,0\,0\,1\,0\,1\,0\,1\,1 \end{pmatrix} \otimes_{max} \begin{pmatrix} 0.8 \\ 0.3 \\ 1 \\ 0.7 \\ 0.1 \\ 0.3 \\ 0.2 \\ 0 \\ 0.4 \\ 0.2 \end{pmatrix} = \begin{pmatrix} 1 \\ 0.4 \\ 1 \\ 1 \\ 0.4 \\ 0.3 \\ 0.4 \\ 0.3 \\ 0.4 \\ 0.4 \end{pmatrix}$$

$$\underline{R_A}\tilde{d} = 1 - M^A \otimes_{max} \tilde{d^c} = \begin{pmatrix} 1 \\ 1 \\ 1 \\ 1 \\ 1 \\ 1 \\ 1 \\ 1 \\ 1 \\ 1 \end{pmatrix} - \begin{pmatrix} 1\,0\,1\,1\,0\,0\,0\,0\,0\,0 \\ 0\,1\,0\,0\,1\,0\,1\,0\,1\,1 \\ 1\,0\,1\,1\,0\,0\,0\,0\,0\,0 \\ 1\,0\,1\,1\,0\,0\,0\,0\,0\,0 \\ 0\,1\,0\,0\,1\,0\,1\,0\,1\,1 \\ 0\,0\,0\,0\,0\,1\,0\,1\,0\,0 \\ 0\,1\,0\,0\,1\,0\,1\,0\,1\,1 \\ 0\,0\,0\,0\,0\,1\,0\,1\,0\,0 \\ 0\,1\,0\,0\,1\,0\,1\,0\,1\,1 \\ 0\,1\,0\,0\,1\,0\,1\,0\,1\,1 \end{pmatrix} \otimes_{max} \begin{pmatrix} 0.2 \\ 0.7 \\ 0 \\ 0.3 \\ 0.9 \\ 0.7 \\ 0.8 \\ 1 \\ 0.6 \\ 0.8 \end{pmatrix} = \begin{pmatrix} 0.7 \\ 0.1 \\ 0.7 \\ 0.7 \\ 0.1 \\ 0 \\ 0.1 \\ 0 \\ 0.1 \\ 0.1 \end{pmatrix}$$

4   Dynamically Maintenance of Approximations in the FDS Under the Variation of Attributes and Objects

In a dynamic information system, the attribute set and object set may be changed simultaneously over time. In this section, we discuss the methods of incremental updating approximations for dynamically changing attributes and objects in FDS. Based on the above analysis, it is found that the relation matrix is the key step to compute rough fuzzy approximations based on matrix. If we can dynamically compute the changed relation matrix with an incremental updating strategy rather than reconstructing it from scratch, then the running-time will be reduced and the lower and upper approximations can be obtained directly. In the following, we discuss how to update the relation matrix incrementally while the attribute and object sets vary.

Let $S^t = \langle U^t, C^t \cup D^t, V^t, f^t\rangle$ be a FDS at time t, $S^{t+1} = \langle U^{t+1}, C^{t+1} \cup D^{t+1}, V^{t+1}, f^{t+1}\rangle$ denote the FDS at time $t+1$, where $U^{t+1} = U^t \cup \Delta U$, $C^{t+1} = C^t \cup \Delta C$, $D^{t+1} = D^t \cup \Delta D$. To incrementally compute the relation matrices, we partition the system S^{t+1} into two subsystems. One is $S^{\Delta U} = \langle \Delta U, C^{t+1} \cup \Delta D, V^{\Delta U}, f^{\Delta U}\rangle$ and the other is $S^{U^t} = \langle U^t, C^{t+1} \cup D^t, V^{U^t}, f^{U^t}\rangle$. Then the $S^{U^t} = \langle U^t, C^{t+1} \cup D^t, V^{U^t}, f^{U^t}\rangle$ is partitioned into two subsystems again: $S^t = \langle U^t, C^t \cup D^t, V^t, f^t\rangle$, $S^{\Delta C} = \langle U^t, \Delta C \cup D^t, V^{\Delta C}, f^{\Delta C}\rangle$. Suppose $|U^{t+1}| = n'$, $|U^t| = n$, $|\Delta U| = n^+$, $|C^{t+1}| = m'$, $|C^t| = m$, $|\Delta C| = m^+$. Then $n' = n + n^+$, $m' = m + m^+$.

Theorem 2. *Let* $M^{C^{t+1}} = (^{U_{t+1}}m_{ij}^{C^{t+1}})_{n' \times n'}$ *be the relation matrix of FDS*
S^{t+1}. *Then* $(M^{C^{t+1}})_{n' \times n'} = \left(\dfrac{(M_{U_t}^{C^{t+1}})_{n \times n} \;\big|\; (M_{U_t, \Delta U}^{C^{t+1}})_{n \times n^+}}{(M_{U_t, \Delta U}^{C^{t+1}})_{n^+ \times n}^T \;\big|\; (M_{\Delta U}^{C^{t+1}})_{n^+ \times n^+}} \right)$, *where* $M_{U_t}^{C^{t+1}}$
denotes the relation matrix of U^t *under* C^{t+1}, $M_{U_t, \Delta U}^{C^{t+1}}$ *denotes the relation*
matrix of U^t *and* ΔU *under* C^{t+1} *and* $M_{\Delta U}^{C^{t+1}}$ *denotes the relation matrix of*
ΔU *under* C^{t+1}.

Proof. According to Definition 3, it is easy to see that the relation matrix $M^{C^{t+1}}$
can be divided into four parts. Each part can be obtained by Definition 3 directly.

To incrementally compute the relation matrices, we partition the relation
matrix to four parts according to Theorem 2, where the first part $(M_{U_t}^{C^{t+1}})_{n \times n}$
can be incrementally computed.

Theorem 3. *Suppose the* $M_{U_t}^{C^{t+1}} = (m_{ij}^{C^{t+1}})_{n \times n}$, $M_{U_t}^{C^t} = (m_{ij}^{C^t})_{n \times n}$, $M_{U_t}^{\Delta C} = (m_{ij}^{\Delta C})_{n \times n}$ *are the relation matrices of the FDS of* S^{U^t}, S^t, $S^{\Delta C}$, *respectively.*
The elements of the relation matrix $M_{U_t}^{C^{t+1}}$ *can be updated as follows.*

(1) if $m_{ij}^{C^t} = 0$, *then* $m_{ij}^{C^{t+1}} = m_{ij}^{C^t}$;
(2) if $m_{ij}^{C^t} = 1$, *and* $m_{ij}^{\Delta C} = 1$, *then* $m_{ij}^{C^{t+1}} = m_{ij}^{C^t}$;
(3) if $m_{ij}^{C^t} = 1$, *and* $m_{ij}^{\Delta C} = 0$, *then* $m_{ij}^{C^{t+1}} = 0$.

Proof. It follows directly from Proposition 2.

According to Theorem 3, we compute the matrix $M_{U_t}^{C^{t+1}}$ only by updating
the case (3) while not recomputing the whole matrix.

Theorem 4. *Given the relation matrices* $M_{\Delta U}^{C^{t+1}} = (m_{ij}^{\Delta U})_{n^+ \times n^+}$, $M_{U_t}^{C^{t+1}} = (m_{ij}^{C^{t+1}})_{n \times n}$. *Then* $M_{U_t, \Delta U}^{C^{t+1}} = (m_{ij}^{U_t, \Delta U})_{n \times n^+}$ *can be updated as follows.*

(1) if x_i *and* x_j *are equivalent under* C^{t+1}, *where* $x_i \in U^t$, $i = \{1, 2, \ldots, n\}$,
$x_j \in \Delta U$, $j = \{n+1, n+2, \ldots, n+n^+\}$, *then* $m_{[i:]}^{U_t, \Delta U} = m_{[(j-n):]}^{\Delta U}$, $m_{[:(j-n)]}^{U_t, \Delta U} = m_{[:i]}^{C^{t+1}}$. *In addition, if* x_i *and* $x_{i'}$ *are equivalent under* C^{t+1}, x_j *and* $x_{j'}$ *are*
equivalent under ΔC, *where* $i' \in \{1, 2, \ldots, n\}$, $j' \in \{n+1, n+2, \ldots, n+n^+\}$,
then $m_{[i':]}^{U_t, \Delta U} = m_{[(j-n):]}^{\Delta U}$, $m_{[:(j'-n)]}^{U_t, \Delta U} = m_{[:i]}^{C^{t+1}}$.
(2) if x_i *and* x_j *do not satisfy the above conditions, then* $m_{i(j-n)}^{U_t, \Delta U} = 0$. *In addi-*
tion, if x_i *and* $x_{i'}$, x_j *and* $x_{j'}$ *satisfy the above conditions, then* $m_{i'(j-n)}^{U_t, \Delta U} = 0$,
$m_{[:(j'-n)]}^{U_t, \Delta U} = m_{[:(j-n)]}^{U_t, \Delta U}$.

where $m_{[i:]}^{U_t, \Delta U}$ *denotes the* ith *row in the matrix* $M_{U_t, \Delta U}^{C^{t+1}}$, $m_{[:j]}^{U_t, \Delta U}$ *denotes the*
jth *column in the matrix* $M_{U_t, \Delta U}^{C^{t+1}}$ *and the others are the same.*

Proof. If x_i and x_j are equivalent, then according to Theorem 2, $m_{[i:]}^{U_t, \Delta U} = m_{[(j-n):]}^{\Delta U}$. Besides, according to Proposition 1, we have $m_{[:(j-n)]}^{U_t, \Delta U} = m_{[:i]}^{C^{t+1}}$. In addition, if x_i and $x_{i'}$, x_j and $x_{j'}$ are in the same the equivalence class, then we have $m_{[i':]}^{U_t, \Delta U} = m_{[i:]}^{U_t, \Delta U} = m_{[(j-n):]}^{\Delta U}$, $m_{[:(j'-n)]}^{U_t, \Delta U} = m_{[:(j-n)]}^{U_t, \Delta U} = m_{[:i]}^{C^{t+1}}$ according to Proposition 1 and the above results. The proof of case (2) is analogous.

5   An Illustrative Example

To incrementally compute the approximations when objects and attributes alter simultaneously, an example is given to illustrate the proposed method. Let $S^t = \langle U^t, C^t \cup D^t, V^t, f^t \rangle$ be a FDS at time t, where $U = \{x_i | i \in \{1, 2, \ldots, 10\}\}$, $C^t = \{c_i, 1 \leq i \leq 4\}$ (see Table 1). At the time $t + 1$, the attributes $\{c_5, c_6\}$ and the objects $\{x_{11}, x_{12}, x_{13}\}$ are added to S^{t+1}. c_5 denotes the runny noses, c_6 denotes the cough, and $V_{c_5} = V_{c_6} = \{No, Yes\} = \{0, 1\}$. Then $\Delta C = \{c_5, c_6\}$, $\Delta U = \{x_{11}, x_{12}, x_{13}\}$, $U^{t+1} = U^t \cup \Delta U$, $C^{t+1} = C^t \cup \Delta C$ (see Table 2).

Table 2. A decision table with fuzzy decision attributes at time t.

U	c_1	c_2	c_3	c_4	c_5	c_6	Flu
x_1	2	1	0	1	1	0	0.8
x_2	1	0	0	1	0	1	0.3
x_3	2	1	0	1	1	0	1
x_4	2	1	0	1	0	0	0.7
x_5	1	0	0	1	0	1	0.1
x_6	0	1	1	0	1	0	0.3
x_7	1	0	0	1	0	1	0.2
x_8	0	1	1	0	0	0	0
x_9	1	0	1	0	0	1	0.4
x_{10}	1	0	1	0	0	1	0.2
x_{11}	1	0	0	1	0	1	0.4
x_{12}	1	0	0	1	0	1	0.1
x_{13}	2	1	0	1	1	1	0.9

Firstly, we compute the relation matrix $M_{U_t}^{\Delta C}$ according to Definition 3 and the relation matrix $M_{U_t}^{C^t}$ is the result of Example 3.

$$M_{U_t}^{\Delta C} = \begin{pmatrix} 1\ 0\ 1\ 0\ 0\ 1\ 0\ 0\ 0\ 0 \\ 0\ 1\ 0\ 0\ 1\ 0\ 1\ 0\ 1\ 1 \\ \mathbf{1}\ 0\ 1\ 0\ 0\ 1\ 0\ 0\ 0\ 0 \\ \mathbf{0}\ 0\ \mathbf{0}\ 1\ 0\ 0\ 0\ 1\ 0\ 0 \\ 0\ \mathbf{1}\ 0\ 0\ 1\ 0\ 1\ 0\ 1\ 1 \\ 1\ 0\ 1\ 0\ 0\ 1\ 0\ 0\ 0\ 0 \\ 0\ 1\ 0\ 0\ \mathbf{1}\ 0\ 1\ 0\ 1\ 1 \\ 0\ 0\ 0\ 1\ 0\ \mathbf{0}\ \mathbf{0}\ 1\ 0\ 0 \\ 0\ 1\ 0\ 0\ 1\ 0\ 1\ 0\ 1\ 1 \\ 0\ 1\ 0\ 0\ 1\ 0\ 1\ 0\ \mathbf{1}\ \mathbf{1} \end{pmatrix} \quad M_{U_t}^{C^t} = \begin{pmatrix} 1\ 0\ 1\ 1\ 0\ 0\ 0\ 0\ 0\ 0 \\ 0\ 1\ 0\ 0\ 1\ 0\ 1\ 0\ 0\ 0 \\ \mathbf{1}\ 0\ 1\ 1\ 0\ 0\ 0\ 0\ 0\ 0 \\ \mathbf{1}\ 0\ \mathbf{1}\ 1\ 0\ 0\ 0\ 0\ 0\ 0 \\ 0\ \mathbf{1}\ 0\ 0\ 1\ 0\ 1\ 0\ 0\ 0 \\ 0\ 0\ 0\ 0\ 0\ 1\ 0\ 1\ 0\ 0 \\ 0\ 1\ 0\ 0\ \mathbf{1}\ 0\ 1\ 0\ 0\ 0 \\ 0\ 0\ 0\ 0\ 0\ \mathbf{1}\ 0\ 1\ 0\ 0 \\ 0\ 0\ 0\ 0\ 0\ 0\ 0\ 0\ 1\ 1 \\ 0\ 0\ 0\ 0\ 0\ 0\ 0\ 0\ \mathbf{1}\ \mathbf{1} \end{pmatrix}$$

Secondly, we compute the relation matrix $M_{U_t}^{C^{t+1}}$. According to Proposition 1, we only compute the elements under the principal diagonal of the matrix $M_{U_t}^{C^{t+1}}$. We judge the elements which values are "1" under the principal diagonal of the matrix $M_{U_t}^{C^t}$ whether change or not according to Theorem 3. Then we can get the matrix $M_{U_t}^{C^{t+1}}$.

$$M_{U_t}^{C^{t+1}} = \begin{array}{c} \begin{array}{cccccccccc} 1\ 2\ 3\ 4\ 5\ 6\ 7\ 8\ 9\ 10 \end{array} \\ \begin{pmatrix} 1\ 0\ 1\ 0\ 0\ 0\ 0\ 0\ 0\ 0 \\ 0\ 1\ 0\ 0\ 1\ 0\ 1\ 0\ 0\ 0 \\ \mathbf{1}\ 0\ 1\ 0\ 0\ 0\ 0\ 0\ 0\ 0 \\ \mathbf{0}\ 0\ \mathbf{0}\ 1\ 0\ 0\ 0\ 0\ 0\ 0 \\ 0\ \mathbf{1}\ 0\ 0\ 1\ 0\ 1\ 0\ 0\ 0 \\ 0\ 0\ 0\ 0\ 0\ 1\ 0\ 0\ 0\ 0 \\ 0\ \mathbf{1}\ 0\ 0\ \mathbf{1}\ 0\ 1\ 0\ 0\ 0 \\ 0\ 0\ 0\ 0\ 0\ \mathbf{0}\ 0\ 1\ 0\ 0 \\ 0\ 0\ 0\ 0\ 0\ 0\ 0\ 0\ 1\ 1 \\ 0\ 0\ 0\ 0\ 0\ 0\ 0\ 0\ \mathbf{1}\ 1 \end{pmatrix} \end{array}$$

where there are eight elements changed under the principal diagonal of the matrix $M_{U_t}^{C^t}$.

Thirdly, we compute the relation matrix $M_{\Delta U}^{C^{t+1}}$ according to Definition 3.

$$M_{\Delta U}^{C^{t+1}} = \begin{array}{c} 1 \\ 2 \\ 3 \end{array} \begin{pmatrix} 1\ & 1\ & 0 \\ 1\ & 1\ & 0 \\ 0\ & 0\ & 1 \end{pmatrix}$$

Then according to Theorem 4,

(1) because x_2 and x_{11} are equivalent, then $m_{[2:]}^{U_t,\Delta U} = m_{[1:]}^{\Delta U}$, $m_{[:1]}^{U_t,\Delta U} = m_{[:1]}^{C^{t+1}}$.

(2) according to $m_{25}^{C^{t+1}} = 1$, $m_{27}^{C^{t+1}} = 1$, we have $m_{[5:]}^{U_t,\Delta U} = m_{[1:]}^{\Delta U}$, $m_{[7:]}^{U_t,\Delta U} = m_{[1:]}^{\Delta U}$, according to the $m_{11,12}^{\Delta U} = 1$, then $m_{[:2]}^{U_t,\Delta U} = m_{[:2]}^{C^{t+1}}$.

(3) because x_2 and x_{13} are not equivalent, then $m_{[13]}^{U_t,\Delta U} = 0$. The others are the same.

We get the relation matrix

$$
M^{C^{t+1}}_{U_t,\Delta U} =
\begin{array}{c c}
& \begin{array}{ccc} \mathbf{1} & \mathbf{2} & \mathbf{3} \end{array} \\
\begin{array}{c} 1 \\ 2 \\ 3 \\ 4 \\ 5 \\ 6 \\ 7 \\ 8 \\ 9 \\ 10 \end{array} &
\left(\begin{array}{ccc}
0 & 0 & 0 \\
1 & 1 & 0 \\
0 & 0 & 0 \\
0 & 0 & 0 \\
1 & 1 & 0 \\
0 & 0 & 0 \\
1 & 1 & 0 \\
0 & 0 & 0 \\
0 & 0 & 0 \\
0 & 0 & 0
\end{array}\right)
\end{array}
$$

where the first and second column are the same to the second column of $M^{C^{t+1}}_{U_t}$, and the second, fifth, seventh row are the same to the first row of $M^{C^{t+1}}_{\Delta U}$. Obviously, it may reduce the computing time than reconstructing the matrix.

Lastly, according to Theorems 1, 2, we obtain the upper and lower approximations of $\tilde{d}^{t+1}$.

$$
\underline{R_{C_{t+1}}\tilde{d}^{t+1}} = \{\frac{0.8}{x_1}, \frac{0.1}{x_2}, \frac{0.8}{x_3}, \frac{0.7}{x_4}, \frac{0.1}{x_5}, \frac{0.3}{x_6}, \frac{0.1}{x_7}, \frac{0}{x_8}, \frac{0.2}{x_9}, \frac{0.2}{x_{10}}, \frac{0.1}{x_{11}}, \frac{0.1}{x_{12}}, \frac{0.9}{x_{13}}\}
$$

$$
\overline{R_{C_{t+1}}\tilde{d}^{t+1}} = \{\frac{1}{x_1}, \frac{0.4}{x_2}, \frac{1}{x_3}, \frac{0.7}{x_4}, \frac{0.4}{x_5}, \frac{0.3}{x_6}, \frac{0.4}{x_7}, \frac{0}{x_8}, \frac{0.4}{x_9}, \frac{0.4}{x_{10}}, \frac{0.4}{x_{11}}, \frac{0.4}{x_{12}}, \frac{0.9}{x_{13}}\}
$$

6   Conclusions

In FDS, the attribute and object sets may alter simultaneously. How to effectively compute the lower and upper rough fuzzy approximations is a crucial problem. In this paper, we presented an incremental method for updating rough fuzzy approximations based on matrix, and gave an example to show the effectiveness of the approach. In the future work, we will develop the algorithm to validate the proposed method and extend them to handle the problems of updating approximations when the attributes, objects and attributes values vary simultaneously with time in FDS. The parallel strategy to improve the algorithm will be taken into account in the future too.

Acknowledgements. This work is supported by the National Science Foundation of China (No. 61175047), NSAF (No. U1230117) and the Young Software Innovation Foundation of Sichuan Province, China (No. 2014-046).

References

1. Błaszczyński, J., Słowiński, R.: Incremental induction of decision rules from dominance-based rough approximations. Electron. Notes Theor. Comput. Sci. **82**(4), 40–51 (2003)
2. Chen, H., Li, T., Luo, C., Horng, S., Wang, G.: A decision-theoretic rough set approach for dynamic data mining. IEEE Trans. Fuzzy Syst. **PP**(99), 1–1 (2015)
3. Chen, H., Li, T., Qiao, S., Ruan, D.: A rough set based dynamic maintenance approach for approximations in coarsening and refining attribute values. Int. J. Intell. Syst. **25**(10), 1005–1026 (2010)
4. Cheng, M., Fang, B., Tang, Y.Y., Zhang, T., Wen, J.: Incremental embedding and learning in the local discriminant subspace with application to face recognition. IEEE Trans. Syst. Man Cybern. Part C Appl. Rev. **40**(5), 580–591 (2010)
5. Cheng, Y.: The incremental method for fast computing the rough fuzzy approximations. Data Knowl. Eng. **70**(1), 84–100 (2011)
6. Dubois, D., Prade, H.: Rough fuzzy sets and fuzzy rough sets. Int. J. Gener. Syst. **17**(2–3), 191–209 (1990)
7. Dy, J.G., Brodley, C.E.: Feature selection for unsupervised learning. J. Mach. Learn. Res. **5**, 845–889 (2004)
8. Karasuyama, M., Takeuchi, I.: Multiple incremental decremental learning of support vector machines. In: Bengio, Y., Schuurmans, D., Lafferty, J.D., Williams, C.K.I., Culotta, A. (eds.) Advances in Neural Information Processing Systems, vol. 22, pp. 907–915. MIT Press, Cambridge (2009)
9. Li, T., Ruan, D., Geert, W., Song, J., Xu, Y.: A rough sets based characteristic relation approach for dynamic attribute generalization in data mining. Knowl. Based Syst. **20**(5), 485–494 (2007)
10. Liu, G.: Axiomatic systems for rough sets and fuzzy rough sets. Int. J. Approx. Reason. **48**(3), 857–867 (2008)
11. Luo, C., Li, T., Chen, H.: Dynamic maintenance of approximations in set-valued ordered decision systems under the attribute generalization. Inf. Sci. **257**, 210–228 (2014)
12. Luo, C., Li, T., Chen, H., Liu, D.: Incremental approaches for updating approximations in set-valued ordered information systems. Knowl. Based Syst. **50**, 218–233 (2013)
13. Luo, C., Li, T., Chen, H., Lu, L.: Fast algorithms for computing rough approximations in set-valued decision systems while updating criteria values. Inf. Sci. **299**, 221–242 (2015)
14. Luo, Q., Wang, G.: Roughness and fuzziness in quantales. Inf. Sci. **271**, 14–30 (2014)
15. Maji, P., Pal, S., Skowron, A.: Preface: pattern recognition and mining. Natural Computing, pp. 1–3 (2015)
16. Pawlak, Z.: Rough sets. Int. J. Comput. Inf. Sci. **11**(5), 341–356 (1982)
17. Pawlak, Z.: Rough Sets: Theoretical Aspects of Reasoning About Data. Kluwer Academic Publishers, Norwell (1992)
18. Polkowski, L., Tsumoto, S., Lin, T.Y. (eds.): Rough Set Methods and Applications: New Developments in Knowledge Discovery in Information Systems. Physica-Verlag GmbH, Heidelberg (2000)
19. Xu, W., Li, W.: Granular computing approach to two-way learning based on formal concept analysis in fuzzy datasets. IEEE Trans. Cybern. **PP**(99), 1–1 (2014)

20. Yao, Y.: A comparative study of fuzzy sets and rough sets. Inf. Sci. **109**(1–4), 227–242 (1998)
21. Zeng, A., Li, T., Zhang, J., Chen, H.: Incremental maintenance of rough fuzzy set approximations under the variation of object set. Fundam. Inform. **132**(3), 401–422 (2014)
22. Zhang, J., Li, T., Ruan, D., Liu, D.: Rough sets based matrix approaches with dynamic attribute variation in set-valued information systems. Int. J. Approx. Reason. **53**(4), 620–635 (2012)

Semi-Supervised Fuzzy-Rough Feature Selection

Richard Jensen[1]([✉]), Sarah Vluymans[2,3], Neil Mac Parthaláin[1],
Chris Cornelis[2,4], and Yvan Saeys[3,5]

[1] Department of Computer Science, Aberystwyth University,
Aberystwyth, Ceredigion, Wales, UK
{rkj,ncm}@aber.ac.uk
[2] Department of Applied Mathematics, Computer Science and Statistics,
Ghent University, Ghent, Belgium
{Sarah.Vluymans,Chris.Cornelis}@ugent.be
[3] VIB Inflammation Research Center, Zwijnaarde, Belgium
{Sarah.Vluymans, Yvan.Saeys}@irc.vib-ugent.be
[4] Department of Computer Science and AI CITIC-UGR,
University of Granada, Granada, Spain
Chris.Cornelis@decsai.ugr.es
[5] Department of Respiratory Medicine, Ghent University, Ghent, Belgium
Yvan.Saeys@ugent.be

Abstract. With the continued and relentless growth in dataset sizes in recent times, feature or attribute selection has become a necessary step in tackling the resultant intractability. Indeed, as the number of dimensions increases, the number of corresponding data instances required in order to generate accurate models increases exponentially. Fuzzy-rough set-based feature selection techniques offer great flexibility when dealing with real-valued and noisy data; however, most of the current approaches focus on the supervised domain where the data object labels are known. Very little work has been carried out using fuzzy-rough sets in the areas of unsupervised or semi-supervised learning. This paper proposes a novel approach for semi-supervised fuzzy-rough feature selection where the object labels in the data may only be partially present. The approach also has the appealing property that any generated subsets are also valid (super)reducts when the whole dataset is labelled. The experimental evaluation demonstrates that the proposed approach can generate stable and valid subsets even when up to 90 % of the data object labels are missing.

Keywords: Fuzzy-rough sets · Feature selection · Semi-supervised learning

1 Introduction

Supervised learning operates on labelled data and attempts to learn the underlying functional relationships in that data. It is the most common paradigm in machine learning and is concerned with the learning of classifiers which can accurately reflect the predictive regularities of the underlying model from the feature

© Springer International Publishing Switzerland 2015
Y. Yao et al. (Eds.): RSFDGrC 2015, LNAI 9437, pp. 185–195, 2015.
DOI: 10.1007/978-3-319-25783-9_17

values and decision class labels. For unsupervised learning, on the other hand, there are no decision class labels and the task is to construct or reconstruct class information from some inherent structure in the data. These techniques attempt to find groups in the data such that objects in the same group are similar to each other in some way and those in different groups are dissimilar. The notion of similarity is however subjective and as such, unsupervised learning approaches are forced to make assumptions about groupings as well as the number of groups into which data objects belong. The semi-supervised learning (SSL) paradigm lies between that of supervised learning and unsupervised learning. It is typically employed when some (but not all) of the data is labelled. The primary aim of SSL is to try to utilise both labelled and unlabelled data and it has therefore attracted much interest due to the abundance of unlabelled data which is available for many real-world problems. The main obstacle for traditional learning methods is that they cannot utilise unlabelled data for knowledge discovery. This has led to a growth in the number of SSL approaches.

Rough sets [7] and fuzzy-rough sets [3] have recently enjoyed much attention particularly for the task of feature selection (FS), due to their domain independence and, in the case of fuzzy-rough sets, the additional ability to handle real-valued data. The vast majority of work carried out in the areas of rough sets and fuzzy-rough sets has been focused on supervised learning approaches, i.e. where all of the class labels are known. There has been very little work in the area of unsupervised learning for fuzzy-rough sets and even less still for semi-supervised learning. The motivation for a fuzzy-rough based semi-supervised feature selection approach is based on the success of the supervised approaches [5,6] and the fact that the subsets produced by the proposed approaches are also shown to be valid (super)reducts for fully labelled data.

The remainder of the paper is structured as follows. In Sect. 2, the preliminaries for fuzzy-rough set theory and FS are covered. In Sect. 3, the proposed approach for semi-supervised fuzzy-rough set FS is presented. Section 4 details an experimental evaluation of the technique, where its performance is assessed through the random removal of class labels from a number of benchmark datasets and using non-parametric statistical analysis. Finally, in Sect. 5, the paper is concluded and some directions for future work are suggested.

2   Rough and Fuzzy-Rough Set Theory

In rough set analysis [7], data is represented as an *information system* $(X, \mathcal{A})$, where $X = \{x_1, \ldots, x_n\}$ and $\mathcal{A} = \{a_1, \ldots, a_m\}$ are finite, non-empty sets of objects and features, respectively. Each $a \in \mathcal{A}$ corresponds to a mapping from X to V_a, which is the value set of a over X. For every subset B of $\mathcal{A}$, the B-indiscernibility relation[1] R_B is defined as

$$R_B = \{(x, y) \in X^2 \mid (\forall a \in B)(a(x) = a(y))\}. \tag{1}$$

Clearly, R_B is an equivalence relation. Its equivalence classes $[x]_{R_B}$ can be used to approximate concepts, i.e., subsets of the universe X. Given $A \subseteq X$, its lower

---

[1] When $B = \{a\}$, i.e., B is a singleton, R_a is written rather than $R_{\{a\}}$.

and upper approximation w.r.t. R_B are respectively defined as

$$R_B{\downarrow}A = \{x \in X \mid [x]_{R_B} \subseteq A\} \tag{2}$$

$$R_B{\uparrow}A = \{x \in X \mid [x]_{R_B} \cap A \neq \emptyset\}. \tag{3}$$

An element x belongs to the lower approximation if it belongs to A and all other instances in its equivalence class do so as well. It belongs to the upper approximation when it does not necessarily belong to A itself, but there is at least one element in its equivalent class that does.

A *decision system* $(X, \mathcal{A} \cup \{d\})$ is a special kind of information system, used in the context of classification. Attribute d $(d \notin \mathcal{A})$ is called the decision feature. Its equivalence classes $[x]_{R_d}$ (or $[x]_d$) are called decision classes. Given $B \subseteq \mathcal{A}$, the B-positive region POS_B contains those objects from X for which the values of B allow to predict the decision class unequivocally. This can be modeled using the lower approximation (2), i.e.,

$$POS_B = \bigcup_{x \in X} R_B{\downarrow}[x]_{R_d}. \tag{4}$$

Indeed, if $x \in POS_B$, it means that whenever an object has the same values as x for the features in B, it will also belong to the same decision class as x. The predictive ability w.r.t. d of the features in B is measured by the following value (degree of dependency of d on B):

$$\gamma_B = \frac{|POS_B|}{|X|}. \tag{5}$$

$(X, \mathcal{A} \cup \{d\})$ is called *consistent* if $\gamma_\mathcal{A} = 1$. A subset B of $\mathcal{A}$ is called a *decision reduct* if it satisfies $POS_B = POS_\mathcal{A}$, i.e., B preserves the decision making power of $\mathcal{A}$, and moreover it cannot be further reduced, i.e., there exists no proper subset B' of B such that $POS_{B'} = POS_\mathcal{A}$. When the latter constraint is removed, i.e. B is not necessarily minimal, this is then termed a decision superreduct.

Fuzzy-rough set theory extends the above notions. A subset B of $\mathcal{A}$ can be defined using the fuzzy B-indiscernibility relation:

$$R_B(x, y) = \underbrace{\mathcal{T}(R_a(x, y))}_{a \in B}, \tag{6}$$

in which $\mathcal{T}$ represents a t-norm[2]. It can easily be seen that if only qualitative features (possibly originating from discretisation) are used, then the traditional concept of the B-indiscernibility relation is recovered. For the lower and upper approximation of a fuzzy set A in X by means of a fuzzy tolerance relation R, the definitions proposed in [8] are adopted and defined, for all x in X,

$$(R{\downarrow}A)(x) = \inf_{y \in X} \mathcal{I}(R(x, y), A(y)) \tag{7}$$

$$(R{\uparrow}A)(x) = \sup_{y \in X} \mathcal{T}(R(x, y), A(y)). \tag{8}$$

---

[2] A t-norm $\mathcal{T}$ is an increasing, commutative, associative $[0, 1]^2 \to [0, 1]$ mapping satisfying $\mathcal{T}(x, 1) = x$ for x in $[0, 1]$.

Here, $\mathcal{I}$ is an implicator[3]. When d is crisp, the fuzzy positive region can be defined as follows [2]:

$$POS_B(x) = (R_B \!\downarrow\! [x]_d)(x). \tag{9}$$

Using the Łukasiewicz implicator $((\forall a, b \in [0,1])(\mathcal{I}(a,b) = \min(1 - a + b, 1))$, it is found for each data instance $x \in X$:

$$
\begin{aligned}
POS_B(x) &= (R_B \downarrow [x]_d)(x) \\
&= \inf_{y \in X} \mathcal{I}(R_B(x,y), [x]_d(y)) \\
&= \min[\inf_{y \in [x]_d} \mathcal{I}(R_B(x,y), [x]_d(y)), \inf_{y \notin [x]_d} \mathcal{I}(R_B(x,y), [x]_d(y))] \\
&= \min[\inf_{y \in [x]_d} \mathcal{I}(R_B(x,y), 1), \inf_{y \notin [x]_d} \mathcal{I}(R_B(x,y), 0)] \\
&= \min[\inf_{y \in [x]_d} \min(1 - R_B(x,y) + 1, 1), \inf_{y \notin [x]_d} \min(1 - R_B(x,y) + 0, 1)] \\
&= \min[1, \inf_{y \notin [x]_d} (1 - R_B(x,y))] \\
&= \inf_{y \notin [x]_d} (1 - R_B(x,y)). \tag{10}
\end{aligned}
$$

From this, an increasing $[0,1]$-valued measure to gauge the degree of dependency of a subset of features on another subset of features can be defined. For FS, it is useful to phrase this in terms of the dependency of the decision feature on a subset of the conditional features:

$$\gamma_B = \frac{|POS_B|}{|X|} = \frac{\sum\limits_{x \in X} POS_B(x)}{|X|}. \tag{11}$$

This measure is used in the fuzzy-rough FS method (FRFS) of [5]. This technique is a hill-climbing algorithm to determine a (super)reduct $B \subseteq \mathcal{A}$. It initialises B as an empty set and iteratively adds the attribute $a \in \mathcal{A} \setminus B$ that leads to the largest increase in the value (11). The algorithm halts when $\gamma_B = \gamma_{\mathcal{A}}$.

3   Semi-Supervised Fuzzy-Rough Feature Selection

One of the primary motivating factors for semi-supervised approaches is the abundance of unlabelled data. Indeed, it is often expensive and time-consuming for domain experts to label data and this is where semi-supervised techniques can take advantage of small amounts of labelled data and (larger) amounts of unlabelled data in order to learn about the underlying predictive regularities.

Using the definitions described in Sect. 2, the original FRFS approach can be altered to handle both labelled and unlabelled data. Consider a feature subset $B \subseteq \mathcal{A}$, the membership degree to the positive region is computed as defined in

---

[3] An implicator $\mathcal{I}$ is a $[0,1]^2 \rightarrow [0,1]$ mapping that is decreasing in its first and increasing in its second argument, satisfying $\mathcal{I}(0,0) = \mathcal{I}(0,1) = \mathcal{I}(1,1) = 1$ and $\mathcal{I}(1,0) = 0$.

Sect. 2. For the semi-supervised paradigm, some data instances in X are missing. Each of these data instances is considered to belong to its own decision class, which contains only that data instance. This impacts the calculation of the positive region (10).

Theorem 1. *Let L be a set of labelled instances and U a set of unlabelled instances, with $L \cap U = \emptyset$ and $L \cup U = X$. The membership degree of an instance $x \in X$ to the positive region in the system can be defined as:*

$$POS_B^{ssl}(x) = \begin{cases} \inf_{y \neq x} \left(1 - R_B(x,y)\right) & \text{if } x \in U \\ \inf_{y \in (U \cup co([x]_d^L))} \left(1 - R_B(x,y)\right) & \text{if } x \in L, \end{cases}$$

where $[x]_d^L$ denotes the set of labelled instances with the same decision value as x and $co(\cdot)$ is the complement operator.

Proof. First consider $x \in U$. Its decision class contains only x, implying that (10) immediately simplifies to

$$POS_B^{ssl}(x) = \inf_{y \neq x} \left(1 - R_B(x,y)\right).$$

Now assume $x \in L$. The decision class of x consists of all labelled instances y with $d(x) = d(y)$. All unlabelled instances are not part of it, as they each belong to their own, distinct classes. In (10), the infimum is therefore taken over $U \cup co([x]_d^L)$, i.e.,

$$POS_B^{ssl}(x) = \inf_{y \in (U \cup co([x]_d^L))} \left(1 - R_B(x,y)\right).$$

$\square$

When $L \neq \emptyset$ and unlabelled data is considered, the dependency degree is defined as

$$\gamma_B^{ssl} = \frac{\sum\limits_{x \in X} POS_B^{ssl}(x)}{|X|}.$$

Theorem 2. *For every $B \subseteq A$, $\gamma_B^{ssl} \leq \gamma_B$.*

Proof. In general, given a function f and sets S and T with $S \subseteq T$,

$$\inf_{x \in T} f(x) \leq \inf_{x \in S} f(x)$$

always holds, as the infimum on the left-hand side is taken over a larger set.

Consider $x \in X$. In the semi-supervised paradigm, either $x \in U$ or $x \in L$ holds. In the first case, it can be found that:

$$POS_B^{ssl}(x) = \inf_{y \neq x} \left(1 - R_B(x,y)\right)$$

$$= \inf_{y \in (X \setminus \{x\})} \left(1 - R_B(x,y)\right)$$

$$\leq \inf_{y \in co([x]_d)} \left(1 - R_B(x,y)\right) \tag{12}$$

$$= \inf_{y \notin [x]_d} \left(1 - R_B(x,y)\right) = POS_B(x)$$

where $[x]_d$ is determined within the fully labelled system. Step (12) holds since $co([x]_d) \subseteq (X \setminus \{x\})$. Secondly, if $x \in L$, then

$$
\begin{aligned}
POS_B^{ssl}(x) &= \inf_{y \in (U \cup co([x]_d^L))} (1 - R_B(x, y)) \\
&\leq \inf_{y \in co([x]_d)} (1 - R_B(x, y)) \\
&= \inf_{y \notin [x]_d} (1 - R_B(x, y)) = POS_B(x)
\end{aligned} \tag{13}
$$

where $[x]_d$ is determined within the fully labelled data. Step (13) holds since $co([x]_d) \subseteq (U \cup co([x]_d^L))$. Indeed, every data instance $y \in co([x]_d)$ remains either labelled in the semi-supervised paradigm ($y \in co([x]_d^L)$) or has no label ($y \in U$). It can therefore be concluded that

$$(\forall x \in X)(POS_B^{ssl}(x) \leq POS_B(x)). \tag{14}$$

Finally, it can be found that:

$$
\gamma_B^{ssl} = \frac{\sum_{x \in X} POS_B^{ssl}(x)}{|X|} \overset{(14)}{\leq} \frac{\sum_{x \in X} POS_B(x)}{|X|} = \gamma_B.
$$

$\square$

This dependency measure can be used in the same way as the original definition as a basis for guiding the search toward optimal subsets. The modified greedy hillclimbing search method ssFRFS is presented in Algorithm 1. The algorithm iteratively adds the best feature (based on γ^{ssl}) to the current subset until the maximal value is attained for the dataset. At this point, the subset is a (super)reduct, because Theorem 2 shows that maximizing γ^{ssl} implies maximizing γ as well.

4   Experimental Evaluation

This section details the experiments conducted and the results obtained for the novel ssFRFS approach. The proposed method was applied to 12 benchmark datasets of different sizes. The results presented here relate to the performance in terms of quality of subsets obtained: classification accuracy and subset size. The effect of the random removal of 10–90 % of the class labels from the datasets on the proposed approach is also assessed.

4.1   Experimental Setup

The 12 different benchmark datasets are drawn from [4]. The class labels are randomly removed from 10 %, 30 %, 50 %, 70 % and 90 % of the labelled data for each dataset in order to simulate the semi-supervised problem domain.

Algorithm 1. The ssFRFS algorithm

---

input : $\mathcal{A}$, the set of all conditional features
output: B, the reduct

1 $B \leftarrow \{\}; \gamma_{best}^{ssl} = 0$
2 **while** $\gamma_{best}^{ssl} \neq \gamma_{\mathcal{A}}^{ssl}$ **do**
3 $\quad T \leftarrow B$
4 $\quad$ **foreach** $a \in (\mathcal{A} \setminus B)$ **do**
5 $\quad\quad$ **if** $\gamma_{B\cup\{a\}}^{ssl} > \gamma_T^{ssl}$ **then**
6 $\quad\quad\quad T \leftarrow B \cup \{a\}$
7 $\quad\quad\quad \gamma_{best}^{ssl} = \gamma_T^{ssl}$
8 $\quad B \leftarrow T$
9 **return** B

---

ssFRFS is compared with the original fuzzy-rough feature selection using a greedy hill-climbing search [5] and applied to the original, fully-labelled data. For all approaches, the Łukasiewicz t-norm $(\max(x + y - 1, 0))$ and the Łukasiewicz fuzzy implicator $(\min(1 - x + y, 1))$ are adopted to implement the fuzzy connectives in (7) and (8).

For the generation of classification results, two different classifier learners have been employed: the rule-based classifier JRip [1] and a nearest-neighbour classifier IBk (with $k = 3$). Five stratified randomisations of 10-fold cross-validation were employed in generating the classification results. It is important to point out here that FS is performed as part of the cross-validation and each fold results in a new selection of features. For the comparison of ssFRFS in terms of model performance, average classification accuracy is used. In addition, a non-parametric Wilcoxon test is performed [10] (significance level $\alpha = 0.05$) in order to evaluate whether any differences in performance are statistically significant. In particular, FRFS is compared with ssFRFS, in order to verify whether the presence of unlabelled data implies a reduction in performance.

4.2 Results

The results of the experimental evaluation are shown in Tables 1, 2 and 3. The results in Table 1 show that the approach achieves a reduction in the subset sizes for almost all cases, with a few small notable exceptions of the *australian* and *glass* datasets when the number of missing labels is greater than 70 %. The *segment* dataset also shows a similar pattern starting at 30 %. It should be noted however that the corresponding level of reduction using FRFS with no missing labels is very small - only a single feature, so this is expected to some degree.

Tables 2 and 3 detail the classification results for the JRip and IBk classifier learners respectively. Examining these results, it is clear that even for those cases with high levels of label removal, ssFRFS can still return comparable classification accuracies to the case where no labels have been removed. Clearly, it

Table 1. Average subset size in terms of original number of features.

Dataset	No. of features	No. of objects	FRFS	Missing class labels				
				10%	30%	50%	70%	90%
Australian	14	690	12.90	13.12	13.70	13.96	14.00[a]	14.00[a]
Cleveland	13	303	7.64	8.32	9.20	9.78	10.08	10.24
Ecoli	7	336	6.00	6.00	6.00	6.00	6.00	6.00
Glass	9	214	9.00	8.84	8.92	8.98	9.00[a]	9.00[a]
Heart	13	270	7.06	7.96	8.90	9.72	10.14	10.32
Ionosphere	34	351	7.04	12.42	16.18	17.32	18.48	18.60
Olitos	25	120	5.00	5.10	5.44	5.74	5.86	5.94
Segment	18	2310	16.00	17.98	18.00[a]	18.00[a]	18.00[a]	18.00[a]
Vehicle	18	946	8.42	9.74	11.04	11.46	11.70	11.84
Water2	38	390	6.00	6.46	6.92	7.00	7.04	7.10
Water3	38	390	6.00	6.26	6.94	7.00	7.02	7.00
Wine	13	178	8.42	5.80	6.34	6.56	6.74	6.80

[a] denotes that no reduction took place

Table 2. Classification accuracy using JRip

Dataset	Unred	FRFS	Missing class labels				
			10%	30%	50%	70%	90%
Australian	85.16	85.16	84.87	85.13	85.10	85.1	85.16
Cleveland	54.23	54.48	54.34	55.10	54.63	54.69	54.42
Ecoli	82.26	81.13	81.13	81.13	81.13	81.13	81.13
Glass	67.17	67.17	66.41	66.99	66.54	67.17	67.17
Heart	72.96	74.15	74.37	74.67	75.04	75.04	75.93
Ionosphere	87.57	87.74	85.13	86.52	85.39	86.96	86.61
Olitos	68.50	62.83	59.83	59.83	63.67	62.00	61.67
Segment	95.31	95.11	95.31	95.28	95.28	95.28	95.28
Vehicle	68.65	63.50	65.72	65.58	66.58	66.10	67.07
2-completed	82.15	83.28	82.97	83.08	82.56	81.54	82.67
3-completed	82.72	81.23	79.79	79.85	78.36	77.64	78.72
Wine	93.54	91.46	89.44	87.08	87.01	86.91	87.31

would be unrealistic to expect that this approach would be able to maintain the robustness of the fully supervised data (no missing labels). After all, there is much discriminative information encoded in class labels, but overall ssFRFS performs well. There is a small decrease in performance when compared to the unreduced data, but the statistical analysis in Tables 4 and 5 shows that no significant differences are observed when the results are compared with the fully supervised approach, regardless of the level of missing labels.

Table 3. Classification accuracy using IBk

Dataset	Unred	FRFS	Missing class labels				
			10 %	30 %	50 %	70 %	90 %
Australian	81.62	79.57	79.39	80.81	81.22	81.62	81.62
Cleveland	55.11	49.24	51.93	53.08	55.23	56.51	56.78
Ecoli	80.78	80.49	80.49	80.49	80.49	80.49	80.49
Glass	70.68	70.68	70.40	70.49	70.49	70.68	70.68
Heart	77.11	72.96	75.19	75.11	76.89	78.52	78.89
Ionosphere	85.30	86.43	84.52	85.22	84.43	85.57	84.52
Olitos	76.33	63.67	63.33	63.33	64.17	64.00	63.17
Segment	97.13	97.57	97.13	97.13	97.13	97.13	97.13
Vehicle	70.00	63.24	64.82	66.22	66.05	66.64	67.06
2-completed	85.03	83.59	81.23	80.31	79.90	79.95	78.77
3-completed	80.77	78.56	76.92	76.36	76.46	73.64	74.72
Wine	95.07	95.41	92.56	92.47	92.48	92.05	91.48

Table 4. Comparison of FRFS with ssFRFS for IBk

FRFS+IBk (0 %)	R^+	R^-	P-value
ssFRFS+IBk (10 %)	43.0	23.0	≥ 0.2
ssFRFS+IBk (30 %)	35.0	31.0	≥ 0.2
ssFRFS+IBk (50 %)	31.0	35.0	≥ 0.2
ssFRFS+IBk (70 %)	36.5	41.5	≥ 0.2
ssFRFS+IBk (90 %)	40.5	37.5	≥ 0.2

Table 5. Comparison of FRFS with ssFRFS for JRip

FRFS+JRip (0 %)	R^+	R^-	P-value
ssFRFS+JRip (10 %)	52.0	14.0	0.10156
ssFRFS+JRip (30 %)	44.0	22.0	≥ 0.2
ssFRFS+JRip (50 %)	38.0	28.0	≥ 0.2
ssFRFS+JRip (70 %)	41.5	24.5	≥ 0.2
ssFRFS+JRip (90 %)	42.5	23.5	≥ 0.2

5   Conclusion

This paper has presented a new approach to feature selection for semi-supervised data. One of the primary motivations behind the development of the approach was to propose a means of carrying out FS when some of the data class labels are missing. Indeed, there is no bound on the number of labelled data objects required and even if all labels are missing, the approach will still return a valid fuzzy-rough (super)reduct. This appealing property means that no consideration needs to be given to any tunable parameters in order to account for differing levels of supervision of the data. Indeed the experimental evaluation has demonstrated that the approach is particularly useful even with high levels of missing labels.

The ideas described in this paper offer some new directions for further development. In particular, other supervised fuzzy-rough approaches could be extended in a similar way to that proposed in this paper, including the fuzzy discernibility matrix-based approaches described in [5]. Also, the idea of object weighting could be incorporated [9], where objects with missing labels are considered to be less important in the calculations when compared to those that are labelled.

The basic semi-supervised concepts could be further generalised to the situation where both conditional and decision features may be missing from the data. This could form the basis for a number of approaches for performing feature selection on sparse data. For the work described here, the fuzzy connectives and similarities are the same as those used for the typical supervised approaches, but it would be interesting to investigate the influence of different choices in this regard.

Acknowledgment. Neil Mac Parthaláin would like to acknowledge the financial support for this research through NISCHR (*National Institute for Social Care and Health Research*) Wales, Grant reference: RFS-12-37. Sarah Vluymans is supported by the Special Research Fund (BOF) of Ghent University. Chris Cornelis was partially supported by the Spanish Ministry of Science and Technology under the project TIN2011-28488 and the Andalusian Research Plans P11-TIC-7765, P10-TIC-6858 and P12-TIC-2958.

References

1. Cohen, W.W.: Fast effective rule induction. In: Proceedings of the 12th International Conference on Machine Learning, pp. 115–123 (1995)
2. Cornelis, C., Jensen, R., Hurtado Martín, G., Ślęzak, D.: Attribute selection with fuzzy decision reducts. Inf. Sci. **180**(2), 209–224 (2010)
3. Dubois, D., Prade, H.: Putting rough sets and fuzzy sets together. In: Słowiński, R. (ed.) Intelligent Decision Support, pp. 203–232. Springer, Dordrecht (1992)
4. Frank, A., Asuncion, A.: UCI Machine Learning Repository. School of Information and Computer Science, University of California, Irvine, CA (2010). http://archive.ics.uci.edu/ml
5. Jensen, R., Shen, Q.: New approaches to fuzzy-rough feature selection. IEEE Trans. Fuzzy Syst. **17**(4), 824–838 (2009)

6. Jensen, R., Tuson, A., Shen, Q.: Finding rough and fuzzy-rough set reducts with SAT. Inf. Sci. **255**, 100–120 (2014)
7. Pawlak, Z.: Rough Sets: Theoretical Aspects of Reasoning About Data. Kluwer Academic Publishing, Dordrecht (1991)
8. Radzikowska, A.M., Kerre, E.E.: A comparative study of fuzzy rough sets. Fuzzy Sets Syst. **126**(2), 137–155 (2002)
9. Widz, S., Ślęzak, D.: Attribute Subset Quality Functions over a Universe of Weighted Objects. In: Kryszkiewicz, M., Cornelis, C., Ciucci, D., Medina-Moreno, J., Motoda, H., Raś, Z.W. (eds.) RSEISP 2014. LNCS, vol. 8537, pp. 99–110. Springer, Heidelberg (2014)
10. Wilcoxon, F.: Individual comparisons by ranking methods. Biometrics Bull. **1**(6), 80–83 (1945)

Modified Generalised Fuzzy Petri Nets for Rule-Based Systems

Zbigniew Suraj[✉]

Chair of Computer Science, Faculty of Mathematics and Natural Sciences,
University of Rzeszów, Prof. St. Pigonia Street 1, 35-310 Rzeszów, Poland
zbigniew.suraj@ur.edu.pl

Abstract. In [10], the generalised fuzzy Petri nets were proposed. This class extends the existing fuzzy Petri nets by introducing three input/output operators in the form of triangular norms, which are supposed to function as substitute for the classical *min*, *max*, and * (the algebraic product) operators. In this paper, we describe so called modified generalised fuzzy Petri nets. A functional interpretation of transitions based on inverted fuzzy implications is added to the model. The proposed net model is not only more comfortable in terms of knowledge representation, but most of all it is more effective in the modelling process of approximate reasoning as in the new class of fuzzy Petri nets the user has the chance to define both the input/output operators as well as transition operators. To demonstrate the power and the usefulness of this model, an application of the modified generalised fuzzy Petri nets in the domain of train traffic control is provided. The proposed approach can be used for knowledge representation and reasoning in decision support systems.

Keywords: Fuzzy Petri net · Fuzzy logic · Knowledge representation · Approximate reasoning · Decision support system

1   Introduction

Petri nets [7] have become an important computational paradigm to represent and analyse a broad class of systems [3,6]. They have been gaining a growing interest among people in Artificial Intelligence due to their adequacy to represent knowledge and reasoning processes graphically in decision support systems [2].

In 1988, Carl G. Looney proposed in [5] so called *fuzzy Petri nets* (*FP*-nets). In his model logical propositions can be associated with Petri nets allowing for logical reasoning about the modelled system. In this class of Petri net models not only crisp but also imprecise, vague and uncertain information is admissible and taken into account. Several authors have proposed different classes of fuzzy Petri nets [2]. These models are based on different approaches combining Petri nets and fuzzy sets introduced by Lotfi A. Zadeh in 1965 [11].

Recently, a new class of *FP*-nets called *generalised fuzzy Petri nets* (*GFP*-nets) [10] has been introduced. This net class extends the existing *FP*-nets by

© Springer International Publishing Switzerland 2015
Y. Yao et al. (Eds.): RSFDGrC 2015, LNAI 9437, pp. 196–206, 2015.
DOI: 10.1007/978-3-319-25783-9_18

introducing two operators: t-norms and s-norms, which are supposed to function as substitute for the *min* and *max* operators used in (classical) *FP*-nets. The latter ones generalize naturally AND and OR logical operators with the Boolean values 0 and 1 which are used in so called *binary Petri nets*.

The aim of this paper is to describe an extended class of *GFP*-nets called modified generalised fuzzy Petri nets (*mGFP*-nets). The main difference between these two classes concerns the definition of the operator binding function δ which connects similarly as in case of *GFP*-nets transitions with triples of operators (In, Out_1, Out_2). The meaning of the first and third operator is the same as in case of *GFP*-nets. However, in the *mGFP*-net model, the meaning of the second operator in the triple (called transition operator) is now significantly different. In the *GFP*-net model the operator Out_1 belongs to the class of t-norms, whereas in the *mGFP*-net model we assume that it belongs to the class of inverted fuzzy implications [9]. The latter one generalizes naturally existing interpretation of transition firing rule in *FP*-nets. Since there exist uncountably many fuzzy implications in the field of fuzzy logic, and the nature of the marking changes in given *mGFP*-nets variously depending on the implication function to be used, it is very difficult to select a suitable implication function for actual applications. But taking into account the methodologies proposed in the papers [8,9] one can reduce the efforts related to a selection of a suitable implication function. In the new net model the user has the chance to define both the input/output operators as well as transition operators depending on own needs. Due to the method described in [9] the user can optimize the truth values (markings) of all output places connected with a given transition t (more precisely, statements corresponding to those places). The choice of suitable operators for a given reasoning process and the speed of reasoning process are very important, especially in real-time decision support systems described by incomplete, imprecise and/or vague information. In order to demonstrate the power and the usefulness of this model, an application of the *mGFP*-nets in the domain of train traffic control is provided. The proposed approach can be used for knowledge representation and reasoning in decision support systems.

The structure of this paper is as follows. In Sect. 2 basic concepts and notation concerning triangular norms and fuzzy implications are provided. Section 3 provides a brief introduction to *GFP*-nets. In Sect. 4 *mGFP*-nets formalism is presented. In order to illustrate the proposed approach, a simple example coming from the domain of train traffic control in Sect. 5 is given. Section 6 includes the summary of our research and some remarks on directions for further investigations related to the presented methodology.

2   Preliminaries

Before giving a formal definition of new Petri net model we remind basic concepts and notation concerning triangular norms and fuzzy implications. As they are used for defining *mGFP*-nets in this paper, we recall their definitions.

2.1 Triangular Norms

Triangular norms (the t-norms and s-norms) generalize naturally classical logical operators AND and OR, respectively.

Let $[0, 1]$ be the interval of real numbers from 0 to 1 (0 and 1 are included).

A function $t : [0, 1] \times [0, 1] \rightarrow [0, 1]$ is said to be a t-norm if it satisfies, for all $a, b, c \in [0, 1]$, the following conditions: (1) it has 1 as the unit element, (2) it is monotone, commutative, and associative.

More relevant examples of t-norms are: the minimum $t(a, b) = min(a, b)$, the algebraic product $t(a, b) = a * b$, and the Łukasiewicz t-norm $t(a, b) = max(0, a + b - 1)$.

A function $s : [0, 1] \times [0, 1] \rightarrow [0, 1]$ is said to be an s-norm if it satisfies, for all $a, b, c \in [0, 1]$, the following conditions: (1) it has 0 as the unit element, (2) it is monotone, commutative, and associative.

More relevant examples of s-norms are: the maximum $s(a, b) = max(a, b)$, the probabilistic sum $s(a, b) = a + b - a * b$, and the Łukasiewicz s-norm $s(a, b) = min(a + b, 1)$.

The set of all triangular norms will be denoted by TN.

For more detailed information about the triangular norms the reader is referred to [4].

2.2 Fuzzy Implications

Fuzzy implications are one of the main operations in fuzzy logic.

A function $I : [0, 1] \times [0, 1] \rightarrow [0, 1]$ is said to be a *fuzzy implication* if it satisfies, for all $x, x_1, x_2, y, y_1, y_2 \in [0, 1]$, the following conditions: (1) $I(., y)$ is decreasing, (2) $I(x, .)$ is increasing, (3) $I(0, 0) = 1$, $I(1, 1) = 1$, and $I(1, 0) = 0$. The set of all fuzzy implications will be denoted by FI.

There exist uncountably many fuzzy implications. Table 1 contains a sample of basic fuzzy implications. For their extended list the reader is referred to ([1], page 4).

Table 1. A sample of basic fuzzy implications

Name	Year	Formula of basic fuzzy implication
Łukasiewicz	1923	$I_{LK}(x, y) = min(1, 1 - x + y)$
Kleene-Dienes	1938	$I_{KD}(x, y) = max(1 - x, y)$
Goguen	1969	$I_{GG}(x, y) = \begin{cases} 1 & if \ x \leq y \\ \frac{y}{x} & if \ x > y \end{cases}$

In the family of all fuzzy implications FI the partial order induced from $[0,1]$ interval exists. Pairs of incomparable fuzzy implications can generate new fuzzy implications by using $min(inf)$ and $max(sup)$ operations. As a result the structure of lattice is created ([1], page 186). This leads to the following question:

how to choose the correct functions among basic fuzzy implications and other generated as described above. In the paper [9], a method for choosing suitable fuzzy implications has been proposed. The method assumes that there is given a basic fuzzy implication $z = I(x, y)$, where x, y belong to [0,1]. x is the truth value of the antecedent and is known. z is the truth value of the implication and is also known. In order to determine the truth value of the consequent y it is needed to compute the inverse function $InvI(x, z)$. In other words, the inverse function $InvI(x, z)$ has to be determined. Of course, not every of basic fuzzy implications can be inverted. The function can be inverted only when it is injective. Moreover, this method allows to compare two fuzzy implications. If the truth value of the antecedent and the truth value of the implication are given, by means of inverse fuzzy implications we can easily optimize the truth value of the implication consequent. In other words, one can choose the fuzzy implication, which has the greatest truth value of the implication consequent or which has greater truth value than another implication. This method is used for defining a new model of fuzzy Petri nets proposed in this paper.

Table 2 lists inverse fuzzy implications and their domains for the fuzzy implications from Table 1.

Table 2. Inverted fuzzy implications

Formula of inverted fuzzy implication	Domain of inverted fuzzy implication
$InvI_{LK}(x, z) = z + x - 1$	$1 - x \leq z < 1, x \in (0, 1]$
$InvI_{KD}(x, z) = z$	$1 - x < z \leq 1, x \in (0, 1]$
$InvI_{GG}(x, z) = xz$	$0 \leq z < 1, x \in (0, 1]$

The resulting inverse functions can be compared with each other so that it is possible to order them. However, in general case, some of those functions are incomparable in the whole domain. Therefore, the domain is divided into separable areas where this property is fulfilled.

For more detailed information about this method the reader is referred to [9].

3   Generalised Fuzzy Petri Nets

In this paper, we assume that the reader is familiar with the basic notions of Petri nets [6].

Definition 1. *[10] A generalised fuzzy Petri net (GFP-net) is said to be a tuple* $N = (P, T, S, I, O, \alpha, \beta, \gamma, Op, \delta, M_0)$, *where:*

- $P = \{p_1, p_2, \ldots, p_n\}$ *is a finite set of places, $n > 0$;*
- $T = \{t_1, t_2, \ldots, t_m\}$ *is a finite set of transitions, $m > 0$;*
- $S = \{s_1, s_2, \ldots, s_n\}$ *is a finite set of statements;*

- *the sets P, T, S are pairwise disjoint and card(P) = card(S);*
- $I: T \rightarrow 2^P$ *is the input function;*
- $O: T \rightarrow 2^P$ *is the output function;*
- $\alpha: P \rightarrow S$ *is the statement binding function;*
- $\beta: T \rightarrow [0,1]$ *is the truth degree function;*
- $\gamma: T \rightarrow [0,1]$ *is the threshold function;*
- *Op is a finite set of t-norms and s-norms called the set of operators;*
- $\delta: T \rightarrow Op \times Op \times Op$ *is the operator binding function;*
- $M_0: P \rightarrow [0,1]$ *is the initial marking, and 2^P denotes a family of all subsets of the set P.*

As for the graphical interpretation, places are denoted by circles and transitions by rectangles. The function I describes the oriented arcs connecting places with transitions, and the function O describes the oriented arcs connecting transitions with places. If $I(t) = \{p\}$ then a place p is called an *input place* of a transition t, and if $O(t) = \{p'\}$, then a place p' is called an *output place* of t. The initial marking M_0 is an initial distribution of numbers in the places. It can be represented by a vector of dimension n of real numbers from $[0,1]$. For $p \in P$, $M_0(p)$ can be interpreted as a truth value of the statement s bound with a given place p by means of the statement binding function α. Pictorially, the tokens are represented by means of grey "dots" together with the suitable real numbers placed inside the circles corresponding to appropriate places. We assume that if $M_0(p) = 0$ then the token does not exist in the place p. The numbers $\beta(t)$ and $\gamma(t)$ are placed in a net picture under the transition t. The first number is usually interpreted as the truth degree of an implication corresponding to a given transition t. Whereas the role of the second one is to limit the possibility of transition firings, i.e., if the input operator In value for all values corresponding to input places of the transition t is less than a threshold value $\gamma(t)$ then this transition cannot be fired (activated). The operator binding function δ connects transitions with triples of operators (In, Out_1, Out_2). The first operator in the triple is called the input operator, and two remaining ones are the output operators. The input operator In concerns the way in which all input places are connected with a given transition t (more precisely, statements corresponding to those places). However, the output operators Out_1 and Out_2 concern the way in which the next marking is computed after firing the transition t. In the case of the input operator we assume that it can belong to one of two classes, i.e., t- or s-norm. In FP-nets the operators: minimum, algebraic product and maximum are usually used. As we know, the first two belong to the class of t-norms, whereas the third belongs to the class of s-norms.

A marking of *GFP*-net N is any function $M: P \rightarrow [0,1]$.

A transition $t \in T$ is *enabled* for marking M, if the value of input operator In for all input places of the transition t by M is positive and greater than, or equal to, the value of threshold function γ corresponding to the transition t.

If M is a marking of N enabling a transition t and M' is the next marking derived from M by firing t, then for each $p \in P$ a procedure for computing the next marking M' is as follows:

1. Numbers from all input places of the transition t are removed.
2. Numbers in all output places of t are modified in the following way:
   - at first the value of input operator In for all input places of t is computed,
   - next the value of output operator Out_1 for the value of In and for the value of truth degree function $\beta(t)$ is determined, and finally,
   - a value corresponding to $M'(p)$ for each $p \in O(p)$ is obtained as a result of output operator Out_2 for the value of Out_1 and the current marking $M(p)$.
3. Numbers in the remaining places of net N are not changed.

Example 1. Consider a *GFP*-net in Fig. 1(a). For the net we have: the set of places $P = \{p_1, p_2, p_3, p_4, p_5\}$, the set of transitions $T = \{t_1, t_2\}$, the input function I and the output function O in the form: $I(t_1) = \{p_1, p_2\}$, $I(t_2) = \{p_2, p_3\}$, $O(t_1) = \{p_4\}$, $O(t_2) = \{p_5\}$, the set of statements $S = \{s_1, s_2, s_3, s_4, s_5\}$, the statement binding function α: $\alpha(p_1) = s_1$, $\alpha(p_2) = s_2$, $\alpha(p_3) = s_3$, $\alpha(p_4) = s_4$, $\alpha(p_5) = s_5$, the truth degree function β: $\beta(t_1) = 0.7$, $\beta(t_2) = 0.8$, the threshold function γ: $\gamma(t_1) = 0.4$, $\gamma(t_2) = 0.3$, and the initial marking $M_0 = (0.6, 0.4, 0.7, 0, 0)$. Moreover, there are: the set of operators $Op = \{max, min, *\}$ and the operator binding function δ defined as follows: $\delta(t_1) = (max, *, max)$, $\delta(t_2) = (min, *, max)$. Transitions t_1 and t_2 are enabled by the initial marking M_0. Firing transition t_1 by M_0 transforms M_0 to the marking $M' = (0, 0, 0.7, 0.42, 0)$ (Fig. 1(b)). However, firing transition t_2 by M_0 results in the marking $M'' = (0.6, 0, 0, 0, 0.32)$.

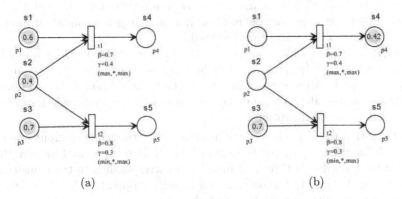

Fig. 1. A fuzzy Petri net with: (a) the initial marking, (b) the marking after firing t_1

For more detailed information about the *GFP*-nets the reader is referred to [10].

4   Modified Generalised Fuzzy Petri Nets

This section presents the main contribution to the paper. Using the notion of the inverted fuzzy implication we reformulated the definition of *GFP*-net as follows:

Definition 2. *A modified generalised fuzzy Petri net (mGFP-net) is said to be a tuple* $N' = (P, T, S, I, O, \alpha, \beta, \gamma, Op, \delta, M_0)$, *where:*

- *$P, T, S, I, O, \alpha, \beta, \gamma, M_0$ have the same meaning as in Definition 1;*
- *$Op = TN \cup FI$ is a union of triangular norms and inverted fuzzy implications called the set of operators;*
- *$\delta : T \to Op \times Op \times Op$ is the operator binding function.*

The operator binding function δ connects similarly as in case of GFP-nets transitions with triples of operators (In, Out_1, Out_2). The meaning of the first and third operator is the same as in case of GFP-nets. However, in this net model, the second operator in the triple (i.e., Out_1) is now called the transition operator and its meaning is significantly different. In the GFP-net model the operator Out_1 belongs to the class of t-norms, whereas in the $mGFP$-net we assume that it belongs to the class of inverted fuzzy implications. In the new net model the user has the chance to define both the input/output operators as well as transition operators depending on own needs. Moreover, taking into account the method described in [9] (see also Subsect. 2.2) the user can optimize the truth values (markings) of all output places connected with a given transition t (more precisely, statements corresponding to those places).

In principle, the dynamics of $mGFP$-nets is defined in an analogous way to the case of GFP-nets. Nevertheless, it is worth emphasising that the definition of a net marking is analogous to GFP-nets, although the definitions of a transition rule and a next marking are substantially modified.

Let N' be a $mGFP$-net. A marking of N' is a function $M : P \to [0, 1]$.

The $mGFP$-net dynamics defines how new markings are computed from the current marking when transitions are fired.

Let $N' = (P, T, S, I, O, \alpha, \beta, \gamma, Op, \delta, M_0)$ be a $mGFP$-net, M be a marking of N', $t \in T$, $I(t) = \{p_{i1}, p_{i2}, \ldots, p_{ik}\}$ be a set of input places for a transition t and $\beta(t) \in (0, 1]$. Moreover, let $\delta(t) = (In, Out_1, Out_2)$ and D be the domain of a transition operator Out_1, i.e., the domain of an inverted fuzzy implication corresponding to the transition t.

A transition $t \in T$ is *enabled* for marking M, if the value of input operator In for all input places of the transition t by M is positive and greater than, or equal to, the value of threshold function γ corresponding to the transition t, and the value belongs to the domain of a transition operator Out_1 of t, i.e., the following two conditions must be satisfied:

- $In(M(p_{i1}), M(p_{i2}), \ldots, M(p_{ik})) \geq \gamma(t) > 0$,
- $In(M(p_{i1}), M(p_{i2}), \ldots, M(p_{ik})) \in D$.

Only enabled transitions can be fired. Firing the enabled transition t in practice consists in either removing or not the tokens from its input places $I(t)$ and adding the tokens to all its output places $O(t)$ without any alteration of the tokens in other places. Each of the actions depends on the firing mode, which will be the subject of discussion further on. We define two operating modes of the $mGFP$-nets in the paper.

Mode 1. If M is a marking of N' enabling transition t and M' the marking derived from M by firing transition t, then for each $p \in P$:

$$M'(p) = \begin{cases} 0 \text{ if } p \in I(t), \\ Out_2(Out_1(In(M(p_{i1}), M(p_{i2}), \ldots, M(p_{ik})), \beta(t)), M(p)) \\ \text{if } p \in O(t), \\ M(p) \text{ otherwise.} \end{cases}$$

In this mode, a procedure for computing the marking M' is described analogously to the case of GFP-nets presented in Sect. 3.

Mode 2. If M is a marking of N' enabling transition t and M' the marking derived from M by firing transition t, then for each $p \in P$:

$$M'(p) = \begin{cases} Out_2(Out_1(In(M(p_{i1}), M(p_{i2}), \ldots, M(p_{ik})), \beta(t)), M(p)) \\ \text{if } p \in O(t), \\ M(p) \text{ otherwise.} \end{cases}$$

The main difference in the definition of a marking M' presented above (*Mode 1*) concerns input places of the fired transition t. In *Mode 1* tokens are removed from all input places of the fired transition t (the first condition of the definition, *Mode 1*), whereas in *Mode 2* all tokens are copied from input places of the fired transition t (the second condition of the definition, *Mode 2*).

Example 2. Consider again a GFP-net in Fig. 1(a). Let a $mGFP$-net be defined as follows. We assume that in the net: the sets P, T, S, and the functions I, O, α, β, γ, M_0 are described analogously to Example 1. However, the set of operators $Op = \{max, min\} \cup \{InvI_{KD}\}$ (for the formula of $InvI_{KD}$ and its domain, see Table 2) and the operator binding function δ defined as follows: $\delta(t_1) = (max, InvI_{KD}, max)$, $\delta(t_2) = (min, InvI_{KD}, max)$. Transitions t_1 and t_2 are enabled by the initial marking M_0. Firing transition t_1 by the marking M_0 according to *Mode 1* transforms M_0 to the marking $M' = (0, 0, 0.7, 0.7, 0)$, where t_2 is disabled. However, firing transition t_2 by the initial marking M_0 according to *Mode 2* results in the marking $M'' = (0.6, 0.4, 0.7, 0, 0.8)$, where t_1 and t_2 are enabled (*Mode 2*). We omit a graphical representation of this net with respect to a limited space of this paper.

5 Example

In order to illustrate our methodology, let us describe a simple example coming from the domain of train traffic control. We consider the following situation: a train B waits at a certain station for a train A to arrive in order to allow some passengers to change train A to train B. Now, a conflict arises when the train A is late. In this situation, the following alternatives can be taken into account:

– train B departs in time, and an additional train is employed for the train A passengers;

- train B departs in time. In this case, passengers disembarking train A have to wait for a later train;
- train B waits for train A to arrive. In this case, train B will depart with delay.

In order to describe the traffic conflict, we propose to consider the following rules:

- r_1: IF s_2 OR s_3 THEN s_6,
- r_2: IF s_1 AND s_4 AND s_6 THEN s_7,
- r_3: IF s_4 AND s_5 THEN s_8,

where:

- s_1: 'Train B is the last train in this direction today',
- s_2: 'The delay of train A is huge',
- s_3: 'There is an urgent need for the track of train B',
- s_4: 'Many passengers would like to change for train B',
- s_5: 'The delay of train A is short',
- s_6: 'Let train B depart according to schedule',
- s_7: 'Employ an additional train C (in the same direction as train B)', and
- s_8: 'Let train B wait for train A'.

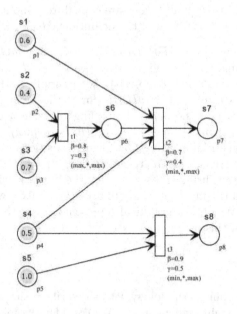

Fig. 2. An example of $mGFP$-net model of train traffic control

In the further considerations we accept the assumptions as in Fig. 2, i.e., (1) the logical operators OR, AND we interpret as max and min fuzzy operators, respectively; (2) the statements s_1, s_2, s_3, s_4 and s_5 we assign the fuzzy values 0.6, 0.4, 0.7, 0.5 and 1, respectively; (3) the truth-values of transitions t_1, t_2 and t_3 are equal to 0.8, 0.7, 0.9, respectively; (4) the threshold values for these three transitions are equal to 0.3, 0.4, 0.5, respectively; (5) each of transitions t_1, t_2 and t_3, firstly, we interpret as the Goguen implication (more precisely, the transition operators Out_{1_i}, i=1,2,3, corresponding to those transitions are interpreted as the inverted Goguen implication, see Table 2). Assessing the statements from s_1 up to s_5, we observe that the transitions t_1 and t_3 can be fired. If one chooses a sequence of transitions $t_1 t_3$ then they obtain the final value, corresponding to the statement s_7, equal to 0.35. In the other case (i.e., for the transition t_3 only), the final value, this time corresponding to the statement s_8, equals 0.45. The detailed computations performed in this case look as follows: (1) for transition t_1 - $InvI_{GG}(x, z) = x * z = max(s_2, s_3) * z = max(0.4, 0.7) * 0.8 = 0.56$; (2) for transition t_2 - $InvI_{GG}(x, z) = x*z = min(s_1, s_4, s_6)*0.7 = min(0.6, 0.5, 0.5)*0.7 = 0.35$; (3) for transition t_3 - $InvI_{GG}(x, z) = x*z = max(s_4, s_5)*0.9 = min(0.5, 1)*0.9 = 0.45$. Secondly, if we interpret these three transitions as the Łukasiewicz implications, and if we choose the same sequences of transitions as above we obtain the final values for the statements s_7 and s_8 equal to 0.2, 0.4, respectively. Thirdly, if we execute the similar simulation of approximate reasoning for three transitions considered above and, if we interpret the transitions as the Kleene-Dienes implications we obtain the final values for s_7 and s_8 equal to 0.7, 0.8, respectively.

This example shows clearly that different interpretations for the transitions may lead to quite different decision results. Choosing a suitable interpretation for fuzzy implications we may apply the approach presented in the paper [9] (see also Subsect. 2.2). The rest in this case certainly depends on the experience of the decision support system designer to a significant degree.

6   Concluding Remarks

The $mGFP$-net model combining the graphical power of Petri nets and capability of fuzzy logic to model rule-based expert knowledge in a decision support system have been described in the paper. $mGFP$-net formalism followed the basic notions from the fuzzy Petri net theory. Using a simple real-life example the suitability and the usefulness of the proposed approach for the design and implementation of decision support systems have been proved. The elaborated approach looks promising with regard to alike application problems that could be solved similarly. Our next problem to be solved is how to adapt the $mGFP$-nets for modelling backward reasoning in decision support systems.

Acknowledgments. This work was partially supported by the Center for Innovation and Transfer of Natural Sciences and Engineering Knowledge at the University of Rzeszów. The author is grateful to the anonymous referees for their helpful comments.

References

1. Baczyński, M., Jayaram, B.: Fuzzy Implications. Studies in Fuzziness and Soft Computing, vol. 231. Springer, Heidelberg (2008)
2. Cardoso, J., Camargo, H. (eds.): Fuzziness in Petri Nets. Studies in Fuzziness and Soft Computing, vol. 22. Springer, Heidelberg (1999)
3. Jensen, K., Rozenberg, G.: High-level Petri Nets, 1st edn. Springer, Heidelberg (1991)
4. Klement, E.P., Mesiar, R., Pap, E.: Triangular Norms. Kluwer Academic Publishers, Dordrecht (2000)
5. Looney, C.G.: Fuzzy petri nets for rule-based decision-making. IEEE Trans. Syst. Man Cybern. **18**(1), 178–183 (1988)
6. Peterson, J.L.: Petri Net Theory and The Modeling of Systems. Prentice-Hall Inc, Englewood Cliffs (1981)
7. Petri, C.A.: Kommunikation mit Automaten. Schriften des IIM Nr. 2, Institut für Instrumentelle Mathematik, Bonn (1962)
8. Suraj, Z., Lasek, A.: Toward optimization of approximate reasoning based on rule knowledge. In: Proceedings of the 2nd International Conference on Systems and Informatics (ICSAI), pp. 281–285 (2014)
9. Suraj, Z., Lasek, A., Lasek, P.: Inverted fuzzy implications in approximate reasoning. Fundamenta Informaticae (special issue after CSP'2014) (2014, accepted)
10. Suraj, Z.: A new class of fuzzy Petri nets for knowledge representation and reasoning. Fundam. Inform. **128**(1–2), 193–207 (2013)
11. Zadeh, L.A.: Fuzzy sets. Inf. Control **8**, 338–353 (1965)

Fuzzy Rough Decision Trees for Multi-label Classification

Xiaoxue Wang, Shuang An, Hong Shi, and Qinghua Hu[✉]

School of Computer Science and Technology, Tianjin University,
Tianjin 300072, China
huqinghua@tju.edu.cn

Abstract. *Multi-label classification exists widely in medical analysis or image annotation. Although there are some algorithms to train models for multi-label classification, few of them are able to extract comprehensible rules. In this paper, we propose a multi-label decision tree algorithm based on fuzzy rough sets, named ML-FRDT. This method can tackle with symbolic, continuous and fuzzy data. We conduct experiments on two multi-label datasets. And the experiment results show that ML-FRDT achieves good performance than some well-established multi-label classification algorithms.*

Keywords: Multi-label learning · Fuzzy rough sets · Decision tree

1 Introduction

In recent years, multi-label learning has been widely discussed in machine learning. In multi-label learning, each instance is associated with one or multiple labels. Simultaneously, multi-label learning widely exists in various real-world applications, such as text classification [1–3], image annotation [4–6], and automatic video understanding [7,8].

The current methods for multi-label classification can be divided into two main categories [9]: problem transformation and algorithm adaptation. The problem transformation methods transform multi-label problems into multiple single-label problems or multi-class problems, such as Binary Relevance (BR) and Label Powerset (LP). This kind of methods are simple and understandable, but they result in high computational cost. The algorithm adaptation methods adapt some algorithms to deal with multi-label problems, rather than transforming the data. For example, Multi-label Decision Tree (ML-DT) [10] adapts decision tree; Multi-label k-Nearest Neighbor (ML-kNN) [11] adapts lazy learning, and Multi-label Radial Basis Function (ML-RBF) [12] adapts radial basis function (RBF). However, ML-kNN and ML-RBF can learn functions for multi-label classification. It's worth noting that only ML-DT can extract comprehensible and interpretable rules from training samples, which is desirable in some multi-label learning tasks, such as medical diagnosis. ML-DT can deal with symbolic

© Springer International Publishing Switzerland 2015
Y. Yao et al. (Eds.): RSFDGrC 2015, LNAI 9437, pp. 207–217, 2015.
DOI: 10.1007/978-3-319-25783-9_19

and continuous attributes. However, ML-DT assumes labels are independent and overlooks correlation and co-occurrence between labels.

Fuzzy rough set theory [13] is a mathematical tool for describing uncertainty of fuzzy information. This theory has been used in building decision trees. Bhatt and Gopal [14] used fuzzy rough dependency as a measure of evaluating attributes in constructing fuzzy-rough classification trees (FRCT). An and Hu [15] proposed a fuzzy rough decision tree algorithm to address a classification problem described with symbolic, continuous or fuzzy attributes. This algorithm can directly build a classification tree without discretization or fuzzification of continuous attributes when splitting branches. While these methods are only applicable to two-class or multi-class classification, they are not suitable for multi-label classification.

In this paper, we propose a fuzzy rough decision tree algorithm to deal with multi-label classification tasks described by symbolic, continuous or fuzzy attributes. Splitting attributes are selected with fuzzy dependency for multi-label feature evaluation, which is proposed by Zhang, et al. in [16]. We redefine the information entropy to adapt multi-label tasks for evaluating the pureness of a node. This paper is organized as follows. In Sect. 2, we review ML-DT and multi-label feature evaluation based on fuzzy rough sets. In Sect. 3, we introduce the fuzzy rough decision tree for multi-label classification. In Sect. 4, numerical experiments are described. Finally, conclusions are given in Sect. 5.

2   Related Works

2.1   Multi-label Decision Trees

Multi-label Decision Tree algorithm (ML-DT) [10] adapts C4.5 [17] to deal with multi-label learning tasks. Based on multi-label entropy, ML-DT defines an information gain criterion to select the optimal attribute of each node.

Given a multi-label classification problem $< U, C, D >$, where U is a multi-label dataset with N samples, C is an attribute set and D is a label set with L labels. To realize ML-DT, the multi-label entropy is defined as

$$MLEnt(U) = -\sum_{j=1}^{L} ((p(d_j) \log p(d_j)) + (q(d_j) \log q(d_j))), \qquad (1)$$

where $p(d_j)$ is the proportion of samples in U with label d_j and $q(d_j)$ represents the proportion of samples without d_j.

We assume that C_k is the optimal attribute of the current node and C_k is a continuous attribute. The information gain obtained by dividing U along attribute C_k at value v is computed as

$$\text{gain}(U, C_k, v) = MLEnt(U) - \sum_{\rho \in \{-,+\}} \frac{|U^\rho|}{|U|} \cdot MLEnt(U^\rho), \qquad (2)$$

where

$$U^- = \{x_i | C_k(x_i) \leq v, 1 \leq i \leq N\},$$

$$U^+ = \{x_i | C_k(x_i) > v, 1 \leq i \leq N\}.$$

Similarly, information gain about discrete attributes also can be computed as Eq. (2).

Just the same as C4.5, ML-DT is constructed from top to down. The attribute and corresponding splitting value with the maximal information gain in Eq. (2) is chosen as the current node and the optimal split of the current node, respectively. Then the samples at the current node are divided into two groups: U^- and U^+. The above process is invoked recursively by treating U^- and U^+ as two subtrees until a certain stopping criterion is met.

Multi-label entropy shown in Eq. (1) represents the total of information needed to describe membership or non-membership of each class. However, it is unnecessary to consider non-membership of certain class in classification problems. Furthermore, it ignores the correlations and co-occurrence between labels. Consequently, we do not use the second term in Eq. (1) and we treat a combination of labels as a new class when computing the entropy in our algorithm.

2.2 Multi-label Feature Evaluation with Fuzzy Rough Sets

Based on fuzzy rough sets theory, Zhang et al. [16] proposed a novel dependency function for multi-label feature evaluation.

Given a multi-label learning task $< U, C, D >$, R_B is a fuzzy equivalence relation on U induced by feature subset B. The membership of sample x to the lower approximation of D with respect to B is defined as the following three forms:

$$\underline{R_B}^1 D(x) = \max_{\{d|d(x)=1, d \in D\}} (\underline{R_B}d(x)) \quad \text{or} \tag{3}$$

$$\underline{R_B}^2 D(x) = \min_{\{d|d(x)=1, d \in D\}} (\underline{R_B}d(x)) \quad \text{or} \tag{4}$$

$$\underline{R_B}^3 D(x) = \underset{\{d|d(x)=1, d \in D\}}{mean} (\underline{R_B}d(x)), \tag{5}$$

where $\underline{R_B}d(x)$ represents the membership degree of sample x certainly belonging to d in terms of B, and $d(x) = 1$ if sample x has label d. Then the dependency function of D in term of B is defined as

$$FD_B(D) = \frac{|\underline{R_B}D|}{|U|} = \frac{\sum_{x \in U} \underline{R_B}D(x)}{|U|}. \tag{6}$$

The fuzzy dependency reflects the consistency of classification when utilizing feature B to classify samples. And in multi-label classification tasks, the classification ability of features is related with multiple labels. So the final result should be induced by the result of each label, which is reflected in the redefined lower approximation and dependency function. It is remarkable that there are three different fusion methods for feature evaluation. The first method in Eq. (3) is an optimistic method, which selects the best feature from the loosest situation by max function. The second method in Eq. (4) is a pessimistic method, which selects the best feature from the most strict situation by the minimum function. And the third method in Eq. (5) selects the best feature from the average situation by mean function. Theorem 1. below indicates that the lower approximation and fuzzy dependency based on these three fusion methods is monotonous with features. It is understandable that as new features are included, the information about classification increases.

Theorem 1. [16] Given a multi-label learning task $< U, C, D >$, $B_1 \subseteq B_2 \subseteq C$, R_1 and R_2 are two fuzzy equivalance relations on U induced by features B_1 and B_2, respectively. Then we get

$$\underline{R_{B_1}}^1 D(x) \leq \underline{R_{B_2}}^1 D(x); \tag{7}$$

$$\underline{R_{B_1}}^2 D(x) \leq \underline{R_{B_2}}^2 D(x); \tag{8}$$

$$\underline{R_{B_1}}^3 D(x) \leq \underline{R_{B_2}}^3 D(x); \tag{9}$$

$$FD_{B_1}(D) \leq FD_{B_2}(D). \tag{10}$$

3   Multi-label Fuzzy Rough Decision Trees

Now, we introduce the multi-label fuzzy rough decision tree (ML-FRDT) algorithm. ML-FRDT is a binary tree. As splitting attribute selection, splitting point selection and stopping criteria are three key factors in building a decision tree. Next, we discuss our algorithm on these three aspects.

Split Attribute Selection: In ML-FRDT, we adopt the fuzzy dependency in Eq. (6) as the criterion to select a splitting attribute, and select the attribute with the maximal dependency as the optimal splitting attribute at the current node.

However, if an attribute has a great influence on multi-label classification, its fuzzy dependency may always be the maximum. Thus this attribute will be the splitting attribute at multiple nodes of the tree, which may cause overfitting. To avoid this, the fuzzy dependency of the selected split attribute will be multiplied by a penalty factor in the next round.

Splitting Point Generation: Suppose that C_0 is selected as the splitting attribute at the current node with dataset U_0 containing n samples. Possible division points of a continuous attribute are determined by the median between two adjacent attribute values, like C4.5. We assume that there are s possible division points of C_0. In order to obtain the optimal splitting point, we redefine the entropy as

$$H\_ML(U_0) = -\sum_{j=1}^{m} p(D_j) \log p(D_j), \tag{11}$$

where D_j is a subset of samples with the same subset of decision labels, and it is regarded as a new class, $p(D_j)$ represents the proportion of samples in current dataset U_0 belonging to class D_j. We set $p(D_j) \log p(D_j) = 0$ if $p(D_j) = 0$. Then the multi-label information gain obtained by dividing U_0 along attribute C_0 at splitting value v is defined as:

$$MLGain(U_0, C_0, v) = H\_ML(U_0) - \sum_{\rho \in \{-,+\}} \frac{|U_0^\rho|}{|U_0|} \cdot H\_ML(U_0^\rho), \quad (12)$$

where

$$U_0^- = \{x_i | C_0(x_i) \leq v, 1 \leq i \leq n\},$$

$$U_0^+ = \{x_i | C_0(x_i) > v, 1 \leq i \leq n\}.$$

Similarly, multi-label information gain of discrete attributes can be calculated as Eq. (12).

Stopping Criteria: After the optimal split being chosen, the dataset U_0 is divided into two groups. If all the samples at a node have the same label set, or the number of samples at a node is smaller than a given threshold λ, this node is regarded as a leaf node.

Given an unknown sample x, it traverses along the path of the learned multi-label decision tree until reaching a leaf node. And the predicted labels of x are determined by the training samples $\chi \subseteq U$ in the leaf node, which is shown as follows:

$$D' = \{d_j | p(d_j) > \tau, 1 \leq j < q\}. \quad (13)$$

Here, τ is a given threshold. $p(d_j)$ represents the proportion of samples in χ with label d_j. Namely, for one leaf node, if the majority of training samples dropping into it have label d_j, the test sample allocated within it will be marked as d_j.

The pseudo-code of ML-FRDT is formulated in Table 1. The method we proposed can tackle multi-label classification with symbolic, continuous and fuzzy features. As selecting feature has three different methods, our algorithm would have three different forms of growing multi-label trees. We mark optimistic method as MLFRDT-opt, negative method as MLFRDT-neg and average method as MLFRDT-ave. Consequently, we have three forms of our algorithm. In Sect. 4, we will show some experiments to test the performance of these forms.

4   Experiments

In this section, we describe the experiments we conducted with ML-FRDT. Firstly, an example on an artificial dataset is taken to demonstrate the process of constructing a fuzzy rough decision tree. Next, we introduce the datasets and evaluation metrics we used in the experiments. Then we discuss the effect of the parameters in our algorithm on the experimental results. In the end, a set of comparative experiments are designed to test the effectiveness of our methods.

Table 1. Pseudo-code of ML-FRDT

Fuzzy Rough Decision Tree for Multi-label classification (ML-FRDT)
Input: U - Training dataset,
$\quad\quad C$ - Attribute set,
$\quad\quad \varsigma$ - Stopping criterion.
Output: T - A multi-label fuzzy rough decision tree,
1. Initialize decision tree T and put the whole training set U at the root node;
2. **if** stopping criterion ς is met **then**
3. $\quad$ break and go to step 16;
4. **else**
5. $\quad$ **for** $i = 1, 2, \ldots, t$ **do**
6. $\quad\quad$ compute the fuzzy dependency in Eq.(6) of c_i;
7. $\quad\quad$ **if** c_i is the selected split attribute of the parent node
8. $\quad\quad\quad$ fuzzy dependency of c_i multiply by a penalty value;
9. $\quad\quad$ **end**
10. $\quad$ **end**
11. $\quad$ Choose c_0 which has the maximum fuzzy dependency as the split attribute of the current node N;
12. $\quad$ Choose v which maximize the value of Eq.(12) as the optimal split point;
13. $\quad$ Divide S (the sample set of the current node) into S^- and S^+ by attribute -value (c_0, v);
14. $\quad$ Set S^- and S^+ as the left subtree and right subtree of N respectively, and construct the left subtree and right subtree recursively
15. **end**
16. Return T.

4.1 A Toy Example

In order to demonstrate the details of building a multi-label fuzzy rough decision tree, we first give a toy example. We adopt MLFRDT-ave method to construct the tree. Table 2 shows some samples of a multi-label task, which totally contains 90 samples described by 4 attributes. C_1 and C_2 are symbolic variables, C_3 and C_4 are continuous variables.

First, we compute the fuzzy dependency of each attribute utilizing Eq. (6). And we get the dependency of each attribute is 0.322, 0.3944, 0 and 0.00043, respectively. So C_2 is selected as the attribute at the root node. C_2 has three discrete values on these samples, so there are two possible split point to choose. The value of Eq. (12) is 0.2647 and 0.3085 if the value of split point is 1 and 2, respectively. So we take 2 as the split point and the dataset is divided into two groups by the split point. These two groups are treated as two subtrees to grow. The above process is invoked recursively until the stopping criterion is met. Figure 1 shows the decision tree trained by ML-FRDT. In Fig. 1, $leaf_i(d, k)(i = 1, 2, 3, 4, 5, 6)$ are leaf nodes, where d is the label set and k is the sample number on the ith leaf node.

According to Fig. 1, six rules are extracted from the decision tree. These rules are shown as follows.

Table 2. Multi-label Dataset

Samples	C_1	C_2	C_3	C_4	Labels
x_1	1	1	0.5	0.12	d_1, d_2
x_2	1	3	0.6	0.24	d_2
x_3	0	3	0.1	0.08	d_2, d_3
x_4	1	2	0.1	0.08	d_1
x_5	0	2	0.7	0.12	d_3
x_6	1	3	0.8	0.33	d_2
x_7	0	3	0.6	0.28	d_2, d_3
x_8	1	1	0.5	0.12	d_1
x_9	0	3	0.3	0.08	d_2, d_3
x_{10}	1	3	0.9	0.24	d_2

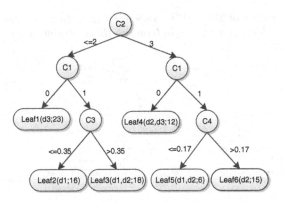

Fig. 1. A Multi-label Fuzzy Rough Decision Tree.

R_1: IF $C_2 \leq 2$ AND $C_1 = 0$ THEN $d_1 = 0, d_2 = 0, d_3 = 1$;
R_2: IF $C_2 \leq 2$, $C_1 = 1$ AND $C_3 \leq 0.35$ THEN $d_1 = 1, d_2 = 0, d_3 = 0$;
R_3: IF $C_2 \leq 2$, $C_1 = 1$ AND $C_3 > 0.35$ THEN $d_1 = 1, d_2 = 1, d_3 = 0$;
R_4: IF $C_2 = 3$ AND $C_1 = 0$ THEN $d_1 = 0, d_2 = 1, d_3 = 1$;
R_5: IF $C_2 = 3$, $C_1 = 1$ AND $C_4 \leq 0.17$ THEN $d_1 = 1, d_2 = 1, d_3 = 0$;
R_6: IF $C_2 = 3$, $C_1 = 1$ AND $C_4 > 0.17$ THEN $d_1 = 0, d_2 = 1, d_3 = 0$;

Referring to the samples in Table 2, it can be concluded that some of the rules are of higher accuracy. More importantly, the rules extracted from the decision tree make the multi-label classification more simple and understandable.

4.2 Numerical Experiments

In this paper, we conduct the experiments of different evaluation metrics [18] on two multi-label datasets [19, 20]. The details of these two datasets are shown in

Table 3. Details of Multi-label Datasets

Name	Instances	Attributes	Nominal attributes	Numeric attributes	Labels
Flags	194	19	9	10	7
Birds	645	260	2	258	19

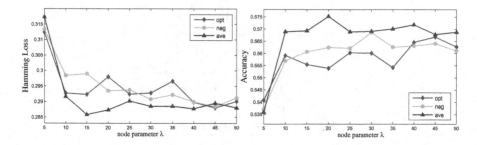

Fig. 2. The performance of ML-FRDT varies in terms of *Hamming Loss* and *Accuracy* as λ increases from 5 to 50 with step of 5 on the **flags** dataset, if τ is fixed at 0.5.

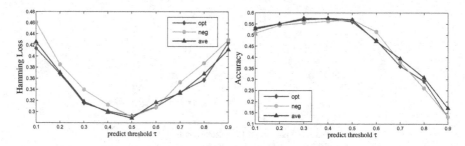

Fig. 3. The performance of ML-FRDT varies in terms of *Hamming Loss* and *Accuracy* as τ on the **flags** dataset, if λ is fixed at 30.

Table 3. And the evaluation metrics we used in this paper are Hamming Loss, F1score, Recall, Precision and Accuracy. We assume that the true label set of sample x_i is D, and the predicted label set is D'. Then the evaluation metrics are computed as: HammingLoss $= \frac{1}{N} \sum_{i=1}^{N} \frac{|D \oplus D'|}{L}$, F1score $= \frac{1}{N} \sum_{i=1}^{N} \frac{2|D \cap D'|}{|D| + |D'|}$, Recall $= \frac{1}{N} \sum_{i=1}^{N} \frac{|D \cap D'|}{|D|}$, Precision $= \frac{1}{N} \sum_{i=1}^{N} \frac{|D \cap D'|}{|D'|}$, Accuracy $= \frac{1}{N} \sum_{i=1}^{N} \frac{|D \cap D'|}{|D \cup D'|}$. And $\oplus$ stands for symmetric difference of two sets.

There are two main parameters in the ML-FRDT algorithm: λ and τ. Figures 2 and 3 show the influence of these parameters on the classification accuracy. In Figs. 2 and 3, MLFRDT-opt, MLFRDT-neg and MLFRDT-ave is marked as opt, neg and ave, respectively.

As described in Sect. 3, if the number of samples at a node reaches λ, it is regarded as a leaf node. And the predict threshold τ determines the labels of

Table 4. Performance comparison of different algorithms on *flags* dataset

Algorithms	HL(↓)	F1score(↑)	Recall(↑)	Precision(↑)	Accuracy(↑)
ML-DT	0.2897	0.6788	0.6865	0.7037	0.5525
ML-kNN	0.2943	0.6776	0.6936	0.6949	0.5515
ML-RBF	0.2720	0.7054	0.7321	0.7181	0.5838
MLFRDT-opt	0.2680	0.7246	0.8755	0.7621	0.5948
MLFRDT-neg	0.2852	0.7127	0.8345	**0.7650**	0.5810
MLFRDT-ave	**0.2551**	**0.7296**	**0.8766**	0.7358	**0.6192**

Table 5. Performance comparison of different algorithms on *birds* dataset

Algorithms	HL(↓)	F1score(↑)	Recall(↑)	Precision(↑)	Accuracy(↑)
ML-DT	0.0962	0.2664	0.2755	0.2700	0.2541
ML-kNN	0.0545	0.4691	0.4537	0.5092	0.4505
ML-RBF	**0.0515**	0.5177	0.5025	0.5629	0.4970
MLFRDT-opt	0.0566	**0.5665**	**0.6995**	**0.5786**	**0.5377**
MLFRDT-neg	0.0580	0.5556	0.6923	0.5470	0.5242
MLFRDT-ave	0.0585	0.5536	0.6804	0.5533	0.5178

leaf nodes. Both of them are of great influence with the classification performance of our algorithm. As shown in Fig. 2, MLFRDT-opt and MLFRDT-neg obtain optimal Hamming Loss when λ is 45, and MLFRDT-ave obtains optimal Hamming Loss when λ is 15. Consequently, different forms of MLFRDT obtain optimal Hamming Loss when λ takes different values, which is the same as the performances in terms of Accuracy. In this case, we should estimate appropriate value range of λ according to the dataset. As shown in Fig. 3, ML-FRDT produces better performance in terms of Hamming Loss when the value of τ is 0.5. However, ML-FRDT is better in terms of Accuracy when τ ranges from 0.3 to 0.5. Consequently, ML-FRDT needs take different values of τ with different evaluation metrics.

4.3 Experiment Results

ML-FRDT is compared with Multi-label Decision Tree (ML-DT) [10], Multi-label k-Nearest Neighbor (ML-kNN) [11] and Multi-label Radial Basis Function (ML-RBF) [12]. Ten-fold cross-validation is performed on each dataset. Tables 4 to 5 present the results on these datasets, where the best result on each metric is shown in bold face. For different evaluation metrics, ↓ indicates the smaller the better and ↑ indicates the larger the better.

As shown in Table 4, ML-FRDT performs best in comparing with others on flags dataset. On birds dataset, ML-FRDT is much better in terms of F1score, Recall, Precision and Accuracy, except Hamming Loss. Although ML-FRDT can't achieve the best performance on all evaluation metrics, ML-FRDT can extract comprehensible and interpretable rules from training samples, which is its main merit.

5   Conclusions

In this paper, we propose a multi-label algorithm ML-FRDT. The new algorithm can generate a decision tree from symbolic, continuous or fuzzy datasets. Experiments on two multi-label datasets show the effectiveness of the proposed algorithm. And the main merit of the proposed algorithm is that it can extract comprehensible and interpretable rules from training samples. Though this method considers the correlations and co-occurrence between labels, it will be time-consuming or cause out-fitting if there are many labels in the training dataset. This problem will be discussed in future work.

References

1. McCallum., A.: Multi-label text classification with a mixture model trained by EM. In: AAAI'99 Workshop on Text Learning, pp. 1–7 (1999)
2. Lewis, D.D., Yang, Y., Rose, T.G., Li, F.: Rcv1: A new benchmark collection for text categorization research. J. Mach. Learn. Res. **5**, 361–397 (2004)
3. Gao, S., Wu, W., Lee C.H., et al.: A MFoM learning approach to robust multiclass multi-label text categorization. In: Proceedings of the Twenty-first International Conference on Machine Learning (ICML 2004), pp. 42. ACM (2004)
4. Boutell, M.R., Luo, J., Shen, X., Brown, C.M.: Learning multi-label scene classification. Pattern Recogn. **37**, 1757–1771 (2004)
5. Wu, F., Han, Y., Tian, Q., Zhuang, Y.: Multi-label boosting for image annotation by structural grouping sparsity. In: Proceedings of the International Conference on Multimedia (MM 2010), pp. 15–24. ACM (2010)
6. Wang, Z., Hu, Y., Chia, L.T.: Multi-label learning by image-to-class distance for scene classification and image annotation. In: Proceedings of the ACM International Conference on Image and Video Retrieval (CIVR 2010), pp. 105–112. ACM (2010)
7. Snoek, C.G., Worring, M., Van Gemert, J.C., Geusebroek, J.M., Smeulders, A.W.: The challenge problem for automated detection of 101 semantic concepts in multimedia. In: Proceedings of the 14th Annual ACM International Conference on Multimedia, pp. 421–430. ACM (2006)
8. Sanden, C., Zhang, J.Z.: Enhancing multi-label music genre classification through ensemble techniques. In: Proceedings of the 34th International ACM SIGIR Conference on Research and Development in Information Retrieval (SIGIR 2011), pp. 705–714. ACM (2011)
9. Zhang, M., Zhou, Z.: A review on multi-label learning algorithms. IEEE Trans. Knowl. Data Eng. **26**, 1819–1837 (2013)
10. Clare, A.J., King, R.D.: Knowledge discovery in multi-label phenotype data. In: Siebes, A., De Raedt, L. (eds.) PKDD 2001. LNCS (LNAI), vol. 2168, pp. 42–53. Springer, Heidelberg (2001)
11. Zhang, M.L., Zhou, Z.H.: ML-KNN: A lazy learning approach to multi-label learning. Pattern Recogn. **40**, 2038–2048 (2007)

12. Zhang, M.L.: ML-RBF: RBF neural networks for multi-label learning. Neural Process. Lett. **29**, 61–74 (2009)
13. Dubois, D., Prade, H.: Rough fuzzy sets and fuzzy rough sets*. Int. J. Gen. Syst. **17**, 191–209 (1990)
14. Bhatt, R.B., Gopal, M.: FRCT: fuzzy-rough classification trees. Pattern Anal. Appl. **11**, 73–88 (2008)
15. An, S., Hu, Q.: Fuzzy rough decision trees. In: Yao, J.T., Yang, Y., Słowiński, R., Greco, S., Li, H., Mitra, S., Polkowski, L. (eds.) RSCTC 2012. LNCS, vol. 7413, pp. 397–404. Springer, Heidelberg (2012)
16. Zhang, L., Hu, Q., Duan, J., Wang, X.: Multi-label feature selection with fuzzy rough sets. In: Miao, D., Pedrycz, W., Slezak, D., Peters, G., Hu, Q., Wang, R. (eds.) RSKT 2014. LNCS, vol. 8818, pp. 121–128. Springer, Heidelberg (2014)
17. Quinlan, J.R.: C4. 5: Programs for Machine Learning. Elsevier, Burlington (1993)
18. Godbole, S., Sarawagi, S.: Discriminative methods for multi-labeled classification. In: Dai, H., Srikant, R., Zhang, C. (eds.) PAKDD 2004. LNCS (LNAI), vol. 3056, pp. 22–30. Springer, Heidelberg (2004)
19. Correa, G.E., Plastino, A., Freitas, A.A.: A genetic algorithm for optimizing the label ordering in multi-label classifier chains. In: IEEE 25th International Conference on Tools with Artificial Intelligence (ICTAI) 2013, pp. 469–476. IEEE (2013)
20. Briggs, F., Huang, Y., Raich, R., Eftaxias, K., Lei, Z., Cukierski, W., et al.: The 9th annual MLSP competition: new methods for acoustic classification of multiple simultaneous bird species in a noisy environment. In: IEEE International Workshop on Machine Learning for Signal Processing (MLSP) 2013, pp. 1–8. IEEE (2013)

Axiomatic Characterizations of Reflexive and $\mathcal{T}$-Transitive $\mathcal{I}$-Intuitionistic Fuzzy Rough Approximation Operators

Wei-Zhi Wu[1,2]($\boxtimes$), You-Hong Xu[1,2], Tong-Jun Li[1,2], and Xia Wang[1,2]

[1] School of Mathematics, Physics and Information Science,
Zhejiang Ocean University, Zhoushan 316022, Zhejiang, China
wuwz@zjou.edu.cn

[2] Key Laboratory of Oceanographic Big Data Mining & Application of Zhejiang
Province, Zhejiang Ocean University, Zhoushan 316022, Zhejiang, China
{xyh,litj}@zjou.edu.cn, bblylm@126.com

Abstract. Axiomatic characterizations of approximation operators are important in the study of rough set theory. In this paper, axiomatic characterizations of relation-based intuitionistic fuzzy rough approximation operators determined by an intuitionistic fuzzy implication operator $\mathcal{I}$ are investigated. We present a set of axioms of lower/upper $\mathcal{I}$-intuitionistic fuzzy set-theoretic operator which is necessary and sufficient for the existence of an intuitionistic fuzzy relation producing the same operator. We show that the lower and upper $\mathcal{I}$-intuitionistic fuzzy rough approximation operators generated by an arbitrary intuitionistic fuzzy relation can be described by single axioms. Moreover, the $\mathcal{I}$-intuitionistic fuzzy rough approximation operators generated by reflexive and $\mathcal{T}$-transitive intuitionistic fuzzy relations can also be characterized by single axioms.

Keywords: Approximation operators · Intuitionistic fuzzy implicators · Intuitionistic fuzzy rough sets · Intuitionistic fuzzy sets · Rough sets

1 Introduction

One of the main directions of research in rough set theory is naturally the generalization of the Pawlak rough set approximations. There are mainly two approaches for the development of rough set approximations, namely the constructive and axiomatic approaches. In the constructive approach, binary relations on the universe of discourse, partitions (or coverings) of the universe of discourse, neighborhood systems, and Boolean algebras are all primitive notions. The lower and upper approximation operators are constructed by means of these notions. Unlike the constructive approach, the axiomatic approach, also called algebraic approach, regards the abstract lower and upper approximation operators as primitive notions. A basic problem in the axiomatic approach is to seek for minimal axiom set of a lower/upper approximation operator, which is necessary and sufficient for the existence of an approximation space producing the

© Springer International Publishing Switzerland 2015
Y. Yao et al. (Eds.): RSFDGrC 2015, LNAI 9437, pp. 218–229, 2015.
DOI: 10.1007/978-3-319-25783-9_20

same abstract approximation operator. Many authors explored and developed the axiomatic approach in the study of rough sets in both of the crisp and fuzzy environments (see e.g. [6–10, 12, 14, 16–19]).

The concept of intuitionistic fuzzy (IF for short) sets, which was originated by Atanassov [1], has played a useful role in the research of uncertainty theories. Unlike a fuzzy set, which gives a degree of which element belongs to a set, an IF set gives both a membership degree and a non-membership degree. Thus, the combination of IF set theory and rough set theory is a new hybrid model to describe the uncertain information and has become an interesting research issue over the years (see e.g. [2, 3, 5, 11, 13, 20–23]). A general family of IF rough set models is determined by an IF implicator $\mathcal{I}$ in which the lower and upper IF rough approximation operators are dual with each other [15].

The main objective of this paper is to present the axiomatic study of relation-based IF rough sets determined by an IF implicator $\mathcal{I}$ in infinite universes of discourse. We will first review the constructive definitions and basic properties of lower and upper $\mathcal{I}$-IF rough approximation operators. We will then show that the $\mathcal{I}$-IF rough approximation operators generated by an arbitrary IF relation can be described by single axioms. Furthermore, we will present that the $\mathcal{I}$-IF rough approximation operators generated by reflexive and $\mathcal{T}$-transitive IF relations can also be characterized by single axioms.

2 Preliminaries

In this section we recall some basic notions and previous results which will be used in the later parts of this paper.

2.1 Intuitionistic Fuzzy Logical Operators

Throughout this paper, U will be a nonempty set called the universe of discourse. The class of all subsets of U will be denoted by $\mathcal{P}(U)$. In what follows, 1_y will denote the fuzzy singleton with value 1 at y and 0 elsewhere; 1_M will denote the characteristic function of a crisp set $M \in \mathcal{P}(U)$.

We first review a lattice on $[0, 1] \times [0, 1]$ originated by Cornelis *et al.* [4].

Definition 1. *Denote*

$$L^* = \{(x_1, x_2) \in [0, 1] \times [0, 1] \mid x_1 + x_2 \le 1\}. \tag{1}$$

A relation $\le_{L^*}$ on L^* is defined as follows: $\forall (x_1, x_2), (y_1, y_2) \in L^*$,

$$(x_1, x_2) \le_{L^*} (y_1, y_2) \iff x_1 \le y_1 \text{ and } x_2 \ge y_2. \tag{2}$$

The relation $\le_{L^*}$ is a partial ordering on L^* and the pair $(L^*, \le_{L^*})$ is a complete lattice with the smallest element $0_{L^*} = (0, 1)$ and the greatest element $1_{L^*} = (1, 0)$. The meet operator $\wedge$ and the join operator $\vee$ on $(L^*, \le_{L^*})$ linked to the ordering $\le_{L^*}$ are, respectively, defined as follows: $\forall (x_1, x_2), (y_1, y_2) \in L^*$,

$$(x_1, x_2) \wedge (y_1, y_2) = (\min(x_1, y_1), \max(x_2, y_2)),$$
$$(x_1, x_2) \vee (y_1, y_2) = (\max(x_1, y_1), \min(x_2, y_2)). \tag{3}$$

and

$$x = y \iff x \leq_{L^*} y \text{ and } y \leq_{L^*} x. \tag{4}$$

Definition 2. *An IF negator on L^* is a decreasing mapping $\mathcal{N} : L^* \to L^*$ satisfying $\mathcal{N}(0_{L^*}) = 1_{L^*}$ and $\mathcal{N}(1_{L^*}) = 0_{L^*}$. If $\mathcal{N}(\mathcal{N}(x)) = x$ for all $x \in L^*$, then $\mathcal{N}$ is called an involutive IF negator.*

The mapping $\mathcal{N}_s$, defined as $\mathcal{N}_s(x_1, x_2) = (x_2, x_1), \forall (x_1, x_2) \in L^*$, is called the *standard IF negator.*

Definition 3. *An IF t-norm on L^* is an increasing, commutative, associative mapping $\mathcal{T} : L^* \times L^* \to L^*$ satisfying $\mathcal{T}(1_{L^*}, x) = x$ for all $x \in L^*$.*

Definition 4. *An IF t-conorm on L^* is an increasing, commutative, associative mapping $\mathcal{S} : L^* \times L^* \to L^*$ satisfying $\mathcal{S}(0_{L^*}, x) = x$ for all $x \in L^*$.*

Obviously, the greatest IF t-norm (respectively, the smallest IF t-conorm) with respect to (w.r.t.) the ordering $\leq_{L^*}$ is min (respectively, max), defined by $\min(x, y) = x \wedge y$ (respectively, $\max(x, y) = x \vee y$) for all $x, y \in L^*$.

An IF t-norm $\mathcal{T}$ and an IF t-conorm $\mathcal{S}$ on L^* are said to be *dual* w.r.t. an IF negator $\mathcal{N}$ if

$$\mathcal{T}(\mathcal{N}(x), \mathcal{N}(y)) = \mathcal{N}(\mathcal{S}(x, y)), \forall x, y \in L^*,$$
$$\mathcal{S}(\mathcal{N}(x), \mathcal{N}(y)) = \mathcal{N}(\mathcal{T}(x, y)), \forall x, y \in L^*. \tag{5}$$

Definition 5. *A mapping $\mathcal{I} : L^* \times L^* \to L^*$ is referred to as an IF implicator on L^* if it is decreasing in its first component (left monotonicity), increasing in its second component (right monotonicity), and satisfies following conditions:*

$$\mathcal{I}(1_{L^*}, 0_{L^*}) = 0_{L^*}, \mathcal{I}(0_{L^*}, 0_{L^*}) = \mathcal{I}(0_{L^*}, 1_{L^*}) = \mathcal{I}(1_{L^*}, 1_{L^*}) = 1_{L^*}. \tag{6}$$

Remark 1. According to the left monotonicity of $\mathcal{I}$, it is easy to verify that $\mathcal{I}((\alpha, \beta), 1_{L^*}) = 1_{L^*}$ for all $(\alpha, \beta) \in L^*$, similarly, by the right monotonicity of $\mathcal{I}$, one can conclude that $\mathcal{I}(0_{L^*}, (\alpha, \beta)) = 1_{L^*}$ for all $(\alpha, \beta) \in L^*$.

Definition 6. *Let $\mathcal{S}$ be an IF t-conorm and $\mathcal{N}$ an IF negator on L^*. An IF $\mathcal{S}$-implicator generated by the $\mathcal{S}$ and $\mathcal{N}$ is a mapping $\mathcal{I}_{\mathcal{S}, \mathcal{N}}$ defined as follows:*

$$\mathcal{I}_{\mathcal{S}, \mathcal{N}}(x, y) = \mathcal{S}(\mathcal{N}(x), y), \quad \forall x, y \in L^*. \tag{7}$$

Definition 7. *Let $\mathcal{T}$ be an IF t-norm on L^*. An IF R-implicator generated by the $\mathcal{T}$ is a mapping $\mathcal{I}_{\mathcal{T}}$ defined as follows:*

$$\mathcal{I}_{\mathcal{T}}(x, y) = \sup\{\gamma \in L^* \mid \mathcal{T}(x, \gamma) \leq_{L^*} y\}, \quad \forall x, y \in L^*. \tag{8}$$

Definition 8. [4] *A mapping $\mathcal{I} : L^* \times L^* \to L^*$ is said to be satisfied, respectively, axiom*

(A1) *if $\mathcal{I}(\cdot, y)$ is decreasing in L^* for all $y \in L^*$ and $\mathcal{I}(x, \cdot)$ is increasing in L^* for all $x \in L^*$ (monotonicity laws);*

(A2) *if $\mathcal{I}(1_{L^*}, x) = x$ for all $x \in L^*$ (neutrality principle);*

(A3) *if $\mathcal{I}(x, y) = \mathcal{I}(\mathcal{N}_{\mathcal{I}}(y), \mathcal{N}_{\mathcal{I}}(x))$ for all $x, y \in L^*$ (contrapositivity);*

(A4) *if $\mathcal{I}(x, \mathcal{I}(y, z)) = \mathcal{I}(y, \mathcal{I}(x, z))$ for all $x, y, z \in L^*$ (exchangeability principle);*

(A5) *if $x \leq_{L^*} y \Longleftrightarrow \mathcal{I}(x, y) = 1_{L^*}$ for all $x, y \in L^*$ (confinement principle);*

(A6) *if $\mathcal{I} : L^* \times L^* \to L^*$ is a continuous mapping (continuity).*

Remark 2. In axiom (A3), the mapping $\mathcal{N}_{\mathcal{I}}$, defined by $\mathcal{N}_{\mathcal{I}}(x) = \mathcal{I}(x, 0_{L^*})$, $x \in L^*$, is an IF negator on L^*, and it is called the negator induced by $\mathcal{I}$. Moreover, it can easily be verified that if axioms (A2) and (A3) hold, then necessarily $\mathcal{N}_{\mathcal{I}}$ is involutive. An IF implicator $\mathcal{I}$ on L^* is called a *border IF implicator* (respectively, EP, CP) if it satisfies axiom (A2) (respectively, (A4), (A5)); an IF implicator $\mathcal{I}$ on L^* is called a *model IF implicator* if it satisfies axioms (A2), (A3) and (A4); an IF implicator on L^* is called a *Łukasiewicz IF implicator* if it satisfies axioms (A2)–(A6).

Theorem 1. [4] *An IF S-implicator $\mathcal{I}_{S,\mathcal{N}}$ on L^* defined by Definition 6 is a model IF implicator on the condition that $\mathcal{N}$ is an involutive IF negator; An IF R-implicator $\mathcal{I}_{\mathcal{T}}$ on L^* defined by Definition 7 is a border IF implicator.*

Given an IF negator $\mathcal{N}$ and a border IF implicator $\mathcal{I}$, we define an $\mathcal{N}$-dual operator of $\mathcal{I}$, $\theta_{\mathcal{I},\mathcal{N}} : L^* \times L^* \to L^*$, as follows:

$$\theta_{\mathcal{I},\mathcal{N}}(x, y) = \mathcal{N}(\mathcal{I}(\mathcal{N}(x), \mathcal{N}(y))), \quad x, y \in L^*. \tag{9}$$

According to Eq. (9), we can conclude following

Theorem 2. [15] *For a border IF implicator $\mathcal{I}$ and an IF negator $\mathcal{N}$, we have*

(1) $\theta_{\mathcal{I},\mathcal{N}}(1_{L^*}, 0_{L^*}) = \theta_{\mathcal{I},\mathcal{N}}(1_{L^*}, 1_{L^*}) = \theta_{\mathcal{I},\mathcal{N}}(0_{L^*}, 0_{L^*}) = 0_{L^*}$.

(2) $\theta_{\mathcal{I},\mathcal{N}}(0_{L^*}, 1_{L^*}) = 1_{L^*}$.

(3) *If $\mathcal{N}$ is involutive, then $\theta_{\mathcal{I},\mathcal{N}}(0_{L^*}, x) = x$ for all $x \in L^*$.*

(4) *$\theta_{\mathcal{I},\mathcal{N}}$ is left monotonic (respectively, right monotonic) whenever $\mathcal{I}$ is left monotonic (respectively, right monotonic).*

(5) *If $\mathcal{I}$ is left monotonic, then $\theta_{\mathcal{I},\mathcal{N}}(x, 0_{L^*}) = 0_{L^*}$ for all $x \in L^*$; and if $\mathcal{I}$ is right monotonic, then $\theta_{\mathcal{I},\mathcal{N}}(1_{L^*}, x) = 0_{L^*}$ for all $x \in L^*$.*

(6) *If $\mathcal{I}$ is an EP IF implicator, then $\theta_{\mathcal{I},\mathcal{N}}$ satisfies the exchange principle, i.e.*

$$\theta_{\mathcal{I},\mathcal{N}}(x, \theta_{\mathcal{I},\mathcal{N}}(y, z)) = \theta_{\mathcal{I},\mathcal{N}}(y, \theta_{\mathcal{I},\mathcal{N}}(x, z)), \quad \forall x, y, z \in L^*. \tag{10}$$

(7) *If $\mathcal{I}$ is a CP IF implicator, then $x \leq y$ iff $\theta_{\mathcal{I},\mathcal{N}}(x, y) = 0_{L^*}$.*

2.2   Intuitionistic Fuzzy Sets

Definition 9. [1] *Let a set U be fixed. An IF set A in U is an object having the form*

$$A = \{\langle x, \mu_A(x), \gamma_A(x)\rangle \mid x \in U\},$$

where $\mu_A : U \to [0,1]$ and $\gamma_A : U \to [0,1]$ satisfy $0 \le \mu_A(x) + \gamma_A(x) \le 1$ for all $x \in U$, and $\mu_A(x)$ and $\gamma_A(x)$ are, respectively, called the degree of membership and the degree of non-membership of the element $x \in U$ to A. The family of all IF subsets in U is denoted by $\mathcal{IF}(U)$. The complement of an IF set A is defined by $\sim A = \{\langle x, \gamma_A(x), \mu_A(x)\rangle \mid x \in U\}$.

For an IF set A and $x \in U$, we denote $A(x) = (\mu_A(x), \gamma_A(x))$, then it is clear that $A \in \mathcal{IF}(U)$ iff $A(x) \in L^*$ for all $x \in U$.

Some basic operations on $\mathcal{IF}(U)$ are introduced as follows [1]: for $A, B, A_i \in \mathcal{IF}(U)$, $i \in J$, J is an index set,

- $A \subseteq B$ iff $\mu_A(x) \le \mu_B(x)$ and $\gamma_A(x) \ge \gamma_B(x)$ for all $x \in U$,
- $A \supseteq B$ iff $B \subseteq A$,
- $A = B$ iff $A \subseteq B$ and $B \subseteq A$,
- $\bigcap_{i \in J} A_i = \{\langle x, \bigwedge_{i \in J} \mu_{A_i}(x), \bigvee_{i \in J} \gamma_{A_i}(x)\rangle \mid x \in U\}$,
- $\bigcup_{i \in J} A_i = \{\langle x, \bigvee_{i \in J} \mu_{A_i}(x), \bigwedge_{i \in J} \gamma_{A_i}(x)\rangle \mid x \in U\}$.

For $(\alpha, \beta) \in L^*$, $\widehat{(\alpha, \beta)}$ will be denoted by the constant IF set: $\widehat{(\alpha, \beta)}(x) = (\alpha, \beta)$, for all $x \in U$. For any $y \in U$ and $M \in \mathcal{P}(U)$, IF sets 1_y, $1_{U-\{y\}}$, and 1_M are, respectively, defined as follows: for $x \in U$,

$$\mu_{1_y}(x) = \begin{cases} 1, & \text{if } x = y, \\ 0, & \text{if } x \ne y. \end{cases} \qquad \gamma_{1_y}(x) = \begin{cases} 0, & \text{if } x = y, \\ 1, & \text{if } x \ne y. \end{cases}$$

$$\mu_{1_{U-\{y\}}}(x) = \begin{cases} 0, & \text{if } x = y, \\ 1, & \text{if } x \ne y. \end{cases} \qquad \gamma_{1_{U-\{y\}}}(x) = \begin{cases} 1, & \text{if } x = y, \\ 0, & \text{if } x \ne y. \end{cases}$$

$$\mu_{1_M}(x) = \begin{cases} 1, & \text{if } x \in M, \\ 0, & \text{if } x \notin M. \end{cases} \qquad \gamma_{1_M}(x) = \begin{cases} 0, & \text{if } x \in M, \\ 1, & \text{if } x \notin M. \end{cases}$$

The IF universe set is $U = 1_U = \widehat{(1,0)} = \widehat{1_{L^*}} = \{\langle x, 1, 0\rangle \mid x \in U\}$ and the IF empty set is $\emptyset = \widehat{(0,1)} = \widehat{0_{L^*}} = \{\langle x, 0, 1\rangle \mid x \in U\}$.

By using L^*, IF sets on U can be represented as follows: for $A, B, A_j \in \mathcal{IF}(U)(j \in J, J$ is an index set), $x, y \in U$, and $M \in \mathcal{P}(U)$

- $A(x) = (\mu_A(x), \gamma_A(x)) \in L^*$,
- $U(x) = (1,0) = 1_{L^*}$,
- $\emptyset(x) = (0,1) = 0_{L^*}$,
- $x = y \Longrightarrow 1_y(x) = 1_{L^*}$ and $1_{U-\{y\}}(x) = 0_{L^*}$,
- $x \ne y \Longrightarrow 1_y(x) = 0_{L^*}$ and $1_{U-\{y\}}(x) = 1_{L^*}$,
- $x \in M \Longrightarrow 1_M(x) = 1_{L^*}$,
- $x \notin M \Longrightarrow 1_M(x) = 0_{L^*}$,

- $A \subseteq B \Longleftrightarrow A(x) \leq_{L^*} B(x), \forall x \in U \Longleftrightarrow B(x) \geq_{L^*} A(x), \forall x \in U,$
- $(\bigcap\limits_{j \in J} A_j)(x) = \bigwedge\limits_{j \in J} A_j(x) = (\bigwedge\limits_{j \in J} \mu_{A_j}(x), \bigvee\limits_{j \in J} \gamma_{A_j}(x)) \in L^*,$
- $(\bigcup\limits_{j \in J} A_j)(x) = \bigvee\limits_{j \in J} A_j(x) = (\bigvee\limits_{j \in J} \mu_{A_j}(x), \bigwedge\limits_{j \in J} \gamma_{A_j}(x)) \in L^*.$

Given an IF implicator $\mathcal{I}$, an involutive IF negator $\mathcal{N}$, and two IF sets A, $B \in \mathcal{IF}(U)$, we define two IF sets $A \Rightarrow_\mathcal{I} B$ and $\theta_{\mathcal{I},\mathcal{N}}(A, B)$ as follows:

$$\begin{aligned}(A \Rightarrow_\mathcal{I} B)(x) &= \mathcal{I}(A(x), B(x)), \ x \in U, \\ \theta_{\mathcal{I},\mathcal{N}}(A, B)(x) &= \theta_{\mathcal{I},\mathcal{N}}(A(x), B(x)), \ x \in U.\end{aligned} \tag{11}$$

It can easily be verified that

$$\theta_{\mathcal{I},\mathcal{N}}(A, B) = {\sim}_\mathcal{N} (({\sim}_\mathcal{N} A) \Rightarrow_\mathcal{I} ({\sim}_\mathcal{N} B)), \tag{12}$$

where $({\sim}_\mathcal{N} A)(x) = \mathcal{N}(A(x)), \ x \in U.$

3   Constructive Definitions of $\mathcal{I}$-Intuitionistic Fuzzy Rough Approximation Operators

In this section, by employing an IF implicator $\mathcal{I}$ on L^*, we will review the constructive definitions of lower and upper approximations of IF sets w.r.t. an arbitrary IF approximation space and present some basic properties of $\mathcal{I}$-IF rough approximation operators.

Definition 10. *Let U and W be two nonempty universes of discourse. A subset $R \in \mathcal{IF}(U \times W)$ is referred to as an IF binary relation from U to W, namely, R is given by*

$$R = \{\langle (x, y), \mu_R(x, y), \gamma_R(x, y)\rangle \mid (x, y) \in U \times W\}, \tag{13}$$

where $\mu_R : U \times W \to [0, 1]$ and $\gamma_R : U \times W \to [0, 1]$ satisfy $0 \leq \mu_R(x, y) + \gamma_R(x, y) \leq 1$ for all $(x, y) \in U \times W$. If R is an IF relation from U to W, then the triple (U, W, R) is called a generalized IF approximation space. If $R \in \mathcal{IF}(U \times U)$ and $R(x, x) = 1_{L^*}$ for all $x \in U$, then R is referred to as a reflexive IF relation on U; if $\bigvee\limits_{y \in U} \mathcal{T}(R(x, y), R(y, z)) \leq_{L^*} R(x, z)$ for all $(x, z) \in U \times U$ (where $\mathcal{T}$ is an IF t-norm on L^*), then R is called a $\mathcal{T}$-transitive IF relation on U.

Definition 11. [15] *Let (U, W, R) be a generalized IF approximation space, $\mathcal{I}$ an IF implicator and $\mathcal{N}$ an IF negator on L^*, for $A \in \mathcal{IF}(W)$, the $\mathcal{I}$-lower and $\mathcal{I}$-upper approximations of A w.r.t. the approximation space (U, W, R), denoted as $\underline{R}_\mathcal{I}(A)$ and $\overline{R}_\mathcal{I}(A)$ respectively, are IF sets of U and are defined as follows:*

$$\begin{aligned}\underline{R}_\mathcal{I}(A)(x) &= \bigwedge\limits_{y \in W} \mathcal{I}(R(x, y), A(y)), \ x \in U, \\ \overline{R}_\mathcal{I}(A)(x) &= \bigvee\limits_{y \in W} \theta_{\mathcal{I},\mathcal{N}}(\mathcal{N}(R(x, y)), A(y)), \ x \in U.\end{aligned} \tag{14}$$

The operators $\underline{R}_\mathcal{I}, \overline{R}_\mathcal{I} : \mathcal{IF}(W) \to \mathcal{IF}(U)$ are, respectively, referred to as lower and upper $\mathcal{I}$-IF rough approximation operators of (U, W, R), and the pair $(\underline{R}_\mathcal{I}(A), \overline{R}_\mathcal{I}(A))$ is called the $\mathcal{I}$-IF rough set of A w.r.t. (U, W, R).

The following theorem shows that the lower and upper $\mathcal{I}$-IF rough approximation operators determined by an IF implicator $\mathcal{I}$ and an involutive IF negator $\mathcal{N}$ are dual with each other.

Theorem 3. [15] *Let (U, W, R) be a generalized IF approximation space, if $\mathcal{I}$ is an IF implicator and $\mathcal{N}$ an involutive IF negator on L^*, then*

$$
\begin{aligned}
&\text{(DIFL) } \underline{R_{\mathcal{I}}}(A) =\sim_{\mathcal{N}} \overline{R_{\mathcal{I}}}(\sim_{\mathcal{N}} A), \forall A \in \mathcal{IF}(W), \\
&\text{(DIFU) } \overline{R_{\mathcal{I}}}(A) =\sim_{\mathcal{N}} \underline{R_{\mathcal{I}}}(\sim_{\mathcal{N}} A), \forall A \in \mathcal{IF}(W).
\end{aligned}
\tag{15}
$$

The next theorem gives some basic properties of $\mathcal{I}$-IF rough approximation operators.

Theorem 4. [15] *Let (U, W, R) be an IF approximation space. IF $\mathcal{I}$ is a continuous, hybrid monotonic and border IF implicator on L^* and $\mathcal{N}$ an involutive IF negator on L^*, then the lower and upper $\mathcal{I}$-IF rough approximation operators satisfy the following properties: For all $A \in \mathcal{IF}(W), A_j \in \mathcal{IF}(W)(\forall j \in J, J$ is an index set), and all $(\alpha, \beta) \in L^*$,*

(IFL1) $\underline{R_{\mathcal{I}}}(\widehat{(\alpha, \beta)} \Rightarrow_{\mathcal{I}} A) = \widehat{(\alpha, \beta)} \Rightarrow_{\mathcal{I}} \underline{R_{\mathcal{I}}}(A)$, *provided that $\mathcal{I}$ is an EP IF implicator.*

(IFU1) $\overline{R_{\mathcal{I}}}(\theta_{\mathcal{I}, \mathcal{N}}(\widehat{(\alpha, \beta)}, A)) = \theta_{\mathcal{I}, \mathcal{N}}(\widehat{(\alpha, \beta)}, \overline{R_{\mathcal{I}}}(A))$, *provided that $\mathcal{I}$ is an EP IF implicator.*

(IFL2) $\underline{R_{\mathcal{I}}}(\bigcap\limits_{j \in J} A_j) = \bigcap\limits_{j \in J} \underline{R_{\mathcal{I}}}(A_j)$.

(IFU2) $\overline{R_{\mathcal{I}}}(\bigcup\limits_{j \in J} A_j) = \bigcup\limits_{j \in J} \overline{R_{\mathcal{I}}}(A_j)$.

Theorems 5 and 6 below show that a reflexive (and, respectively, $\mathcal{T}$-transitive) IF relation can be characterized by properties of $\mathcal{I}$-IF rough approximation operators.

Theorem 5. [15] *Let (U, R) be an IF approximation space (i.e. R is an IF relation on U), $\mathcal{I}$ a border and CP IF implicator, and $\mathcal{N}$ an involutive IF negator. Then*

$$
\begin{aligned}
R \text{ is reflexive} &\Longleftrightarrow \text{(IFLR) } \underline{R_{\mathcal{I}}}(A) \subseteq A, \quad \forall A \in \mathcal{IF}(U), \\
&\Longleftrightarrow \text{(IFUR) } A \subseteq \overline{R_{\mathcal{I}}}(A), \quad \forall A \in \mathcal{IF}(U).
\end{aligned}
$$

Theorem 6. [15] *Let (U, R) be an IF approximation space, $\mathcal{N}$ an involutive IF negator, $\mathcal{I}$ an IF implicator and $\mathcal{T}$ an IF t-norm satisfying*

$$
\mathcal{I}(a, \mathcal{I}(b, c)) = \mathcal{I}(\mathcal{T}(a, b), c), \quad \forall a, b, c \in L^*.
\tag{16}
$$

(1) *If R is $\mathcal{T}$-transitive, then*

$$
\begin{aligned}
&\text{(IFLT) } \underline{R_{\mathcal{I}}}(A) \subseteq \underline{R_{\mathcal{I}}}(\underline{R_{\mathcal{I}}}(A)), \quad \forall A \in \mathcal{IF}(U). \\
&\text{(IFUT) } \overline{R_{\mathcal{I}}}(\overline{R_{\mathcal{I}}}(A)) \subseteq \overline{R_{\mathcal{I}}}(A), \quad \forall A \in \mathcal{IF}(U).
\end{aligned}
$$

(2) *If $\mathcal{I}$ is a CP and border IF implicator, then*

$$
\text{(IFLT)} \Longleftrightarrow \text{(IFUT)} \Longrightarrow R \text{ is } \mathcal{T}\text{-transitive.}
\tag{17}
$$

4 Axioms of $\mathcal{I}$-Intuitionistic Fuzzy Rough Approximation Operators

In this section, we will investigate the use of abstract IF set-theoretic operators to characterize $\mathcal{I}$-IF rough approximation operators.

Definition 12. *Let* $L, H : \mathcal{IF}(W) \rightarrow \mathcal{IF}(U)$ *be two operators.* L *and* H *are called dual operators w.r.t. an involutive IF negator* $\mathcal{N}$ *if they satisfy* (IFDL) *or* (IFDU):

(IFDL) $L(A) = \sim_{\mathcal{N}} H(\sim_{\mathcal{N}} A), \quad \forall A \in \mathcal{IF}(W)$.
(IFDU) $H(A) = \sim_{\mathcal{N}} L(\sim_{\mathcal{N}} A), \quad \forall A \in \mathcal{IF}(W)$.

If the lower and upper IF approximation operators are dual with each other, then we only need to define one of the two operators, and the other can be derived from the dual property. Furthermore, properties of one operator can be obtained from the corresponding properties of its dual operator. Similar to the lower IF approximation operators in [23] and by using the dual approach, we can obtain following theorem which shows that a set of axioms of lower/upper $\mathcal{I}$-IF set-theoretic operator is necessary and sufficient for the existence of an IF relation producing the same operator.

Theorem 7. *Let* $L, H : \mathcal{IF}(W) \rightarrow \mathcal{IF}(U)$ *be a dual pair of IF operators, if* $\mathcal{I}$ *is an EP IF implicator on* L^*, *then there exists an IF relation* R *from* U *to* W *such that*

$$L(A) = \underline{R_{\mathcal{I}}}(A), H(A) = \overline{R_{\mathcal{I}}}(A), \quad \forall A \in \mathcal{IF}(W) \tag{18}$$

iff L *satisfies properties* (AIFL1) *and* (AIFL2):

(AIFL1) $L(\widehat{(\alpha,\beta)} \Rightarrow_{\mathcal{I}} A) = \widehat{(\alpha,\beta)} \Rightarrow_{\mathcal{I}} L(A), \forall A \in \mathcal{IF}(W), \forall (\alpha,\beta) \in L^*$,

(AIFL2) $L(\bigcap_{j \in J} A_j) = \bigcap_{j \in J} L(A_j), \forall A_j \in \mathcal{IF}(W)(j \in J, J$ *is any index set*).

or, equivalently, H *obeys properties* (AIFU1) *and* (AIFU2):

(AIFU1) $H(\theta_{\mathcal{I},\mathcal{N}}(\widehat{(\alpha,\beta)}, A)) = \theta_{\mathcal{I},\mathcal{N}}(\widehat{(\alpha,\beta)}, H(A)), \forall A \in \mathcal{IF}(W), \forall (\alpha,\beta) \in L^*$.

(AIFU2) $H(\bigcup_{j \in J} A_j) = \bigcup_{j \in J} H(A_j), \forall A_j \in \mathcal{IF}(W)(j \in J, J$ *is any index set*).

Theorem 7 shows that $\{$(AIFL1), (AIFL2)$\}$ (respectively, $\{$(AIFU1), (AIFU2)$\}$) is the basic set of axioms to characterize the generalized lower (respectively, upper) $\mathcal{I}$-IF rough approximation operator. The next theorem indicates that $\{$(AIFL1), (AIFL2)$\}$ (respectively, $\{$(AIFU1), (AIFU2)$\}$) can be replaced by a single axiom.

Theorem 8. *Let* $L, H : \mathcal{IF}(W) \rightarrow \mathcal{IF}(U)$ *be a dual pair of IF operators, if* $\mathcal{I}$ *is an EP IF implicator on* L^*, *then there exists an IF relation* R *from* U *to* W *such that Eq.* (18) *holds iff* L *satisfies following* (AIFL):

(AIFL) $\forall A_j \in \mathcal{IF}(W), \forall (\alpha_j, \beta_j) \in L^*, j \in J,$ *where* J *is any index set,*

$$L(\bigcap_{j \in J}((\widehat{\alpha_j, \beta_j}) \Rightarrow_{\mathcal{I}} A_j)) = \bigcap_{j \in J}((\widehat{\alpha_j, \beta_j}) \Rightarrow_{\mathcal{I}} L(A_j)). \tag{19}$$

or, equivalently, H obeys following (AIFU):

(AIFU) $\forall A_j \in \mathcal{IF}(W), \forall (\alpha_j, \beta_j) \in L^*, j \in J$, *where J is any index set,*

$$H\left(\bigcup_{j \in J} \theta_{\mathcal{I},\mathcal{N}}((\widehat{\alpha_j, \beta_j}), A_j)\right) = \bigcup_{j \in J} \theta_{\mathcal{I},\mathcal{N}}((\widehat{\alpha_j, \beta_j}), H(A_j)). \tag{20}$$

By Theorems 4, 5, and 7, Theorem 16 in [23] and the dual approach, we can conclude following

Theorem 9. *Let* $L, H : \mathcal{IF}(U) \to \mathcal{IF}(U)$ *be a dual pair of IF operators, if* $\mathcal{I}$ *is a border, EP and CP IF implicator on* L^*, *then there exists a reflexive IF relation R on U such that*

$$L(A) = \underline{R}_{\mathcal{I}}(A), H(A) = \overline{R}_{\mathcal{I}}(A), \quad \forall A \in \mathcal{IF}(U) \tag{21}$$

iff L satisfies properties {(AIFL1), (AIFL2), (AIFLR)}, *or, equivalently, H obeys properties* {(AIFU1), (AIFU2), (AIFUR)}, *where*

(AIFLR) $L(A) \subseteq A, \quad \forall A \in \mathcal{IF}(U).$

(AIFUR) $A \subseteq H(A), \quad \forall A \in \mathcal{IF}(U).$

Remark 3. Theorem 9 shows that {(AIFL1), (AIFL2), (AIFLR)} (respectively, {(AIFU1), (AIFU2), (AIFUR)}) is the basic set of axioms to characterize the lower (respectively, upper) $\mathcal{I}$-IF rough approximation operator generated by a reflexive IF relation. The next theorem indicates that the set of axioms {(AIFL1), (AIFL2), (AIFLR)} (respectively, {(AIFU1), (AIFU2), (AIFUR)}) can be replaced by a single axiom.

Theorem 10. *Let* $L, H : \mathcal{IF}(U) \to \mathcal{IF}(U)$ *be a dual pair of IF operators, if* $\mathcal{I}$ *is a border, EP and CP IF implicator on* L^*, *then there exists a reflexive IF relation R on U such that Eq.* (21) *holds iff L satisfies the following* (ALR):

(ALR) $\forall A_j \in \mathcal{IF}(U), \forall (\alpha_j, \beta_j) \in L^*, j \in J$, *where J is any index set,*

$$L\left(\bigcap_{j \in J}((\widehat{\alpha_j, \beta_j}) \Rightarrow_{\mathcal{I}} A_j)\right) = \left(\bigcap_{j \in J}((\widehat{\alpha_j, \beta_j}) \Rightarrow_{\mathcal{I}} A_j)\right) \cap \left(\bigcap_{j \in J}((\widehat{\alpha_j, \beta_j}) \Rightarrow_{\mathcal{I}} L(A_j))\right).$$

or, equivalently, H obeys the following (AUR):

(AUR) $\forall A_j \in \mathcal{IF}(U), \forall (\alpha_j, \beta_j) \in L^*, j \in J$, *where J is any index set,*

$$H\left(\bigcup_{j \in J} \theta_{\mathcal{I},\mathcal{N}}((\widehat{\alpha_j, \beta_j}), A_j)\right) = \left(\bigcup_{j \in J} \theta_{\mathcal{I},\mathcal{N}}((\widehat{\alpha_j, \beta_j}), A_j)\right) \cup \left(\bigcup_{j \in J} \theta_{\mathcal{I},\mathcal{N}}((\widehat{\alpha_j, \beta_j}), H(A_j))\right).$$

By Theorems 4, 6, and 7, Theorem 18 in [23] and the dual approach, we can conclude following

Theorem 11. *Let* $L, H : \mathcal{IF}(U) \to \mathcal{IF}(U)$ *be a dual pair of IF operators, if* $\mathcal{I}$ *is a border, EP and CP IF implicator, and* $\mathcal{T}$ *an IF t-norm on* L^* *satisfying Eq.* (16), *then there exists a* $\mathcal{T}$-*transitive IF relation R on U such that Eq.* (21)

holds iff L *satisfies properties* {(AIFL1), (AIFL2), (AIFLT)}, *or, equivalently,* H *obeys properties* {(AIFU1), (AIFU2), (AIFUT)}, *where*

(AIFLT) $L(A) \subseteq L\big(L(A)\big), \quad \forall A \in \mathcal{IF}(U)$.

(AIFUT) $H\big(H(A)\big) \subseteq H(A), \quad \forall A \in \mathcal{IF}(U)$.

Remark 4. Theorem 11 shows that {(AIFL1), (AIFL2), (AIFLT)} (respectively, {(AIFU1), (AIFU2), (AIFUT)}) is the basic set of axioms to characterize the lower (respectively, upper) $\mathcal{I}$-IF rough approximation operator generated by a $\mathcal{T}$-transitive IF relation. The next theorem indicates that the set of axioms {(AIFL1), (AIFL2), (AIFLT)} (respectively, {(AIFU1), (AIFU2), (AIFUT)}) can be replaced by a single axiom.

Theorem 12. *Let* $L, H : \mathcal{IF}(U) \to \mathcal{IF}(U)$ *be a dual pair of IF operators, if* $\mathcal{I}$ *is a border, EP and CP IF implicator, and* $\mathcal{T}$ *an IF t-norm on* L^* *satisfying Eq. (16), then there exists a* $\mathcal{T}$-*transitive IF relation* R *on* U *such that Eq. (21) holds iff* L *satisfies the following* (ALT):

(ALT) $\quad \forall A_j \in \mathcal{IF}(U), \forall (\alpha_j, \beta_j) \in L^*, j \in J,$ *where* J *is any index set,*

$$L\big(\bigcap_{j \in J} ((\widehat{\alpha_j, \beta_j}) \Rightarrow_{\mathcal{I}} A_j)\big) = \big(\bigcap_{j \in J} ((\widehat{\alpha_j, \beta_j}) \Rightarrow_{\mathcal{I}} L(A_j))\big)$$
$$\cap \big(\bigcap_{j \in J} ((\widehat{\alpha_j, \beta_j}) \Rightarrow_{\mathcal{I}} L(L(A_j)))\big).$$

or, equivalently, H *obeys the following* (AUT):

(AUT) $\quad \forall A_j \in \mathcal{IF}(U), \forall (\alpha_j, \beta_j) \in L^*, j \in J,$ *where* J *is any index set,*

$$H\big(\bigcup_{j \in J} \theta_{\mathcal{I}, \mathcal{N}}((\widehat{\alpha_j, \beta_j}), A_j)\big) = \big(\bigcup_{j \in J} \theta_{\mathcal{I}, \mathcal{N}}((\widehat{\alpha_j, \beta_j}), H(A_j))\big)$$
$$\cup \big(\bigcup_{j \in J} \theta_{\mathcal{I}, \mathcal{N}}((\widehat{\alpha_j, \beta_j}), H(H(A_j)))\big).$$

Remark 5. According to Theorems 9 and 11, we observe that {(AIFL1), (AIFL2), (AIFLR), (AIFLT)} (respectively, {(AIFU1), (AIFU2), (AIFUR), (AIFUT)}) is the basic set of axioms for characterizing the lower (respectively, upper) $\mathcal{I}$-IF rough approximation operators generated by a reflexive and $\mathcal{T}$-transitive IF relation. In the next theorem, we will show that {(AIFL1), (AIFL2), (AIFLR), (AIFLT)} (respectively, {(AIFU1), (AIFU2), (AIFUR), (AIFUT)}) can be replaced by a single axiom.

Theorem 13. *Let* $L, H : \mathcal{IF}(U) \to \mathcal{IF}(U)$ *be a dual pair of IF operators, if* $\mathcal{I}$ *is a border, EP and CP IF implicator, and* $\mathcal{T}$ *an IF t-norm on* L^* *satisfying Eq. (16), then there exists a reflexive and* $\mathcal{T}$-*transitive IF relation* R *on* U *such that Eq. (21) holds iff* L *satisfies the following* (ALRT):

(ALRT) $\quad \forall A_j \in \mathcal{IF}(U), \forall (\alpha_j, \beta_j) \in L^*, j \in J,$ *where* J *is any index set,*

$$L\big(\bigcap_{j \in J} ((\widehat{\alpha_j, \beta_j}) \Rightarrow_{\mathcal{I}} A_j)\big) = \big(\bigcap_{j \in J} ((\widehat{\alpha_j, \beta_j}) \Rightarrow_{\mathcal{I}} A_j)\big) \cap \big(\bigcap_{j \in J} ((\widehat{\alpha_j, \beta_j}) \Rightarrow_{\mathcal{I}} L(A_j))\big)$$
$$\cap \big(\bigcap_{j \in J} ((\widehat{\alpha_j, \beta_j}) \Rightarrow_{\mathcal{I}} L(L(A_j)))\big).$$

or, equivalently, H obeys the following (AURT)*:*

(AURT) $\forall A_j \in \mathcal{IF}(U), \forall(\alpha_j, \beta_j) \in L^*, j \in J,$ *where J is any index set,*

$$H(\bigcup_{j \in J} \theta_{\mathcal{I},\mathcal{N}}(\widehat{(\alpha_j, \beta_j)}, A_j)) = (\bigcup_{j \in J} \theta_{\mathcal{I},\mathcal{N}}(\widehat{(\alpha_j, \beta_j)}, A_j)) \cup (\bigcup_{j \in J} \theta_{\mathcal{I},\mathcal{N}}(\widehat{(\alpha_j, \beta_j)}, H(A_j)))$$
$$\cup (\bigcup_{j \in J} \theta_{\mathcal{I},\mathcal{N}}(\widehat{(\alpha_j, \beta_j)}, H(H(A_j)))).$$

5   Conclusion

Axiomatic characterization of approximation operators, which aims to investigate the mathematical structures of rough sets, is an important issue in the study of rough set theory. In this paper, we have studied the axiomatic approach to relation-based $\mathcal{I}$-intuitionistic fuzzy rough sets determined by an IF implicator $\mathcal{I}$. We have found the necessary and sufficient properties of abstract lower/upper $\mathcal{I}$-intuitionistic fuzzy approximation operator which guarantee the existence of an intuitionistic fuzzy relation which produces the same operator. We have shown that the lower (and, respectively, upper) $\mathcal{I}$-intuitionistic fuzzy rough approximation operator generated by a generalized intuitionistic fuzzy relation can be described by only one axiom. We have also presented that each of $\mathcal{I}$-IF rough approximation operators corresponding to reflexive and $\mathcal{T}$-transitive intuitionistic fuzzy relations can be characterized by single axioms.

Acknowledgement. This work was supported by grants from the National Natural Science Foundation of China (Nos. 61272021, 61202206, and 61173181), and the Zhejiang Provincial Natural Science Foundation of China (Nos. LZ12F03002 and LY14F030001).

References

1. Atanassov, K.: Intuitionistic Fuzzy Sets: Theory and Applications. Studies in Fuzziness and Soft Computing, vol. 35. Physica-Verlag, Heidelberg (1999)
2. Chakrabarty, K., Gedeon, T., Koczy, L.: Intuitionistic fuzzy rough set. In: Proceedings of 4th Joint Conference on Information Sciences (JCIS), pp. 211–214, Durham, NC (1998)
3. Cornelis, C., Cock, M.D., Kerre, E.E.: Intuitionistic fuzzy rough sets: at the crossroads of imperfect knowledge. Expert Syst. **20**, 260–270 (2003)
4. Cornelis, C., Deschrijver, G., Kerre, E.E.: Implication in intuitionistic fuzzy and interval-valued fuzzy set theory: construction, classification, application. Int. J. Approx. Reason. **35**, 55–95 (2004)
5. Jena, S.P., Ghosh, S.K.: Intuitionistic fuzzy rough sets. Notes Intuit. Fuzzy Sets **8**, 1–18 (2002)
6. Liu, G.L.: Axiomatic systems for rough sets and fuzzy rough sets. Int. J. Approx. Reason. **48**, 857–867 (2008)
7. Liu, G.L.: Using one axiom to characterize rough set and fuzzy rough set approximations. Inf. Sci. **223**, 285–296 (2013)

8. Liu, X.D., Pedrycz, W., Chai, T.Y., Song, M.L.: The development of fuzzy rough sets with the use of structures and algebras of axiomatic fuzzy sets. IEEE Trans. Knowl. Data Eng. **21**, 443–462 (2009)

9. Mi, J.-S., Zhang, W.-X.: An axiomatic characterization of a fuzzy generalization of rough sets. Inf. Sci. **160**, 235–249 (2004)

10. Morsi, N.N., Yakout, M.M.: Axiomatics for fuzzy rough sets. Fuzzy Sets Syst. **100**, 327–342 (1998)

11. Radzikowska, A.M.: Rough approximation operations based on IF sets. In: Rutkowski, L., Tadeusiewicz, R., Zadeh, L.A., Żurada, J.M. (eds.) ICAISC 2006. LNCS (LNAI), vol. 4029, pp. 528–537. Springer, Heidelberg (2006)

12. Radzikowska, A.M., Kerre, E.E.: A comparative study of fuzzy rough sets. Fuzzy Sets Syst. **126**, 137–155 (2002)

13. Samanta, S.K., Mondal, T.K.: Intuitionistic fuzzy rough sets and rough intuition-istic fuzzy sets. J. Fuzzy Math. **9**, 561–582 (2001)

14. Thiele, H.: On axiomatic characterisation of crisp approximation operators. Inf. Sci. **129**, 221–226 (2000)

15. Wu, W.-Z., Gao, C.-J., Li, T.-J., Xu, Y.-H.: On dual intuitionistic fuzzy rough approximation operators determined by an intuitionistic fuzzy implicator. In: Ciucci, D., Inuiguchi, M., Yao, Y., Ślęzak, D., Wang, G. (eds.) RSFDGrC 2013. LNCS, vol. 8170, pp. 138–146. Springer, Heidelberg (2013)

16. Wu, W.-Z., Leung, Y., Mi, J.-S.: On characterizations of $(\mathcal{I}, \mathcal{T})$-fuzzy rough approx-imation operators. Fuzzy Sets Syst. **154**, 76–102 (2005)

17. Wu, W.-Z., Zhang, W.-X.: Constructive and axiomatic approaches of fuzzy approx-imation operators. Inf. Sci. **159**, 233–254 (2004)

18. Yao, Y.Y.: Constructive and algebraic methods of the theory of rough sets. J. Inf. Sci. **109**, 21–47 (1998)

19. Yao, Y.Y.: Two views of the theory of rough sets in finite universes. Int. J. Approx. Reason. **15**, 291–317 (1996)

20. Zhang, X.H., Zhou, B., Li, P.: A general frame for intuitionistic fuzzy rough sets. Inf. Sci. **216**, 34–49 (2012)

21. Zhou, L., Wu, W.-Z.: On generalized intuitionistic fuzzy approximation operators. Inf. Sci. **178**, 2448–2465 (2008)

22. Zhou, L., Wu, W.-Z.: Characterization of rough set approximations in Atanassov intuitionistic fuzzy set theory. Comput. Math. Appl. **62**, 282–296 (2011)

23. Zhou, L., Wu, W.-Z., Zhang, W.-X.: On characterization of intuitionistic fuzzy rough sets based on intuitionistic fuzzy implicators. Inf. Sci. **179**, 883–898 (2009)

Granular Computing

Knowledge Supported Refinements for Rough Granular Computing: A Case of Life Insurance Industry

Kao-Yi Shen[1](✉) and Gwo-Hshiung Tzeng[2]

[1] Department of Banking and Finance, Chinese Culture University (SCE),
Taipei 10659, Taiwan
kyshen@sce.pccu.edu.tw, atrategy@gmail.com
[2] Graduate Institute of Urban Planning, College of Public Affairs,
National Taipei University, 151 University Rd., San Shia District,
New Taipei City 23741, Taiwan
ghtzeng@gm.ntpu.edu.tw

Abstract. Dominance-based rough set approach (DRSA) has been adopted in solving various multiple criteria classification problems with positive outcomes; its advantage in exploring imprecise and vague patterns is especially useful concerning the complexity of certain financial problems in business environment. Although DRSA may directly process the raw figures of data for classifications, the obtained decision rules (i.e., knowledge) would not be close to how domain experts comprehend those knowledge—composed of granules of concepts—without appropriate or suitable discretization of the attributes in practice. As a result, this study proposes a hybrid approach, composes of DRSA and a multiple attributes decision method, to search for suitable approximation spaces of attributes for gaining applicable knowledge for decision makers (DMs). To illustrate the proposed idea, a case of life insurance industry in Taiwan is analyzed with certain initial experiments. The result not only improves the classification accuracy of the DRSA model, but also contributes to the understanding of financial patterns in the life insurance industry.

Keywords: Dominance-based rough set approach (DRSA) · Multiple attribute decision making (MADM) · DEMATEL-based ANP (DANP) · Approximation space (AS) · Granular computing · Financial performance (FP) · Life insurance industry

1 Introduction

In modern business world, a management team or decision maker (DM) has to make complicated judgments to attain satisfactory prospects. However, considering the complex and imprecise relationships among the criteria/attributes/variables (these three terms are used interchangeably in this paper) for reaching superior financial performance (FP), how to obtain intuitive or obvious guidance from historical patterns—for FP prediction—is still challenging. Conventional research mainly relies on statistical analysis, e.g., regression, time series and principle component analyses; nevertheless,

© Springer International Publishing Switzerland 2015
Y.Yao et al. (Eds.): RSFDGrC 2015, LNAI 9437, pp. 233–244, 2015.
DOI: 10.1007/978-3-319-25783-9_21

statistical analyses stand on the ground of unpersuasive assumptions: such as the independence of variables (in regressions) and certain probabilistic distributions of data. Analytical tools/techniques with fewer assumptions or constraints, that may model the imprecise concepts and relationships in a real business environment, are highly required [6]. In addition, to resolve real world problems often has to consider a large amount of variables in practice. Consequently, based on the advantages of soft computing techniques (e.g., fuzzy set and rough set theories) and multiple attribute decision making (MADM), the combination or integration of those two disciplines (i.e., soft computing and MADM) shed lights on solving this kind of complex and yet valuable problems [13, 14].

Among various soft computing techniques, the rough set approach (RSA) proposed by Pawlak [8], has strength in handling imprecise attributes/variables even while the number of variables is large. Classical RSA categorizes objectives/alternatives by analyzing the indiscernibility of data, and the extended dominance-based rough set approach (DRSA) [5] further considers the dominance relationship in variables and attribute values, which is more suitable to solve MADM problems. Although the DRSA has been applied in certain financial applications with positive outcomes [13–15], while applying DRSA for modeling vague/imprecise concepts, there is a need to divide an attribute into several intervals (i.e., granulized concepts)—to be closer to how domain experts process those concepts or to obtain superior classification accuracy —for the subsequent induction of decision rules. As claimed in the previous work [18], the search for (semi-)optimal approximation space (AS) would be a problem of high complexity. Until recently, the mainstream for searching the optimal approximation spaces is either machine learning (e.g., genetic algorithm based searching strategy) or heuristic based approach; however, the machine learning based approach often ends up with unexplainable discretization result (i.e., far from how human experts comprehend those concepts), and the heuristic approach would encounter obstacles while the number of attributes/variables is too large for experts to give precise opinions. Knowledge supported discretization (retrieved from domain experts and transformed by reliable MADM methods) to obtain appropriate granular concepts for financial applications is still underexplored.

As a result, the present study proposes a hybrid approach—the combination of DRSA and a MADM method—to find suitable approximation spaces for obtaining applicable guidance/knowledge (i.e., decision rules) for the FP prediction problem. The major motivation of this study originated from the idea that granulized concepts of financial applications/problems should be close to how domain experts comprehend/ process the approximation reasoning. Furthermore, the present study presumes that an attribute should be divided into more values (i.e., as smaller granules) to reflect its importance in constructing a DRSA information system (IS). This idea is intuitive to comprehend; if a_1 attribute weighted more than a_2 in a model, then the change of a_1 should be more sensitive than the change of a_2 to influence the classification outcome. To find suitable approximation spaces for predicting FP, a group of life insurance companies was analyzed as an example, and the experimental result supports the aforementioned presumption in this case; in addition, understandable decision rules were obtained to provide insights regarding the FP prediction of life insurance industry in a real case.

The remainder is structured as below. Section 2 provides a brief discussion regarding the involved research methods, including DRSA and DEMATEL-based ANP (termed DANP). In Sect. 3, the proposed approach is explained with needed steps. Section 4 uses the openly accessed financial data of the life insurance companies in Taiwan as an empirical case, with experimental results and discussions. Section 5 concludes this study with the research findings and future directions of the proposed approach.

2 Preliminary

The section briefly reviews the concept of RSA, DRSA, and the adopted MADM method (i.e., DANP) and certain applications in finance. The relationship among granular computing, RSA and MADM methods are also mentioned.

2.1 Rough Set Approach (RSA) and Extended Applications

The classical RSA [8] was based on solid logical and mathematical foundations, which could discern objects considering a group of attributes while the indiscernibility of attributes is preserved for inductions. Compared with the logical operations on the crisp sets, the RSA is more capable to tackle social science problems with vague or imprecise patterns. However, while using the RSA to support certain decision making problems, there is a need to consider the preferential order of attribute values; for example, in evaluating credit loads, a company with higher cash flows from operations is often preferred. Therefore, the subsequent DRSA was proposed by Greco et al. [5] to enhance the RSA—by incorporating the dominance relationship—for solving complex decision making problems. Certain RSA or DRSA extended approaches were developed to consider the precision or dominance relationship in attributes for making approximations, such as variable precision rough set model (VPRSM) [20], near set approach (NSA) [9], tolerance rough set (TRS) [17], quasi dominance rough set approach (QDRSA) [2], and variable-consistency dominance-based rough set approach (VC-DRSA) [1, 4]. For example, the QDRSA was devised to addressed preference-ordered attributes in the application of bioinformatics [2]; nevertheless, it is constraint to the two-value domain of the decision attribute, which might not be suitable to evaluate decision problems in a complicated business environment. Therefore, owing to the popularity of DRSA in MCDM research [3, 21], the present study intends to illustrate the proposed approach by adopting the DRSA.

Although the DRSA has been adopted in various applications with positive outcomes, we mainly focus on the discussion of financial applications in here; for example, the DRSA was recently applied on predicting the FP of banks [12, 13], diagnosing the FP of the high tech industry [15], and retrieving knowledge from technical analysis for investments [16]. Nevertheless, previous works mainly adopted intuitive discretization for the involved attributes; it is our hope to examine how domain experts' knowledge may refine the approximation spaces for the addressed issue in the present study.

2.2   DEMATEL-Based ANP (DANP) Method

Various MADM methods may support DMs to obtain or retrieve the relative weight of each criterion/attribute for the addressed issue/problem; among those MADM methods, the ANP (Saaty 2004) [10] was extended from the analytic hierarchy process (AHP) to allow for interdependence among the considered criteria, which is one of the most prevailing MADM methods. The ANP decomposes problem into clusters (dimensions), and each cluster contains multiple attributes/criteria for evaluation. To improve the equal weighting assumption of the typical ANP, the DEMATEL technique [7] was introduced to combine with the basic concept of ANP for retrieving the influential weights of attributes, called DANP (DEMATEL-based ANP) [11]. This hybrid MADM method (combining DEMATEL with ANP) has at least two advantages in here: reliable knowledge retrieval based on pairwise comparison and adjusted dimensional weights for forming the DANP model. The finalized influential weights of DANP can thus indicate the experts' knowledge to support the refinements of approximation spaces based on the aforementioned presumption.

3   Research Approach

Section 3 provides the background information of the incorporated methods in the proposed approach. The proposed approach comprises of two building blocks: the DRSA and DANP methods. The basic idea of DRSA is discussed with the suggested refinement of approximation spaces. Also, the essential idea of DANP is discussed.

3.1   Dominance-Based Rough Set Approach (DRSA)

A typical DRSA model begins from a 4-tuple information system (IS) that $IS = (U, Q, V, f)$; in which, U is a finite set of universe, Q is a finite set of z attributes (i.e., $Q = \{q_1, q_2 \ldots, q_z\}$), V is the value domain of attribute (i.e., $V = \bigcup_{q \in Q} V_q$), and f denotes a total function (i.e. $f : U \times Q \rightarrow V$). Considering an attribute with continuous property (e.g., debt ratio), the attribute would need to be discretized as several intervals/values/granules to denote certain concepts (e.g., very high debt ratio); if the ith attribute was discretized as h_i values, then the jth granule of the attribute i can be denoted as G_i^j (for $i = 1, \ldots, z$; $1 \leq j \leq h_i$). The set Q comprises of two main parts: condition attributes set C and a decision attribute d ($d \in D$, where D is the decision attribute set), for most of the decision problems. In the universe of U, a complete outranking relation can be defined as $\succeq_q$ with respect to a criterion $q \in Q$.

For any two objects x and y (alternatives) in U, a complete outranking relation with regard to a criterion q can be defined as $\succeq_q$; if $x \succeq_q y$, it denotes that "x is at least as good as y regarding criterion q". For a decision attribute d that belongs to D, which categorizes the objects in U into a finite number (e.g., m) of decision classes (DCs), denoted as $Cl = \{Cl_t : Cl_1, Cl_2, \ldots, Cl_m\}$. For each $x \in U$, object x belongs to only one $Cl_t(Cl_t \in Cl)$. For Cl with predefined preferential order (i.e. for all $r, s = 1, \ldots, m$, if

$r \succ s$, the decision class Cl_r is preferred to Cl_s), an downward union $Cl_t^{\leq}$ and upward union $Cl_t^{\geq}$ of classes can be defined as Eq. (1):

$$Cl_t^{\leq} = \bigcup_{s \leq t} Cl_s \text{ and } Cl_t^{\geq} = \bigcup_{s \geq t} Cl_s \qquad (1)$$

For brevity, only the upward union is discussed; the downward union could be reasoned in analogy. For $x, y \in U$, if x dominates y with respect to the attribute set P (for $P \subseteq C$), it can be denoted as $xD_P y$ to denote x P-dominates y. Then, for a set of objects that dominate x with regard to P, it can be denoted as $D_P^+(x) = \{y \in U : yD_P x\}$, the P-dominating set. On the other side, a set of objects that are dominated by x with regard to P can be denoted as $D_P^-(x) = \{y \in U : xD_P y\}$(i.e., the P-dominated set).

In the next, the P-lower and P-upper approximation of an upward union with respect to $P \subseteq C$ can be define by $\underline{P}(Cl_t^{\geq}) = \{x \in U : D_P^+(x) \subseteq Cl_t^{\geq}\}$ and $\bar{P}(Cl_t^{\geq}) = \{x \in U : D_P^-(x) \cap Cl_t^{\geq} \neq \emptyset\}$. In DRSA, $\underline{P}(Cl_t^{\geq})$ denotes all of the objects $x \in U$ that are for sure to be included in the upward union $Cl_t^{\geq}$; the P-upper approximation $\bar{P}(Cl_t^{\geq})$ can be interpreted as all of the objects possibly belongs to $Cl_t^{\geq}$ (i.e., uncertain). Then, the boundary region $Bn_P = \bar{P}(Cl_t^{\geq}) - \underline{P}(Cl_t^{\geq})$ can thus be defined ($t = 2, \ldots, m$). Considering each subset set $P \subseteq C$, the quality of approximation of Cl (by using a set of condition attributes P) is defined as Eq. (2). If an object x P-dominating object y on all attributes in P ($P \subseteq C$), then x should also dominate y on the decision attribute, denoted as $xD_P y$. The quality of approximation defines the ratio of the objects P-consistent with the dominance relationship divided by the total number of objects in U, and $|\bullet|$ denotes the cardinality in Eq. (2)

$$\gamma_P(Cl) = \left| U - \left(\bigcup_{t \in \{2, \ldots, m\}} Bn_P(Cl_t^{\geq}) \right) \right| \Big/ |U| \qquad (2)$$

Each minimal subset P ($P \subseteq C$) that may satisfy $\gamma_P(Cl) = \gamma_C(Cl)$ is called a REDUCT of Cl, and the intersection of all REDUCTs represent the indispensable attributes to maintain the quality of approximation, termed $CORE_{Cl}$. The proposed approach adopts the CORE attributes for the subsequent DANP analysis, and the CORE attributes with higher influential weights would be discretized into smaller granules based on the presumption in this study. In addition, the induction of dominance-based approximations may generate a set of decision rules, in the form of "**if** *antecedents* **then** *consequence*". The obtained decision rules with refined granules of knowledge are expected to deliver insightful knowledge, concerning the FP prediction problem for the life insurance industry.

3.2 DEMATEL-Based ANP Method (DANP)

The CORE attributes obtained from the initial DRSA could be used to design a questionnaire, to collect domain expert's opinions by pairwise comparisons of each two attributes. Experts are asked to judge the direct effect that they feel attribute (criterion) k would have on attribute (criterion) l, indicated as a_{kl}. The arithmetic mean of experts' opinions are used for forming the initial average influence relation matrix A in the DEMATEL analysis (with assumed n criteria), as Eq. (3).

$$A = \begin{bmatrix} a_{11} & \cdots & a_{1l} & \cdots & a_{1n} \\ \vdots & & \vdots & & \vdots \\ a_{k1} & \cdots & a_{kl} & \cdots & a_{kn} \\ \vdots & & \vdots & & \vdots \\ a_{n1} & \cdots & a_{nl} & \cdots & a_{nn} \end{bmatrix} \tag{3}$$

Then, the direct influence relation matrix D can be obtained by normalize the initial average influence relation matrix A referring to Eqs. (4)–(5).

$$D = eA \tag{4}$$

$$e = \min\left\{ 1 \Big/ \max_k \sum_{l=1}^{n} a_{kl}, \; 1 \Big/ \max_l \sum_{k=1}^{n} a_{kl} \right\}, \; k, l \in \{1, 2, \ldots, n\} \tag{5}$$

And the total influence relation matrix T can be decomposed as Eq. (6), while $w \to \infty$, $D^w = [0]_{n \times n}$, and I denotes the identity matrix in Eqs. (6)–(7).

$$T = D + D^2 + \ldots + D^w = D(I - D^w)(I - D)^{-1} \tag{6}$$

$$T = D(I - D)^{-1} = [t_{kl}]_{n \times n}, \; \text{when } \lim_{w \to \infty} D^w \cong [0]_{n \times n} \tag{7}$$

The total influence relation matrix T (obtained from DEMATEL analysis) can be normalized to be T_C^N; similarly, the total influence relation matrix of the dimensions matrix as the normalized T_D^N. Then, the total influence relation matrix T can be normalized, and then transpose into an un-weighted super-matrix $W = (T_C^N)'$. The DEMATEL weighted super-matrix W^{DW} can be obtained by multiple T_D^N with W, i.e., $W^{DW} = T_D^N W$. The weighted super-matrix is multiple by itself several times to obtain the stable super-matrix. The influential weight of each attribute/criterion in the DANP can thus be obtained. The detail formulae regarding the transformation from T to the stable super-matrix may refer to the previous works [7].

3.3   Suggested Steps for the Proposed Approach

The suggested steps to implement the hybrid model are as below: (1) Define condition and decision attributes; (2) Discretize each attribute by intuition (e.g., as three values to denote high, middle, and low); (3) Conduct DRSA analysis for identifying the CORE attributes; (4) Collect expert's opinions and calculate the influential weight of each CORE attribute by the DANP method; (5) Discretize those attributes with higher influential weights into more values (i.e., as smaller granules) and examine the classification outcomes of DRSA by experiments; and (6) Obtain decision rules by DRSA from the refined granules of knowledge for gaining insights.

The collection of experts' opinions might need certain explanations in here. Since the DANP is a pairwise comparison approach, a questionnaire will be designed that contains the core attributes. Experts would be asked for questions like: "Considering the future FP of a life insurance company, what do you feel about the influence that attribute i might have on the other attribute j?" Opinions would range from 0 (no influence) to 4 (very high influence), based on the experience/knowledge of experts. The detailed discussions and required steps for the DANP calculations can be found in the previous studies [7, 11].

4   Empirical Case of Life Insurance Industry in Taiwan

To examine the proposed approach in a real business environment, the life insurance industry in Taiwan was analyzed as an empirical case. Considering the importance of life insurance industry on strengthening citizens' confidence of a nation's economy, the present study attempts to retrieve the implicit and yet critical knowledge regarding the change of FP in this industry, for those concerned stakeholders: the potential or existing customers, the government authority, the investors, the creditors (e.g., banks), and the management teams of the insurance companies.

4.1   Data

All the registered life insurance companies in Taiwan have to report their latest financial statements for the government and the public to review/monitor, and the authority of insurance industry in Taiwan periodically releases all of those registered life insurance companies' major financial information on the internet; those openly-accessed financial data were available from 2009, and all of those historical data were included for DRSA modeling. In this empirical case, a one-period lagged model was constructed, and each company's condition attributes at time t were associated with its decision attribute at time $t + 1$. There were total 114 observations, and the observations in the recent time frame (27 observation, from 2012 to 2013) were used as the testing set, and the others the training set. The data were retrieved from the website of Taiwan Insurance Institute [19] at the end of 2014. The original released ratios (i.e., criteria) comprises of 19 criteria in five dimensions: *Capital Structure* (D_1), *Payback* (D_2), *Operational Efficiency* (D_3), *Revenue Quality* (D_4), and *Capital Efficiency* (D_5) (this dimension was renamed from *"Profitability"*, because the profitability indicator *ROA* in this dimension was used as the decision attribute d at time $t + 1$).

At the first stage, as no specific guidance was provided for the approximation spaces, all the condition attributes were discretized into three intervals to denote "high," "middle," and "low" of each attribute (at time t) by intuition; similarly, "Good", "Mediocre", and "Bad" for the decision attribute (i.e., *ROA* at time $t + 1$). A 5-fold cross validation was repeated five times—on the training set—for the DRSA model; furthermore, the support vector machine (SVM) and the variable-consistency dominance-based rough set approach (VC-DRSA) classifiers (while consistency level = 0.95 and 0.90, i.e., CL = 0.95 and 0.90) were calculated and summarized in Table 1. The averaged classification accuracy (CA) of the DRSA model reached 65.39 %, which is the highest one among the compared classifiers. In the next, the whole training set was used for inducting decision rules, and the testing set was then validated. The study adopted 5-fold cross validation on the training set by further dividing the training set as sub-training and sub-testing sets; those two sets were randomly divided by jMAF in each cross validation, to calculate the CAs in Tables 1 and 5. The CA of the testing set reached 62.96 % in the DRSA model. The DRSA model at this stage formed a set of CORE attributes (13 condition attributes). Those CORE set attributes/criteria (Table 2) denote the minimal and dispensable attributes that may classify alternatives/observations without losing the quality of approximation accuracy.

Table 1. Classification accuracy of various classifiers by all the attributes (unit:%)

	DRSA	VC-DRSA (CL = 0.95)	VC-DRSA (CL = 0.90)	SVM (RBF kernel)
1	65.21	60.87	62.61	56.14
2	69.57	63.48	67.83	62.61
3	66.96	60.00	69.57	60.87
4	60.87	67.83	62.61	56.82
5	64.35	63.48	59.13	59.52
Average	**65.39**	63.13	64.35	59.19
Testing set	62.96	51.85	59.26	55.55

In the next, the presumption that higher influential criteria (attributes) should be divided into more values (i.e., granules of knowledge) was examined at this stage (i.e., Step 5 in Subsect. 3.3). The knowledge from domain experts' perceived influential weights of the CORE attributes regarding the future FP of the life insurance industry was obtained. Although the original number of condition attributions was reduced to 13, the complexity of multiple criteria still impedes experts to give direct opinions for the importance of each criterion on these 13 criteria (in five dimensions). Therefore, the prevailing DANP method was adopted to collect domain experts' opinions, and transformed those opinions (in the form of pairwise comparisons by questionnaires) to form the initial average relation matrix A (refer Eq. (1), and the matrix A was then turned into the total influence relation matrix T (Table 3).

All the five domain experts have more than 15 years' experience in the insurance industry; among those experts, there is a retired government official from Insurance Bureau. The other four experts' titles include Manager, Senior Manager (Unit Manager), Sales Director and Assistant Vice President. As stated in Subsect. 3.2, the initial average relation matrix A was preprocessed by the DEMATEL technique to adjust the influential weight of each dimension before reaching the finalized influential weight of each criterion. For brevity, the calculation details of DANP (refer to the Appendix of [15]) are not provided in here.

Table 2. Descriptions of the CORE attributes

Dimensions	Criteria		Definitions/Descriptions
Capital Structure (D_1)	Debt	C_1	Total debt/total assets
	Δ Provision	C_2	Change rate of Provision for life insurance reserve
Payback (D_2)	1st Y-Premium	C_3	First year premium ratio
	RY-Premium	C_4	Renewable premium ratio
Operational Efficiency (D_3)	N-Cost	C_5	New contract cost/new contract revenue
	Δ Equity	C_6	Change rate of share holders' equity
	Δ NetProfit	C_7	Change rate of net profit
	CapInvest	C_8	Total invested capital/total assets
Earning Quality (D_4)	Persistency	C_9	Persistency of the valid contracts in the 25th month
Capital Efficiency (D_5)	NetProfit	C_{10}	Net profitability of capital utilization
	ROI	C_{11}	Return on investment ratio
	O-Profit	C_{12}	Operational profits/operational incomes
	RealEstate	C_{13}	Investment and loan on real estate/total assets

Table 3. Total influence relation matrix T

	C_1	C_2	C_3	C_4	C_5	C_6	C_7	C_8	C_9	C_{10}	C_{11}	C_{12}	C_{13}
C_1	0.10	0.15	0.11	0.12	0.13	0.18	0.21	0.21	0.12	0.17	0.16	0.17	0.14
C_2	0.10	0.05	0.09	0.10	0.09	0.12	0.15	0.12	0.10	0.10	0.07	0.10	0.10
C_3	0.11	0.12	0.08	0.19	0.19	0.13	0.22	0.17	0.19	0.12	0.10	0.22	0.12
C_4	0.12	0.11	0.14	0.10	0.19	0.14	0.24	0.17	0.21	0.12	0.10	0.24	0.11
C_5	0.15	0.12	0.17	0.20	0.12	0.15	0.25	0.17	0.20	0.13	0.12	0.26	0.12
C_6	0.11	0.06	0.05	0.06	0.07	0.06	0.14	0.11	0.07	0.10	0.11	0.11	0.09
C_7	0.16	0.09	0.13	0.14	0.16	0.14	0.15	0.17	0.14	0.18	0.15	0.21	0.13
C_8	0.18	0.13	0.13	0.14	0.18	0.16	0.29	0.16	0.16	0.23	0.21	0.27	0.20
C_9	0.14	0.14	0.12	0.19	0.16	0.15	0.25	0.19	0.11	0.15	0.13	0.25	0.13
C_{10}	0.19	0.11	0.13	0.15	0.19	0.16	0.29	0.23	0.16	0.14	0.22	0.28	0.20
C_{11}	0.20	0.11	0.14	0.15	0.18	0.17	0.30	0.27	0.17	0.21	0.13	0.28	0.22
C_{12}	0.19	0.11	0.16	0.18	0.20	0.16	0.27	0.24	0.20	0.18	0.15	0.17	0.13
C_{13}	0.12	0.06	0.08	0.08	0.10	0.12	0.20	0.20	0.10	0.18	0.16	0.19	0.09

4.2 Knowledge-Supported Refinements of Approximation Spaces

The obtained final influential weight for each criterion (attribute) from DANP is shown in Table 4 with the corresponding discretized values of each attribute in the experiments. Refer to the widely accepted 5-point Likert-scale in social science; the attribute at the highest level of influential weight was discretized into five intervals, from

1 (lowest) to 5 (highest). The attributes as the second level of importance were discretized into four values, and the others as the original three values (i.e., 1, 2, 3).

Table 4. DANP weight and discretized values of each attribute/criterion

	C_1	C_2	C_3	C_4	C_5	C_6	C_7	C_8	C_9	C_{10}	C_{11}	C_{12}	C_{13}
^1w_i	0.10	0.07	0.08	0.10	0.05	0.05	0.08	0.06	0.20	0.05	0.04	0.07	0.04
The 1st group experiment													
^2d_i	4	4	4	4	3	3	4	3	5	3	3	4	3
The 2nd group experiment													
d_i	4	3	4	4	3	3	3	3	5	3	3	3	3

Note: w_i denotes the influential weight of the ith criterion/attribute, and d_i denotes the discretized values (intervals) of the ith criterion.

In Table 5, both the 1st and 2nd group of experiments—based on the refined approximation spaces—revealed superior averaged CAs compared with the original 3-level discretization of the DRSA model. The averaged CA (repeated five times by the 5-fold cross validation) of the DRSA model (same as in Table 1) in the 1st group experiment reached 69.74 %, and the CA of the testing set was 70.37 %; the experimental outcomes thus support the proposed presumption in this study. Even though, the reported CAs in here might not be good enough; compared with those works based on evolutional optimization techniques to increase CAs, the proposed approach was devised to retrieve rough knowledge (i.e., granulized concepts in decision rules) that are close to how domain experts comprehend this problem.

Table 5. Classification accuracy of the 1st and 2nd group discretization in DRSA (unit: %)

	Original AS	1st group AS*	2nd group AS*
Average	65.39	69.74	67.65
Testing set	62.96	70.37	66.67

Note: Refer Table 4 for the approximation spaces (AS*) of the 1st/2nd group on each attribute.

5   Conclusion and Remarks

The present study is based on the essential idea: the granulized concepts for modeling complex social science problems (e.g., the FP prediction) should be close to how domain experts comprehend those concepts, and the attributes with higher influential weights should be discretized as smaller granules for the DRSA modeling for superior classification outcome. This presumption was examined by using a group of life insurance companies in Taiwan with supportive findings. Compared with previous research, the present study suggested a new approach to incorporate domain experts' perception/feeling/knowledge, regarding the FP prediction problem, for searching suitable (rather than semi-optimal or optimal) approximation spaces. This approach may bridge the two disciplines—soft computing and MADM—for modeling complex and yet valuable financial problems in practice. The strong decision rules from the refined DRSA model may further provide implications that are easy-to-understand, e.g., "**if** $C_3 \geq$ moderately low & $C_9 \geq$ somewhat low & $C_{11} \geq$ high & $C_{12} \geq$ moderately high & $C_{13} \geq$ high, **then** DC $\geq$ Good FP in the next period", the descriptive scale for each attribute is based on the discretized values (e.g., discretized as four values: "very low," "moderately low," "moderately high," and "very high").

Although the proposed approach has gained certain supportive evidence in a real case, there are still several limitations. First, the external influences to the changes of FP in the life insurance industry might exist (e.g., financial crises or new policies), which might cause less satisfactory CA results. Second, owing to the length limitation, only two groups of refined approximation spaces were analyzed and reported; future research may devise more systematic experiments to enhance the approximation spaces for the addressed problem. Last, the obtained result is limited to the analyzed life insurance industry; nevertheless, the proposed approach could be referred by the other interested researchers to work on this direction in the future.

References

1. Błaszczyński, J., Greco, S., Słowiński, R., Szelg, M.: Monotonic variable consistency rough set approaches. Int. J. Approx. Reason. **50**(7), 979–999 (2009)
2. Cyran, K.A.: Quasi dominance rough set approach in testing for traces of natural selection at molecular level. In: Cyran, K.A., Kozielski, S., Peters, J.F., Stańczyk, U., Wakulicz-Deja, A. (eds.) Man-Machine Interactions. AISC, vol. 59, pp. 163–172. Springer, Heidelberg (2009)
3. Figueira, J., Greco, S., Ehrgott, M.: Multiple Criteria Decision Analysis: State of the Art Surveys, vol. 78. Springer Science & Business Media, New York (2005)
4. Greco, S., Matarazzo, B., Słowiński, R., Stefanowski, J.: Variable consistency model of dominance-based rough sets approach. In: Ziarko, W.P., Yao, Y. (eds.) RSCTC 2000. LNCS (LNAI), vol. 2005, pp. 170–181. Springer, Heidelberg (2001)
5. Greco, S., Matarazzo, B., Słowiński, R.: Multicriteria classification by dominance-based rough set approach. In: Kloesgen, W., Zytkow, J. (eds.) Handbook of Data Mining and Knowledge Discovery. Oxford University Press, New York (2002)
6. Liou, J.J.H., Tzeng, G.H.: Comments on "Multiple criteria decision making (MCDM) methods in economics: an overview". Technol. Econ. Dev. Econ. **18**(4), 672–695 (2012)
7. OuYang, Y.P., Shieh, H.M., Tzeng, G.H.: A VIKOR technique based on DEMATEL and ANP for information security risk control assessment. Inf. Sci. **232**, 482–500 (2013)
8. Pawlak, Z.: Rough sets. Int. J. Comput. Inf. Sci. **11**(5), 341–356 (1982)
9. Peters, J.F.: Near sets. General theory about nearness of objects. Appl. Math. Sci. **1**(53), 2609–2629 (2007)
10. Saaty, T.L.: Decision making with Dependence and Feedback: The Analytic Network Process. RWS Publications, Pittsburgh (1996)
11. Shen, K.Y., Yan, M.R., Tzeng, G.H.: Combining VIKOR-DANP model for glamor stock selection and stock performance improvement. Knowl. -Based Syst. **58**, 86–97 (2014)
12. Shen, K.Y., Tzeng, G.H.: DRSA-based neuro-fuzzy inference system for the financial performance prediction of commercial bank. Int. J. Fuzzy Syst. **16**(2), 173–183 (2014)
13. Shen, K.Y., Tzeng, G.H.: A decision rule-based soft computing model for supporting financial performance improvement of the banking industry. Soft. Comput. **19**(4), 859–874 (2015)

14. Shen, K.Y., Tzeng, G.H.: Combining DRSA decision-rules with FCA-based DANP evaluation for financial performance improvements. Technol. Econ. Dev. Econ. (2015, in press). doi:10.3846/20294913.2015.1071295
15. Shen, K.Y., Tzeng, G.H.: A new approach and insightful financial diagnoses for the IT Industry based on a hybrid MADM model. Knowl. -Based Syst. **85**, 112–130 (2015)
16. Shen, K.Y., Tzeng, G.H.: Fuzzy inference enhanced VC-DRSA model for technical analysis: investment decision aid. Int. J. Fuzzy Syst. **17**(3), 375–389 (2015). doi:10.1007/s40815-015-0058-8
17. Skowron, A., Stepaniuk, J.: Tolerance approximation spaces. Fundam. Inform. **27**, 245–253 (1996)
18. Skowron, A., Stepaniuk, J., Swiniarski, R.: Modeling rough granular computing based on approximation spaces. Inf. Sci. **184**(1), 20–43 (2012)
19. Taiwan Insurance Institute. http://www.tii.org.tw/eindex.asp. Accessed 2014
20. Ziarko, W.: Variable precision rough sets model. J. Comput. Syst. Sci. **46**(1), 39–59 (1993)
21. Zopounidis, C., Galariotis, E., Doumpos, M., Sarri, S., Andriosopoulos, K.: Multiple criteria decision aiding for finance: an updated bibliographic survey. Eur. J. Oper. Res. **247**(2), 339–348 (2015). doi:10.1016/j.ejor.2015.05.032

Building Granular Systems - from Concepts to Applications

Marcin Szczuka[1]([⊠]), Andrzej Jankowski[2], Andrzej Skowron[1,3],
and Dominik Ślęzak[1,4]

[1] Institute of Mathematics, University of Warsaw,
Banacha 2, 02-097 Warsaw, Poland
szczuka@mimuw.edu.pl
[2] Knowledge Technology Foundation, Nowogrodzka 31, 00-511 Warsaw, Poland
andrzej.adgam@gmail.com
[3] Systems Research Institute, Polish Academy of Sciences,
Newelska 6, 01-447 Warsaw, Poland
skowron@mimuw.edu.pl
[4] Infobright Inc., Krzywickiego 34 pok. 219, 02-078 Warsaw, Poland
slezak@mimuw.edu.pl

Abstract. Granular Computing (GrC) is a domain of science aiming at
modeling computations and reasoning that deals with imprecision, vagueness and incompleteness of information. Computations in GrC are performed on granules which are obtained as a result of information granulation. Principal issues in GrC concern processes of representation, construction, transformation and evaluation of granules. It also requires align-
ing with some of the fundamental computational issues concerning, e.g.,
interaction and adaptation. The paper outlines the current status of GrC
and provides the general overview of the process of building granular solutions to challenges posed by various real-life problems involving granularity. It discusses the steps that lead from raw data and imprecise/vague
specification towards a complete, useful application of granular paradigm.

Keywords: Granular computing · Information granulation ·
Vagueness · Computing with words · Soft computing · Cyber-physical
systems

1 Introduction

In recent years, one can observe a growing interest in Granular Computing (GrC)
as a methodology for modeling and conducting complex computations, in various
domains of Artificial Intelligence (AI) and Information Technology (IT). GrC is

This work was partially supported by the Polish National Science Centre (NCN) grant
DEC-2012/05/B/ST6/03215 as well as by the Polish National Centre for Research and
Development (NCBiR) grants O ROB/0010/03/001 and PBS2/B9/20/2013.

The original version of this chapter was revised: The acknowledgement was
modified. The correction to this chapter is available at https://doi.org/10.1007/
978-3-319-25783-9_45

© Springer International Publishing Switzerland 2015
Y. Yao et al. (Eds.): RSFDGrC 2015, LNAI 9437, pp. 245–255, 2015.
DOI: 10.1007/978-3-319-25783-9_22

already a well-established area with some encyclopedia entries [1–3], research monographs [4–9], edited collections of articles [10–21], conference series[1,2] and scientific journals[3,4].

GrC is founded on information granulation, a notion which is inherent in human thinking and reasoning. The underlying realization in GrC is that achieving full precision is sometimes expensive and not very meaningful in modeling and controlling complex information systems. When a task at hand involves dealing with incomplete, uncertain, and vague information, it may be difficult to discern distinct objects and one may find it convenient to consider granules for its handling. At the level of data granules can be composed of objects that are drawn together by indiscernibility, similarity, functional or physical adjacency and so on [22]. At the level of conceptual modelling each of granules, according to its structure and level of granularity, may reflect a specific aspect of the problem or form a portion of the system domain.

Modern computations are often performed on complex structures and their features, which need to be discovered from data and domain knowledge. Granules gathering such complex structures are expected to correspond to primitives expressed in some kind of high level language, which need to be approximated step by step, starting from low-level concepts extractable from raw data. Such granules can be treated as computational building blocks aiming at comprehension of complex situations.

Granular Systems (GrS) is an umbrella term that is used to describe those complex, intelligent systems that originate from general, frequently vague and imprecise specification and employ information granularity at their basis. They use various granule models as basic pieces and various GrC models in the role of an engine that processes them.

GrC became an effective framework in the design and implementation of intelligent systems for various real life applications. The developed systems exploit the tolerance for imprecision, uncertainty and partial truth under soft computing framework, in order to achieve tractability, robustness and resemblance with human-like (natural) decision-making [7].

This article is structured as follows. We begin with introduction of the information granulation paradigm (Sect. 2) and show how the general granular concepts translate to models of granules (Sect. 3). Then we present an overview of granular computation mechanisms (Sect. 4), discuss the ways of evaluating their properties (Sect. 5) and conclude the paper (Sect. 6).

2   Information Granulation

The term *information granulation* (or *information granularity*) is used in different contexts and numerous domains of applications. The pioneering studies

---

[1] http://dblp.uni-trier.de/db/conf/grc/index.html.

[2] http://dblp.uni-trier.de/db/conf/rsfdgrc/index.html.

[3] http://www.inderscience.com/jhome.php?jcode=ijgcrsis.

[4] http://www.springer.com/engineering/computational+intelligence+and+complexity/journal/41066.

by Zadeh have led toward envisioning information granulation as one of the foundations of Computational Intelligence (CI) [22,23].

There are two important human abilities, *graduation* and *granulation* of perceptions, that are desirable for approaches involving information granulation. Zadeh presented it as follows [22,24]:

> *Informally, a granule is a clump of values of a perception (e.g., perception of age), which are drawn together by proximity, similarity, or functionality. More concretely, a granule may be interpreted as a restriction on the values that a variable can take. In this sense, words in a natural language are, in large measure, labels of granules. A linguistic variable is a variable whose values are words or, equivalently, granules.*

Information granulation is naturally linked to the idea of computing with words and working with *vague concepts*. Mathematics requires that all notions (including set) must be exact, otherwise precise reasoning would be impossible. However, philosophers (see, e.g., [25]) and recently computer scientists as well as other researchers have become interested in *vague* (imprecise) concepts. Throughout the XX century one there occurred a significant drift of paradigms in modern science from dealing with precise concepts to vague concepts, especially in the case of complex systems (e.g., in economy, biology, psychology, sociology, quantum mechanics). In classical set theory, a set is uniquely determined by its elements. This means that every element must be uniquely classified as belonging to the set or not. That is to say the notion of a set is a *crisp* (precise) one. In contrast, all concepts that we are using in natural language are vague. Vagueness is actually an essential feature of natural language [26]. Moreover, a natural language is not a calculus with rigid rules matching all possible circumstances.

The impreciseness and vagueness of the objects of interest is intrinsic to some of the well established areas of science. The uncertainty of quantum mechanics or the mathematical objects such as *infinitely small* in non-standard analysis or a *neighborhood* of a point in some topological spaces (V-spaces) are just some of the prominent examples.

Approximate reasoning based on natural language should be based on vague concepts and not on bi-valent classical logic. One should also note that vagueness also relates to insufficient specificity, as the result of lack of feasible searching methods for sets of features adequately describing concepts.

Discussion on vague (imprecise) concepts in philosophy includes the following characteristic features of them [25]: (i) the presence of borderline cases, (ii) boundary regions of vague concepts are not crisp, (iii) vague concepts are susceptible to sorites paradoxes. From this it follows that at a given situation or moment of time we can only induce some approximations of vague concepts which should be adaptively changed when data and/or knowledge are changing.

Information granulation is also closely related to Computing with Words (CW) [23,24,27,28]. The main distinction between GrC and CW is that in CW the labels (or, more precisely syntax) of granules are defined by words (or expressions) from natural language. This is strongly related to linguistic variables [29].

3   Models of Granules

Granules can be regarded as abstract entities, corresponding to basic portions of information in our processing scheme. A granule is determined by a pair containing an expression formulated in some kind of language (formal, natural, basically any form of abstraction) and a meaning (semantics), which is actually a mechanism of classification whether a given object (or rather the data that we can see in relation to the object) satisfies the expression. It is also interesting to note that the way in which information about a given object is gathered can be parameterized and tuned adaptively.

Granules may have actually two kinds of labels representing, in a sense, their syntax. The first kind refers to formal languages (e.g., propositional formulæ) and is more directly related to rough sets, fuzzy sets, intervals and so on, which reflect "image" of labels at the level of empirical data. The second kind, representing vague concepts, requires utilization of natural languages (e.g., using words such as *safe* or *risky*) and can be approximated at the level of data by means of interaction between semantic and data-related layers of granule specifications.

In order to work with granules in practice, one needs at least two layers of their representation. The first, is the layer of semantic description of atomic or vague concepts. The second layer consists of the corresponding subsets of data elements that make it possible to translate computing with words to computing with actual data. We can say that atomic concepts are expected to be translated to a data level using more formalized language, such as e.g. decision logic or formal concept analysis, while vague concepts (referring, e.g., to some classes of complex situations) require more advanced, approximate constructions to reflect their meaning in data. Consequently, the area of GrC is strongly related to different approaches to dealing with imperfect information such as rough sets [30–32], fuzzy sets [24,33] and interval analysis [34]. These methodologies, discussed in more detail below, are just a few prominent representatives of granular representation.

The rough set philosophy is founded on the assumption that with every object of the universe of discourse we associate some information (data, knowledge). Objects characterized by the same information are indiscernible (similar) in view of the available information about them. The *indiscernibility relation* generated in this way is the mathematical basis of rough set theory. Each rough set has *borderline cases* (*boundary line*), i.e., objects which cannot be classified with certainty as members of either the set or its complement. This means that borderline cases cannot be properly classified by employing available knowledge. From this perspective, rough sets are related to the idea of vagueness mentioned in previous section.

Fuzzy sets admit partial membership of an element to a given granule [22–24,27,33]. Fuzzy membership functions are taking values in interval [0,1]. The approach is making it possible to work with the notions where the principle of dichotomy is neither justified nor advantageous. Problem solving by using fuzzy sets is based on selecting some elementary granules (i.e., fuzzy sets) and next on construction of fuzzy granules leading to the problem solution(s). The constructions of fuzzy granules are based on fuzzy logic. Let us also note that by using, e.g., shadowed sets [35] one can construct interesting granules by distinguishing which objects fully belong to the concept, which are excluded from it, and whose belongingness is completely unknown.

Interval analysis has emerged as soon as digital computers were used for dealing with problems caused by the representation of any number on a digital computer by a finite number of bits [34]. This interval nature of the arguments (variables) implies that the results are also intervals. Hence, the results have also the interval character. One can also observe that interval analysis is used due to aleatoric uncertainty, i.e., uncertainty occurring when for the same object the measurement results differ with time because the object changes or these results differ because we may select different boundaries (change specification) for the object. Interval analysis is instrumental in the analysis of propagation of granularity embedded in the original arguments (intervals). A lot of applications of interval analysis in combination with rough sets, fuzzy sets or probability methods were developed. Combination of these approaches leads to construction of higher order granules, e.g., in type-2 fuzzy sets, membership grades are quantified as fuzzy sets in [0,1] or subintervals of the unit interval. The elements of interval computing were also considered within rough set frameworks for analysis of incomplete and non-deterministic data sets [36,37], as well as for representing and computing with summarized statistics of granules or concepts [38,39].

All of the approaches to impreciseness discussed above are useful and each of them has its strengths. That is why in real applications they are usually intertwined in a way that boost they collective ability to address the task at hand. The synergy that results from expert use of various approaches together is a sought-after quality of any granular system.

4 Models of Computing with Granules

The developed GrC methods concern representation, construction, and processing of granules [7]. The label *Granular Computing* was suggested by Lin in late 1990s [40]. In GrC we deal with calculi of granules defined by elementary granules (e.g., indiscernibility or similarity classes) and some operations allowing us to construct new granules from already defined ones by their amalgamation and aggregation [41,42]. Graduation and granulation, as mentioned here, play an important role in differentiating between traditional and granular computing.

When thinking about a problem-solving environment inspired by natural language expert descriptions, we need a phase of design (including optimization and tuning) and a phase of its utilization in practice, for newly observed situations. Let us assume for a while that the system is already tuned, therefore, the vague and often very complex concepts reflecting not only the goals but also, e.g., risky situations are already mapped by appropriate granules. A given specific goal can be then paired with a number of strategies for its reaching, described by vague concepts as well. Then, starting from the initial information, the problem solver begins interacting with the environment, performing some actions and gathering feedback in terms of available (raw and then transformed) measurements and human observations. We can then say that an agent – a model of a human in the world of granular actions – compares the current state with historical data, in order to better understand the current situation, suggest some actions or ask for more specific information.

One can see that such requirements can be reflected by some widely known notions, such as induction, deduction and abduction, what-if analysis, reasoning and judgement. Quite often, like in control systems, the final goal is also combined with a constrained-specific goal of keeping the overall environment in a stable, safe state, which requires multi-level granular modeling too. Just like in the application of granular computing in building a support system for a commander of the fire brigade [43]. The plethora of possible readings from sensors and reports from firemen as well as background knowledge (building blueprints, engineering experience, etc.) is fed into a granular system that ultimately provides an advice to the commander in such a way that the resulting action scenario better fulfills vague constraints for concepts like "safe move" or "life threat avoidance".

One of the ways to arrive at the proper granular configuration is by applying the *adaptive judgment*. Adaptive judgment can be seen as the basic tool in discovering relevant granules of different complexity used for approximation of complex vague concepts and inducing approximate reasoning schemes on such approximations in the dynamically changing environment. In a nutshell, this approach makes it possible to build multi-layer hierarchies of granules accounting for both the high-level requirements and low-level limitations of data sources at the sensory level.

It all does not mean that GrC-specific techniques are the only way to address such problems. However, they are often useful to understand and somewhat decompose the physical and mathematical complexities of real-world processes using some analogies with natural-language-based hierarchical modeling. From this perspective, operating with granules is quite convenient, as they allow more imprecise mappings between concepts and subsets of variously represented objects satisfying those concepts. Human way of describing, combining and reasoning about objects can be then often naturally translated onto data manipulation operations known from relational databases, statistical transformations or spatio-temporal operators. Operations of the synthesis of objects described vaguely in natural language can be then mapped onto aggregations, amalgamations, similarities and gradual satisfiability at the level of data.

Given the above way of understanding how a granular system can be utilized, we can also think about its learning. It is certainly a very complicated, open and iterative process if we need to tune all layers of a granular system deployment, i.e., ontology of involved concepts' descriptions, their data-based meanings together with stability of data operations modelling concepts' synthesis, and physical parameters responsible for data acquisition from devices. Having all those degrees of freedom as components of the learning process, we can conduct a kind of supervised, semi-supervised or completely unsupervised learning against available empirical data. Still, the ways of data acquisition, preparation and granulation can play an important role in the learning process as well.

5   Evaluation of Granular Systems

The number of ways the GrS can be constructed given specification, constraints and the underlying data set(s) is practically infinite. Granulation itself may be introduced into universe of discourse in great many ways, as exemplified by the diversity of approaches presented in previous sections. The granules can be created as snapshots of information, such as sets and intervals, fuzzy and rough subsets, cluster and pattern descriptions, as well as various other constructs over the original universe of investigated objects. They can also take a form of the sets of original or granulated objects gathered together according to some constraints, with various types of links and hierarchical structures spanned over them.

The granules, whatever their origin, may provide the means to simplify, enhance, and speed up the computational tasks such as: searching, mining, or reasoning. We expect that by replacing original objects with granules we may obtain more compact outcome and avoid too detailed, hence non-transparent results, as well as – what is equally important in many applications – unnecessarily complex calculations leading to those results.

One of the principal issues in intelligent systems is related to control of interactive computations over granules for achieving the target goals. In granular environments we are often faced with considerable imprecision or vagueness which, if not handled carefully, can cause inefficiencies of computational processes, both with respect to their speed and their accuracy. The key to success is to properly manage the quality associated with each given level of granularity. In particular, one needs to take care of controlling the growth of imprecision along the chain of granular computations (see e.g. [44]) and the specifics of those computations, as well as optimizing the quality of granular systems both at the point of their construction and application-specific deployment.

As mentioned before, the measures we are after should assess the quality of granular system on the application level. They should provide the means for judging the quality of the outcome from the calculations (algorithms) that work with granules as an input.

The ultimate criterion of GrS' success should be related to the tangible requirements of particular domains of applications, treating the previously introduced measures rather as the means for heuristic estimate of the efficiency of granular solutions. In order to better grasp it, a high level quality measure is needed, one that assesses the overall performance of the granular system in the way similar to, e.g., empirical risk function known from the classical statistical learning theory.

The measures and criteria that we refer to are very varied, and not all of them are simply the sets of numerical values. There are two kinds (or rather two levels) of quality measures we are concerned with. First type of these measures is used to select and control the granule generators. These measures are used on the basic level to evaluate generated granules without looking at the way they will be further utilized. In particular, Pedrycz in [45] consider the principle of *justifiable granularity*. In a nutshell, the justifiable granularity principle require granule generators (algorithms for creation of granules) to work in such a way that the resulting granules are strongly supported by experimental evidence (*justified* by original data) and semantically *meaningful* (of sufficient specificity, sound level of detail). This principle mostly pertains the creation of granules, i.e., finding the proper granular representation. It can be materialized in the form of numerical evaluation functions for various types of granular environments, e.g., fuzzy-set-based or interval-based [46].

Our focus extends the above investigations or, in other words, treats their outcomes as one of necessary components of the granular system evaluation framework. One may say that the above-mentioned numerical evaluation functions are heuristic measures that – if appropriately designed and tuned – should anticipate whether a granular system is going to perform accurately and efficiently in practice. While the approach of Pedrycz ([45]) postulates the creation of granules to be treated as a two-way trade-off between justifiability (support) and meaningfulness (specificity), we want to include more factors, in particular, usefulness and applicability. The optimization of the granular system in our view is a three-way wager between justifiability, usefulness, and versatility. The versatility is understood by means of the algorithms that will operate on granules on the application level. We would like to have a granular system that nicely balances the effort needed to obtain the result using granules and the (possibly reduced) precision of the final result.

The granules are subjected to algorithms for creation of a model or a computational scheme (a computation on granules) that we are looking for (e.g., classification, information retrieval, or query processing). The resulting model is then evaluated with respect to quality, accuracy and applicability. In our view the final score should be established by calculating an analog of the empirical risk measure. Such risk-like measure is taken as a summarized expectation for creating a loss (or gain) due to use of particular granular system in a given application. It is quite common, and can be argued as commonsense, to make assessment of the quality of solution by hypothesizing the situations in which the gain/loss can be generated in our system, and then weighting them by the likelihood of their occurrence. The necessary step in this "granular risk assessment" is the calculation of gain (or loss). The proper choice of loss/gain function is a challenging task in itself, but this is the point where we may make a clever use of previously calculated "standard" measures such as accuracy, computational cost, etc.

6   Conclusions

We have outlined some basic issues concerning Granular Systems and Granular Computing. It is only a brief overview, but can serve as a start point for anyone interested in deepening the knowledge about computational approaches to tackling the problems intrinsic in modern computational intelligence systems such as: vagueness, incompleteness, inconsistency, or imprecision.

It is worth mentioning that the importance of this approach will grow in the future for development of intelligent systems, where we are dealing with with data and knowledge making it possible to obtain only imperfect (i.e., imprecise, uncertain, incomplete, or unreliable) information about perceived objects.

References

1. Lin, T.Y., et al.: Granular computing - topical section. In: Meyers, R.A. (ed.) Encyclopedia of Complexity and Systems Science, pp. 4283–4435. Springer, New York (2009)
2. Yao, Y., Zhong, N.: Granular computing. In: Wah, B., Wah, B.M. (eds.) Wiley Encyclopedia of Computer Science and Engineering. Wiley, New York (2008)
3. Pedrycz, W.: History and development of granular computing. In: UNESCO-EOLSS Joint Committee, (ed.) Encyclopedia of Life Support Systems (EOLSS). Eolss Publishers, Paris (2012)
4. Apolloni, B., Pedrycz, W., Bassis, S., Malchiodi, D.: The Puzzle of Granular Computing. Studies in Computational Intelligence, vol. 138. Springer, Heidelberg (2008)
5. Bargiela, A., Pedrycz, W.: Granular Computing: An Introduction. Kluwer Academic Publishers, Dordrecht (2003)
6. Bello, R., Falcón, R., Pedrycz, W.: Granular Computing: At the Junction of Rough Sets and Fuzzy Sets. Studies in Fuzziness and Soft Computing, vol. 234. Springer, Heidelberg (2010)
7. Pedrycz, W.: Granular Computing Analysis and Design of Intelligent Systems. CRC Press, Taylor and Francis, Boca Raton (2013)
8. Polkowski, L., Artiemjew, P.: Granular Computing in Decision Approximation: An Application of Rough Mereology. Intelligent Systems Reference Library. Springer, Switzerland (2015)
9. Stepaniuk, J.: Rough-Granular Computing in Knowledge Discovery and Data Mining. Springer, Heidelberg (2008)
10. Inuiguchi, M., Hirano, S., Tsumoto, S. (eds.): Rough Set Theory and Granular Computing. Studies in Fuzziness and Soft Computing, vol. 125. Springer, Heidelberg (2003)
11. Lin, T.Y., Yao, Y., Zadeh, L.A. (eds.): Rough Sets, Granular Computing and Data Mining. Studies in Fuzziness and Soft Computing. Physica-Verlag, Heidelberg (2001)
12. Pal, S.K., Skowron, A. (eds.): Rough Fuzzy Hybridization: A New Trend in Decision-Making. Springer, Singapore (1999)
13. Pal, S.K., Polkowski, L., Skowron, A. (eds.): Rough-Neural Computing: Techniques for Computing with Words. Cognitive Technologies. Springer, Heidelberg (2004)
14. Pedrycz, W. (ed.): Granular Computing: An Emerging Paradigm. Studies in Fuzziness and Soft Computing, vol. 70. Physica-Verlag, Heidelberg (2001)
15. Pedrycz, W. (ed.): Knowledge-Based Clustering. From Data to Information Granules. Wiley, New York (2005)
16. Bargiela, A., Pedrycz, W. (eds.): Human-Centric Information Processing Through Granular Modelling. Studies in Computational Intelligence, vol. 182. Springer, Heidelberg (2009)

17. Pedrycz, W., Chen, S.M. (eds.): Granular Computing and Intelligent Systems Design with Information Granules of Higher Order and Higher Type. Studies in Computational Intelligence, vol. 502. Springer, Heidelberg (2011)

18. Pedrycz, W., Chen, S.M. (eds.): Information Granularity, Big Data, and Computational Intelligence. Studies in Big Data, vol. 8. Springer, Heidelberg (2015)

19. Yao, J.T. (ed.): Novel Developments in Granular Computing: Applications for Advanced Human Reasoning and Soft Computation. IGI Global, Hershey (2010)

20. Zadeh, L.A., Kacprzyk, J. (eds.): Computing with Words in Information/Intelligent Systems. Physica-Verlag, Heidelberg (1999)

21. Zhang, L., Zhang, B. (eds.): Quotient Space Based Problem Solving: A Theoretical Foundation of Granular Computing. Elsevier, Amsterdam (2014)

22. Zadeh, L.A.: Toward a theory of fuzzy information granulation and its centrality in human reasoning and fuzzy logic. Fuzzy Sets Syst. **90**, 111–127 (1997)

23. Zadeh, L.A.: Generalized theory of uncertainty (GTU) - principal concepts and ideas. Comput. Stat. Data Anal. **51**, 15–46 (2006)

24. Zadeh, L.A. (ed.): Computing with Words: Principal Concepts and Ideas. Studies in Fuzziness and Soft Computing, vol. 277. Springer, Heidelberg (2012)

25. Keefe, R.: Theories of Vagueness. Cambridge Studies in Philosophy. Cambridge University Press, Cambridge (2000)

26. Baker, G., Hacker, P.: Wittgenstein: Understanding and Meaning. Analytical Commentary on the Philosophical Investigations, Part II: Exegesis 1–184, vol. 1, 2nd edn. Wiley-Blackwell Publishing, Oxford (2004)

27. Zadeh, L.A.: Fuzzy logic = Computing with words. IEEE Trans. Fuzzy Syst. **2**, 103–111 (1996)

28. Zadeh, L.A.: From computing with numbers to computing with words - from manipulation of measurements to manipulation of perceptions. IEEE Trans. Circuits Syst. **45**, 105–119 (1999)

29. Zadeh, L.A.: Outline of a new approach to the analysis of complex systems and decision processes. IEEE Trans. Syst. Man Cybern. **SMC–3**, 28–44 (1973)

30. Pawlak, Z.: Rough Sets: Theoretical Aspects of Reasoning about Data. System Theory, Knowledge Engineering and Problem Solving, vol. 9. Kluwer Academic Publishers, Dordrecht (1991)

31. Pawlak, Z., Skowron, A.: Rudiments of rough sets. Inf. Sci. **177**(1), 3–27 (2007)

32. Pawlak, Z., Skowron, A.: Rough sets: some extensions. Inf. Sci. **177**(1), 28–40 (2007)

33. Zadeh, L.A.: Fuzzy sets. Inf. Control **8**, 338–353 (1965)

34. Moore, R., Kearfott, R.B., Cloud, M.J.: Introduction to Interval Analysis. SIAM, Philadelphia (2009)

35. Pedrycz, W.: From fuzzy sets to shadowed sets: interpretation and computing. Int. J. Intell. Syst. **24**(1), 48–61 (2009)

36. Sakai, H., Okuma, H., Nakata, M., Ślęzak, D.: Stable rule extraction and decision making in rough non-deterministic information analysis. Int. J. Hybrid Intell. Syst. **8**(1), 41–57 (2011)

37. Sakai, H., Wu, M., Nakata, M.: Apriori-based rule generation in incomplete information databases and non-deterministic information systems. Fundamenta Informaticae **130**(3), 343–376 (2014)

38. Ślęzak, D., Synak, P., Wojna, A., Wróblewski, J.: Two database related interpretations of rough approximations: data organization and query execution. Fundamenta Informaticae **127**(1–4), 445–459 (2013)

39. Pankratieva, V.V., Kuznetsov, S.O.: Relations between proto-fuzzy concepts, crisply generated fuzzy concepts, and interval pattern structures. Fundamenta Informaticae **115**(4), 265–277 (2012)
40. Lin, T.Y.: Data mining and machine oriented modeling: a granular computing approach. Appl. Intell. **13**(2), 113–124 (2000)
41. Polkowski, L., Skowron, A.: Rough mereological calculi of granules: a rough set approach to computation. Comput. Intell. **17**(3), 472–492 (2001)
42. Skowron, A., Stepaniuk, J., Peters, J.F., Świniarski, R.W.: Calculi of approximation spaces. Fundamenta Informaticae **72**, 363–378 (2006)
43. Krasuski, A., Jankowski, A., Skowron, A., Ślęzak, D.: From sensory data to decision making: a perspective on supporting a fire commander. In: 2013 IEEE/WIC/ACM International Conferences on Web Intelligence and Intelligent Agent Technology, Atlanta, Georgia, USA, 17–20 November 2013, Workshop Proceedings, pp. 229–236. IEEE Computer Society (2013)
44. Szczuka, M.S., Skowron, A., Stepaniuk, J.: Function approximation and quality measures in rough-granular systems. Fundamenta Informaticae **109**(3), 339–354 (2011)
45. Pedrycz, W.: The principle of justifiable granularity and an optimization of information granularity allocation as fundamentals of granular computing. J. Inf. Process. Syst. **7**(3), 397–412 (2011)
46. Apolloni, B., Pedrycz, W., Bassis, S., Malchiodi, D.: The Puzzle of Granular Computing. Studies in Computational Intelligence, vol. 138. Springer, Heidelberg (2008)

The Rough Granular Approach to Classifier Synthesis by Means of SVM

Jacek Szypulski and Piotr Artiemjew[✉]

Department of Mathematics and Computer Science,
University of Warmia and Mazury, Sloneczna 54, 10-710 Olsztyn, Poland
{jszypulski,artem}@matman.uwm.edu.pl

Abstract. In this work we exploit the effects of applying methods for constructions of granular reflections of decision systems developed up to now in the framework of rough mereology, along with kernel methods for the building of classifiers. In this preliminary report we present results obtained with the SVM classification with use of the RBF kernel. The approximation metod we use is the optimized ε concept dependent granulation. We experimentally verify the validity of this new approach with test data: Wisconsin Diagnostic Breast Cancer, Fertility Diagnosis, Parkinson Disease and the Prognostic Wisconsin Breast Cancer Database. The results are very promising as the obtained accuracy is not diminished but the size of the granular decision system is radically diminished.

Keywords: Rough sets · Decision systems · SVM · Granular rough computing

1 Introduction

The basic way of approximations of decision system considered in this work was proposed by Polkowski in [5–7]. Such approximation methods turn out to work in an effective way with many classifiers, including classic rough set methods, classifiers based on weak variants of rough inclusions, rule-based classifiers, Bayes classifier, k-nn classifier and so on - see among others [8,9]. The classic method - normal granulation [5–7] - was later extended to many variants, among others into granulation in decision concepts with discernibility ratio of descriptors (the ε - granulation) - see [1,2]. This variant is used in this paper in context of SVM classification. Additionally, the process of granulation is optimized by parallel approximation and covering. The detail description of our method is provided in Sect. 3.

The rest of the paper is as follows. In Subsects. 1.1, 1.2, 1.3 and 1.4 we have the motivation, the methodology and theoretical background for the used approximation method, as well as the brief description of used classifier. In Sect. 2 we have the detail description of our modification of granulation method. In Sect. 3 we have the experimental session, and finally in Sect. 4 we have conclusion and future work description.

© Springer International Publishing Switzerland 2015
Y. Yao et al. (Eds.): RSFDGrC 2015, LNAI 9437, pp. 256–263, 2015.
DOI: 10.1007/978-3-319-25783-9_23

1.1 Motivation

The main reason for this work was to check the effectiveness of the SVM classifier in classification of granulated data. This is the basic result, which gives us the first insight on the behaviour of such classifier on the approximation based on the ε concept-dependent granulation - see [1,2].

1.2 Methodology

We optimize the granulation method by parallel computing of granules and a covering, see Sect. 3. We use classification by SVM with RBF kernel and default parameters from LIBSVM tool. For evaluation of results we use multiple Cross Validation 5 method.

Let us to provide the theoretical background for our approach.

1.3 Granulation in Rough Mereology

Rough mereology is a theory of the predicate $\mu(x, y, r)$ read: "x is a part of y to a degree r", and called a *rough inclusion*, see [8].

We recall that an *information system* (a *data table*) is represented as a pair (U, A) where U is a finite set of things and A is a finite set of *attributes*; each attribute $a : U \to V$ maps the set U into the *value set V*. For an attribute a and a thing v, $a(v)$ is the value of a on v.

We apply a particular form of a rough inclusion defined as follows.

For an attribute a, we let a_{max}, resp. a_{min}, to be the maximal, resp., the minimal value of the attribute a on objects in the decision system, and then $span(a) = |a_{max} - a_{min}|$ is the span of a.

Given a parameter ε, we call two values $a(u), a(v)$ of the attribute a ε–*similar* if the inequality $|a(u) - a(v)| \leq \varepsilon$ holds, in symbol $sim_\varepsilon(a(u), a(v))$. For a given object u, we define the *granule about u and of the granulation radius r*, $g_\varepsilon(u, r)$ as the set,

$$g_\varepsilon(u, r) = \{v : \frac{|\{a \in A : IND_\varepsilon(u, v)\}|}{|A|} \geq r\}, \tag{1}$$

$$where \ IND_\varepsilon(u, v) = \{a \in A : sim_\varepsilon(a(u), a(v))\}.$$

where $|.|$ denotes the size of a set. Having granules defined, we continue with the granulation procedure. We apply the sequential covering method by selecting an object, building a granule around it, removing the granule from the universe of objects and repeating until all objects are covered. Each of the obtained granules in the covering is factorized by selecting for each attribute the representative value for the granule by majority voting with random tie resolution. By this process, each granule is replaced with a vector of attribute values. The obtained reflection of the original decision system is then subject to classification by means of C–SVC, with the radial basis kernel RBF, see [3,4].

1.4 Support Vector Machine Classifier

In this work for the experiments we use the LIBSVM tool for classification based on SVM with RBF kernel [4]. The SVM classifier is used based on scripts with generated approximated data. During the classification the test and training data are scaled into the interval $[-1, 1]$, and after that the model is created based on training data, and classification of test data is performed.

In the next section we have brief description of our approach to granulation, investigated recently in the [1,2].

2 Optimized Concept Dependent ε-Granulation

In general the process consists of the following steps. We get the random training object and form around it, the ε granule. In the next step we get the object outside the covering, and we form the next granule. The process of classification for the most of parameters converges fastly, and there is no need to compute the approximations for all objects, with exception of the granules which contain only their centers. The process stops where the covering is equal to the universe of training object.

The detail procedure of covering is the following,
(i) from the decision system (U, A, d), we form the TRN and TST data sets,
(ii) $U_{cover} = \emptyset$,
(iii) we set the granulation radius r_{gran} and the discernibility ratio of attributes ε,
(iv) for given $TRN = \{u_1, u_2, ..., u_{|TRN|}\}$, we form the $TRN_{temp} = TRN - U_{cover}$, we get in random way the object $u \in TRN_{temp}$, and form the granule

$$g_{r_{gran}}^{\varepsilon,cd}(u) = \{v \in U, \frac{|IND_\varepsilon(u,v)|}{|A|} \geq r_{gran} \text{ and } d(u) = d(v)\}$$

$$IND_\varepsilon(u,v) = \{a \in A : \frac{|a(u) - a(v)|}{max_a - min_a} \leq \varepsilon\},$$

(v) $U_{cover} \leftarrow g_{r_{gran}}^{\varepsilon,cd}(u)$,
(vi) if the U_{cover} is equal TRN, we go to the point (vii), otherwise to the point
(vii) we form the granular reflections of the original TRN system based on the granules from U_{cover} with use of majority voting where the discernibility ratio of attribute values is considered and the ties are resolved randomly.

3 Experimental Session

The experimental session is designed based on five times the Cross Validation 5 method. The original decision system is split into five parts, and we conduct five Train and Test experiments, where each time the other part is treated as test set and the rest as training set. In the next step the training decision systems are

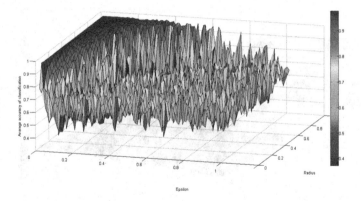

Fig. 1. 5xCV5 - Classification result for Wisconsin Diagnostic Breast Cancer data set

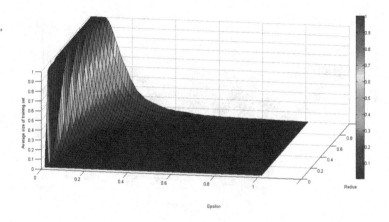

Fig. 2. Size of the training set of granular Wisconsin Diagnostic Breast Cancer data

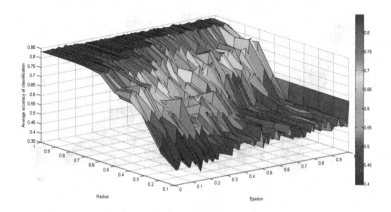

Fig. 3. 5xCV5 - Classification result for Fertility Diagnosis data set

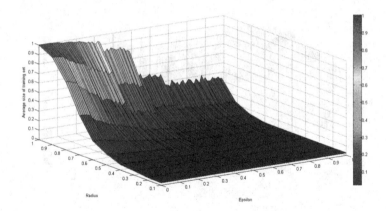

Fig. 4. Size of the training set of granular Fertility Diagnosis data

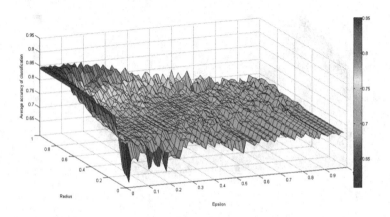

Fig. 5. 5xCV5 - Classification result for Parkinson Disease data set

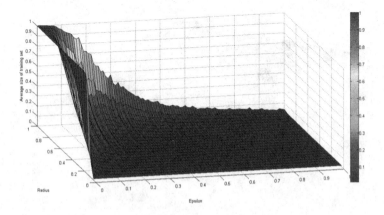

Fig. 6. Size of the training set of granular Parkinson Disease data

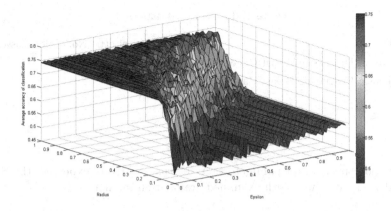

Fig. 7. 5xCV5 - Classification result for WPBC data set

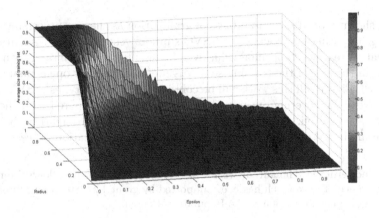

Fig. 8. The size of the training set of granular WPBC data

granulated based on ε concept dependent granulation, and we form the granular reflections of data for all radii $(r_g ran)$ and epsilons (ε). And the test sets are classified based on SVM by TRN sets. The process is repeated five times and we get the average result of accuracy and percentage size of granulated training decision systems.

3.1 Results of Experiments

We have carried out experiments for the data sets from UCI repository listed in the Table 1.

In the Figs. 1, 3, 5 and 7, we have the Average of accuracy of classification for all the spectrum of granulation radii and parameters ε, and in the Figs. 2, 4, 6 and 8, respective percentage size of granulated training sets. The results show the high approximation level with maintenance of the good accuracy of classification

Table 1. Table of used data sets

Name	No. of obj	No. of attr	No. of dec classes
Wisconsin Diagnostic Breast Cancer	569	32	2
Wisconsin Prognostic Breast Cancer	198	34	2
Fertility	100	10	2
Parkinsons	197	23	2

for the wide spectrum of parameters. As it was intuitively expected, the SVM method works very well with examined granulation method.

4   Conclusions

Results shown indicate the usefulness of granulation as the pre–processing stage in classifier synthesis by means of C–SVC support vector machine. Accuracy of classification reaches its optimal value even for small granulation radii with large decrease in the size of the training set. This preliminary work opens many ways of research. First of all this work will be continued with other types of SVM's and kernels. Next the covering optimization method will be compared with the case without optimization. Another way of research will be the comparison between the covering based on joint and disjoint granules in optimization process.

Acknowledgement. The author wishes to thank Professor Lech Polkowski for kind help and advice. The research has been supported by grant 1309-802 from Ministry of Science and Higher Education of the Republic of Poland.

References

1. Artiemjew, P.: On strategies of knowledge granulation and applications to decision systems, Ph.D. Dissertation, Polish Japanese Institute of Information Technology. L. Polkowski, Supervisor, Warsaw (2009)
2. Artiemjew, P.: A review of the knowledge granulation methods: discrete vs continuous algorithms. In: Skowron, A., Suraj, Z. (eds.) Rough Sets and Intelligent Systems. ISRL, vol. 43, pp. 41–59. Springer, Heidelberg (2013)
3. Boser, B.E., Guyon, I., Vapnik, V.: A training algorithm for optimal margin classifiers. In: Proceedings of Vth International Workshop on Computational Learning Theory, pp. 144–152. ACM Press (1992)
4. Chang, C.-C., Lin, C.-J.: LIBSVM, a library for support vector machines. ACM Trans. Intell. Syst. Technol. **2**, 27:1–27:27 (2011). http://www.csie.ntu.edu.tw/~cjlin/libsvm
5. Polkowski, L.: Formal granular calculi based on rough inclusions (a feature talk). In: Proceedings 2005 IEEE International Conference on Granular Computing GrC 2005, pp. 57–62. IEEE Press (2005)

6. Polkowski, L.: Granulation of knowledge in decision systems: the approach based on rough inclusions. The method and its applications. In: Kryszkiewicz, M., Peters, J.F., Rybiński, H., Skowron, A. (eds.) RSEISP 2007. LNCS (LNAI), vol. 4585, pp. 69–79. Springer, Heidelberg (2007)
7. Polkowski, L.: Formal granular calculi based on rough inclusions (a feature talk). In: Proceedings 2006 IEEE International Conference on Granular Computing GrC 2006, pp. 57–62. IEEE Press (2006)
8. Polkowski, L.: Approximate Reasoning by Parts. An Introduction to Rough Mereology. ISRL, vol. 20. Springer, Heidelberg (2011)
9. Polkowski, L., Artiemjew, P.: Granular Computing in Decision Approximation - An Application of Rough Mereology. ISRL, vol. 77. Springer, Switzerland (2015). ISBN 978-3-319-12879-5, pp. 1–422

Data Mining and Machine Learning

The Boosting and Bootstrap Ensemble for Classifiers Based on Weak Rough Inclusions

Piotr Artiemjew[(✉)]

Department of Mathematics and Computer Science,
University of Warmia and Mazury, Olsztyn, Poland
`artem@matman.uwm.edu.pl`

Abstract. In the recent works we have investigated the classifiers based on weak rough inclusions, especially the 8v1.1 - 8v1.5 algorithms. These algorithms in process of weights forming for classification dynamically react on the distance between the particular attributes. Our results show the effectiveness of these methods and the wide application in many contexts, especially in the context of classification of DNA Microarray data. In this work we have checked a few methods for classifier stabilisation, such as the Bootstrap Ensemble, Boosting based on Arcing, and Ada-Boost with Monte Carlo split. We have performed experiments on selected data from the UCI Repository. The results show that the committee of weak classifiers stabilised our algorithms in the context of accuracy of classification. The Boosting based on Arcing turned out to be the most promising method among those examined.

Keywords: Rough sets · Decision systems · Bootstrap ensemble · Ada-Boost · Arcing

1 Introduction

The main motivation to conduct this experimental work was the necessity of search for methods, which would stabilise the classification process of our classifiers, making the classification model less influenced by the particular split of the data during model creation. Seeing the effectiveness of ensemble scheme of classification, in the context of rough set methods - see [16–20] we decided to check the selected methods on our classifiers.

We have chosen for our experiments a group of classifiers based on weak variants of rough inclusions Alg8v1.1-Alg8v1.5 - see [1–5,8,9]. These classifiers turned out to be among the best we developed in their family of methods.

In this work we make a preliminary search for the best stabilisation method in the context of the chosen data sets. The Boosting is the family of algorithms for the sequential production of classifiers, where each classifier is dependent on the previous one and focuses on the previous one's error [10,12,13]. Examples that are incorrectly predicted in previous classifiers are chosen more often, or weighted more heavily. The Ensemble of Bootstraps [11] forms the group of

© Springer International Publishing Switzerland 2015
Y. Yao et al. (Eds.): RSFDGrC 2015, LNAI 9437, pp. 267–277, 2015.
DOI: 10.1007/978-3-319-25783-9_24

classifiers, where each classifier is dependent on the previous one, but there is no influence of classification accuracy.

We have prepared three types of experiment, where the first one consists of the stabilisation of classification based on the mentioned Committee of Bootstraps [11], the second on Boosting based on Arcing [14,15], and the third on Ada-Boost with Monte Carlo split [10,12,13]. The detailed description is given in Sect. 3.

The rest of the paper has the following content. In Subsect. 1.1 we have the theoretical introduction to rough mereology classifiers. In Sect. 2 we have the detailed description of our classifiers. In Sect. 3 we have described the Boosting methods used in this work. In Sect. 4 we show the results of the experiments, and we conclude the paper in Sect. 5.

We now describe background information about how the classifiers are considered in terms of rough set theory.

1.1 Theoretical Background of Our Classifiers

Firstly we introduce the residual rough inclusions. Considering the decision system in the form of triple (U, A, d), where the U is the set of objects, the A is a set of conditional attributes, and $d \notin A$ is the decision attribute, the rough inclusion in terms of residuum of a t–norm can be outlined in the following way. The t–norm, see, e.g., [6], is a function $T : [0, 1] \times [0, 1] \rightarrow [0, 1]$ which is symmetrical, associative, increasing in each coordinate, and subject to boundary conditions: $T(x, 0) = 0, T(x, 1) = x$. The more popular t–norms are, the Łukasiewicz t–norm $L(x, y) = max\{0, x+y-1\}$, the Product t–norm $P(x, y) = x \cdot y$, and the Minimum t–norm $M(x, y) = min\{x, y\}$. The *residuum* $x \Rightarrow_T y$ of a t–norm T is meant as a function, defined in the following way, $x \Rightarrow_T y \geq r$ *if and only if* $T(x, r) \leq y$. All t–norms L, P, M are continuous, in their cases, the residual implication is given by the formula, $x \Rightarrow_T y = max\{r : T(x, r) \leq y\}$. Residual rough inclusions on the interval $[0, 1]$ are defined, see, e.g., [7] as,

$$\mu_T(x, y, r) \; if \; and \; only \; if \; x \Rightarrow_T y \geq r \tag{1}$$

To transfer the residual rough inclusions from the unit interval (1) to decision systems, see [7], we use the following factors, $dis_\varepsilon(u, v) = \frac{|\{a \in A : ||a(u) - a(v)|| \geq \varepsilon\}|}{|A|}$ and $ind_\varepsilon(u, v) = \frac{|\{a \in A : ||a(u) - a(v)|| < \varepsilon\}|}{|A|}$, where ε is in the interval $[0, 1]$, $|A|$ is the number of attributes in A, and $||a(u) - a(v)||$ means the distance between the attribute values. A variant of weak rough inclusion, see [7], is given as, $\mu_T^\varepsilon(v, u, r)$ *if and only if* $dis_\varepsilon(u, v) \Rightarrow_T ind_\varepsilon(u, v) \geq r$.

And now, we show the way of voting for decision assignment. To synthesize the classifier, the rough inclusion, e.g., induced by the Łukasiewicz t–norm L, is used as follows. Considering splitting of original data set into a training (U_{trn}, A, d), and a test (U_{tst}, A, d) sets. For all training objects v we form the following weight with respect to a test object u. $w(v, u, \varepsilon) = dis_\varepsilon(u, v) \Rightarrow_T ind_\varepsilon(u, v)$. And finally, all decision classes compete with respect to parameter,

$Param(c) = \sum_{\{v \in U_{trn}:d(v)=c\}} w(v, u, \varepsilon)$. The class with the smallest parameter assigns the decision to test object u.

After this introduction into the theoretical background we show applications of the presented scheme.

2   8_v1.1-8_v1.5 Classifiers

The way of voting shown in the previous section is here extended along the lines of [8]. Now, we show the procedure of Alg. 8_v1.1-8_v1.5 - for more detail see [1–5, 8, 9]. In general, the classification process consists of the following steps. We form the training and test systems. The table of maximal and minimal attribute values is computed from the training data set. The outliers from the test set are mapped into the interval of founded minimal and maximal values. The parameter $\varepsilon \in [0, 1]$ is fixed. After the preliminary steps, test object u can be classified in the following ways. In case of Alg. 8v1.1 we compute the weight $w(u, v)$ for all training objects v as follows, $w(u, v) \leftarrow w(u, v) + \frac{|a(u)-a(v)|}{max_{trn\ set}a - min_{trn\ set}a}, a \in A$. For the rest of algorithms we have two subcases, in which, respectively, the $\frac{|a(u)-a(v)|}{max_{trn\ set}a - min_{trn\ set}a} \geq \varepsilon \rightarrow$, and $\frac{|a(u)-a(v)|}{max_{trn\ set}a - min_{trn\ set}a} < \varepsilon \rightarrow$. For Alg. 8v1.2, respective to subcases we have weights,

$$w(u, v) \leftarrow w(u, v) + \frac{|a(u) - a(v)|}{max_{trn\ set}a - min_{trn\ set}a} * \frac{(1 + \varepsilon)}{\varepsilon}.$$

$$w(u, v) \leftarrow w(u, v) + \frac{|a(u) - a(v)|}{max_{trn\ set}a - min_{trn\ set}a} * \frac{1}{\varepsilon}.$$

For Alg. 8v1.3, respectively,

$$w(u, v) \leftarrow w(u, v) + \frac{|a(u) - a(v)|}{max_{trn\ set}a - min_{trn\ set}a} * (1 + \varepsilon).$$

$$w(u, v) \leftarrow w(u, v) + \frac{|a(u) - a(v)|}{max_{trn\ set}a - min_{trn\ set}a} * \frac{1}{1 + \varepsilon}.$$

In case of Alg. 8v1.4,

$$w(u, v) \leftarrow w(u, v) + \frac{|a(u) - a(v)|}{max_{trn\ set}a - min_{trn\ set}a}.$$

$$(\varepsilon + \frac{|a(u) - a(v)|}{max_{trn\ set}a - min_{trn\ set}a})$$

$$w(u, v) \leftarrow w(u, v) + \frac{|a(u) - a(v)|}{(max_{trn\ set}a - min_{training\ set}a) \cdot \varepsilon}$$

And finally for Alg. 8v1.5 we have

$$w(u, v) \leftarrow w(u, v) + \frac{|a(u) - a(v)|}{max_{trn\ set}a - min_{trn\ set}a}$$

$$w(u, v) \leftarrow w(u, v) + \frac{|a(u) - a(v)|}{(max_{trn\ set}a - min_{trn\ set}a) \cdot \varepsilon}$$

After the weights are computed for all training objects v with respect to test object u, the following parameter is computed for all decision classes c. $Param(c) = \sum_{\{v \in U_{trn}: d(v) = c\}} w(u_q, v_p)$. The class with the minimal value of $Param$ assigns the decision to u.

In the next section we introduce the selected methods for classifiers stabilisation.

3   Classifiers Stabilisation Methods

There are many, methods which let us model classification problems and check the effectiveness of classifiers in a particular context. Among the more popular methods are the Cross Validation method, Monte Carlo Cross Validation, Bagging, Committee of classifiers, and Boosting. Due to lack of space, in this work we have chosen only a few methods to check the boosting, stabilisation effect on our classifiers. Let us list the methods with brief descriptions below.

3.1   Bootstrap Ensembles

In other words the random committee of bootstraps [11], it is a method in which the original decision system - the basic knowledge - is split into (TRN) training data set, and ($TSTvalid$) validation test data set. And from the TRN system, for a fixed number of iterations, we form new Training systems ($NewTRN$) by random choice with returning of $card\{TRN\}$ objects. In all iterations we classify the TRNvalid system in two ways: the first based on the actual $NewTRN$ system and the second based on the committee of all performed classifications. In the committee majority voting is performed and the ties are resolved randomly.

3.2   Boosting Based on Arcing

In the general context, this method is similar to the first one, but here the TRN is split into two data sets $NewTRN$ and $NewTST$ - see [14,15]. The split is based on Bootstraps and the NewTRN system is formed by weights assigned to training objects. The objects are chosen for the NewTRN with a fixed probability determined by weights. The weights are initially equal, but after the first classification of the NewTST system based on NewTRN, the weights are modified in such a way that well-classified objects have lowered weights, and after that the normalisation of weights is performed. This method of forming Bootstraps is called Arcing. After the classification, the $NewTRN$ classifies the $TSTvalid$ in a single iteration and as the committee of classifiers. In Arcing the factor for weights modification is equal $\frac{1-Accuracy}{Accuracy}$.

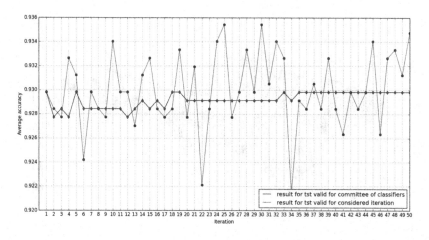

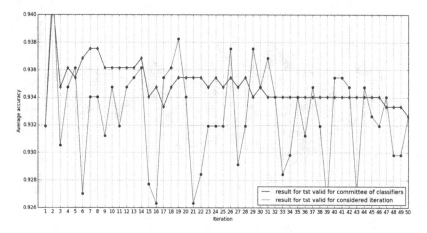

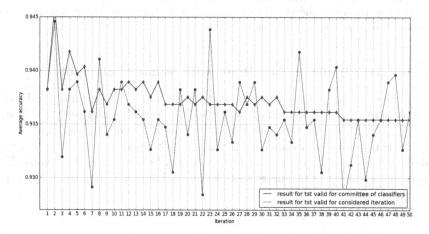

Fig. 1. The result for 5×50 iterations of learning for Wisconsin Diagnostic Breast Cancer for 8v1.1 algorithm. The first picture show the result for Ensemble of Bootstraps, the second for Arcing, and the last one for Ada-Boost

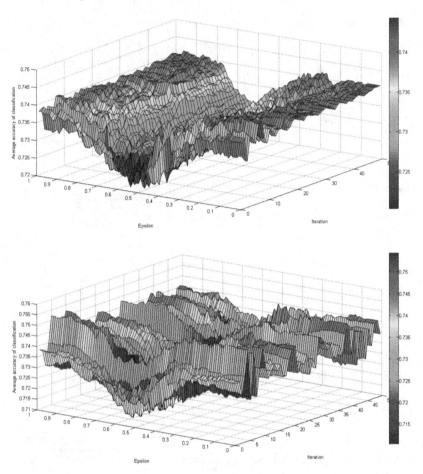

Fig. 2. The result for 5 x Ensemble of Bootstraps with 50 iterations of learning for Pima Indians Diabetes and 8v1.2 algorithm. The higher result is for committee of classifiers, the lower one for particular iterations

3.3   Boosting Based on Ada-Boost with Monte Carlo Split

In this case we use a similar method of classification to the one previously described, but here we use a different method for NewTRN and NewTST forming - see [10,12,13]. We split the TRN data set according to the fixed ratio, and choose the objects in the NewTRN based on weights. The good split ratio is about 0.6, because it is close to the approximate size of the distinguishable objects in the bootstraps. The rest of the algorithm works in a similar way to the previous one.

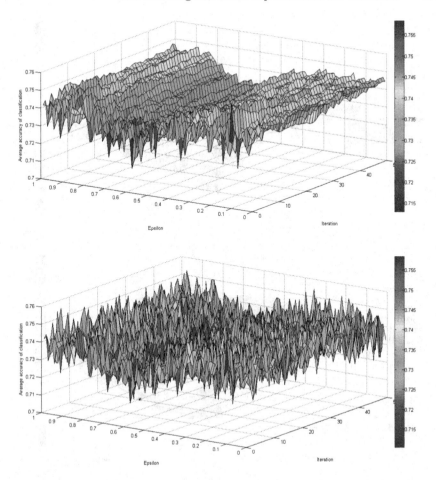

Fig. 3. The result for 5 x Boosting based on Arcing with 50 iterations of learning for Pima Indians Diabetes and 8v1.3 algorithm. The higher result is for committee of classifiers, the lower one for particular iterations

4   Experimental Session

In the experimental part we have carried out experiments on the real data from the UCI Repository. We use Australian, Pima Indians Diabetes, Heart Disease, and Wisconsin Breast Cancer data sets. The boosting, stabilisation methods are performed five times and the average results are presented on the plots. The classifier methods we use are mentioned Alg1v1.1-Alg1v1.5. The experimental session was extensive but due to lack of space we show only a few selected results.

4.1   The Results of Experiments

For the three described methods we used our five classifiers Alg1v1.1-Alg1v1.5. In the Figs. 1, 2, 3, 4 and 5 we have exemplary results from these experiments.

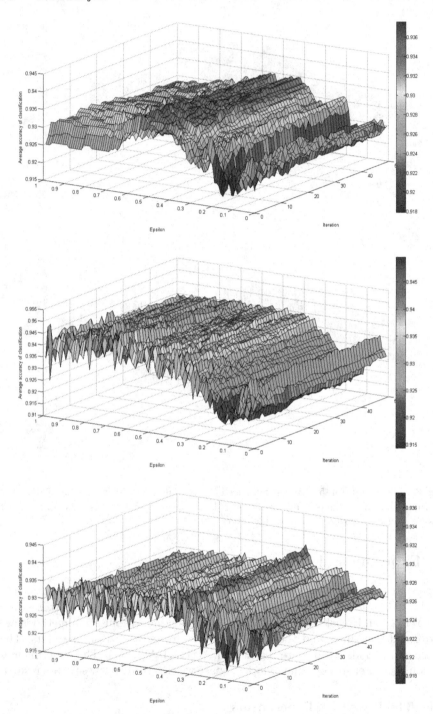

Fig. 4. The result for 5×50 iterations of learning for Wisconsin Diagnostic Breast Cancer for 8v1.4 algorithm. The first picture show the result for Ensemble of Bootstraps, the second for Arcing, and the last one for Ada-Boost

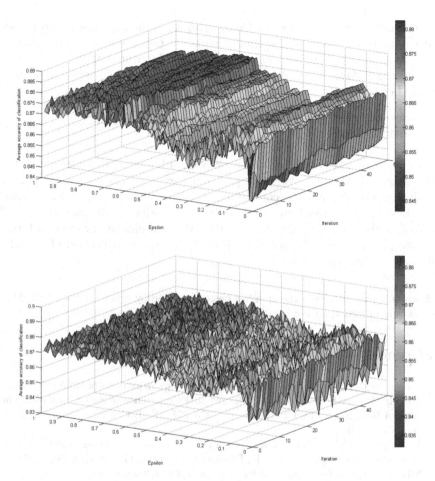

Fig. 5. The result for 5 x Ensemble of Bootstraps with 50 iterations of learning for Pima Indians Diabetes and 8v1.2 algorithm. The higher result is for committee of classifiers, the lower one for particular iterations

Particularly in the Fig. 1 we have the result for Alg8v1.1 for three stabilisation methods. In the Figs. 2, 3 and 5, we have the demonstration of stabilization effect for Diabetes and Australian data with Ensemble of Bootstraps for Alg8v1.2, Arcing for Alg8v1.3 and Ada-Boost with Monte Carlo split for Alg8v1.5 respectively. In the Fig. 4, we have the result Alg8v1.4 and all the stabilisation methods.

Considering all results, it turned out that the stabilisation of classification for the committee started from about 20 iterations of learning, and after that the standard deviation of results was less than 0.005. Obviously the results for the committee are more stable than for the single classifiers for particular iterations. For the selected results the results for the committee of bootstraps can be treated as a base line, and after that the boosting based on the Arcing and the Monte Carlo Split shows slightly better and more stable results than for the pure committee without boosting. Considering all of the experiments we can see that the boosting based on the Arcing works for the investigated data slightly better that the other methods for parameterized algorithms Alg8v1.2-Alg8v1.5. Only in case of Alg8v1.1 an Ada-Boost with Monte Carlo split seems to be work slightly better than the Arcing.

5   Conclusions

In this paper we have compared three methods of boosting, stabilisation of classifiers Alg1v1.1-Alg1v1.5. We have performed experiments on real data from the UCI repository, and the conclusion is that the Boosting based on Arcing works best in most cases for our parameterized classifiers. We observed that all methods in the form of committee stabilised the classification process. In future work we plan to check the boosting effect during the classification of granular reflections of decision systems.

Acknowledgements. The research has been supported by grant 1309-802 from Ministry of Science and Higher Education of the Republic of Poland.

References

1. Artiemjew, P.: Classifiers based on rough mereology in analysis of DNA microarray data. In: Proceedings of Second International Conference of Soft Computing and Pattern Recognition (SoCPar2010), pp. 273–278. IEEE Computer Society, Cergy Pontoise (2010)
2. Artiemjew, P.: The extraction method of DNA microarray features based on experimental A statistics. In: Yao, J.T., Ramanna, S., Wang, G., Suraj, Z. (eds.) RSKT 2011. LNCS, vol. 6954, pp. 642–648. Springer, Heidelberg (2011)
3. Artiemjew, P.: The extraction method of DNA microarray features based on modified F statistics vs. classifier based on rough mereology. In: Kryszkiewicz, M., Rybinski, H., Skowron, A., Raś, Z.W. (eds.) ISMIS 2011. LNCS (LNAI), vol. 6804, pp. 33–42. Springer, Heidelberg (2011)
4. Artiemjew, P.: Review of the extraction methods of DNA microarray features based on central decision class separation vs rough set classifier. Found. Comput. Decis. Sci. **37**(4), 241–254 (2012)
5. Artiemjew, P.: Rough mereology classifier vs simple DNA microarray gene extraction methods. Int. J. Data Min. Model. Manage. Spec. Issue Pattern Recogn. **6**(2), 110–126 (2014)
6. Hájek, P.: Metamathematics of Fuzzy Logic. Kluwer, Dordrecht (1998)
7. Polkowski, L.: A unified approach to granulation of knowledge and granular computing based on rough mereology: a survey. In: Pedrycz, W., Skowron, A., Kreinovich, V. (eds.) Handbook of Granular Computing, pp. 375–401. Wiley, New York (2008)
8. Polkowski, L., Artiemjew, P.: On classifying mappings induced by granular structures. In: Peters, J.F., Skowron, A., Rybiński, H. (eds.) Transactions on Rough Sets IX. LNCS, vol. 5390, pp. 264–286. Springer, Heidelberg (2008)

9. Polkowski, L., Artiemjew, P.: Granular Computing in Decision Approximation - An Application of Rough Mereology. Intelligent Systems Reference Library, vol. 77. Springer, Switzerland (2015). ISBN 978-3-319-12879-5, pp. 1–422

10. Ohno-Machado, L.: Cross-validation and Bootstrap Ensembles, Bagging, Boosting, Harvard-MIT Division of Health Sciences and Technology, HST.951J: Medical Decision Support, Fall (2005). http://ocw.mit.edu/courses/health-sciences-and-technology/hst-951j-medical-decision-support-fall-2005/lecture-notes/hst951_6.pdf

11. Zhou, Z.-H.: Ensemble Methods: Foundations and Algorithms. Chapman and Hall/CRC, p. 23. ISBN 978-1439830031. The term boosting refers to a family of algorithms that are able to convert weak learners to strong learners (2012)

12. Schapire, R.E.: The boosting approach to machine learning: an overview. In: MSRI (Mathematical Sciences Research Institute) Workshop on Nonlinear Estimation and Classification (2003)

13. Zhou, Z.-H.: Boosting 25 years, CCL 2014 Keynote (2014)

14. Breiman, L.: Arcing classifier (with discussion and a rejoinder by the author). Ann. Statist. **26**(3), 801–849 (1998). Accessed 18 January 2015. Schapire (1990) proved that boosting is possible, p. 823

15. Schapire, R.E.: A Short Introduction to Boosting (1999)

16. Hu, X.: Construction of an ensemble of classifiers based on rough sets theory and database operations. In: Proceedings of the IEEE International Conference on Data Mining (ICDM2001) (2001)

17. Hu, X.: Ensembles of classifiers based on rough sets theory and set-oriented database operations. In: Presented at the 2006 IEEE International Conference on Granular Computing, Atlanta, GA (2006)

18. Saha, S., Murthy, C.A., Pal, S.K.: Rough set based ensemble classifier for web page classification. Fundamenta Informaticae **76**(1–2), 171–187 (2007)

19. Shi, L., Weng, M., Ma, X., Xi, L.: Rough set based decision tree ensemble algorithm for text classification. J. Comput. Inf. Syst. **6**(1), 89–95 (2010)

20. Murthy, C.A., Saha, S., Pal, S.K.: Rough set based ensemble classifier. In: Kuznetsov, S.O., Ślęzak, D., Hepting, D.H., Mirkin, B.G. (eds.) RSFDGrC 2011. LNCS, vol. 6743, pp. 27–27. Springer, Heidelberg (2011)

Extraction of Off-Line Handwritten Characters Based on a Soft K-Segments for Principal Curves

Na Jiao[✉]

Department of Information Science and Technology,
East China University of Political Science and Law,
Shanghai 201620, People's Republic of China
jiaonaecupl@gmail.com

Abstract. Principal curves are nonlinear generalizations of principal components analysis. They are smooth self-consistent curves that pass through the middle of the distribution. By analysis of existed principal curves, we learn that a soft k-segments algorithm for principal curves exhibits good performance in such situations in which the data sets are concentrated around a highly curved or self-intersecting curves. Extraction of features are critical to improve the recognition rate of off-line handwritten characters. Therefore, we attempt to use the algorithm to extract structural features of off-line handwritten characters. Experiment results show that the algorithm is not only feasible for extraction of structural features of characters, but also exhibits good performance. The proposed method can provide a new approach to the research for extraction of structural features of characters.

Keywords: Off-line handwritten characters features · A soft k-segments algorithm for principal curves · Structural features · Features extraction

1 Introduction

Automatic fingerprint identification system typically consists of fingerprint acquisition, fingerprint image preprocessing, minutiae extraction, fingerprint matching, fingerprint classification and recognition. Fingerprint feature extraction results directly affect the final result of fingerprint matching and fingerprint classification. A thinning algorithm is usually used to obtain fingerprint skeleton. A large number of scanning, traversal and mathematical methods are required in thinning methods, so there are inevitably false minutia. Furthermore, the thinned image is lossy compression image of the original, therefore it cannot preserve accurate information from the original image. Moreover, the thinned image is stored in bitmap format, which need a great deal of storage space. It is not effective for the fingerprint feature point extraction and distinguishing the true from the false.

To avoid the above shortcomings, we should choose principal curves to extract fingerprint skeleton. Hastie and Stuetzle (hereafter HS) originally proposed the notion of principal curves to solve the problems in traditional machine learning and multivariate data analysis in 1984. The definition of principal curves

© Springer International Publishing Switzerland 2015
Y. Yao et al. (Eds.): RSFDGrC 2015, LNAI 9437, pp. 278–285, 2015.
DOI: 10.1007/978-3-319-25783-9_25

is based on the concept of self-consistency. Self-consistency means that each point of the curves is the average of all points that project there. Self-consistent smooth curves which pass through the middle of a data set. Moreover, we hope to find curves passing through the middle of the datasets, which can truly reflect the shape of the data. Principal curves are non-linear generalizations of principal components, the basic idea of which is to seek low-dimensional manifolds embedded in the multi-dimensional space. Due to all the properties and advantages of principal curves which have gained its rapid development since 1990s with various definitions of principal curves. Banfield and Raftery defined BR principal curve in 1992 [3]. Kegl gave his principal curve definition called PL principal curve in 2000 [4]. Verbeek proposed K-segment principal curve in 2002 [5], and Delicado defined D principal curve in 2003 [8]. There are a lot of achievements on the applications of principal curves, such as shape detection [3], speech recognition [7], image skeletonization, character feature extraction [6], bill recognition [9], pattern classification [10] and intelligent transportation analysis [2].

There are several problems in HS principal curve, PL principal curve, BR principal curve and so on. A common feature of aforementioned methods is that they consist of a combination of local models that are related by a fixed topology. When the dataset are concentrated around a highly curved or self-intersecting curve, these methods show poor performance. The reason is the fixed topology among the local model and the bad initialization. One often does not make sure the number of local models and one has to make a guess for a priori of local model. In addition, the definition of HS principal curve explicitly excludes self-intersecting curves as principal curves. Due to bad initialization in cases where the dataset are concentrated around highly curved or self-intersecting curves, PL principal curve also fail. Figure 1. shows the results for PL principal curve and HS principal curve. The dataset of fingerprint image is highly curved or self-intersecting, HS principal curve and PL principal curve are not fit for fingerprint skeletonization extraction. A soft K-segments algorithm for principal curves can be used for fingerprint skeletonization extraction. The soft K-segments algorithm for principal curves is able to overcome some of these local optima. And it can jumps in the space of curves.

The rest of this paper is organized as follows. Section 2 describes the relevant primary concept of principal curves and a soft K-segments algorithm for principal curves. Section 3 proposes structural extraction of off-line handwritten characters based on a soft K-segments principal curves. Finally, Sect. 4 gives a summary and future research.

2   A Soft K-segments Algorithm for Principal Curves

In this section, we briefly introduce the basic concepts of principal curves and a soft K-segments algorithm for principal curves.

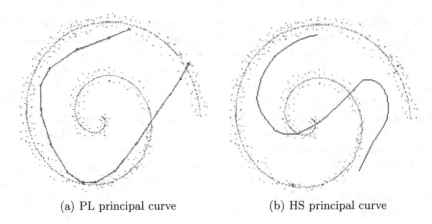

(a) PL principal curve   (b) HS principal curve

Fig. 1. PL principal curve and HS principal curve

2.1 Principal Curves

Let Y denote a random vector in R^d, and $f(s)$ be a smooth curve in R^d parameterized by $s \in R$. The curve is a principal curve if the following self-consistent criterion is satisfied:

$$f(s) = E(Y \,|s_f(y) = s) \,,$$

$$s_f(y) = \sup \{s : \|y - f(s)\| = \inf_\tau \|y - f(\tau)\|\},$$

where $s_f(y)$ is the projection index function which maps any value of $Y = y$ to the value of s for which $Y = y$ is closest to y. $\|.\|$ denotes the Euclidean distance in R^d.

In brief, principal curves are smooth one-dimensional curves which pass through the middle of a set of d-dimensional data points [1]. The goal is to provide smooth and low dimensional summaries of dataset. The definition of a principal curve indicates that any point of a principal curve is the condition expectation of those points that project to this point, and a principal curve satisfies the property of self-consistency. Principal curves form a nonlinear generalization of principal component analysis. The theoretical foundations of these curves are a low-dimensional nonlinear manifold embedded in a high-dimensional space [1,8]. Figure 2 shows a first principal component line and a principal curve. Compared with the corresponding principal component, two obvious advantages of a principal curve can be observed: a principal curve can retain more information of dataset. Furthermore, it can describe the outline of original information better.

(a) The first principal component   (b) The principal curve

Fig. 2. The comparison between first principal component and principal curve

2.2 A Soft K-segments Algorithm for Principal Curves

The algorithm contains the following steps:

Step 1: The initialization step

(1) Given a date set $X = (x_1, x_2, ..., x_n)$.

(2) Compute a first principal component of the dataset. Set the length of initial segment $s_1 = 3\sigma$, the region $Voronoi$ of segment s_1 is $V_1 = \{x_1, ...x_n\}$, $k=1$. σ^2 is the variance of the first principal component.

(3) Set $k\_$max.

Step 2: Insert a new segment

If $k > k\_$max, then go to end.

Else

(1) Compute the data point which satisfies:

$$x_q = \inf \left\{ x_t : \left| \sum_{i=1}^{n} g(x_i, x_t) = \max(\sum_{i=1}^{n} g(x_i, x_j)) \right. \right\},$$

where

$$g(x_i, x_j) = \begin{cases} dist(x_i) - d(x_i, x_j) & dist(x_i) - d(x_i, x_j) > 0 \\ 0 & dist(x_i) - d(x_i, x_j) \leq 0 \end{cases},$$

$1 \leq i, j \leq n$, $dist(x_i) = \min\limits_{j=1,2,...,k} d(x_i, s_j)$, $d(x_i, x_j) = ||x_i - x_j||^2$. Define the $Voronoi$ region of point x_q,

$V_q = \{x \in X| \, ||x - x_q|| \leq \min d(x, s_j), j = 1, 2, ...k\}$, V_q contains all points for with the j th is the closest.

(2) Compute a first principal component of the dataset. Set the length of initial segment $s_1 = 3\sigma$, σ^2 is the variance.

(3) $k = k + 1$, the new segment is denoted as s_k. Set the region $Voronoi$ of s_k which is $V_k = \Phi$.

Step 3: The adjusting step

Adjust the new segment and other segments. The method is outlined as follows:

(1) Set original region $Voronoi$ of all the segments $(V_1, V_2, ..., V_k)$.

(2) Compute new region $Voronoi$ of all the segments. $\forall s_i$, $i = 1, ...k$,

$$V_i' = \{x_j \in X \, | ||x_j - s_i|| = \min\limits_{t=1,2,...,k} ||x_j - s_t||\}.$$

(3) If $\left(V_1', V_2', \cdots, V_k'\right)$ and $(V_1, V_2, \cdots, V_k)$ are identical, then the adjusting step end.

Else compute a first principal component of V_j' and set $= \left(V_1', V_2', \cdots, V_k'\right) = (V_1, V_2, \cdots, V_k)$.

Step 4: Combining segments Step

Link the segments together to form a 'Hamiltonian path' (HP).

(1) Set $p = k$ (p is the number of sub-HP), p sub-HP consists of the $2p$ end-points and 2^p edges of p - segments.

(2) If $p < 2$, then go to 3). Else compute the total cost of the edge $c(e_i) = l(e_i) + \lambda_1 a(e_i)$, $0 \leq \lambda_1 \in \Re$ being a parameter to be set by the user. The term $l(e_i)$ denotes the length of the edge. The length of an edge $e_i = (v_l, v_m)$ is taken as the Euclidean distance between its vertices (v_l, v_m). $a(e_i)$ is a penalty term equal to the sum of the angles between adjacent edges, $a(e_i) = \alpha + \beta$. Figure 3. illustrates the angle penalty. Connect the vertices whose $c(e_i)$ is minimum. $p = p - 1$. Return to 2).

(3) Use the $2 - opt\ TSP$ optimization scheme to improve our initial HP.

(4) Compute the objective function $OF = n \log l + \sum\limits_{i=1}^{k} \sum\limits_{x \in V_I} d(s_i, x)^2/(2\sigma^2)$, l is length of HP. If the objective function reaches its minimum, then go to end. Else return Step 2.

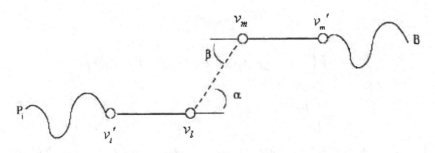

Fig. 3. The angle penalty of $(v_l,\ v_m)$ is $\alpha + \beta$

The time complexity of a soft K-segments algorithm for principal curves is $O\left(kn^2\right) \approx O\left(n^2\right)$ ($k \ll n$) (k is the number of segments, n is the number of data) (Table 1).

The flow-chart of a soft K-segments algorithm for principal curves is Fig. 4.

3 Structural Extraction of Off-Line Handwritten Characters Based on a Soft K-segments Principal Curves

In this section, we extract structural features of off-line handwritten characters based on a soft K-segments principal curves. We conduct experiments with six kinds of off-line handwritten character. Experimental results are shown in Fig. 5.

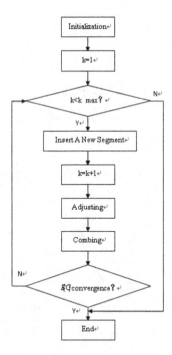

Fig. 4. The flow-chart of a soft K-segments algorithm for principal curves

Table 1. Comparison of average running time

Time	2	6	8	e	q	w
HL	0.45	0.38	0.67	0.41	0.43	0.36
PL	1.28	1.11	2.26	1.23	1.26	1.06
Propose method	0.51	0.42	0.63	0.48	0.47	0.37

From Fig. 5, we can see that the soft K principal curve algorithm can be applied in great curved or intersection of off-line handwritten character. From the experiments we know there are several curves in each character. A curve is composed of a series of points, and each pair of adjacent points have an edge. For example, the first letter is composed of 3 curves.

The algorithms were implemented on Intel CORE 2, WinXP, 2.47 GHz. The average running time of HS was faster than other algorithms. The running time of PL algorithm is almost 2.5 times that of our algorithm. The algorithm proposed in this paper is less efficient than HS. The reason is that the initialization step of the algorithm is time-consuming, but the rest of the proposed algorithm help reduce the running time.

We analyze that the number of strokes and the number of loop is very important to extract structural features of off-line handwritten characters. We put

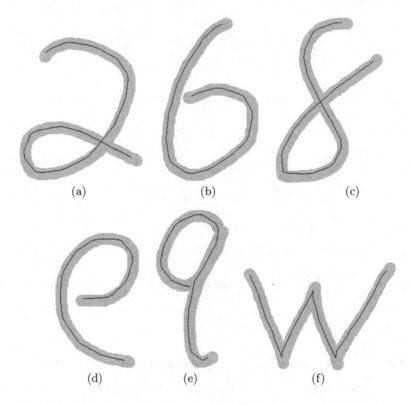

(a)　　　　　　　　(b)　　　　　　　　(c)

(d)　　　　　　　　(e)　　　　　　　　(f)

Fig. 5. Structural extraction based on a soft K-segments principal curves

the number of strokes as the overall features of the classification, for example, (l, l) (g, g), (m, m), (h, h), ($2, 2$), and so on. For the same number of strokes of each class, we use the circuit to the off-line handwritten character further classification. For example, the number of strokes is 2, there are three categories ($a\, b\, de\, gp\, q\, g$), (8), ($ij\, mv\, wz\, 23$). According to the position of the intersection, the details of the characteristics of the off-line handwritten character can be further distinguished.

4　Conclusions

In this paper, we apply the soft K-segments principal curves algorithm to extract structural features of off-line handwritten characters. Experiment results show that the algorithm is feasible for extraction of structural features of characters. Our future work will focus on comparison with other algorithms.

Acknowledgements. This paper is supported by the National Social Science Fund (Granted No. 13CFX049), Shanghai University Young Teacher Training Program

(Granted No. hdzf10008) and the Research Fund for East China University of Political science and Law (Granted No. 11H2K034).

References

1. Hastie, T.: Principal curves and surfaces. Laboratory for Computational Statistics, Stanford University, Department of Statistics, Technical report 11 (1984)
2. Banfield, J.D., Raftery, A.E.: Ice floe identification in satellite images using mathematical morphology and clustering about principal curves. J. Am. Stat. Assoc. **87**, 7–16 (1992)
3. Kegl, B., KrzyzaK, A., et al.: A polygonal line algorithm for constructing principal curves. In: Proceedings of Neural Information Processing System, pp. 501–507. Computer Press, Denver Colorado, USA (1999)
4. Zhang, J.P., Chen, D.W., Kruger, U.: Adaptive constraint k-segment principal curves for intelligent transportation systems. IEEE Trans. Intell. Transp. Syst. **9**, 666–677 (2008)
5. VerbeeK, J.J., Vlassis, N., Krose, B.: A K-segments algorithm for finding principal curves. Technical report: Pattern Recognition Letters 23, Computer Science of Institute, University of Amsterdam, pp. 1009–1017 (2002)
6. Verbeek, J.J., Vlassis, N., Kröse, B.J.A.: A soft k-segments algorithm for principal curves. In: Dorffner, G., Bischof, H., Hornik, K. (eds.) ICANN 2001. LNCS, vol. 2130, pp. 450–456. Springer, Heidelberg (2001)
7. Delicado, P.: Another look at principal curves and surfaces. J. Multivar. Anal. **7**, 84–116 (2001)
8. Zhang, J.P., Wang, J.: An overview of principal curves. Chin. J. Comput. **26**, 129–146 (2003)
9. Bas, E., Erdogmus, D.: Principal curves as skeletons of tubular objects: locally characterizing the structures of axons. Neuroinformatics **9**, 181–191 (2011)
10. Wang, H.N., Lee, T.C.M.: Extraction of curvilinear features from noisy point patterns using principal curves. Pattern Recogn. Lett. **29**, 2078–2084 (2008)

A Knowledge Acquisition Model Based on Formal Concept Analysis in Complex Information Systems

Xiangping Kang[1]($\boxtimes$), Duoqian Miao[1], and Na Jiao[2]

[1] The Key Laboratory of Embedded System and Service Computing,
Ministry of Education, Tongji University,
Shanghai 200092, People's Republic of China
tongji_kangxp@sina.com, miaoduoqian@163.com
[2] Department of Information Science and Technology,
East China University of Political Science and Law, Shanghai 201620,
People's Republic of China
jiaonaecupl@gmail.com

Abstract. Normally, in some complex information systems, the binary relation on domain of any attribute is just a kind of ordinary binary, which does not meet some common properties such as reflexivity, transitivity or symmetry. In view of the above-mentioned facts this paper attempts to employ FCA(Formal Concept Analysis), proposes a rough set model based on FCA, in which equivalence relations, dominance relations, similarity relations(or tolerance relations) and neighborhood relations on universe are expanded to general binary relations and problems in rough set theory are discussed based on FCA. Particularly, from the above description of complex information systems, we can see that the relation in domain of any attribute may be extremely complex, which often leads to high time complexity and space complexity in the process of knowledge acquisition. For above reason this paper introduces granular computing(GrC), which can effectively reduce the complexity to a certain extent.

Keywords: Rough set · Formal concept analysis · Granular computing

1 Introduction

Rough set theory, introduced by Pawlak in 1982 [6], is a mathematical theory which can be used to deal with vague and uncertain problems. The theory of FCA, proposed by Wille in the same year [10], is a tool for concept discovery from data, in which the relationship of concepts is embodied by concept lattice. As two active relevant research fields in artificial intelligence and information science, they have many common characteristics, such as they have common research backgrounds and aims, and they are both closely related to topology, algebra and logic. Therefore, the study of combination of two theories has fundamental significance, in recent years scholars have done a lot of research in this area [3,5,9].

© Springer International Publishing Switzerland 2015
Y. Yao et al. (Eds.): RSFDGrC 2015, LNAI 9437, pp. 286–297, 2015.
DOI: 10.1007/978-3-319-25783-9_26

An information system is an quadruple $IS = (U, AT, V, f)$, where U is a finite nonempty set of objects, called a universe, and AT is a finite nonempty set of attributes, $V = \bigcup_{m \in AT} V_m$ and V_m is a domain of attribute m; $f : U \times AT \to V$ is a function such that $f(x, m) \in V_m$ for every $x \in U$, $m \in AT$, called an information function.

Pawlak rough set theory is defined using the indiscernibility relation, which implies that values in any V_m are independent of one another. However, in the real world, we may face cases that there are complex internal relationship in any V_m, that is, values describing attributes may exist some special relationship, such as some attribute values are ordinal or similar. Therefore, Pawlak rough set theory is inapplicable in dealing with above information systems. To overcome this insufficiency, scholars have done a lot of extended research work, in which rough set models based on dominance relations or similarity relations are the most common and widely used models [2,4,8].

However, in some complex information systems, the relationship between different values in any V_m is more complex rather than ordinal or similar, meanwhile, the corresponding relation on V_m often does not satisfies common properties such as reflexivity, transitivity or symmetry. Obviously, in above situation, for finding potential, valuable, simple information from chaotic, strong interference and large data, we need to extend the rough set theory further. For all that, on the basis of previous research this paper attempts to introduce advantages of FCA into rough set theory.

Extremely complex relations of domains of attributes often lead to complex lattice structure and huge concepts, and cause high time complexity and space complexity for further calculation and analysis of problems. For above-mentioned reasons this paper introduces GrC which has unique advantages in modeling and analysis of the large and complex data, and it can effectively simplify the complex structure and reduce the scale of concepts to a certain extent.

In general, this paper not only provides an useful method for applying FCA and GrC to complex information systems, but also offers a new idea for the extension of the rough set model. And above result is also the innovation of this thesis.

The rest of the paper is organized: FCA is briefly introduced in Sect. 2; Sect. 3 discusses the classification problem in the domain of any attribute based on GrC; Sect. 4 translates complex information systems into one-valued formal contexts; Sect. 5 discusses algebraic structures in complex information systems; Sect. 6 offers solutions to the problem of reduct, core and dependency etc.; conclusions and the discussion of further work close the paper in Sect. 7.

2   Basic Notions of FCA

A formal context is a triple $K = (G, M, I)$, where G and M are sets, and $I \subseteq G \times M$ is a binary relation. In the case, members of G are called objects and members of M are called attributes, and I is viewed as an incidence relation between objects and attributes. Accordingly, $(g, m) \in I$ denotes "the object g has the attribute m".

Definition 1. *[1] Let $K = (G, M, I)$ be a formal context, for any $A \subseteq G$ and $B \subseteq M$, we define:*

$$A' = \{\, m \in M \mid (g, m) \in I, \forall g \in A \,\}; \quad B' = \{\, g \in G \mid (g, m) \in I, \forall m \in B \,\}$$

If $A' = B$ and $B' = A$, then (A, B) is called a concept. In this case, A is called the extent, B is called the intent. The order "$\leq$" between concepts (A_1, B_1) and (A_2, B_2) is defined as

$$(A_1, B_1) \leq (A_2, B_2) \Leftrightarrow A_1 \subseteq A_2 \Leftrightarrow B_1 \supseteq B_2$$

The ordered set $(\mathscr{B}(K), \leq)$ is a complete lattice, where $\mathscr{B}(K)$ is the set of all concepts.

Proposition 1. *[1] If $K = (G, M, I)$ is a formal context, $A, A_1, A_2 \subseteq G$ are sets of objects and $B, B_1, B_2 \subseteq M$ are sets of objects, then*

(1) $A_1 \subseteq A_2 \Rightarrow A_2' \subseteq A_1'$ (2) $B_1 \subseteq B_2 \Rightarrow B_2' \subseteq B_1'$

(3) $A \subseteq A''; B \subseteq B''$ (4) $A' = A'''; B' = B'''$

Proposition 2. *[1] The ordered set $(\mathscr{B}(K), \leq)$ is a complete lattice, its corresponding infimum and supremum are:*

$$(1) \bigwedge_{t \in T} (A_t, B_t) = \left(\bigcup_{t \in T} A_t, \left(\bigcup_{t \in T} B_t \right)'' \right) \quad (2) \bigvee_{t \in T} (A_t, B_t) = \left(\left(\bigcup_{t \in T} A_t \right)'', \bigcup_{t \in T} B_t \right)$$

3   Classification Analysis in Domain of Attribute Based On GrC

Let V_m be the domain of attribute m, R_m be a relation on V_m. If R_m does not meet some common properties such as reflexivity, transitivity or symmetry, that we say R_m is just a general binary relation on V_m. For example, If there exists some type of relationship "$\perp$" between values in V_m, then we can deduce a binary relation $R_m = \{(v, w) \mid v \perp w, \ v, w \in V_m\}$ on V_m. In fact, R_m objectively reflects the relationship between values in V_m, if $(v, w) \notin R_m$, then there exists no relationship "$\perp$" between v and w; if $(v, w) \in R_m$, then there exists relationship "$\perp$" between v and w.

In view of the above-mentioned facts, for any attribute $m \in AT$, if the corresponding relation R_m on V_m is general, then we say $IS = (U, AT, V, f)$ is a complex information system. For convenience IS mentioned above is formalized as $IS = (U, AT, \Re)$, where $\Re = \{R_m \mid m \in AT\}$.

For example, Table 1 is a complex information system about cars, where $U = \{1, 2, \ldots, 8\}$ is the set of various type of cars and $AT = \{a, b, c, d, e\}$ is the set of attributes with a="the evaluation of price", b="the evaluation of size", c="the evaluation of engine", d="the evaluation of maximum speed", and e="the evaluation of performance/ price ratio". $R_a, R_b, \ldots, R_e$ in Table 2

Table 1. A complex information system

	a	b	c	d	e
1	u_1	v_1	w_1	x_1	z_1
2	u_2	v_2	w_2	x_2	z_2
3	u_3	v_3	w_3	x_3	z_3
4	u_4	v_4	w_4	x_4	z_4
5	u_5	v_5	w_5	x_5	z_5
6	u_6	v_6	w_6	x_6	z_6
7	u_7	v_7	w_7	x_7	z_7
8	u_8	v_8	w_8	x_8	z_8

Table 2. The $\Re$ of complex information system in Table 1

(a) R_a

	u_1	u_2	u_3	u_4	u_5	u_6	u_7	u_8
u_1	×	×	×	×	×	×		
u_2	×	×	×	×	×	×		
u_3	×	×	×		×	×	×	×
u_4	×	×		×	×	×	×	×
u_5	×	×	×		×	×	×	×
u_6	×	×	×	×	×		×	×
u_7		×	×	×	×	×	×	×
u_8		×	×	×	×	×	×	×

(b) R_b

	v_1	v_2	v_3	v_4	v_5	v_6	v_7	v_8
v_1	×	×	×	×				
v_2	×	×	×	×				
v_3	×	×	×	×	×	×	×	×
v_4	×	×		×	×	×		×
v_5			×	×	×	×	×	×
v_6			×	×	×	×	×	×
v_7		×			×	×	×	×
v_8		×			×	×	×	×

(c) R_c

	w_1	w_2	w_3	w_4	w_5	w_6	w_7	w_8
w_1	×		×		×		×	
w_2		×	×	×		×	×	
w_3		×	×	×		×		×
w_4		×	×		×	×	×	
w_5	×	×	×	×		×		×
w_6	×			×	×		×	×
w_7	×	×	×	×	×	×	×	
w_8	×			×				

(d) R_d

	x_1	x_2	x_3	x_4	x_5	x_6	x_7	x_8
x_1	×		×				×	×
x_2		×		×	×	×	×	×
x_3	×		×		×			
x_4		×						×
x_5	×	×	×					×
x_6		×			×	×	×	×
x_7	×	×			×	×	×	
x_8	×		×	×			×	×

(e) R_e

	z_1	z_2	z_3	z_4	z_5	z_6	z_7	z_8
z_1	×	×			×	×	×	×
z_2	×	×					×	×
z_3	×	×	×	×	×			
z_4	×	×	×	×		×		
z_5	×			×	×		×	×
z_6	×	×	×	×	×	×	×	×
z_7		×	×			×	×	
z_8		×	×			×	×	

are corresponding binary relations on $V_a, V_b, \ldots, V_e$ in Table 1. In fact, for any $m \in AT$, R_m in Table 2 reflects relationships among various evaluations in V_m (in Table 1).

From the above description of complex information systems, we can see the relation in domain of any attribute may be extremely complex, that often leads to high time complexity and space complexity in the process of knowledge acquisition. For above mentioned reasons this paper introduces GrC which has unique advantage in modeling and analysis of the large and complex data, and discusses the classification problem of domain in the different granulation, it can effectively reduce the complexity of knowledge acquisition to a certain extent. In recent years, GrC plays an important role in knowledge discovery, data mining and soft computing, which helps to solve the problem more scientific, rational and easy. For example, when the problem is too complex or costly, in order to better understand and solve problems rather than submerging in unnecessary details, larger granulations help to identify useful information and hide some specific details, and the problem can be solved from the overall picture. It can be said GrC has the unique advantage in large, complex data modeling and analysis.

Typically, attribute values can simply be classified as 'numeric' and 'non-numeric'. For any attribute $m \in AT$, if the type of values in V_m is 'non-numeric', we need to determine whether any two values are similar and how to measure the degree of the similarity. For example, We can use the method similar to the AHP theory [7], which employs integers $1, \ldots, 9$ as the metric values. In particular, human's subjective judgments usually conclude following levels: no similarity, weak similarity, similarity, strong similarity and complete similarity, which were denoted as 1,3,5,7,9 correspondingly. and then another four levels between above levels are denoted as 2,4,6,8 respectively.

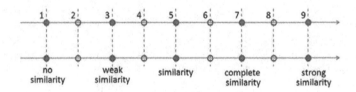

Fig. 1. Similarities at different levels.

If the type of values in V_m is 'non-numeric', then based on human's subjective judgment the degree of similarity between $v_i, v_j \in V_m$ is:

$$\tilde{r}_{ij} = \frac{1}{8} \cdot (s_{ij} - 1), where \ s_{ij} \in \{1, 2, \ldots, 9\}$$

If the type of values in V_m is 'numeric', then the degree of the similarity between $v_i, v_j \in V_m$ is defined as follows:

$$\tilde{r}_{ij} = 1 - \frac{1}{\max\{v_1, v_2, \ldots, v_n\}} |v_i - v_j|$$

Based on above discussion, we can define a similarity relation matrix $\tilde{R}_m = (\tilde{r}_{ij})_{n \times n}$. Obviously, the fuzzy clustering algorithm can be used to calculate the fuzzy equivalence relation matrix R_m of $\tilde{R}_m$. And further by introducing parameter $\lambda \in [0, 1]$, we can obtain the λ-cut equivalence relation matrix $R_m^\lambda = (r_{ij})_{n \times n}$. In fact R_m^λ is a equivalence relation on V_m, and V_m / R_m^λ is a partition of V_m. Let $V_m^\lambda = V_m/R_m^\lambda = \{X_1, X_2, \cdots, X_l\}$, then the knowledge granulation of R_m^λ is defined as:

$$\nu(R_m^\lambda) = \frac{1}{n^2} \sum_{i=1}^{l} |X_i|^2$$

$X_i \in V_m^\lambda$ is called a granule or a class.

The process of the classification of domain of any attribute is not the keystone of the paper, so we only give a brief account. In the following, we assume the classification result is known, and overlook the calculating process. In this paper, the classification results of $V_a \ldots V_e$ in Table 1 are shown as follows separately.

- $V_a^\lambda : u_1, u_2, u_3u_6, u_4u_5, u_7, u_8;$
- $V_b^\lambda : v_1, v_2, v_3v_4, v_5, v_6, v_7, v_8;$
- $V_c^\lambda : w_1w_2, w_3w_4w_5w_6, w_7w_8;$
- $V_d^\lambda : x_1x_2, x_3, x_4, x_5, x_6x_8, x_7;$
- $V_e^\lambda : z_1, z_2, z_3, z_4, z_5z_6, z_7, z_8$

4   One-Valued Formal Contexts

As we know that operators in Definition 1 only can be used for one-valued contexts. Based on this, before applying operators in Definition 1 we need to translate complex information systems into one-valued formal contexts.

Definition 2. *For every attribute* $m \in AT$, *we can further expand* R_m *to* R_m^λ, *where* R_m^λ *is described as follows: for any* $v, w \in V_m$, *if*

$$\frac{1}{|X| \times |Y|} \times |(X \times Y) \cap R_m| \geq \delta, \ where \ v \in X \in V_m^\lambda \ and \ w \in Y \in V_m^\lambda$$

then $(v, w) \in R_m^\lambda$. *In the following, we say* R_m^λ *is a variable-precision relation of* R_m.

For example, let $\delta = 0.6$, then base on Definition 3 and the classification result in Sect. 3, variable-precision relations shown in Table 3 can be obtained from binary relations in Table 2. In addition, we can easily discover from Fig. 2 and Fig. 3 that R_c^λ is simpler and more intuitive than R_c.

From above discussion, some conclusions can be inferred immediately as follows:

- when λ is invariable, if $\delta \downarrow$, then $|R_m^\lambda| \uparrow$.
- if $\lambda \uparrow$, then $|V_m^\lambda| \uparrow$.

In fact, in $IS = (U, AT, \Re)$, for any $v \in V_m$, this paper only care about the relationship between it and other value in V_m rather than the size of it. Based on this we try to translate IS into an one-valued formal context, which has filtered out some redundant information(we do not care about in IS). There are some differences between the above procedure with the classic scaling procedure in FCA, such as compared to classic scaling procedure the above procedure removes some redundant information, which we do not care about.

Table 3. The $\Re$ of complex information system in Table 1

(a) R_a^λ

	u$_1$	u$_2$	u$_3$	u$_4$	u$_5$	u$_6$	u$_7$	u$_8$
u$_1$	x	x	x	x	x	x		
u$_2$	x	x	x	x	x	x		
u$_3$	x	x	x	x	x	x	x	x
u$_4$	x	x	x	x	x	x	x	x
u$_5$	x	x	x	x	x	x	x	x
u$_6$	x	x	x	x	x	x	x	x
u$_7$		x	x	x	x	x	x	
u$_8$		x	x	x	x	x	x	

(b) R_b^λ

	v$_1$	v$_2$	v$_3$	v$_4$	v$_5$	v$_6$	v$_7$	v$_8$
v$_1$	x	x	x	x				
v$_2$	x	x	x	x				
v$_3$	x	x	x	x	x	x	x	
v$_4$	x	x	x	x	x	x	x	
v$_5$			x	x	x	x	x	x
v$_6$			x	x	x	x	x	x
v$_7$					x	x	x	x
v$_8$					x	x	x	x

(c) R_c^λ

	w$_1$	w$_2$	w$_3$	w$_4$	w$_5$	w$_6$	w$_7$	w$_8$
w$_1$			x	x	x	x		
w$_2$			x	x	x	x		
w$_3$	x	x	x	x	x	x	x	x
w$_4$	x	x	x	x	x	x	x	x
w$_5$	x	x	x	x	x	x	x	x
w$_6$	x		x	x	x	x	x	x
w$_7$			x	x	x	x		
w$_8$			x	x	x	x		

(d) R_d^λ

	x$_1$	x$_2$	x$_3$	x$_4$	x$_5$	x$_6$	x$_7$	x$_8$
x$_1$					x	x	x	x
x$_2$					x	x	x	x
x$_3$								
x$_4$								
x$_5$	x	x			x	x	x	x
x$_6$	x	x			x	x	x	x
x$_7$	x	x			x	x	x	x
x$_8$	x	x			x	x	x	x

(e) R_e^λ

	z$_1$	z$_2$	z$_3$	z$_4$	z$_5$	z$_6$	z$_7$	z$_8$
z$_1$	x	x			x	x	x	x
z$_2$	x	x			x	x	x	x
z$_3$	x	x	x	x				
z$_4$	x	x	x	x				
z$_5$			x	x	x	x	x	x
z$_6$			x	x	x	x	x	x
z$_7$	x	x			x	x		
z$_8$	x	x			x	x		

Definition 3. *Let $IS = (U, AT, \Re)$ be a complex information system, $0 \leq \lambda \leq 1$, $K_\lambda = (U \times U, AT, J_\lambda)$ is called an one-valued context deduced from IS, where J_λ is described as: for any $x, y \in U$ and $m \in AT$, there exists*

$$(x,y)J_\lambda m \Leftrightarrow (v,w) \in R_m^\lambda, \text{where } f(x,m) = v,\ f(y,m) = w$$

By the above translation rule the one-valued formal context can be deduced from Table 1, which is shown in Table 4.

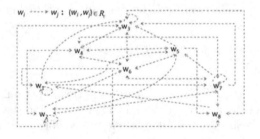

Fig. 2. The binary relation R_c in Table 2

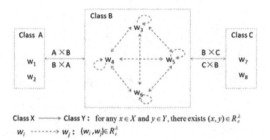

Class X ⟶ Class Y : for any $x \in X$ and $y \in Y$, there exists $(x, y) \in R_c^{\lambda}$

w_i ⤏ w_j : $(w_i, w_j) \in R_c^{\lambda}$

Fig. 3. The variable-precision relation R_c^{λ} of R_c in Table 2

Table 4. The one-valued context deduced from Table 1

	a	b	c	d	e
(1, 1)	×	×			×
(1, 2)	×	×			×
(1, 3)	×	×	×		
(1, 4)	×	×	×		
(1, 5)	×		×	×	×
(1, 6)	×		×	×	×
(1, 7)				×	×
⋮	⋮	⋮	⋮	⋮	⋮
(8, 6)	×	×	×		×
(8, 7)	×	×		×	×
(8, 8)	×	×		×	×

5 Algebraic Structure in the Complex Information System

In recent years many scholars discussed the granularity model based on equivalence relation, dominance relation, similarity relation, tolerance relation, neighborhood relation and other complex relations. Based on above discussions this paper expands above common binary relations to the more general binary relation

$$R_B^{\lambda} = \{(x,y) \in U \times U \mid \forall m \in B, \ f(x,m) = v, f(y,m) = w, \ (v, \ w) \in R_m^{\lambda}\}$$

In K_{λ}, it's obvious that operators in Definition 1 are described as: for any $R \subseteq U \times U$,

$$R' = \{m \in M \mid ((x,y), m) \in J_{\lambda}, \forall (x,y) \in R\}$$

and for any $B \subseteq AT$,

$$B' = \{(x,y) \in U \times U \mid ((x,y), \ m) \in J_{\lambda}, \ \forall m \in B \}$$

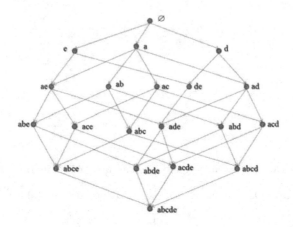

Fig. 4. A λ relational concept lattice with respect to Table 4

if $R' = B$ and $B' = R$, then we say (R, B) is a λ-relational concept. Let (R_1, B_1) and (R_2, B_2) be λ-relational concepts, we define

$$(R_1, B_1) \leq_\lambda (R_2, B_2) \Leftrightarrow R_1 \subseteq R_2 \Leftrightarrow B_2 \subseteq B_1$$

Obviously, the ordered set $(\mathscr{B}(K_\lambda), \leq_\lambda)$ is a complete lattice, which is called a λ relational concept lattice. For example, from Table 4 we can obtain a lattice shown in Fig. 4 (the extent is not easy to be expressed, so only the intent of concept is given).

Theorem 1. *In $K_\lambda = (U \times U, AT, \ J_\lambda)$, the following statements are true*

(1) *Let $B \subseteq AT$, then $B' = R_B^\lambda$;*
(2) *If $(R, B) \in \mathscr{B}(K_\lambda)$, then $R = R_B^\lambda$;*
(3) *Let $B, D \in AT$, if $B' = D'$, then $R_B^\lambda = R_D^\lambda$.*

Proof. The conclusion (1) can be immediately inferred by the definition of operators in K_λ. (2) If $(R, B) \in \mathscr{B}(K_\lambda)$, then we can obtain $R = B'$. In addition, based on conclusion (1) there exists $B' = R_B^\lambda$. Together with $R = B'$ we can see $R = R_B^\lambda$. Therefore (2) is true. (3) We can see from conclusion (1) that $R_B^\lambda = R_D^\lambda$ is true.

In summary, $(\mathscr{B}(K_\lambda), \leq_\lambda)$ can organize all the general relations on U in the form of lattice, which can be viewed as the algebraic structure contained in the complex information system IS.

6   Knowledge Acquisition in Complex Information Systems

We know that solutions to problems of reduct, core and dependency etc. in FCA occupy important position. Based on this, in complex information systems, we propose a solution based on operators in K_λ.

Table 5. A set of dependencies in Table 1

$bc \rightarrow abc$	$bcde \rightarrow abcde$	$cde \rightarrow acde$	$bde \rightarrow abde$
$b \rightarrow ab$	$c \rightarrow ac$	$bd \rightarrow abd$	$ce \rightarrow ace$
$cd \rightarrow acd$	$be \rightarrow abe$	$bce \rightarrow abce$	$bcd \rightarrow abcd$

Definition 4. *In $IS = (U, AT, \Re)$, let $B \subseteq AT$, if $m \in B$ and $R_B^\lambda \neq R_{B-m}^\lambda$, we say m is indispensable in B; if any $m \in B$ is indispensable, we say B is independent, the set of all independent sets is denoted by IND_λ; let $C \subseteq B$, if C is independent and satisfies $R_B^\lambda = R_C^\lambda$, then C is called a reduct of B, the set of all reducts in B is denoted as $RED_\lambda(B)$; let $B \subseteq AT$, the set of all indispensable attributes in B is called the core of B denoted as $CORE_\lambda(B)$; let $B, D \subseteq AT$, for $\forall x, y \in U$, if*

$$(\forall m \in B, (f(x, m), f(y, m)) \in R_m^\lambda) \Rightarrow (\forall n \in D, (f(x, n), f(y, n)) \in R_n^\lambda)$$

we say D is dependent on B, and $B \rightarrow D$ is a dependency. If D is not dependent on B, then the corresponding degree of dependence is defined as follows:

$$\gamma_B(D) = \frac{1}{|B'|} \times |B' \cap D'|$$

For example, when $B = \{a, b\}$ and $D = \{c, d, e\}$, it's not hard to calculate D is not dependent on B, and $\gamma_B(D) = 0.4$.

Theorem 2. *In above Definition there exist following conclusions*

(1) $IND_\lambda = \{B \subseteq AT \mid \forall a \in B, (B - a)' \neq B'\}$;
(2) $CORE_\lambda(B) = \{a \in B \mid (B - a)' \neq B'\}$;
(3) *if C is the minimal set in B satisfying $C' = B'$, then $C \in RED_\lambda(B)$.*
(4) $B \rightarrow D \Leftrightarrow B' \subseteq D'$.

Proof. Conclusions can be immediately inferred by Definition 4 and Theorem 1.

Theorem 3. *Let $B, D \subseteq AT$, $B \rightarrow D$. If $B \subseteq B_1$ and $D_1 \subseteq D$, then $B_1 \rightarrow D_1$.*

Proof. $B' \subseteq D'$ can be inferred from $B \rightarrow D$. In addition, $B'_1 \subseteq B'$ and $D' \subseteq D'_1$ can be deduced from $B \subseteq B_1$ and $D_1 \subseteq D$. Hence $B'_1 \subseteq D'_1$, that is, $B_1 \rightarrow D_1$ is true.

Definition 5. *If $B \rightarrow D$ can be inferred from $B_1 \rightarrow D_1$ by some inference rule ξ, we say $B \rightarrow D$ can be ξ-inferred from $B_1 \rightarrow D_1$. In this case, we call $B \rightarrow D$ is relatively redundant. In addition, if $B = D$, we say $B \rightarrow D$ is absolutely redundant.*

Normally, the number of dependencies is quite large. In the following, to find valuable dependencies we need to remove some redundant dependencies. In fact, from Theorem 3 we can define a inference as follows:

$$\frac{B \subseteq B_1, \ D_1 \subseteq D, \ B \rightarrow D}{B_1 \rightarrow D_1}$$

and then based on above inference, we can remove all relatively redundant dependencies. For example, in Table 1, if we remove all relatively redundant dependencies based on above inference and absolutely redundant dependencies, and then can obtain a smaller set of dependencies shown in Table 5.

7   Conclusions

As a tool for data analysis, FCA possesses good mathematical properties, and has attracted great concerns in recent years. With the research background of complex information systems this paper proposes a rough set model based on FCA, in which common relations on universe such as equivalence relation, dominance relation, similarity relation, tolerance relation and neighborhood relation are expanded to a general binary relation and problems of algebraic structure, reduct, core, dependency are discussed. In fact, the algebraic structure in the complex information system discussed in this paper is a lattice structure that can organize binary relations on universe in the form of lattice organically. Finally, how to eliminate redundant dependencies is studied. Particularly, since the relation on domain of any attribute may be extremely complex, often cause the high time complexity and space complexity to solve above problems. For above reason this paper introduces granular computing(GrC), which can effectively reduce the complexity to a certain extent. In general, this paper not only provides a useful method for applying the formal concept analysis to the complex information system, but also offers a new idea for the extension of the rough set model. Exploration of wider combinations of FCA and rough set theory will also be one focus of our future research.

Acknowledgments. We would like to thank anonymous reviewers very much for their professional comments and valuable suggestions. This work was supported by the National Postdoctoral Science Foundation of China (No. 2014M560352) and the National Natural Science Foundation of China (No. 61273304).

References

1. Ganter, B., Wille, R.: Formal Concept Analysis: Mathematical Foundations. Springer, Berlin (1999)
2. Greco, S., Matarazzo, B., Slowinski, R.: Rough approximation of a preference relation by dominance relation. Eur. J. Oper. Res. **117**, 63–83 (1999)
3. Kang, X.P., Li, D.Y., Wang, S.G., Qu, K.S.: Rough set model based on formal concept analysis. Inf. Sci. **222**, 611–625 (2013)
4. Leung, Y., Li, D.Y.: Maximal consistent block technique for rule acquisition in incomplete information systems. Inf. Sci. **153**, 85–106 (2003)
5. Mi, J.S., Leung, Y., Wu, W.Z.: Approaches to attribute reduct in concept lattices induced by axialities. Knowl.-Based Syst. **23**, 504–511 (2010)
6. Pawlak, Z.: Rough sets. Int. J. Comput. Inform. Sci. **11**, 341–356 (1982)
7. Saaty, T.L.: A scaling method for priorities in hierarchical structures. J. Math. Psychol. **15**(3), 234–281 (1977)

8. Slowinski, R., Vanderpooten, D.: A generalized definition of rough approximations based on similarity. Knowl. Data Eng. **12**, 331–336 (2000)

9. Wei, L., Qi, J.J.: Relation between concept lattice reduct and rough set reduct. Knowl.-Based Syst. **23**, 934–938 (2010)

10. Wille, R.: Restructuring lattice theory: an approach based on hierarchies of concepts. In: Rival, I. (ed.) Ordered Sets, pp. 445–470. Reidel, Dordrecht, Boston (1982)

The Borda Count, the Intersection and the Highest Rank Method in a Dispersed Decision-Making System

Małgorzata Przybyła-Kasperek[✉]

Institute of Computer Science, University of Silesia,
Będzińska 39, 41-200 Sosnowiec, Poland
malgorzata.przybyla-kasperek@us.edu.pl
http://www.us.edu.pl

Abstract. The main aim of the article is to compare the results obtained using three different methods of conflict analysis in a dispersed decision-making system. The conflict analysis methods, used in the article, are discussed in the paper of Ho, Hull and Srihari [6] and in the book of Black [2]. All these methods are used if the individual classifiers generate rankings of classes instead of unique class choices. The first method is the Borda count method, which is a generalization of the majority vote. The second is the intersection method, which belong to the class set reduction method. The third one is the highest rank method, which belong to the methods for class set reordering. All of these methods were used in a dispersed decision-making system which was proposed in the paper [12].

Keywords: Decision-making system · Global decision · Conflict analysis · Borda count method · Intersection method · Highest rank method

1 Introduction

Decision making based on dispersed knowledge is an extremely important issue. Information from one subject is collected by different units. In this way we get access to the knowledge that is stored in separate databases. In order to make decisions we want to use all available knowledge, that is, we must use the dispersed knowledge.

In this paper an approach proposed in the article [12] is used. The aim of the paper is to investigate the use of three selected conflict analysis methods in the system with dynamically generated clusters. The problem of conflict analysis arises because the inference is being conducted in groups of knowledge bases. By a conflict, we mean a situation in which conflicting decisions are taken for the specified set of conditions on the basis of knowledge stored in different groups of knowledge bases. Three methods known from the literature [2,6] were used to conflict analysis: the Borda count method, the intersection method and the highest rank method.

© Springer International Publishing Switzerland 2015
Y. Yao et al. (Eds.): RSFDGrC 2015, LNAI 9437, pp. 298–309, 2015.
DOI: 10.1007/978-3-319-25783-9_27

The concept of distributed decision making is widely discussed in the paper [13]. The concept of taking a global decision on the basis of local decisions is also used in issues concerning the multiple model approach. Examples of the application of this approach can be found in the literature [1,16]. Also in many other papers [3,14], the problem of using distributed knowledge is considered. This paper describes a different approach to the global decision-making process. We assume that the set of local knowledge bases that contain information from one domain is pre-specified. The only condition which must be satisfied by the local knowledge bases is to have common decision attributes.

2   A Brief Overview of Decision-Making System Using Dispersed Knowledge

The concept of a dispersed decision-making system is being considered by the author for several years. In the first stage of studies the considerations were directed to a system with a static structure [10,17]. In recent papers a system with a dynamic structure has been proposed [11,12]. During the construction of this system's structure a negotiation stage is used. The main assumptions, notations and definitions of the system are described below. A detailed discussion can be found in the paper [12].

We assume that the knowledge is available in a dispersed form, which means in a form of several decision tables. The set of local knowledge bases that contain data from one domain is pre-specified. The only condition which must be satisfied by the local knowledge bases is to have common decision attributes. We assume that each local knowledge base is managed by one agent, which is called a resource agent. We call ag in $Ag = \{ag_1, \ldots, ag_n\}$ a resource agent if he has access to resources represented by a decision table $D_{ag} := (U_{ag}, A_{ag}, d_{ag})$, where U_{ag} is a finite nonempty set called the universe; A_{ag} is a finite nonempty set of conditional attributes, V_a^a is a set of attribute a values; d_{ag} is referred to as a decision attribute, V_{ag}^d is called the value set of d_{ag}. We want to designate homogeneous groups of resource agents. The agents who agree on the classification for a test object into the decision classes will be combined in the group. It is realized in two steps. At first initial coalitions are created. Then the negotiation stage is implemented. These two steps are based on the test object classification carried out by the resource agents. For each agent the classification is represented as a vector of values, whose dimension is equal to the number of decision classes. This vector will be defined on the basis of certain relevant objects. That is the objects from the decision tables of agents that carry the greatest similarity to the test object. From decision table of resource agent $D_{ag}, ag \in Ag$ and from each decision class, the smallest set containing at least m_1 objects for which the values of conditional attributes bear the greatest similarity to the test object is chosen. The value of the parameter m_1 is selected experimentally. Then for each resource agent $i \in \{1, \ldots, n\}$ and the test object x, a c-dimensional vector $[\bar{\mu}_{i,1}(x), \ldots, \bar{\mu}_{i,c}(x)]$ is generated. The value $\bar{\mu}_{i,j}(x)$ is equal to the average value of the similarity of the test object to the relevant objects of agent ag_i, belonging to the decision class v_j. In the experimental part of this paper the Gower similarity measure [11] was used. This measure enables the analysis of data sets that have qualitative, quantitative and binary attributes. On the basis of the vector of values defined above, a vector of the rank is specified. The vector of rank is defined as follows: rank 1 is assigned to the values of the decision attribute that are taken with the maximum level of certainty. Rank 2 is assigned to the next most certain decisions, etc. Proceeding in this way for each resource agent $ag_i, i \in \{1, \ldots, n\}$, the vector of rank $[\bar{r}_{i,1}(x), \ldots, \bar{r}_{i,c}(x)]$ will be defined.

In order to create clusters of agents, relations between the agents are defined. Definitions of the relations of friendship and conflict as well as the method for determining the intensity of conflicts were taken from the papers of Pawlak [8,9]. Relations between agents are defined by their views on the classification of the test object x to the decision class. We define the function $\phi^x_{v_j}$ for the test object x and each value of the decision attribute $v_j \in V^d$; $\phi^x_{v_j} : Ag \times Ag \to \{0,1\}$

$$\phi^x_{v_j}(ag_i, ag_k) = \begin{cases} 0 & \text{if } \bar{r}_{i,j}(x) = \bar{r}_{k,j}(x) \\ 1 & \text{if } \bar{r}_{i,j}(x) \neq \bar{r}_{k,j}(x) \end{cases} \text{ where } ag_i, ag_k \in Ag. \text{ We also define the}$$

intensity of conflict between agents using a function of the distance between agents. We define the distance between agents ρ^x for the test object x: ρ^x : $Ag \times Ag \to [0,1]$, $\rho^x(ag_i, ag_k) = \frac{\sum_{v_j \in V^d} \phi^x_{v_j}(ag_i, ag_k)}{card\{V^d\}}$, where $ag_i, ag_k \in Ag$.

Definition 1. *Let p be a real number, which belongs to the interval $[0, 0.5)$. We say that agents $ag_i, ag_k \in Ag$ are in a friendship relation due to the object x, which is written $R^+(ag_i, ag_k)$, if and only if $\rho^x(ag_i, ag_k) < 0.5 - p$. Agents $ag_i, ag_k \in Ag$ are in a conflict relation due to the object x, which is written $R^-(ag_i, ag_k)$, if and only if $\rho^x(ag_i, ag_k) > 0.5 + p$. Agents $ag_i, ag_k \in Ag$ are in a neutrality relation due to the object x, which is written $R^0(ag_i, ag_k)$, if and only if $0.5 - p \leq \rho^x(ag_i, ag_k) \leq 0.5 + p$.*

By using the relations defined above we can create groups of resource agents, which are not in conflict relation. The first step involves the creation of groups of agents remaining in the friendship relation.

Definition 2. *Let Ag be the set of resource agents. The initial cluster due to the classification of object x is the maximum, due to the inclusion relation, subset of resource agents $X \subseteq Ag$ such that $\forall_{ag_i, ag_k \in X} R^+(ag_i, ag_k)$.*

In the second stage of clustering, limitations imposed on compatibility of agents are relaxed. We assume that during the negotiation, agents put the greatest emphasis on compatibility of ranks assigned to the decisions with the highest ranks. We define the function ϕ^x_G for the test object x; $\phi^x_G : Ag \times Ag \to [0, \infty)$

$$\phi^x_G(ag_i, ag_j) = \frac{\sum_{v_l \in Sign_{i,j}} |\bar{r}_{i,l}(x) - \bar{r}_{j,l}(x)|}{card\{Sign_{i,j}\}} \text{ where } ag_i, ag_j \in Ag \text{ and } Sign_{i,j} \subseteq V^d$$

is the set of significant decision values for the pair of agents ag_i, ag_j. In the set $Sign_{i,j}$ there are the values of the decision, which the agent ag_i or agent ag_j gave the highest rank.

During the negotiation stage, the intensity of the conflict between the two groups of agents is determined by using the generalized distance. The generalized distance between agents for the test object x is denoted by ρ_G^x; $\rho_G^x : 2^{Ag} \times 2^{Ag} \rightarrow [0, \infty)$. The value of the generalized distance function for two sets of agents X and Y is equal to the average value of the function ϕ_G^x for each pair of agents ag, ag' belonging to the set $X \cup Y$. This value can be interpreted as the average difference of the ranks assigned to significant decisions within the combined group of agents consisting of the sets X and Y.

For each agent ag that has not been included to any initial clusters, the generalized distance value is determined for this agent and all initial clusters, with which the agent ag is not in a conflict relation and for this agent and other agents without coalition, with which the agent ag is not in a conflict relation. Then the agent ag is included to all initial clusters, for which the generalized distance does not exceed a certain threshold, which is set by the system's user. Also agents without coalition, for which the value of the generalized distance function does not exceed the threshold, are combined into a new cluster. After completion of the second stage of the process of clustering we get the final form of clusters. On the basis of the knowledge of agents from one cluster, local decisions are taken. An important problem that occurs is to eliminate inconsistencies in the knowledge stored in different knowledge bases. In previous papers the approximated method of the aggregation of decision tables have been used to eliminate inconsistencies in the knowledge [10–12]. In this paper, we also use this method. In the method for every cluster, a kind of combined information is determined. This combined information is in the form of aggregated decision table. Object of this table are constructed by combining relevant object from decision tables of the resource agents that belong to one cluster. Based on the aggregated decision tables global decisions are taken using the methods of conflict analysis.

3   Methods of Conflict Analysis

In this article, we use three different methods of conflict analysis: the Borda count method, the intersection method and the highest rank method. These methods are discussed in the paper of Ho, Hull and Srihari [6] and in the book of Black [2]. All these methods are used if the individual classifiers generate rankings of classes instead of unique class choices. However, this restriction can be relaxed. These methods can also be used when the individual classifiers generate the probability distributions over different decision. On the basis of such probability distribution the ranking of classes can be generated in a simple way. This can be realized in the following way: rank 1 is assigned to the values of the decision attribute that are taken with the maximum level of certainty. Rank 2 is assigned to the next most certain decisions, etc. In the considered system, such transformation is applied.

On the basis of each aggregated decision table a ranking of classes is generated. At first, a c-dimensional vector of values $[\mu_{j,1}(x), \ldots, \mu_{j,c}(x)] \in [0, 1]^c$ is generated for each j-th cluster, where c is the number of all of the decision classes. The value $\mu_{j,i}(x)$ determines the level of certainty with which the decision v_i is taken by agents for a given test object x belonging to the cluster$j$.

This vector will be defined on the basis of relevant objects. From each aggregated decision table and from each decision class, the smallest set containing at least m_2 objects for which the values of conditional attributes bear the greatest similarity to the test object is chosen. The value of the parameter m_2 is selected experimentally. The value $\mu_{j,i}(x)$ is equal to the average value of the similarity of the test object to the relevant objects form j-th aggregated decision table, belonging to the decision class v_i. On the basis of the vector of values defined above, for each cluster a vector of the rank is specified. The vector of rank is defined as follows: rank 1 is assigned to the values of the decision attribute that are taken with the maximum level of certainty. Rank 2 is assigned to the next most certain decisions, etc. Proceeding in this way for each j-th cluster, the vector of rank $[r_{j,1}(x), \ldots, r_{j,c}(x)]$ will be defined. In this way, for each cluster the ranking of classes is generated.

The Borda Count Method

The Borda count method is simple, has low computational complexity and is easy to implement. This method belongs to class set reordering methods. It consists in the designation for each decision class the sum of the number of classes ranked below it by each cluster. Thus for each decision class v_i the value is determined $\sum_{j\text{-th cluster}}(card\{V^d\} - r_{j,i}(x))$. The decision classes sorted according to their Borda count gives the final ranking. For the purposes of dispersed decision-making system it is assumed that the set of decisions taken by the system is the set of classes which have the maximum value of the Borda count. The Borda count method requires no training and all classifiers are treated equally. In some applications, it may be undesirable, because the method does not take into account the differences in the individual classifier capabilities.

The Intersection Method

The intersection method belongs to a group of methods for class set reduction. The purpose of these methods is reducing the number of classes in the output without losing the true class. This is achieved by using two criteria: the size of the result set should be minimized, and the probability of inclusion of the true class should be maximized. The intersection method requires training, so it is more computationally complex than the Borda method. In order to apply the intersection method to the dispersed decision-making system, it had to be a little modified as described below. At first, the classification of objects from the training set is made by each of the individual classifiers. Therefore for each resource agent the leave-one-out method is implemented on the training set. The ranking of classes for resource agent is generated according to the method described at the beginning of this section. So firstly the μ vector is generated using the parameter m_2, and then the vector of rank is determined. After rankings are obtained for all resource agents, the lowest rank ever given by each agent to any true class is determined. Then a threshold on the ranks is determined for each cluster, by calculating the minimum of thresholds on the ranks assigned to the resource agents belonging to the cluster. For the test object the global decisions generated by the dispersed decision-making system is equal to the intersection of large neighborhoods taken from each cluster. The sizes of the neighborhoods are determined by the calculated thresholds on the ranks. The neighborhoods may be very large, since they are determined by the worst-case behavior of the cluster. The intersection method is suitable when classifiers deal well with all objects from the training set. Unfortunately, that happens rarely. Therefore, we use another modification. It was assumed that the worst rank can be considered as the threshold value only when it is assigned to a certain small percentage of cases. The percentage of cases is determined by the parameter u.

The Highest Rank Method

The highest rank method is simple, has low computational complexity, requires no training and belongs to class set reordering methods, just like the Borda count method. The method has the following steps: for a given test object and for each cluster the ranking of classes is determined. Then, for each decision class the minimum (highest) of these ranks is assigned. Thus for each decision class v_i the value $\min_{j\text{-th cluster}} r_{j,i}(x)$ is calculated. The classes are then sorted by these scores to derive a combined ranking for that test object. The set of decisions taken by the dispersed system is the set of classes which have the highest rank. A big disadvantage of this method is that the combined ranking may have many ties. The number of classes sharing the same ranks depends on the number of classifiers used. During the experiments with the dispersed decision-making system it was noted that there are some weak clusters, which assigns the highest rank to several decision classes at the same time. This significantly increases the number of ties, and made uninteresting results. Therefore, for the highest rank method only the individual classifiers were taken into consideration which made unambiguous decisions. That is the highest rank was assigned to one decision class.

Now we will briefly discuss the computational complexity and scalability of the proposed methods. In the paper [12] a computational complexity of the clustering method with a negotiation stage was discussed. The method has an exponential complexity of computing due to the number of resource agents. The number of resource agents in the system is rather small, therefore the algorithm execution time is short. In the paper [17] a computational complexity of the approximated method of the aggregation of decision tables was discussed. The method has an exponential complexity due to the number of relevant objects. This method takes most of the time while generating a global decision by the system. The Borda count and the highest rank methods require no training and they are based on arithmetic operations performed on the rankings of classes, therefore their computational complexity is low. These methods have a polynomial complexity due to the number of synthesis agents and the number of decision classes. The intersection method requires training and therefore has a higher computational complexity. The method has a polynomial complexity due to the number of resource agents and the number of objects in their decision tables. As was mentioned, most of the time while generating global decisions is taken by the method of the aggregation of decision tables. But the complexity of this method depends on the chosen parameter value, which specifies the number of relevant objects. For large (big) data sets lower parameter value can be selected. Experiments on data set, which contains more than 4,000 objects were carried out. However, due to the limited length of the article, results of these experiments are not presented here.

4   Experiments

The aim of the experiments is to compare the quality of the classification made by the decision-making system using three different methods of conflict analysis. The Borda count method, the intersection method and the highest rank method were considered. For the experiments the following data, which are in the UCI repository (http://www.ics.uci.edu/~mlearn/MLRepository.html), were used: Soybean Data Set and Vehicle Silhouettes data set. In order to determine the efficiency of inference of the dispersed decision-making system with respect to the analyzed data, each data set was divided into two disjoint subsets: a training set and a test set. Table 1 gives a numerical summary of the data sets.

Table 1. Data set summary

Data set	# The training set	# The test set	# Conditional attributes	# Decision classes
Soybean	307	376	35	19
Vehicle Silhouettes	592	254	18	4

Because the available data sets are not in the dispersed form, in order to test the dispersed decision-making system the training set was divided into a set of decision tables. Divisions with a different number of decision tables were considered. For each of the data sets used, the decision-making system with five different versions (with 3, 5, 7, 9 and 11 decision tables) were considered. For these systems, we use the following designations: WSD_{Ag1}^{dyn} - 3 decision tables; WSD_{Ag2}^{dyn} - 5 decision tables; WSD_{Ag3}^{dyn} - 7 decision tables; WSD_{Ag4}^{dyn} - 9 decision tables; WSD_{Ag5}^{dyn} - 11 decision tables. Note that the division of the data set was not made in order to improve the quality of the decisions taken by the decision-making system, but in order to store the knowledge in a dispersed form. We consider the situation, that is very common in life, in which data from one domain are collected by different units as separate knowledge bases. For each data set we have 5 versions of the dispersion, therefore it can be said that 10 different dispersed data set have been used for experiments.

All used conflict analysis methods have one disadvantage. The combined ranking may have many ties. The number of classes sharing the same ranks depends on the number of classifiers used. In addition it was noted that the highest rank method generates the largest number of ties and the Borda count method generates the least number of ties. In order to analyze these properties, an additional method of ties resolving has not been applied. But the appropriate classification measures were applied, which are adapted to this situation. The measures of determining the quality of the classification are: *estimator of classification error e* in which an object is considered to be properly classified if the decision class used for the object belonged to the set of global decisions generated by the system; *estimator of classification ambiguity error* e_{ONE} in which object is considered to be properly classified if only one, correct value of the decision was generated to this object; *the average size of the global decisions sets* $\overline{d}_{WSD_{Ag}^{dyn}}$ generated for a test set.

In the description of the results of experiments for clarity some designations for algorithms and parameters have been adopted: m_1 - parameter which determines the number of relevant objects that are selected from each decision class of the decision table and are then used in the process of cluster generation; p - parameter which occurs in the definition of friendship, conflict and neutrality relations; u - parameter of the intersection method; $A(m)$ - the approximated method of the aggregation of decision tables; $C(m_2)$ - the method of conflict analysis (the Borda count method, the intersection method or the highest rank method), with parameter which determines the number of relevant objects that are used to generate ranking of classes.

The results of the experiments with the Soybean data set are presented in Table 2. In the table the following information is given: the name of dispersed decision-making system (System); the selected, optimal parameter values (Parameters); the algorithm's symbol (Algorithm); the three measures discussed earlier $e, e_{ONE}, \overline{d}_{WSD_{Ag}^{dyn}}$; the time t needed to analyse a test set expressed in minutes.

Based on the results of the experiments given in Table 2, the following conclusions can be drawn. The dispersed decision support system with the Borda count method generates in most cases one decision. The quality of the classification increases with increasing number of considered decision tables, in other words - when the knowledge is more dispersed. The system with the intersection method generates much better results but this is achieved at the cost of increased number of generated decision classes. The system with the highest rank method produces the worst results, even though the average size of the global decisions sets is not significant different from the size of the sets generated when using the intersection method. The reason for the fact that the highest rank method produces the worst results, is that this method favors one or a small number of individual classifiers. As was mentioned earlier, this method does not consider clusters that do not take unambiguous decisions. The highest rank method is good for combining a small number of classifiers, each of which specializes on inputs of a particular type. In the Borda count method and the intersection approach all classifiers are treated equally, which turned out to be appropriate for the considered data set.

The results of the experiments with the Vehicle data set are presented in Table 3. Similarly, in the case of the Vehicle data set the Borda count method generates the smallest average size of the global decisions sets. Only in three

Table 2. Summary of experiments results with the Soybean data set

The Borda count method

System	Parameters	Algorytm	e	e_{ONE}	$\bar{d}_{WSD_{Ag}}$	t
WSD_{Ag1}	$m_1 = 6, p = 0.15$	$A(1)C(1)$	0.117	0.189	1.109	0.04
WSD_{Ag2}	$m_1 = 1, p = 0.15$	$A(6)C(11)$	0.112	0.191	1.104	0.10
WSD_{Ag3}	$m_1 = 11, p = 0.05$	$A(1)C(1)$	0.096	0.239	1.189	0.08
WSD_{Ag4}	$m_1 = 11, p = 0.15$	$A(1)C(1)$	0.082	0.160	1.085	0.21
WSD_{Ag5}	$m_1 = 6, p = 0.05$	$A(1)C(1)$	0.077	0.229	1.197	2.36

The intersection method

System	Parameters	Algorytm	e	e_{ONE}	$\bar{d}_{WSD_{Ag}}$	t
WSD_{Ag1}	$m_1 = 1, p = 0.05, u = 1\%$	$A(1)C(1)$	0.043	0.396	1.676	0.04
WSD_{Ag2}	$m_1 = 6, p = 0.05, u = 1\%$	$A(2)C(1)$	0.021	0.316	1.556	0.07
WSD_{Ag3}	$m_1 = 11, p = 0.05, u = 2\%$	$A(1)C(1)$	0.080	0.354	1.556	0.10
WSD_{Ag4}	$m_1 = 11, p = 0.05, u = 3\%$	$A(5)C(1)$	0.045	0.263	1.426	0.25
WSD_{Ag5}	$m_1 = 6, p = 0.05, u = 1\%$	$A(3)C(1)$	0.053	0.364	1.566	2.41

The highest rank method

System	Parameters	Algorytm	e	e_{ONE}	$\bar{d}_{WSD_{Ag}}$	t
WSD_{Ag1}	$m_1 = 1, p = 0.15$	$A(3)C(2)$	0.104	0.351	1.303	0.06
WSD_{Ag2}	$m_1 = 1, p = 0.15$	$A(3)C(3)$	0.069	0.508	1.497	0.07
WSD_{Ag3}	$m_1 = 1, p = 0.15$	$A(10)C(1)$	0.152	0.428	1.372	1.21
WSD_{Ag4}	$m_1 = 1, p = 0.15$	$A(1)C(1)$	0.117	0.527	1.545	0.18
WSD_{Ag5}	$m_1 = 20, p = 0.15$	$A(1)C(1)$	0.130	0.681	2.340	3.02

cases, the systems with 3, 5 and 7 decision tables, the use of the intersection method has improved the efficiency of inference. The highest rank method produces results that are not interesting, because the average size of the global decisions sets is above or close to the value 2, while in the data set there are only 4 decision classes.

The papers [4,5,7,15] also shows the experiments with the Soybean and the Vehicle Silhouettes data set. Data in the non-dispersible form were examined there. Table 4 presents the selected results given in those papers. Presented, in this paper, results can not be compared uniquely with the results shown in Table 4. Because the decision-making system, described in the paper, generates a set of decisions, while Table 4 shows the results of algorithms that generate one decision. Two of the conflict analysis methods (the Borda count method and the intersection method) used in this paper generate a small set of decisions. Whereas the highest rank method generates larger set of decisions, especially for the Vehicle data set. It should be noted that for the Soybean data set the average size of the global decisions sets is close to the value 1 or 1.5, note that there are 19 decision classes. This means that this result may be considered as a quite good result.

Table 3. Summary of experiments results with the Vehicle Silhouettes data set

The Borda count method

System	Parameters	Algorytm	e	e_{ONE}	$\overline{d}_{WSD_{Ag}}$	t
WSD_{Ag1}	$m_1 = 11, p = 0.05$	$A(11)C(16)$	0.193	0.386	1.236	0.15
WSD_{Ag2}	$m_1 = 1, p = 0.05$	$A(1)C(2)$	0.256	0.433	1.220	0.07
WSD_{Ag3}	$m_1 = 11, p = 0.05$	$A(5)C(6)$	0.236	0.350	1.142	0.26
WSD_{Ag4}	$m_1 = 6, p = 0.05$	$A(7)C(7)$	0.276	0.358	1.098	4.14
WSD_{Ag5}	$m_1 = 11, p = 0.05$	$A(2)C(9)$	0.264	0.354	1.106	2.18

The intersection method

System	Parameters	Algorytm	e	e_{ONE}	$\overline{d}_{WSD_{Ag}}$	t
WSD_{Ag1}	$m_1 = 6, p = 0.05, u = 8\%$	$A(3)C(1)$	0.087	0.713	1.650	0.13
WSD_{Ag2}	$m_1 = 1, p = 0.05, u = 8\%$	$A(1)C(6)$	0.138	0.705	1.606	0.15
WSD_{Ag3}	$m_1 = 6, p = 0.05, u = 9\%$	$A(2)C(6)$	0.130	0.634	1.528	0.22
WSD_{Ag4}	$m_1 = 1, p = 0.05, u = 3\%$	$A(1)C(1)$	0.248	0.665	1.508	0.23
WSD_{Ag5}	$m_1 = 1, p = 0.05, u = 3\%$	$A(6)C(1)$	0.272	0.661	1.425	3.55

The highest rank method

System	Parameters	Algorytm	e	e_{ONE}	$\overline{d}_{WSD_{Ag}}$	t
WSD_{Ag1}	$m_1 = 1, p = 0.05$	$A(2)C(1)$	0.154	0.504	1.472	0.06
WSD_{Ag2}	$m_1 = 1, p = 0.05$	$A(8)C(2)$	0.098	0.780	1.953	0.17
WSD_{Ag3}	$m_1 = 1, p = 0.05$	$A(4)C(8)$	0.059	0.811	2.169	0.13
WSD_{Ag4}	$m_1 = 1, p = 0.05$	$A(3)C(5)$	0.039	0.890	2.417	0.20
WSD_{Ag5}	$m_1 = 1, p = 0.05$	$A(7)C(9)$	0.047	0.937	2.594	6.04

Table 4. Results of experiments from other papers

Algorithm	Soybean Error rate	Vehicle Silhouettes Error rate
C4.5	0.126	0.294
Bagged C4.5	0.101	0.257
Adaboosted C4.5	0.076	0.220
Randomized C4.5	0.085	0.248
K-Nearest neighbor	0.160	0.335
Naive bayes	0.106	0.550

In the case of the Soybean data set, the system with the Borda count method generates comparable results to the results shown in Table 4. The efficiency of inference has significantly improved in comparison with the results shown in Table 4 when applying the intersection method. This is achieved at the expense of an increased number of generated decision classes. But as was mentioned earlier, the average size of the global decisions sets close to the value 1.5 can be considered as a quite good result when we have 19 decision classes. Applying the highest rank method did not bring the improvement of the classification quality in comparison with the results given in Table 4.

In the case of the Vehicle Silhouettes data set, the system with the Borda count method generates comparable results to the results shown in Table 4. In the case of dispersed systems with 3, 5 and 7 decision tables the efficiency of inference has significantly improved in comparison with the results shown in Table 4, when applying the intersection method. This is achieved at the cost of an increased number of generated decision classes. Application of the highest rank method does not give interesting results, because the average size of the global decisions sets is above or close to the value 2, while in the data set there are only 4 decision classes. Summarizing the results it should be noted that the biggest advantage of the proposed approach is the possibility of using dispersed knowledge, which is not possible in the approaches presented in other papers.

5   Conclusions

In this article, three different methods of conflict analysis were used in the dispersed decision-making system: the Borda count method, the intersection method and the highest rank method. In the experiments, which are presented, dispersed data have been used: Soybean data set and Vehicle Silhouettes data set. Based on the presented results of experiments it can be concluded that the Borda count method produces acceptable results with the average size of the global decisions sets close to the value 1. The intersection method generates much better results but at the expense of a slightly increased number of generated decision classes. The highest rank method often produces results that are not interesting.

References

1. Bazan, J., Peters, J., Skowron, A., Nguyen, H., Szczuka, M.: Rough set approach to pattern extraction from classifiers. Electron. Notes Theor. Comput. Sci. **82**, 20–29 (2003)
2. Black, D.: The Theory of Committees and Elections. University Press, Cambridge (1958)
3. Delimata, P., Suraj, Z.: Feature selection algorithm for multiple classifier systems: a hybrid approach. Fundamenta Informaticae **85**(1–4), 97–110 (2008)
4. Dietterich, T.G.: An experimental comparison of three methods for constructing ensembles of decision trees: bagging, boosting, and randomization. Mach. Learn. **40**(2), 139–157 (2000)
5. Gou, J., Xiong, T., Kuang, Y.: A novel weighted voting for K-Nearest neighbor rule. J. Comput. **6**(5), 833–840 (2011)
6. Ho, T.K., Hull, J.J., Srihari, S.N.: Decision combination in multiple classifier systems. IEEE Trans. Pattern Anal. Mach. Intell. **16**(1), 66–75 (1994)
7. Kononenko, I., Simec, E., Robnik-Sikonja, M.: Overcoming the myopia of inductive learning algorithms with RELIEFF. Appl. Intell. **7**(1), 39–55 (1997)

8. Pawlak, Z.: On conflicts. Int. J. of Man-Mach. Stud. **21**, 127–134 (1984)
9. Pawlak, Z.: An inquiry anatomy of conflicts. J. Inf. Sci. **109**, 65–78 (1998)
10. Przybyła-Kasperek, M., Wakulicz-Deja, A.: Application of reduction of the set of conditional attributes in the process of global decision-making. Fundamenta Informaticae **122**(4), 327–355 (2013)
11. Przybyła-Kasperek, M., Wakulicz-Deja, A.: Global decision-making system with dynamically generated clusters. Inf. Sci. **270**, 172–191 (2014)
12. Przybyła-Kasperek, M., Wakulicz-Deja, A.: A dispersed decision-making system - the use of negotiations during the dynamic generation of a systems structure. Inf. Sci. **288**, 194–219 (2014)
13. Schneeweiss, C.: Distributed Decision Making. Springer, Berlin (2003)
14. Skowron, A., Wang, H., Wojna, A., Bazan, J.: Multimodal classification: case studies. In: Peters, J.F., Skowron, A. (eds.) Transactions on Rough Sets V. LNCS, vol. 4100, pp. 224–239. Springer, Heidelberg (2006)
15. Soria, D., Garibaldi, J.M., Ambrogi, F., Biganzoli, E., Ellis, I.O.: A 'nonparametric' version of the naive Bayes classifier. Knowl.-Based Syst. **24**(6), 775–784 (2011)
16. Ślęzak, D., Wróblewski, J., Szczuka, M.: Neural network architecture for synthesis of the probabilistic rule based classifiers. Electron. Notes Theor. Comput. Sci. **82**, 251–262 (2003)
17. Wakulicz-Deja, A., Przybya-Kasperek, M.: Application of the method of editing and condensing in the process of global decision-making. Fundamenta Informaticae **106**(1), 93–117 (2011)

An Ensemble Learning Approach Based on Missing-Valued Tables

Seiki Ubukata[1]([✉]), Taro Miyazaki[2], Akira Notsu[1], Katsuhiro Honda[1],
and Masahiro Inuiguchi[2]

[1] Graduate School of Engineering, Osaka Prefecture University,
Gakuencho 1-1, Sakai, Osaka 599-8531, Japan
`{subukata,notsu,honda}@cs.osakafu-u.ac.jp`
[2] Graduate School of Engineering Science, Osaka University,
Toyonaka, Osaka 560-8531, Japan
`miyazaki@inulab.sys.es.osaka-u.ac.jp, inuiguti@sys.es.osaka-u.ac.jp`

Abstract. In classification problems on rough sets, the effectiveness of ensemble learning approaches such as bagging, random forests, and attribute sampling ensemble has been reported. We focus on occurrences of deficiencies in columns on the original decision table in random forests and attribute sampling ensemble approaches. In this paper, we generalize such deficiencies of columns to deficiencies of cells and propose an ensemble learning approach based on missing-valued decision tables. We confirmed the effectiveness of the proposed method for the classification performance through numerical experiments and the two-tailed Wilcoxon signed-rank test. Furthermore, we consider the robustness of the method in absences of condition attribute values of unknown objects.

Keywords: Rough sets · Ensemble learning · MLEM2 · Incomplete data · Missing attribute values

1 Introduction

Rough set theory was proposed by Z. Pawlak in 1982 and has been developed by many researchers [9]. In the field of data mining, rough set approaches have been applied to various issues such as decision rule inductions from decision tables and classification problems by means of decision rules [4,10]. In rough sets, various methods which handle incomplete decision tables including missing values by use of extended binary relations have been proposed and allow rule induction from incomplete data [5–7,11].

In this paper, we deal with classification problems based on rough sets. Miyazaki *et al.* [8] applied some ensemble learning approaches such as bagging, random forests, and attribute sampling ensemble to classification problems based on rough sets and reported its effectiveness. Some ensemble learning approaches are tested under rough set frameworks also in [1,12]. We focus on occurrences of deficiencies of columns on the original decision table in random forest and

© Springer International Publishing Switzerland 2015
Y. Yao et al. (Eds.): RSFDGrC 2015, LNAI 9437, pp. 310–321, 2015.
DOI: 10.1007/978-3-319-25783-9_28

attribute sampling ensemble approaches and generalize them to deficiencies of cells. In this paper, we call an element of a decision table determined by a row and a column a *cell*.

In this paper, we propose an ensemble leaning approach based on missing-valued decision tables and confirm the efficacy of the method through numerical experiments and the two-tailed Wilcoxon signed-rank test. We also consider the robustness of the method in absences of condition attribute values of unknown objects.

2   Preliminaries

In this section, we provide a basic explanation of rough set based classifications and ensemble learning.

2.1   Data Mining Based on Rough Sets

In rough set analyses, a data set is often expressed as a decision table. A decision table is defined by $\langle U, C \cup \{d\}, V, \rho \rangle$, where U is a non-empty finite set of objects, C is a finite set of condition attributes, $\{d\}$ is a set of a decision attribute, $V = \bigcup \{V_a\}_{a \in C \cup \{d\}}$ is a set of attribute values, and $\rho : U \times C \cup \{d\} \to V$ is a information function which assigns an attribute value of an object. In rough sets, binary relations are used for granulation of a universe and approximations of subsets of objects. In typical rough set approaches, a indiscernibility relation is often used. A indiscernibility relation is defined by

$$IND(B) = \{(x, y) \in U \times U \mid \forall a \in B, \rho(x, a) = \rho(y, a)\},$$

where $B \subseteq C \cup \{d\}$. Using this relation, the lower and upper approximations of X are obtained as follows:

$$IND(B)[X] = \{x \in U \mid [x]_{IND(B)} \subseteq X\},$$

$$IND(B)\langle X \rangle = \{x \in U \mid [x]_{IND(B)} \cap X \neq \emptyset\},$$

where $[x]_{IND(B)}$ is the equivalence class of x with respect to $IND(B)$. The indiscernibility relation means that two objects are indistinguishable each other by use of focused attributes.

J. W. Grzymala-Busse has been developed the data mining system LERS and proposed a framework of decision rule induction based on rough sets and classification by means of decision rules [4]. MLEM2 is a subsystem of LERS and an algorithm which induces the minimal set of decision rules which have minimal condition parts from the given decision table. Decision rules which are extracted from lower approximations are certain rules. On the other hand, decision rules which are extracted from upper approximations are possible rules.

2.2   Missing Value Handling in Rough Sets

In rough sets, various methods which handle incomplete decision tables which include missing values have been proposed [5, 6, 11].

In this paper, we introduce methods which handle missing values by extended binary relations. The missing value is denoted by "*". In general data mining, in order to make it possible to analyze incomplete data tables which include missing values, preprocessing such as list-wise deletion and imputation of missing values are usually executed in order to transform incomplete tables to complete tables. On the other hand, rough set based approaches using extended binary relations can handle missing values directly without these preprocessing.

As typical binary relations which handle missing values, there are a similarity relation and a tolerance relation. Rough approximations of subsets of objects using these relations enable rule induction from incomplete decision tables.

Following are the definitions of a similarity relation, a tolerance relation and rough approximations based on these relations. In rough sets, there are various approximation methods such as singleton approximations, subset approximations and concept approximations. In this paper, we use subset approximations, especially, we use subset lower approximations.

A Similarity Relation. A similarity relation is defined as

$$S(B) = \{(x, y) \in U \times U \mid \forall a \in B, \rho(x, a) = \rho(y, a) \vee \rho(y, a) = *\}.$$

In this relation, a link from a first object to a second object is not connected even if they have equal attribute values except for missing values, because they can have different attribute values if different values are substituted into missing values. This relation is reflexive and transitive, and not necessarily symmetric. Accessible sets stand on the relation and the inverse relation are defined as follows:

$$S_B(x) = \{y \in U \mid (x, y) \in S(B)\},$$
$$S_B^{-1}(x) = \{y \in U \mid (y, x) \in S(B)\}.$$

$S_B(x)$ is the set of objects which are accessible from x by $S(B)$, and S_B^{-1} is the set of objects which are accessible to x by $S(B)$. Using these sets, the subset lower and upper approximations of X are defined as follows:

$$S(B)[X] = \bigcup \{S_B^{-1}(x) \mid x \in U, S_B^{-1}(x) \subseteq X\},$$

$$S(B)\langle X \rangle = \bigcup \{S_B(x) \mid x \in X\}.$$

A Tolerance Relation. A tolerance relation is defined as

$$T(B) = \{(x, y) \in U \times U \mid \forall a \in B, \rho(x, a) = \rho(y, a) \vee \rho(x, a) = * \vee \rho(y, a) = *\}.$$

In this relation, two objects which have same attribute values except for missing values are linked because they can have equal attribute values for all attributes if appropriate values are substituted into missing values. This relation is reflexive and symmetric, and not necessarily transitive.

The subset lower and upper approximations of X are defined as follows:

$$T(B)[X] = \bigcup \{T_B(x) \mid x \in U, T_B(x) \subseteq X\},$$

$$T(B)\langle X \rangle = \bigcup \{T_B(x) \mid x \in U, T_B(x) \cap X \neq \emptyset\},$$

where $T_B(x)$ is the set of objects which are accessible from x by $T(B)$.

Missing Value Handling in LERS. MLEM2 is also applied to incomplete decision tables which include missing values which are considered as lost values or "do not care" values, by using a characteristic relation, that is, the combination of a tolerance relation and a similarity relation [5].

2.3 Ensemble Learning

Ensemble learning is useful way to address overfitting problems and improve the classification performance in classification problems. Let us explain basic ensemble learning approaches, that is, bagging, random forests, and attribute sampling ensemble.

Bagging. Bagging (bootstrap aggregating) was proposed by L. Breiman in 1996 [2]. In bagging approach, multiple new tables are generated by the bootstrap sampling from the given data. Using a certain method, weak classifiers are obtained from each generated decision table. Finally, one prediction is determined by aggregating predictions of these weak classifiers.

Random Forests. Random forests was also proposed by L. Breiman in 2001 [3]. In random forests, attribute sampling is carried out in addition to the bootstrap sampling in bagging. Predictions are executed in the same way as bagging.

Attribute Sampling Ensemble. Random forests includes the bootstrap procedure and the attribute sampling procedure. We can consider a method which includes only attribute sampling process. In this paper, we call the method attribute sampling ensemble. This ensemble learning approach was tested in [8].

3 An Ensemble Learning Approach Based on Missing-Valued Decision Tables

In this section, we propose an ensemble learning approach based on missing-valued decision tables. In this method, deficiencies of columns in random forests or attribute sampling ensemble approaches are generalized into deficiencies of cells.

In basic ensemble approaches such as random forests and attribute sampling ensemble approaches, the attribute sampling are executed. The attribute sampling may provide reductions of columns of a original decision table. We can interpret the procedure as a procedure in which attribute values of all cells in columns which are not selected in the attribute sampling procedure are replaced by the missing value "*". We generalize this deficiency procedure to a procedure in which attribute values of cells are replaced by the missing value in increments of cells. In other words, whereas the attribute sampling procedure causes complete deficiencies of columns, proposed method causes partial deficiencies of columns.

Now we consider a method which generates partially losses of attribute values in increments of cells. Let $p \in [0, 1]$ be a missing value rate. In the procedure of the proposed method, each attribute value of cells on the original decision table is replaced by the missing value "*" with a probability p. Executing this procedure in n times to the original decision table, n different incomplete decision tables are generated.

Next, we create weak classifiers from each incomplete decision table which includes missing values by means of MLEM2 rule induction algorithm using lower subset approximations based on extended binary relations, e.g. a similarity relation and a tolerance relation. Thereby, we obtain n classifiers, i.e. n sets of decision rules. Finally, one estimation result is made by a majority voting of estimations of the weak classifiers. Weak classifiers are the decision rules extracted from each incomplete decision tables and classifications by use of extracted decision rules are executed based on the LERS classification framework [4].

In the case that we use data mining techniques which can not handle missing values directly, incomplete decision tables need to be translated to complete decision tables by use of preprocessing which purges missing values by means of list-wise deletions and imputations of missing values. However, analyses including removal of missing values by these processing may not be essential in the proposed approach which appends missing values purposely. Rough set approaches with extended binary relations can handle missing values directly and do not need particular preprocessing against missing values. Therefore, methods based on extended binary relations have a particular affinity for the proposed method. The proposed approach based on missing-valued tables has applicability to not only rough sets or MLEM2 but also other methods which can handle missing values.

Obviously, if $p = 0$, each weak classifier conforms to the classifier stands on the original decision table, and thus the aggregated prediction is same to the predication by the original classifier. If $p = 1$, all objects have no certain attribute values and no rule can be extracted. Therefore, we cannot obtain any prediction.

Table 1. Data set summary.

Data set	#Objects	#Attributes	#Classes	Attribute Type
Balance	625	4	3	Numerical
Blood_Transfusion	748	5	2	Numerical
Dermatology	366	34	6	Numerical
Ecoli	336	7	8	Numerical
Glass	214	9	6	Numerical
Iris	150	4	3	Numerical
Soybean_small	47	35	4	Nominal
Wine	178	13	3	Numerical

4   Numerical Experiments

We carried out numerical experiments in order to determine the classification performance of the proposed methods. In this section, we present the experimental results and considerations. We also conducted the two-tailed Wilcoxon signed-rank test in order to confirm the superiority of proposed method. First, we show the changes in classification error rates of the proposed method by missing value rates. Next, we compare the classification performance of the proposed method with other classification methods such as bagging and random forests.

We used eight data sets shown in Table 1. The data sets were retrieved from the UCI machine learning repository [13]. The application was implemented using Java and a desktop computer equipped with CPU Intel(R) Core(TM) i7-4770 @3.40 GHz.

4.1   Classification Error Rates by Missing Value Rate

We determined the classification performance of the proposed method using a similarity relation and a tolerance relation. The parameter p (missing value rate) was gradually incremented from 0.0 to 1.0 with the increment of 0.1. The weak classifier, that is, the set of decision rules extracted from each missing-valued table is induced from subset lower approximations of each decision class using a similarity relation or a tolerance relation. In all ensemble methods, we set up the ensemble number $n = 10$. Each error rate is the average of 10 times of 10-fold cross-validations. The results are shown in Fig. 1.

Let us consider the results.

- In all data sets except for 'Iris', classification accuracies are improved by adding a certain level of missing values. The appropriate missing value rate depends on data sets, however, in almost data sets, classification accuracies achieve a peak in the case of adding 10 to 20 % missing values.
- If a missing value rate grows on in excess of the peak, the classification performance gradually grows worse. Extremely large missing value rates cause poor

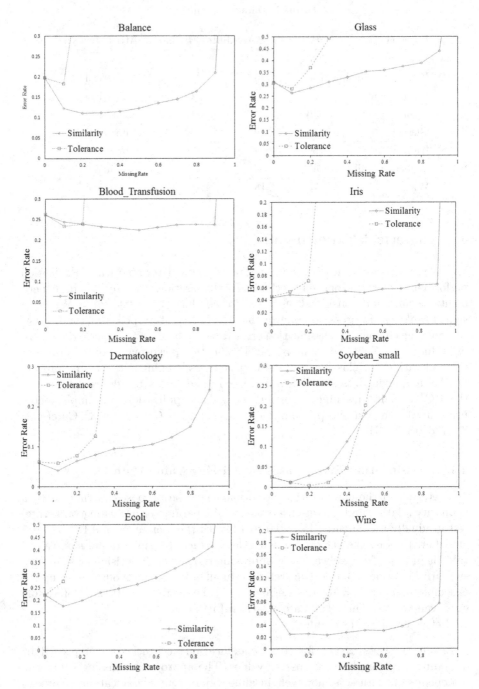

Fig. 1. The changes of error rates of proposed methods by missing value rates for each data set in 10 times of 10-fold cross-validations.

performances. Moreover, if $p = 1$, any classifications are not executed because lower approximations become the empty set. It is interesting that there exist data sets in which classification accuracies are kept even with deficiencies in most of attribute values such as 'Balance', 'Blood_Transfusion', 'Iris', and 'Wine'.

- There exists a data set in which missing value approaches are ineffective in improving performance, such as 'Iris'.
- In all data sets except for 'Soybean_small', a similarity relation provides better results than a tolerance relation.

Let us consider the difference in performance between two relations. In the cases of using a tolerance relation, accessible sets become larger than the cases of using a similarity relation, thus, lower approximations become smaller. On the other hand, in the cases of using a similarity relation, lower approximations become larger. In this research, because decision rules are inducted from lower approximations, smaller lower approximations may cause decreases of supports of decision rules and decreases in the classification performance. Thus, it was predictable results, however, in a data set 'Soybean_small', a tolerance relation provides better performance than a similarity relation exceptionally. Characteristics of data sets which cause such an exception should be investigated in order to use two relations as the situation demands.

4.2   Comparison with Other Methods

We confirmed the effectiveness of the proposed method by comparing with other methods. We compared classification error rates of the proposed method with MLEM2, bagging and random forests applied to MLEM2. The ensemble number of each ensemble approach is 10.

In this paper we abbreviate classification approaches as follows.

- MLEM2: the classifier using the set of decision rules extracted by MLEM2 based on the LERS method.
- MB: the classifier in which bagging is applied to MLEM2 classifier.
- MRF: the classifier in which random forests is applied to MLEM2 classifier.
- MMS: the proposed classifier based on missing-valued tables, MLEM2, and a similarity relation.
- MMT: the proposed classifier based on missing-valued tables, MLEM2, and a tolerance relation.

We compared error rates of these methods. Each error rate is the average of error rates of 10 times 10-fold cross-validation.

Table 2 shows error rates of these methods for each data set. Values of the column 'MRF' are the best values among the numbers of sampling attribute. Values of columns 'MMS' and 'MMT' are the best values among the missing value rates. In each data set, the best value is indicated by boldface. Because of the randomness of cross-validation, each value of the table has statistical dispersion with every experiments.

Table 2. Error rates of each method for each data set in 10 times 10-fold cross-validations.

	MLEM2	MB	MRF	MMS	MMT
Balance	0.199	0.142	0.140	**0.112**	0.184
Blood_Transfusion	0.261	0.247	0.232	**0.224**	0.234
Dermatology	5.95	4.17	**3.6**	4.04	5.82
Ecoli	22.11	17.83	17.74	**17.56**	22.11
Glass	30.11	28.95	30.97	**26.31**	28.16
Iris	4.87	4.73	4.6	**4.4**	4.6
Soybean_small	2.85	3.55	0.65	1.25	**0.45**
Wine	6.67	5.46	3.14	**2.36**	5.39

Table 3. The test statistics of each pair of the classification methods by the Wilcoxon signed-rank test.

	MLEM2	MB	MRF	MMS	MMT
MLEM2	-	4*	4*	0**	0.5**
MB	-	-	6	0**	16
MRF	-	-	-	11	11.5
MMS	-	-	-	-	4*
MMT	-	-	-	-	-

It was shown that a proposed method using a similarity relation (MMS) provides good overall performance. Except for 'Dermatology', the proposed method based on missing-valued tables provides the best performance. Even in 'Dermatology' in which the proposed method worth than MRF, MMS provides better performance than MLEM2 and MB. Hence, we confirmed the efficacy of the proposed method for classification problems.

The Wilcoxon Signed-Rank Test. We conducted the two-tailed Wilcoxon signed-rank test based on Table 2 in order to confirm the superiority of proposed method. Table 3 shows the test statistics of each pair of the classification methods by the Wilcoxon signed-rank test. In the case that the sample size is 8, the critical value at a significance level $\alpha = 0.05$ is $T_{0.05} = 3$, and the critical value at a significance level $\alpha = 0.1$ is $T_{0.1} = 5$. Asterisks * represent that there is a significant difference between the pair at a significance level of $\alpha = 0.05$. Double asterisks ** represent that there is a significant difference between the pair at a significance level of $\alpha = 0.1$.

The following is the results of the test.

- MB and MRF are significantly better than MLEM2 at a significance level of $\alpha = 0.1$.

Table 4. The robustness of each method for the data sets

	MLEM2	MB	MRF	MMS	MMT
Dermatology	0.953	0.959	**0.994**	0.985	0.953
Ecoli	0.795	0.803	0.882	**0.909**	0.795
Glass	0.74	0.81	**0.94**	0.92	0.76
Iris	0.88	0.88	0.91	**0.97**	0.96
Soybean_small	0.9866	0.9804	**0.9983**	0.9978	0.9965
Wine	0.8557	0.8857	0.9971	**0.9975**	0.8863

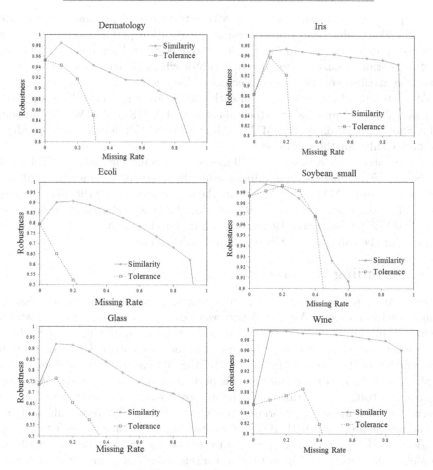

Fig. 2. The changes of the robustness of proposed methods by missing value rates for the data sets

- MMS and MMT are significantly better than MLEM2 at a significance level of $\alpha = 0.05$.
- MMS is significantly better than MB at a significance level of $\alpha = 0.05$.

– MMS is significantly better than MMT at a significance level of $\alpha = 0.1$.
– There is no significant difference between MMS and MRF.

Whereas MMS is significantly better than MLEM2 and MB at the level $\alpha = 0.05$, and better than MMT at the level $\alpha = 0.1$, MRF is not significant better than them at the same level. Hence, we concluded that MMS has the best classification performance.

5 Effects on the Robustness Against Attributes Deficiencies

It is supposable that some condition attribute values of unknown objects are lost. Classifiers which can provide stable classification performance under such circumstances are desired. In this paper, we defined the robustness as the average of classification accuracies for decision tables which is deleted a column of a condition attribute of the original decision table one by one.

We determined the robustness of each classification approach for some data sets in Table 1. Figure 2 shows the robustness of MMS and MMT by missing value rate p. In all data sets, MMS adding 10 to 20 % missing values produce the best results for the robustness.

Table 4 shows the robustness of all methods for some the data sets. Values of the column 'MRF' are the best values among the numbers of sampling attribute. Values of columns 'MMS' and 'MMT' are the best values among the missing value rates. In all data sets, MMS or MRF provide the best performance and almost same level of performance. Hence, we confirmed the efficacy of the proposed method for the robustness against attributes deficiencies.

6 Concluding Remarks

In this paper, we proposed an ensemble learning approach based on missing-valued decision tables. We tested two ways using a similarity relation and a tolerance relation in order to extract decision rules from missing-valued tables. We carried out numerical experiments in order to compare the classification performance of the proposed method with MLEM2, bagging and random forests applied to MLEM2. As a result of the experiments and the two-tailed Wilcoxon signed-rank test, we confirmed that the proposed method provides good overall result if we set up an appropriate missing value rate. We also confirmed the efficacy of the proposed method for the robustness against attributes deficiencies.

Because the classification performance of the proposed method depends on missing value rate p, we have to test various values of p in order to obtain better performance. In terms of future issues, we plan to consider methods which determine the parameter p automatically.

At the present stage, the way to replace certain attribute values to the missing value is a random strategy, however methods which replace them systematically can be considered. In this research, we used only lower approximations, however, in rough sets, various approximations such as upper approximations and parameterized approximations are provided and should be tested.

References

1. Błaszczyński, J., Słowiński, R., Stefanowski, J.: Variable consistency bagging ensembles. Trans. Rough Sets **5946**, 40–52 (2010)
2. Breiman, L.: Bagging predictors. Mach. Learn. **24**, 123–140 (1996)
3. Breiman, L.: Random forests. Mach. Learn. **45**, 5–32 (2001)
4. Grzymala-Busse, J. W., Siddhaye, S.: Rough set approaches to rule induction from incomplete data. In: Proceedings of the IPMU'2004, the 10th International Conference on Information Processing and Management of Uncertainty in Knowledge-Based Systems, Perugia, vol. 2, pp. 923–930 (2004)
5. Grzymala-Busse, J.W., Clarka, P.G., Kuehnhausen, M.: Generalized probabilistic approximations of incomplete data. Int. J. Approximate Reasoning **55**(1), 180–196 (2014). Part 2
6. Kryszkiewicz, M.: Rough set approach to incomplete information systems. Inf. Sci. **112**, 39–49 (1998)
7. Latkowski, R.: On decomposition for incomplete data. Fundamenta Informaticae **54**, 1–16 (2003). IOS Press
8. Miyazaki, T., Ubukata, S., Inuiguchi, M.: Ensemble learning approach based on rough set theory. In: Proceedings of the 59th Annual Conference of the Institute of Systems, Control and Information Engineers (ISCIE), Osaka, vol. 316, no. 6, pp. 1–2 (2015)
9. Pawlak, Z.: Rough sets. Int. J. Comput. Inf. Sci. **11**(5), 341–356 (1982)
10. Skowron, A.: Boolean reasoning for decision rules generation. In: Komorowski, J., Raś, Z.W. (eds.) ISMIS 1993. LNCS, vol. 689, pp. 295–305. Springer, Heidelberg (1993)
11. Stefanowski, J., Tsoukias, A.: Incomplete information tables and rough classification. Comput. Intell. **17**(3), 545–566 (2001)
12. Stefanowski, J.: The bagging and n2-classifiers based on rules induced by MODLEM. In: Tsumoto, S., Słowiński, R., Komorowski, J., Grzymała-Busse, J.W. (eds.) RSCTC 2004. LNCS (LNAI), vol. 3066, pp. 488–497. Springer, Heidelberg (2004)
13. UCI machine learning repository. http://archive.ics.uci.edu/ml/

Multiple-Side Multiple-Learner for Incomplete Data Classification

Yuan-ting Yan, Yan-Ping Zhang[✉], and Xiu-Quan Du

Key Laboratory of Intelligent Computing and Signal Processing of Ministry
of Education, School of Computer Science and Technology,
Anhui University, Hefei 230601, Anhui Province, China
365975632@qq.com, zhangyp2@gmail.com

Abstract. Selective classifier can improve classification accuracy and algorithm efficiency by removing the irrelevant attributes of data. However, most of them deal with complete data. Actual datasets are often incomplete due to various reasons. Incomplete dataset also have some irrelevant attributes which have a negative effect on the algorithm performance. By analyzing main classification methods of incomplete data, this paper proposes a Multiple-side Multiple-learner algorithm for incomplete data (MSML). MSML first obtains a feature subset of the original incomplete dataset based on the chi-square statistic. And then, according to the missing attribute values of the selected feature subset, MSML obtains a group of data subsets. Each data subset was used to train a sub classifier based on bagging algorithm. Finally, the results of different sub classifiers were combined by weighted majority voting. Experimental results on UCI incomplete datasets show that MSML can effectively reduce the number of attributes, and thus improve the algorithm execution efficiency. At the same time, it can improve the classification accuracy and algorithm stability too.

Keywords: Incomplete data · Multiple-side · Feature subset · Multiple-learner

1 Introduction

When solving a problem, human usually ignore the irrelevant details and focus on the major part of the problem, in this way, they can simplify the problem solving. For example, feature selection [1], attribute reduction [2] in knowledge mining, etc. In addition, analyzing problem in several different aspects and then combing their results is another common solution of human problem solving. There are many related researches, such as subspace [3], multiple view learning [4], and so on.

These two ways of problem solving have been widely used in classification problem [5,6]. First of all, ignore irrelevant information can improve the algorithm execution efficiency. Studies have shown that irrelevant attributes have a negative effect on classification accuracy. Secondly, classifying from several

© Springer International Publishing Switzerland 2015
Y. Yao et al. (Eds.): RSFDGrC 2015, LNAI 9437, pp. 322–331, 2015.
DOI: 10.1007/978-3-319-25783-9_29

different views and then combine their results is another effective method to improve classification accuracy. However, most of the researches are deal with complete data. At the same time, in many practical applications, missing values are often inevitable due to various reasons. Such as equipment errors, data loss, manual data input, etc. So, classification on incomplete data is very necessary.

To avoid the negative impact of irrelevant attributes on the classification performance, we propose a multiple-side multiple-learner algorithm (MSML) for incomplete data. MSML first uses chi-square statistic evaluation algorithm to delete some unimportant attributes, and then constructs a group of classifiers according to the missing feature values in the selected feature subset. Finally, the results of different classifiers are combined by weighted majority voting.

The rest of the paper is organized as follows. The research on incomplete data classification is briefly reviewed in Sect. 2. In Sect. 3, we introduce the MSML algorithm. Section 4 gives the numerical experiments on 8 real incomplete datasets form UCI Machine Learning Repository. Section 5 concludes the paper.

2   Related Work

Some scholars have studied the classification on incomplete data. There are two simple methods to deal with incomplete datasets. One way is simply ignore the samples with missing values. However, this may cause loss of potential profitable information, leading to an insufficient amount of samples for investigation [7]. Imputation method is another common solution to replace missing values with a particular value of the individual variables. Both methods are known to incur potential estimation bias [8,9]. One kind of methods can avoid the estimation bias is to use the EM algorithm [10], gradient descent [11], Gibbs sampling [12] or Logistic regression algorithm [13]. But this kind of methods relies on the assumption that data are missing at random and there is no technique to verify this assumption. Meanwhile, this kind of methods will suffer a dramatic decrease in accuracy when this assumption is violated.

To avoid the missing at random assumption, Ramoni and Sebastiani proposed a Robust Bayes Classifier (RBC) [14] that needs no assumption about data missing mechanism. However, similar to Naive Bayes Classifier, RBC also makes the assumption that attributes are independent for each class. Krause et. al [15] introduced an ensemble method to deal with incomplete data, sub classifiers were trained on random feature subsets. The method also assumed that the value of any feature is independent of all others. Chen et.al [16] put forward a noninvasive neural network ensemble (NNNE) method without any assumptions about the data distribution. This method generates a community of base classifiers trained only with known values. But it did not take into account the differences of attribute importance degree. To overcome the limitation, a multi-granulation ensemble method (MGNNE) was proposed [17]. Information entropy was applied to measure attribute importance degree. However, the performance of MGNNE relies on the proportion of samples whit no missing values. Moreover, all the above three methods did not consider the negative effect of irrelevant attributes on classification performance.

Considering the characteristic of incomplete dataset and the negative effect of irrelevant attributes on classification performance. We propose a new algorithm called multiple-side multiple-learner classification algorithm (MSML) to deal with incomplete data.

3   MSML

3.1   Chi-Square Statistic Feature Evaluation Algorithm

We apply chi-square statistic to calculate the importance degree between each attributes and class variable respectively. A feature subset is selected by removing the attributes with cumulative probability distribution (*cdf*) values smaller than threshold α. We first give the method to construct the contingency table of an attribute variable with respect to the class variable.

Given an incomplete dataset D, suppose A is an attribute of D with m values, d is the class variable with l values. Note, we use '?' to denote the missing (unknown) value. The process of constructing contingency table M_A of attribute A with respect to d can be described as follows:

(1) Count the following frequencies:
$$f_{ij} = f(A = a_i, d = d_j), f_{(m+1)j} = f(A =?, d = d_j),$$
$$f_{i(l+1)} = f(A = a_i, d =?) \text{ and } f_{(m+1)(l+1)} = f(A =?, d =?).$$

(2) Allocate $f_{(m+1)j}$, $f_{i(l+1)}$ and $f_{(m+1)(l+1)}$ to f_{ij}.
   To update f_{ij}:

   (2.1) Compute the following summation:
   $$row_i = \sum_{j=1}^{l} f_{ij}, col_j = \sum_{i=1}^{m} f_{ij}, N = \sum_{i=1}^{m}\sum_{j=1}^{l} f_{ij}$$
   (2.2) Update f_{ij}:
   $$f'_{ij} \leftarrow f_{ij} + f_{i(l+1)} \times \frac{col_j}{N} + f_{(m+1)j} \times \frac{row_i}{N} + f_{(m+1)(l+1)} \times \frac{f_{ij}}{N}$$
(3) Obtain the contingency table $M(M_{ij} = f'_{ij})$

According to the above steps, we can get all the contingency tables between each attribute and class variable, respectively. Given a contingency table M of attribute A with respect to d, we use the chi-square statistics to measure the importance degree of attribute A. The chi-square attribute evaluation algorithm of incomplete dataset is as follows:

(1) Construct the contingency table M ($m * n$) of each attribute with respect to class variable, m and n are the number of distinct values (except '?') of attribute A and class variable d, respectively;
(2) For a contingency table M of attribute A with respect to d, compute the chi-square statistic $Chi(A, d)$;

(2.1) Compute the summation of each row of M_A denoted by r_i and each column of M_A denoted by c_j, respectively:

$$r_i = \sum_{j=1}^{n} v_{ij}, (i = 1, 2, ..., m), c_j = \sum_{i=1}^{m} v_{ij}, (j = 1, 2, ..., n);$$

(2.2) For each pair of (i, j), calculate the expected frequency E_{ij}:

$$E_{ij} = \frac{r_i \cdot c_j}{N}(i = 1, 2, ..., m; j = 1, 2, ...n; N = \sum_{i=1}^{m} \sum_{j=1}^{n} f_{ij});$$

(2.3) Compute the chi-square statistic value $Chi(A, d) = \sum_{i=1}^{m} \sum_{j=1}^{n} \frac{(E_{ij} - v_{ij})^2}{E_{ij}}$;

(2.4) Compute the *cdf* value P_A corresponding to $Chi(A, d)$.

(3) Select the feature subset S_1 consist of attributes with *cdf* value bigger than threshold $\alpha(0 < \alpha < 1)$.

According to the above method, we can get a feature subset S_1 of the incomplete dataset. S_1 consists of attributes with *cdf* value bigger than a given threshold α. In general, S_1 still have missing values. We will construct a group of classifiers on S_1.

3.2 Multiple-Side Multiple-Learner for Incomplete Data

Let $D = \{(x_i, y_i) | i = 1, 2, ..., n\}$ be the incomplete dataset. Where n denote the size of the dataset. Suppose there are d features of the input space $X = (X^{(1)}, X^{(2)}, ..., X^{(d)})$. If a value $x_i^{(j)}$ of sample x_i is unknown, it is denoted as $x_i^{(j)} = null$. For convenience, we first give some definitions as follows:

Definition 1: For a sample x_i of D, the missing value set of sample x_i is defined as a feature subset $mset\{i\}$ that x_i is missing for all features in $mset\{i\}$ and is complete for all features in X but not in $mset\{i\}$.

$mset\{i\} = \{X^{(j)} | (\forall X^{(j)} \in mset\{i\} \wedge x_i^{(j)} = null) \wedge (\forall X^{(j)} \notin mset\{i\} \wedge x_i^{(j)} \neq null)\}$.

Definition 2: The missing attribute set (MS) of D is defined as a set of missing value sets, $MS = \{MS_1, ..., MS_k\}$, in which each missing value set is unique.

Definition 3: A complete data subsets X_{mset_R} is defined as a data subset corresponding to a missing attribute set $mset_R$.

$X_{mset_R} = \{x_i^{(j)} | x_i \in D \wedge \forall j \notin mset_R(x_i^{(j)} \neq null)\}$

Note, each complete data subset corresponding to a unique feature subset (or missing attribute set) of the incomplete dataset.

To improve the algorithm performance, each complete data subset is used to train a classifier based on bagging algorithm. For a test sample, the algorithm chooses the classifiers that did not require the missing value of the test sample to predict it. And then weighted majority voting is applied to combine the prediction results of the test sample.

Algorithm 1. Multiple-side Multiple-learner Classification.

Input: Training dataset $D_{train} = \{(x_i, y_i)|i = 1, ..., n\}$.
  Testing dataset $D_{test} = \{(x_i, y_i)|i = 1, ..., m\}$.
Output: Prediction results $Y = \{Y_1, ..., Y_m\}$.
  Initialize $Y \leftarrow \emptyset$, $temp \leftarrow \emptyset$
  Training
  Obtain the missing attribute set $MS = \{MS_1, ..., MS_k\}$ and the complete data subsets $X = \{X_1, ..., X_k\}$ of D_{train}.
  Calculate the mutual information $MI = \{MI_1, ..., MI_k\}$.
  for $i \leftarrow 1$ **to** $length(X)$ **do**
        Generate a classifier h_i on X_i by using bagging algorithm and bp network.
  end
  Testing
  for $j \leftarrow 1$ **to** m **do**
        Obtain the missing value set $mset\{j\}$ of sample j and set $temp \leftarrow \emptyset$;
        for $i \leftarrow 1$ **to** $length(MS)$ **do**
              if $mset\{j\} \subseteq MS_i$ **then**
                    $temp = [temp, h_i(x_j)]$;
              end
        end
        Obtain final result Y_j of sample j by using weighted majority voting.
        $Y \leftarrow [Y, Y_j]$.
  end
  return Y

In this paper, to determine the final prediction of test sample, some factors are concerned to realize the weighted majority voting. First, each complete data subset has a unique feature subset with an relevance degree for prediction the class label. Moreover, the sub classifiers trained on complete data subsets have different prediction accuracies, as is commonly agreed that higher testing accuracy tends to have greater prediction accuracy. Besides, the size of complete data subsets are different, it is also a factor need to be considered. Combining these three factors, each available sub classifier is assigned a weight by the following method.

$$w_i = \frac{MI_i |X_{mset_i}| ACC_i}{\sum MI_i |X_{mset_i}| ACC_i} \tag{1}$$

Here $|X_{mset_i}|$ denote the size of complete data subset X_{mset_i}, MI_j denote the relevance degree (measured by mutual information) between attributes set and class variable on data subset X_{mset_i}, ACC_i denote the testing accuracy of the i_{th} sub classifier. Algorithm 1 gives the MSML algorithm.

4   Experiments

4.1   Experimental Description

To testify the validity of MSML, we carried out experiments on 8 benchmark datasets with missing data from UCI machine learning repository [18]. All our experiments were programming by MatlabR2001a. The implementation was performed on an Intel Core i5 CPU running at 3.2GHz (4CPUs) and 4GB RAM. Table 1 gives the detail information about the datasets used for experiments.

For MSML, MGNNE and NNNE, a faster BP algorithm called Levenberg-Marquardt algorithm which has an efficient implementation provided by Matlab is used in our experiments. The number of input nodes (id) is determined by the number of available attributes on each data subsets, and the number of output nodes (od) is determined by the number of classes. According to the geometric pyramid rule, the number of hidden nodes is $\sqrt{id * od}$. We evaluate the accuracy using ten-folds cross validation approach where a given dataset is randomly partitioned into ten folds of equal size. For each complete data subset, we apply the bagging algorithm to improve the algorithm performance, and set 10 as the number of replicates [19].

Table 1. Summarization of datasets characteristics

Dataset name	Instance	Attributes	Classes
Automobile	205	26	6
Bands	540	39	2
B.cancer	699	10	2
Credit	690	15	2
Heart-h	294	13	2
Vote	435	16	2
L.cancer	32	56	2
Mushroom	8124	22	2

4.2   Experimental Results and Analysis

In our algorithm, the attributes with cdf values smaller than threshold α will be deleted to avoid the adverse effect of irrelevant attributes on algorithm performance. We choose two datasets Bands and L.cancer to study the relationship between algorithm performance and the threshold. We set the threshold to vary from 0.50 to 0.95 with the interval 0.05. Table 2 and Table 3 report the results.

One can see that, with the increase in the number of α, both the number of selected attributes and the algorithm runtime decreased gradually. During the process of α increased from 0.5 to 0.9, algorithm accuracy is basically unchanged. When $\alpha = 0.95$, the algorithm performance on both datasets has an obvious decline (Bands: about 2 % decline, L.cancer: about 46 % decline). From the experimental results, in this paper, we choose $\alpha = 0.9$ as the threshold to delete unimportant attributes, and thus to improve algorithm efficiency.

Table 4 gives the accuracy comparison of our algorithm to MGNNE, NNNE and RBC on 8 datasets. We can see that, overall speaking, NNNE has relatively poor performance. RBC has best accuracy on two datasets B.cancer and

Table 2. Performance of MSML on Bands with the change of α

α	#.Attributes	Accuracy	Runtime(s)
0.50	36	0.769	976
0.55	34	0.775	971
0.60	33	0.776	968
0.65	33	0.778	968
0.70	32	0.778	965
0.75	32	0.776	965
0.80	30	0.776	763
0.85	29	0.779	752
0.90	26	0.779	732
0.95	24	0.758	717

Table 3. Performance of MSML on L.cancer with the change of α

α	#.Attributes	Accuracy	Runtime(s)
0.50	44	0.577	11.2
0.55	41	0.574	10.5
0.60	38	0.579	10.4
0.65	34	0.575	5.1
0.70	32	0.574	5.1
0.75	28	0.573	4.9
0.80	25	0.579	4.9
0.85	23	0.573	4.9
0.90	17	0.576	4.9
0.95	11	0.300	0.5

Heart-h. MSML has best performance on 5 datasets, and MGNNE has a slightly better accuracy than MSML on dataset Vote. It indicates that there are a small amount of relevant attributes been removed from dataset Vote when we set $\alpha = 0.9$. One effective solution is to increase the number of selected attributes by setting a smaller threshold. On four datasets Automobile, Bands, Credit and L.cancer, compared with MGNNE, MSML has a certain improvement on accuracy (1 % ~ 2 %). It suggests that the irrelevant attributes has an adverse impact on algorithm accuracy.

By deleting irrelevant attributes, compared with NNE-based algorithms, the execution efficiency of MSML is greatly improved. Table 5 shows the details of three algorithms MSML, MGNNE and NNE on 8 datasets. Note that the difference between MGNNE and NNE is that MGNNE modified the weighted majority voting method of NNE by applying information entropy to measure

Table 4. The average accuracy of the four classifiers

Datasets	MSML	MGNNE	NNE	RBC
Automobile	**70.05 ± 0.11**	68.18 ± 0.08	66.31 ± 0.10	68.49 ± 3.84
Bands	**77.85 ± 0.06**	75.62 ± 0.05	74.63 ± 0.06	71.36 ± 0.48
B.cancer	94.99 ± 0.02	93.99 ± 0.02	93.85 ± 0.02	**97.11 ± 0.16**
Credit	**86.45 ± 0.04**	85.50 ± 0.04	84.81 ± 0.06	86.18 ± 0.40
Heart-h	80.91 ± 0.07	80.59 ± 0.06	81.69 ± 0.06	**85.88 ± 2.11**
Vote	94.47 ± 0.02	**94.71 ± 0.03**	0.942 ± 0.03	90.25 ± 0.19
L.cancer	**57.83 ± 0.28**	52.17 ± 0.27	49.75 ± 0.28	56.13 ± 1.67
Mushroom	**99.96 ± 0.01**	99.96 ± 0.01	99.86 ± 0.01	95.96 ± 0.02

the importance degree of each sub classifiers. So the number of attributes and the number of data subsets of both methods are equal. Thus, we just list the runtime of MGNNE.

We can see that quite a few irrelevant attributes was deleted on three datasets Bands, Credit and Heart-h, so the number of complete data subsets decreased a lot, thus the algorithm computational time is greatly reduced. At the same time, the runtime of MSML is higher than MGNNE and NNNE on the two datasets Automobile and Mushroom. That is because both datasets has only one attribute was removed, and the number of data subsets is unchanged. However, the chi-square statistic attribute evaluation algorithm is introduced, which increases a certain algorithm execution time. For dataset L.cancer, the algorithm runtime has an apparent decline because its attributes number reduced from 56 to 17. Meanwhile, one can see that there is only one data subsets of MSML, which means that all the attributes with missing values are deleted. Overall, by removing irrelevant attributes, MSML can effectively enhance execution efficiency on the basis of guarantee algorithm accuracy.

Table 5. Runtime, number of selected attributes and number of data subsets

Dataset	Runtime			#.Subsets		#.Attributes	
	MSML	MGNNE	NNNE	MSML	NNE	MSML	NNNE
Automobile	89.5	80.8	77.1	6	6	24	25
Bands	731.6	989	989	40	66	26	39
B.cancer	50.6	111	110.9	2	2	9	10
Credit	77.8	183	183	4	8	11	15
Heart-h	46	123.9	122.5	4.9	12	7	13
Vote	899	961	961	64.7	73	15	16
L.cancer	4.9	19.3	18.9	1	3	17	56
Mushroom	718.6	640.6	636	2	2	21	22

5   Conclusion and Discussion

By removing the irrelevant attributes of dataset, and then building ensemble classifier on the selected attributes set is an effective way to improve algorithm accuracy and execution efficiency. Most current studies require complete data. However, actual datasets are mostly incomplete due to various reasons, thus build classifier can deal with incomplete data is meaningful.

This paper puts forward a multiple-side multiple-learner classification algorithm to deal with incomplete data based on the characteristics of incomplete dataset. MSML first construct the contingency table of all attributes with respect to class variable, and then MSML introduces chi-square statistic evaluation algorithm to select a feature subset by removing the irrelevant attributes. Experiments show that MSML is an effective classification method to deal with incomplete dataset.

Acknowledgments. This work was supported by National Natural Science Foundation of China (Nos.61175046 and 61203290).

References

1. Guyon, I., Elisseeff, A.: An introduction to variable and feature selection. J. Mach. Learn. Res. **3**, 1157–1182 (2003)
2. Qian, Y., Liang, J., Pedrycz, W., Dang, C.: Positive approximation: an accelerator for attribute reduction in rough set theory. Artif. Intell. **174**(9), 597–618 (2010)
3. Kuncheva, L.I., Rodrguez, J.J., Plumpton, C.O., et al.: Random subspace ensembles for fMRI classification. IEEE Trans. Med. Imaging **29**(2), 531–542 (2010)
4. Zhang, J., Zhang, D.: A novel ensemble construction method for multi-view data using random cross-view correlation between within-class examples. Pattern Recogn. **44**(6), 1162–1171 (2011)
5. Sun, S., Zhang, C.: Subspace ensembles for classification. Phys. A Stat. Mech. Appl. **385**(1), 199–207 (2007)
6. Bryll, R., Gutierrez-Osuna, R., Quek, F.: Attribute bagging: improving accuracy of classifier ensembles by using random feature subsets. Pattern Recogn. **36**(6), 1291–1302 (2003)
7. Allison, P.D.: Missing Data. Sage Publications, Thousand Oaks (2001)
8. Roderick L., J A, Rubin, D.B.: Statistical Analysis with Missing Data, vol. 43, no. 4, pp. 364–365. Wiley, New York (2002)
9. Gheyas, I.A., Smith, L.S.: A neural network-based framework for the reconstruction of incomplete data sets. Neurocomputing **73**(16), 3039–3065 (2010)
10. Dempster, A.P., Laird, N.M., Rubin, D.B.: Maximum likelihood from incomplete data via the EM algorithm. J. Roy. Stat. Soc. Ser. B (Methodol.) **39**, 1–38 (1977)
11. Russell, S., Binder, J., Koller, D., Kanazawa, K.: Local learning in probabilistic networks with hidden variables. In: Proceedings of IJCAI 1995, pp. 1146–1152 (1995)
12. Geman, S., Geman, D.: Stochastic relaxation, Gibbs distributions, and the Bayesian restoration of images. IEEE Trans. Pattern Anal. Mach. Intell. **6**, 721–741 (1984)

13. Williams, D., Liao, X., Xue, Y., Carin, L., Krishnapuram, B.: On classification with incomplete data. IEEE Trans. Pattern Anal. Mach. Intell. **29**(3), 427–436 (2007)
14. Ramoni, M., Sebastiani, P.: Robust Bayes classifiers. Artif. Intell. **125**(1), 209–226 (2001)
15. Krause, S., Polikar, R.: An ensemble of classifiers approach for the missing feature problem. In: IEEE Proceedings of the International Joint Conference on Neural Networks, vol. 1, pp. 553–558 (2003)
16. Chen, H., Du, Y., Jiang, K.: Classification of incomplete data using classifier ensembles. In: IEEE International Conference on Systems and Informatics. pp. 2229–2232 (2012)
17. Yan, Y.-T., Zhang, Y.-P., Zhang, Y.-W.: Multi-granulation ensemble classification for incomplete data. In: Miao, D., Pedrycz, W., Slezak, D., Peters, G., Hu, Q., Wang, R. (eds.) RSKT 2014. LNCS, vol. 8818, pp. 343–351. Springer, Heidelberg (2014)
18. UCI Repository of machine learning databases for classification. http://archive.ics.uci.edu/ml/datasets.html
19. Breiman, L.: Bagging predictors. Mach. Learn. **24**(2), 123–140 (1996)

Sparse Matrix Feature Selection
in Multi-label Learning

Wenyuan Yang, Bufang Zhou, and William Zhu$^{(\boxtimes)}$

Lab of Granular Computing, Minnan Normal University, Zhangzhou, China
williamfengzhu@gmail.com

Abstract. High-dimensional data are commonly met in multi-label learning, and dimensionality reduction is an important and challenging work. In this paper, we propose sparse matrix feature selection to reduce data dimension in multi-label learning. First, the feature selection problem is formalized by sparse matrix. Second, an sparse matrix feature selection algorithm is proposed. Third, four feature selection are compared with the proposed methods and parameter optimization analysis is also provide. Experiments reported the proposed algorithms outperform the other methods in most cases of tested datasets.

Keywords: Multi-label learning · feature selection · sparse matrix · machine learning

1 Introduction

Multi-label learning is an important research topic in the field of machine learning. Multi-label learning studies the problem where each training example is represented by a single instance while associated with a set of labels simultaneously and the task is to predict the proper label set for each unseen instance. During the past decade, significant amount of progresses have been made toward this emerging machine learning paradigm [10].

In multi-label learning, high-dimensional data are common. When dealing with label learning problems, high dimensionality calls for more computational time and space requirements. It is an important to reduce the dimensions of high-dimensional data. Feature extraction and feature selection are the two main approaches to dimensionality reduction. The former aims to map the original features into a low-dimensional space via certain transformation and then generates some new features [4], while the latter aims to find an optimal feature subset given a certain predetermined criterion [3].

Feature selection aims to reduces data dimensionality by removing irrelevant and redundant features. It chooses a subset of the original features according to a selection criterion, and arouse the beneficial effects for applications, such as speeding up a data mining algorithm, improving predictive accuracy, and enhancing result comprehensibility. It is an important technique widely used in pattern analysis and machine learning [11].

© Springer International Publishing Switzerland 2015
Y. Yao et al. (Eds.): RSFDGrC 2015, LNAI 9437, pp. 332–339, 2015.
DOI: 10.1007/978-3-319-25783-9_30

Many feature selection methods often select the top ranked features, which are evaluated independently by each feature. However, ranking features underlying the relation between an individual feature and a feature subset is a combinational optimization problem. The matrix factorization method has become popular techniques to deal with this type of relation for feature selection [8]. Feature selection problems are transformed into finding optimal solutions of optimization problems, and then the corresponding matrix update iterative algorithms can be developed, which implies the optimality of a feature subset as a whole, not just some individual features.

In this paper, We provide one efficient feature selection based on sparse matrix for dimensionality reduction, the method can improve processing speed and accuracy. Firstly the feature selection problem is formalized by sparse matrix, and then its equivalent characterizations are presented from a theoretical viewpoint, then an sparse matrix feature selection algorithm is proposed, finally four algorithms are compared with the proposed methods.

The paper is arranged as follows. Some recent works on feature selection is presented in Sect. 2. The feature selection criterion with sparse matrix is provided in Sect. 3. We propose sparse matrix feature selection algorithms for this criterion in Sect. 4. Experimental results are reported and analyzed in Sect. 5. Finally, this paper is concluded in Sect. 6.

2   Related Works

In this part, we review the existing multi-label learning algorithms and feature selection criteria. A kind of effective multi-label learning algorithms is ML-KNN [9]. select features algorithms include ReliefF, KruskalWallis, MRMR [11].

2.1   Multi-Label K-Nearest-Neighbor

Multi-Label k-Nearest-Neighbor(ML-KNN) based on statistical information derived from the label sets of an unseen instance's neighboring instances, ML-KNN utilizes maximum a posteriori principle to determine the label set for the unseen instance. ML-KNN estimates the prior probabilities firstly, then estimates the posterior probabilities, Finally, using the Bayesian rule and compute the algorithm's outputs based on the estimated probabilities [9].

ML-KNN is a binary relevance learner, i.e., it learns a single classifier h_i for each label $\lambda_i \in \mathcal{L}$. However, instead of using the standard K-Nearest-Neighbor(KNN) classifier as a base learner, it implements the h_i by means of a combination of KNN and Bayesian inference: Given a query instance x with unknown multi-label classification $L \in \mathcal{L}$, it finds the k nearest neighbors of x in the training data and counts the number of occurrences of λ_i among these neighbors. Considering this number, y, as information in the form of a realization of a random variable Y , the posterior probability of $\lambda_i \in \mathcal{L}$ is given by [2,9]

$$\mathbf{P}(\lambda_i \in L | Y = y) = \frac{\mathbf{P}(Y=y|\lambda_i \in L) \cdot \mathbf{P}(\lambda_i \in L)}{\mathbf{P}(Y=y)}$$

which leads to the decision rule

$$h_i(x) = \begin{cases} 1 & \text{if } \mathbf{P}(Y = y|\lambda_i \in L)\mathbf{P}(\lambda_i \in L) \geqslant \mathbf{P}(Y = y|\lambda_i \notin L)\mathbf{P}(\lambda_i \notin L \\ 0 & \text{otherwise} \end{cases}$$

The prior probabilities $\mathbf{P}(\lambda_i \in L)$ and $\mathbf{P}(\lambda_i \notin L)$ as well as the conditional probabilities $\mathbf{P}(Y = y|\lambda_i \in L)$ and $\mathbf{P}(Y = y|\lambda_i \notin L)$ are estimated from the training data in terms of corresponding relative frequencies.

2.2   Feature Selection with ReliefF

Relief [5] and its multiclass extension ReliefF [6] are supervised feature weighting algorithms of the filter model. Assuming that p instances are randomly sampled from data, the evaluation criterion of Relief is defined as

$$SC_R(f_i) = \frac{1}{2}\sum_{t=1}^{p} d(f_{t,i} - f_{NM_{(x_t),i}}) - d(f_{t,i} - f_{NH_{(x_t),i}})$$

Where $f_{t,i}$ denotes the value of instance x_t on feature f_i, $f_{NM_{(x_t),i}}$ and $f_{NH_{(x_t),i}}$ denote the values on the ith feature of the nearest points to xt with the same and different class label, respectively, and $d(\cdot)$ is a distance measurement. To handle multi-class problems, the above criterion is extended to the following formulation:

$$SC_R(f_i) = \frac{1}{p}\sum_{t=1}^{p}\{-\frac{1}{m_{x_t}}\sum_{x_j \in NH(x_t)} d(f_{t,i} - f_{j,i})+$$

$$\sum_{y \neq y_{x_t}} \frac{1}{m_{x_t,y}} \frac{P(y)}{1 - P(y_{x_t})} \sum_{x_j \in NM(x_t,y)} d(f_{t,i} - f_{j,i})\}$$

Where y_x is the class label of the instance x_t and $P(y)$ is the probability of an instance being from the class y. $NH(x)$ or $NM(x,y)$ denotes a set of nearest points to x with the same class of x, or a different class (the class y), respectively. m_x and $m_{x_t,y}$ are the sizes of the sets $NH(x_t)$ and $NM(x_t,y)$, respectively. Usually, the size of both $NH(x)$ and $NM(x,y)$, $\forall y \neq y_x$, is set to a prespecified constant k. The evaluation criteria of Relief and ReliefF suggest that the two algorithms select features contributing to the separation of samples from different classes [7].

3   Evaluation Metrics

In multi-label learning, performance evaluation is much more complicated than single-label learning as each instance could have multiple labels simultaneously. One direct solution is to calculate the classic single-label metric (such as precision, recall and F-measure) on each possible label independently, and then combine the outputs from each label through micro- or macro-averaging. However, this intuitive way of evaluation does not consider the correlations between different labels of each instance. Therefore, given a test set $S = \{(x_i, Y_i)|1 \leqslant i \leqslant p\}$ the following evaluation metrics specifically designed for multi-label learning are used in this paper [10].

(1) Hamming Loss: $hloss(h) = \frac{1}{p}\sum\limits_{i=1}^{p}\frac{1}{q}|h(x_i)\Delta Y_i|$

Here, Δ stands for the symmetric difference between two sets. The hamming loss evaluates the fraction of misclassified instance-label pairs, i.e. a relevant label is missed or an irrelevant is predicted. Note that when each example in S is associated with only one label, $hloss_S(h)$ will be $2/q$ times of the traditional misclassification rate.

(2) One-error: $One - error(f) = \frac{1}{p}\sum\limits_{i=1}^{p}[\![\arg\max\limits_{y\in Y} f(x_i, y)] \notin Y_i]\!]$

Where for any predicate π, $[\![\pi]\!]$ equals 1 if π holds and 0 otherwise. The one-error evaluates the fraction of examples whose top-ranked label is not in the relevant label set.

(3) Coverage: $Coverage(f) = \frac{1}{p}\sum\limits_{i=1}^{p}\max\limits_{y\in Y_i} rank_f(x_i, y) - 1$

The coverage evaluates how many steps are needed, on average, to move down the ranked label list so as to cover all the relevant labels of the example.

(4) Ranking Loss:

$$rloss(f) = \frac{1}{p}\sum\limits_{i=1}^{p}\frac{1}{|Y_i||\bar{Y_i}|}|\{(y_1, y_2)|f(x_i, y_1) \leqslant f(x_i, y_2), (y_1, y_2) \in Y_i \times \bar{Y_i}\}|$$

Where $\bar{Y_i}$ denotes the complementary set of Y_i in Y. The ranking loss evaluates the fraction of reversely ordered label pairs, i.e. an irrelevant label is ranked higher than a relevant label. (5) Average Precision:

$$avgprec(f) = \frac{1}{p}\sum\limits_{i=1}^{p}\frac{1}{|Y_i|}\sum\limits_{y\in Y_i}\frac{|\{y'|rank_f(x_i, y') \leqslant rank_f(x_i, y), y' \in Y_i\}|}{rank_f(x_i, y)}$$

The average precision evaluates the average fraction of relevant labels ranked higher than a particular label $y \in Y_i$

The first metric *Hamming loss* is defined based on the multi-label classifier $h(\cdot)$, which evaluates how many times an instance-label pair is misclassified. The other four metrics are defined using the real-valued function $f(\cdot, \cdot)$ concerning the ranking quality of different labels for each instance. *One-error* evaluates how many times the top-ranked label is not in the set of proper labels of the instance; *Coverage* evaluates how many steps are needed, on average, to move down the label list in order to cover all the proper labels of the instance; *Ranking loss* evaluates the average fraction of label pairs that are for the instance *Average precision* evaluates the average fraction of labels ranked above a particular label $y \in Y$ which actually are in Y.

For *Hamming loss, One-error, Coverage* and *Ranking loss*, the smaller the metric value the better the system's performance. For *Average precision*, on the other hand, the bigger the value the better the system's performance.

4   Feature Selection Base on Sparse Representation

4.1   Feature Selection via Sparse Representation

Let $X \in \mathbb{R}^{n \times d}$ and $Y \in \mathbb{R}^{n \times l}$ K is similarity matrix of Y, $\hat{X} \in \mathbb{R}^{n \times k}$, $\hat{K}$ preservation the similarity

$$\mathbb{O} = \arg\min_{W} \frac{1}{2} ||K - \hat{K}||_F \tag{1}$$

We introduce a zero-one matrix to replace the index subset of the selected features. Without loss of generality, we can define an sparse matrix $W \in \mathbb{R}^{d \times |I|}$ from an index set I in the following form:

$$W_{ij} = \begin{cases} 1, & \text{the } j\text{-th element of } I \text{ is } i, \\ 0, & \text{otherwise.} \end{cases} \tag{2}$$

It is noted that the number of the rows of the sparse matrix is equal to the number of all features, and that of the columns is equal to the number of the selected features.

$\hat{K}$ can be presented by sparse matrix W, that is, $\hat{K} = (XW)(XW)^T$, so, Eq.(1) is written as follow.

$$\mathbb{O} = \arg\min_{W} \frac{1}{2} ||K - (XW)(XW)^T||_F \tag{3}$$

Similarity matrix K can be decomposed into matrices multiplying of V and the V transpose, $K = VV^T$

$$\mathbb{O} = \arg\min_{W} \frac{1}{2} ||VV^T - (XW)(XW)^T||_F \tag{4}$$

We introduce parameter λ represents the adjustable variables and change Eq. (4) into Eq. (5)

$$\mathbb{O} = \arg\min_{W} \frac{1}{2} ||V - XW||_F + \lambda ||W|| \tag{5}$$

In order to solve the problem of Eq. (refSimilaritionPara), 1-norm of W are used.

$$\mathbb{O}_1 = \arg\min_{W} \frac{1}{2} ||V - XW||_F + \lambda ||W||_1 \tag{6}$$

1-norm is a kind of sparse representation and can take advantage of the extensive use of Lasso (The least absolute shrinkage and selection operator) algorithm to solve it.

4.2 Algorithm for Feature Selection

Based on the above, write the algorithm of feature selection based on sparse matrix is as follows.

Algorithm 1. Feature selection based on sparse matrix(FSSM)

Input: Data matrix $X \in \mathbb{R}^{n \times d}$, the number of selected features k and parameter λ.
Output: An index set of the selected features $I \subseteq \{1, \cdots, d\}$ and $|I| = k$.
Initialize sparse matrix W ;
Calculate similarity matrix K ;
Decompose K into V where $K = VV^T$;
$W = \arg \min_{W} \frac{1}{2}||V - XW||_F + \lambda||W||_1$; //calculate by lasso
$W = (w_1,, w_d)^T$, indexW=sort($||w_i||_2$) //descending order sort
I=indexW$(1 : k)$ //select I as the top k index set

---

5 Experiments

In this section, the proposed algorithm are tested in benchmark datasets by experiments and compared with four state-of-the-art feature selection algorithms.

5.1 Dataset

Experiments are conducted in publicly available datasets[1], including Computers, Recreation, Science and Society. Datasets covering different data sources serve as a good test bed for the performance of the proposed algorithms [1]. These datasets are described in Table 1.

5.2 Experimental Setting

To verify the effectiveness and efficiency of the proposed algorithms, we compare the proposed algorithms ReliefF, KruskalWallis and MRMR with the following four feature selection algorithms. NSF stands all original features are selected.

Table 1. Data description

DID	Dataset	♯ Instances(train and test)	♯ Features	♯ Classes
1	Computers	5000	681	33
2	Recreation	5000	606	22
3	Science	5000	743	40
4	Society	5000	636	27

---

[1] http://lamda.nju.edu.cn/Data.ashx.

Table 2. Experimental results

	FSSM	NSF	ReliefF	KruskalWallis	MRMR
Ave. Prec. ↑					
Computers	0.6612±7.67E-5	0.6377±3.82E-4	0.6497±5.52E-5	0.6346±1.46E-4	0.6569±1.16E-4
Recreation	0.5529±4.60E-5	0.5212±2.48E-4	0.5176±2.56E-4	0.4367±3.58E-4	0.5427±3.55E-4
Science	0.5070±1.70E-4	0.4958±4.08E-4	0.4980±1.59E-4	0.4385±1.83E-4	0.5155±2.02E-4
Society	0.6046±1.78E-4	0.5938±2.53E-4	0.5920±4.78E-4	0.5599±7.78E-5	0.6010±9.30E-5
Coverage ↓					
Computers	4.0418±0.0702	4.2922±0.0241	4.1202±0.0548	4.2718±0.0826	4.0480±0.1030
Recreation	4.4740±0.0414	4.6898±0.0380	4.6062±0.0940	5.0172±0.0831	4.5244±0.0431
Science	6.5746±0.1097	6.5298±0.2289	6.6500±0.1526	7.2792±0.1540	6.4564±0.0376
Society	5.5348±0.0635	5.6282±0.0800	5.6614±0.1039	5.9462±0.0997	5.5012±0.0467
Hamming loss ↓					
Computers	0.0378±2.94E-6	0.0392±2.76E-6	0.0384±4.07E-6	0.0387±3.21E-6	0.0377±4.87E-6
Recreation	0.0564±6.16E-6	0.0575±3.19E-6	0.0590±7.33E-6	0.0632±1.59E-6	0.0574±7.34E-6
Science	0.0336±7.63E-7	0.0338±7.69E-7	0.0341±8.49E-7	0.0357±1.22E-6	0.0336±1.51E-6
Society	0.0550±1.71E-6	0.0554±5.12E-6	0.0557±5.52E-6	0.0580±8.98E-6	0.0549±3.82E-6
One-error ↓					
Computers	0.4116±1.56E-4	0.4364±1.14E-3	0.4230±1.72E-4	0.4416±3.44E-4	0.4202±3.14E-4
Recreation	0.5646±6.73E-4	0.6106±4.12E-4	0.6180±5.23E-4	0.7368±7.01E-4	0.5798±7.28E-4
Science	0.6020±3.56E-4	0.6284±5.47E-4	0.6150±1.75E-4	0.6954±3.48E-4	0.5916±5.00E-4
Society	0.4374±3.57E-4	0.4474±5.08E-4	0.4526±8.12E-4	0.4942±2.80E-4	0.4414±3.17E-4
Ranking Loss ↓					
Computers	0.0826±2.08E-5	0.0897±1.04E-5	0.0856±8.54E-6	0.0898±3.85E-5	0.0832±3.73E-5
Recreation	0.1623±4.49E-5	0.1739±5.04E-5	0.1696±1.36E-5	0.1923±1.38E-4	0.1648±3.44E-5
Science	0.1264±4.86E-5	0.1272±9.68E-5	0.1285±5.43E-5	0.1451±5.50E-5	0.1234±2.18E-5
Society	0.1379±6.70E-5	0.1415±7.47E-5	0.1426±5.80E-5	0.1521±2.59E-5	0.1372±2.64E-5

5.3   Experimental Result and Analysis

Compared four algorithms are no features select (NSF), ReliefF, KruskalWallis, MRMR, all of these are based on ML-KNN algorithm and its evaluation by 10 fold cross validation, experimental results are shown in Table 2. Experiments report shows that FSSM algorithms outperform the four methods in most cases for tested datasets.

6   Conclusions and Further Studies

In this paper, we proposed a new feature selection base sparse matrix. Experimental results show that our algorithms outperformed the compared feature selection algorithms. And our next work is to research 2-norm that a kind of smooth matrix so we can use trace and derivative method to solve.

Acknowledgments. This work is in part supported by National Nature Science Foundation of China under Grant Nos. 61170128 and 61379049, the Key Project of Education Department of Fujian Province under Grant No. JA13192, the Zhangzhou Municipal Natural Science Foundation under Grant No. ZZ2013J03, and the Minnan Normal University Doctoral Research Foundation under Grant No. 2004L21424.

References

1. Abajo, E., Dinez, A.: Graphs with maximum size and lower bounded girth. Appl. Math. Lett. **25**(3), 575–579 (2012)
2. Cheng, W., Hllermeier, E.: Combining instance-based learning and logistic regression for multilabel classification. Mach. Learn. **76**, 211–225 (2009)
3. Guyon, I., Elisseeff, A.: An introduction to variable and feature selection. J. Mach. Learn. Res. **3**, 1157–1182 (2003)
4. Guyon, I., Elisseeff, A. (eds.): Feature Extraction: Foundations and Applications. Studies in Fuzziness and Soft Computing, vol. 207. Springer, Heidelberg (2006)
5. Kira, K., Rendell, L.A.: A Practical Approach to Feature Selection. In: International Conference on Machine Learning, pp. 249–256 (1992)
6. Kononenko, I.: Estimating attributes: analysis and extensions of RELIEF. In: Bergadano, Francesco, De Raedt, Luc (eds.) ECML 1994. LNCS, vol. 784, pp. 171–182. Springer, Heidelberg (1994)
7. Liu, H., Motoda, H.: Computational Methods of Feature Selection. Chapman & Hall, London (2008)
8. Nie, F., Huang, H., Cai, X., Ding, C.: Efficient and robust feature selection via joint $\ell_{2,1}$-norms minimization. In: Advances in Neural Information Processing Systems, pp. 1813–1821 (2010)
9. Zhang, M., Zhou, Z.: ML-KNN: a lazy learning approach to multi-label learning. Pattern Recogn. **40**, 2038–2048 (2007)
10. Zhang, M., Zhou, Z.: A review on multi-label learning algorithms. IEEE Trans. Knowl. Data Eng. **26**(8), 1819–1837 (2014)
11. Zhao, Z., Wang, L., Liu, H., Ye, J.: On similarity preserving feature selection. IEEE Trans. Knowl. Data Eng. **25**(3), 619–632 (2013)

A Classification Method for Imbalanced Data Based on SMOTE and Fuzzy Rough Nearest Neighbor Algorithm

Weibin Zhao[1(✉)], Mengting Xu[1], Xiuyi Jia[2], and Lin Shang[1]

[1] State Key Laboratory for Novel Software Technology,
Nanjing University, Nanjing 210046, China
{njzhaowb,xumtpark}@gmail.com, shanglin@nju.edu.cn
[2] School of Computer Science and Engineering,
Nanjing University of Science and Technology, Nanjing 210094, China
jiaxy@njust.edu.cn

Abstract. FRNN (Fuzzy Rough Nearest Neighbor) algorithm has exhibited good performance in classifying data with inadequate features. However, FRNN does not perform well on imbalanced data. To overcome this problem, this paper introduces a combination method. An improved SMOTE method is adopted to balance data and FRNN is applied as the classification method. Experiments show that the combination method can obtain a better result rather than classical FRNN algorithm.

Keywords: Imbalanced data · SMOTE · Fuzzy rough set · Nearest neighbor · Classification

1 Introduction

Classification is an important task of machine learning and data mining. Classification modelling is to learn a function from training data, which makes as few errors as possible when being applied to data previously unseen. A range of classification algorithms, such as decision tree, neural network, Bayesian network, nearest neighbor, support vector machines, and the newly reported associative classification, have been developed and successfully applied to many application domains [1,2]. However, reports from both academia and industry indicate that imbalanced class distribution of a data set has posed a serious difficulty to most classifier learning algorithms, which assume a relatively balanced distribution [16]. In fact, imbalanced problem exists in many real-world domains. Particularly for a two-class classification, the imbalanced problem is one in which one class is represented by a large of samples, while the other one is represented by only a few [3,4]. This problem is very common in real-world applications, such as anomaly detection [6], medical applications [7], database marketing [5] and so on. In such applications, the traditional classification algorithms mentioned above often obtain high accuracy over the majority class, while for the minority class the opposite occurs. This happens because the classifier focuses

© Springer International Publishing Switzerland 2015
Y. Yao et al. (Eds.): RSFDGrC 2015, LNAI 9437, pp. 340–351, 2015.
DOI: 10.1007/978-3-319-25783-9_31

on global measures that do not take into account the class data distribution, nevertheless the most interesting information is often found within the minority class [8]. Many techniques and methods for dealing with class imbalance have been proposed [9]. These techniques can be grouped into two main categories: those that modify the data distribution by preprocessing techniques (data level methods) [14,15], and those at the level of the learning algorithm which adapt a base classifier to deal with class imbalance (algorithm level methods) [10,11].

In this paper, we present a combine solution to classify imbalanced data that is based on FRNN (Fuzzy Rough Nearest Neighbor) classifier [12,13]. The reason we use FRNN classifier is that it is good at handling incomplete information system classification and we do not need to do extra preprocessing. However, there are some weaknesses in FRNN algorithm. For instance, FRNN treats the majority class and minority class in a symmetric way and hence makes no provisions for the class imbalance [8]. Therefore, we use an improved SMOTE (synthetic minority over-sampling) technique [14] to balance the original data firstly, then use the FRNN to make classification for the balanced data. Our method is validated by an extensive experimental study, showing statistically better results than FRNN only used.

The paper is organized as follows: Sect. 2 introduces the problem of imbalanced classification and describes two types of approach to solve this problem. Section 3 gives the review of our proposed method which is the combination of the improved SMOT and the FRNN. Section 4 gives some comparative experiments and our analysis, finally end with conclusion in Sect. 5.

2   Background

Imbalanced data always exists in the real world, and classification accuracy of minority class is more valuable than those of majority class, so it is a meaning job to improve the classification accuracy of minority class. In this section, we will introduce the nature of the problem and related solutions.

2.1   Imbalanced Data Problem

Sun et. al argued that in a data set with the class imbalance problem, the most obvious characteristic is the imbalance data distribution between classes [11]. However, the former theoretical and experimental studies indicated that the imbalance data distribution is not the only key factor that influences the performance of a available classifier in identifying rare samples. Other influential factors include small sample size, separability and the existence of within-class sub-concepts.

2.2   Existing Solutions of Imbalanced Data Problem

As mentioned above, much work has been done in addressing the class imbalance problem. There are two types of solutions to tackle the problems. One is using resampling technique to modify the data distribution, the other is modifying the algorithms, which mainly improving accuracy in classification of minority class. These two methods are described as below.

Solutions at the Data Level. Resampling is a major technique to handle imbalanced data classification problems. Oversampling techniques are often chosen in dealing with the imbalanced classification [14,15]. Among those oversampling techniques, SMOTE is an useful oversampling method that creates new minority class samples by interpolating between minority class samples and their nearest neighbors. It can effectively make origin data become more balanced and reduce the overfitting of the classifier. Deficiency of SMOTE is the randomness in the process of synthesis of new samples, it lack precise control for the synthesis of new samples and choose the samples of minority without awareness. Also it often cause overlap problems in sampling, making classification accuracy decrease. In Sect. 3, we will show an improved SMOTE used in the paper.

Solutions at the Algorithm Level. In this case, traditional classifiers are modified to favor minority class, so it can optimize the decision making [16]. Major methods are shown as belows. Adjusting probability density [10]: as the positive sample is known, probability density estimation can be achieved by using an appropriate statistical distribution. Cost-sensitive [11]: these algorithms try to minimize higher cost errors. The kernel idea is that the cost of misclassifying a positive sample should be higher than the cost of misclassifying a negative one.

3 The Proposed Method: A Combination of DB_SMOTE and FRNN

As a kind of classical oversampling method, SMOTE resamples the small class through taking each small class example and introducing synthetic examples along the line segments joining its small class nearest neighbors [15]. For example, assume the amount of over-sampling needed is 200 %, then for each small class example, two nearest neighbors belonging to the same class are identified and one synthetic example is generated in the direction of each. The synthetic example is generated in the following way: take the difference between the attribute vector (example) under consideration and its nearest neighbor. Multiply this difference by a random number between 0 and 1, and add it to the attribute vector under consideration. Approach effectively forces the decision region of the minority class to become more general [8,16]. Based on the NN (nearest neighbor) algorithm, FRNN consider how to calculate the similarity between samples when information systems have both discrete and continuous attributes. Without transforming all attributes into discrete or continuous beforehand, FRNN can construct the upper and lower approximations for the majority and minority respectively, so it reveals more flexible in real-world application [12]. In this paper, we combine an improved SMOTE and FRNN to deal with the imbalanced data classification problem.

3.1 DB_SMOTE (Distance-based Synthetic Minority Over-Sampling Technique)

Liu et. al presented an improved SMOTE called DB_SMOTE based on the distance between samples and aggregation degree of the class [14]. The foundation of DB_SMOTE is the assumption that the sample which is located in the edge of the class does favor to form a classification boundary. By comparing the distance between sample and class center to the class average distance, the seed samples are acquired and new samples can be synthesized sequentially in the connection between the seed sample and class center. The advantage of such method is that its operation cost is relative low and anti-noise performance is good. So we decided to adopt DB_SMOTE to make oversampling in this paper.

Let S be a set of certain kind of samples $S = \{d_i, i = 1, 2, \ldots, n\}$, d_i denotes a m-dimensional vector and m represents the number of attributes.

Definition 1 (Class Center). *Class center (cc) denotes the average center of the samples in data space, a class center cc is a vector having same dimensions as the sample, cc can be computed as follow:*

$$cc = \frac{1}{n} \sum_{i=1}^{n} d_i \tag{1}$$

Definition 2 (Class Average Distance). *Class average distance (cd) refers to the average distance between the sample to the class center, it is a scalar. The cd reflects the degree of class aggregation and can be computed as:*

$$cd = \frac{1}{n} \sum_{i=1}^{n} D(d_i, cc) \tag{2}$$

where D denotes the Euclidean distance between d_i and cc.

Definition 3 (Seed Sample). *Seed sample (ss) represents the sample which has a farther distance to the class center than cd, it can be defined by:*

$$ss = \{d_i \mid D(d_i, cc) > cd\} \tag{3}$$

The set composed by seed samples is called candidate set. In order to avoid introducing too much noise in the process of synthesizing new samples, the DB_SMOTE modifies the classic SMOTE: it appoints the class center as a reference point and forms a segment between the sample and reference point, so new samples are synthesized in this segment. It can ensure that new samples can be located inside of the class.

Definition 4 (Synthetic of New Sample). *Synthetic of new sample (sns) can be defined as below:*

$$sns = ss_i + (ss_i - cc) * r \tag{4}$$

ss_i is a sample of candidate set r is a random number located in [0, 1].

3.2 FRNN (Fuzzy Rough Nearest Neighbor)

The K-nearest neighbor (KNN) algorithm [18] is a well-known classification technique that assigns a test sample to the decision class most common among its K nearest neighbors, i.e. the K training samples that are closest to the test sample. Obviously, the KNN depends on the relative importance (closeness) of each neighbor w.r.t. the test sample. However, when noise is present or the classes are overlapping in the region, the relative closeness will affects the classification result. To address this problem, Jensen et al. [12,13] present an alternative approach (FRNN) based on fuzzy rough set theory, which uses a test sample's nearest neighbors to construct the lower and upper approximation of each decision class, and then computes the membership of the test sample to these approximations. The method is very flexible and more robust in the presence of noisy data. The FRNN is described in detail as follows:

Considering two-class classification, in order to predict the class of a new test sample x, the FRNN algorithm computes the sum of the memberships of x to the fuzzy-rough lower and upper approximation of each class, and assigns the sample to the class for which this sum is higher. More precisely, let $\mathcal{F}$ be an implicator, $\mathcal{T}$ a t-norm and R a fuzzy relation that represents approximate indiscernibility between samples. The membership degrees $\underline{P}(x)$ and $\underline{N}(x)$ of x to the lower approximation of P and N are defined by, respectively [8]:

$$\underline{P}(x) = \min_{y \in U} \mathcal{F}(R(x,y), P(y)) \tag{5}$$

$$\underline{N}(x) = \min_{y \in U} \mathcal{F}(R(x,y), N(y)) \tag{6}$$

The value $\underline{P}(x)$ can be interpreted as the degree to which objects outside P (thus, in N) which are approximately indiscernible from x do not exist. A similar interpretation can be given to the value $\underline{N}(x)$. On the other hand, the membership degrees $\overline{P}(x)$ and $\overline{N}(x)$ of x to the upper approximation of P and N under R are defined by, respectively:

$$\overline{P}(x) = \max_{y \in U} \mathcal{T}(R(x,y), P(y)) \tag{7}$$

$$\overline{N}(x) = \max_{y \in U} \mathcal{T}(R(x,y), N(y)) \tag{8}$$

$\overline{P}(x)$ can be interpreted as the degree to which another element in P close to x exists, and similarly for $\overline{N}(x)$.

$$\mathcal{F}(a,b) = max(1-a, b) \tag{9}$$

$$\mathcal{T}(a,b) = min(a,b) \tag{10}$$

In this paper, we consider $\mathcal{F}$ and $\mathcal{T}$ defined by Eqs. (9) and (10), It can be verified that in this case, Eqs. (5)–(8) can be simplified to

$$\underline{P}(x) = \min_{y \in N} 1 - R(x,y) \tag{11}$$

$$\underline{N}(x) = \min_{y \in P} 1 - R(x, y) \tag{12}$$

$$\overline{P}(x) = \max_{y \in P} R(x, y) \tag{13}$$

$$\overline{N}(x) = \max_{y \in N} R(x, y) \tag{14}$$

In other words, $\underline{P}(x)$ is determined by the similarity to the closest negative (majority) sample, and $\underline{N}(x)$ is determined by the similarity to the closest positive (minority) sample. On the other hand, to obtain $\overline{P}(x)$ and $\overline{N}(x)$, we look for the most similar element to x belonging to the positive, respectively, negative class. Also, the lower and upper approximations are clearly related: $\overline{P}(x) = 1 - \underline{N}(x)$ and $\overline{N}(x) = 1 - \underline{P}(x)$. The FRNN algorithm then determines the classification of the test sample x as follows. We compute

$$\mu_P(x) = \frac{\underline{P}(x) + \overline{P}(x)}{2} = \frac{\underline{P}(x) + 1 - \underline{N}(x)}{2} \tag{15}$$

$$\mu_N(x) = \frac{\underline{N}(x) + \overline{N}(x)}{2} = \frac{\underline{N}(x) + 1 - \underline{P}(x)}{2} \tag{16}$$

x is classified to the positive class if $\mu_P(x) \geq \mu_N(x)$, otherwise it is classified to the negative class. Apart from the FRNN, we also need to make a choice for the fuzzy relation R. In order to determine the approximate indiscernibility between two samples x and y based on the set A of attributes, in this paper we assume the following definitions. Given a continuous attribute a, the fuzzy relationship between the x, y in terms of a is

$$R_a(x, y) = 1 - \frac{|a(x) - a(y)|}{range(a)} \tag{17}$$

while for a discrete attribute a,

$$R_a(x, y) = \begin{cases} 1, & if \ a(x) = a(y) \\ 0, & otherwise \end{cases} \tag{18}$$

The choice of fuzzy relation will affect the accuracy of classification in FRNN, literature [8] proved that among those of fuzzy relation model, choosing Eq. (19) for FRNN can achieve higher accuracy.

$$R(x, y) = \frac{R_{a1}(x, y) + \ldots + R_{am}(x, y)}{m} \tag{19}$$

The main drawback of FRNN for imbalanced classification is that it treats all classes symmetrically, not making a distinction between majority and minority samples. So in this paper, we combine FRNN with DB_SMOTE (DB_SMOTE + FRNN) to make classification for imbalanced data.

3.3   The Algorithm of DB_SMOTE + FRNN

Firstly, we use DB_SMOTE resampling the origin data set in order to improve the data distribution, then make classification by FRNN with a fuzzy relation as Eq. (19) shown.

As we can see from the flowchart of our proposed method in Fig. 1, firstly we apply the DB_SMOTE algorithm to make the data more balance. In this process, we may generate a lot of minority samples according to the data center. The details we can get from the Algorithm 1. Next, ten-fold cross is employed to divided the whole dataset into ten parts, each time we choose nine as train set to train a FRNN classifier, and then the rest part of the data as the test set, finally the classification results obtained. The pseudocode of the FRNN Classification algorithm is shown in Algorithm 2.

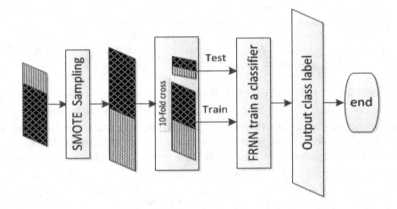

Fig. 1. Flowchart of DB_SMOTE + FRNN algorithm

Algorithm 1. DB_SMOTE resampling

Input: Origin set of minority, $DS = \{Xi \mid i = 1, 2, \ldots, m\}$, the balance factor σ
Output: New set of minority, $DS = \{Xi \mid i = 1, 2, \ldots, n\}$
1: Compute cc and cd according to Eqs. (1) and (2)
2: Create seed sample according to Eq. (3)
3: **for** $i \rightarrow \lfloor \sigma \rfloor$ **do**
4:    **for** $i \rightarrow m$ **do**
5:       Generate random number γ
6:       Generate new sample y according to Eq. (4)
7:       $\mid DS \mid = \mid DS \mid \cup y$
8:    **end for**
9: **end for**
10: Output new set of monority

Algorithm 2. FRNN Classification

Input: Training data set $\{Xi \mid i = 1, 2, \ldots, n\}$, which each sample has m conditional attributes and one decision attribute.

Output: Decision class P or N.

1: Mark the m-dimensional attribute as a continuous or discrete property
2: Create seed sample according to Eq. (3)
3: **for** $i \to n$ **do**
4:     **for** $j \to m$ **do**
5:         **if** the jth attribute is countinuous **then**
6:             Compute $R_{aj}(x_i, j)$ according to Eq. (17)
7:         **else**
8:             Compute $R_{aj}(x_i, j)$ according to Eq. (18)
9:         **end if**
10:    **end for**
11:    Compute $R(x_i, y)$ according to Eq. (19)
12: **end for**
13: Compute $\underline{P}(x), \underline{N}(x)$ according to Eqs. (11) and (12)
14: Compute $\mu_P(x), \mu_N(x)$ according to Eqs. (15) and (16)
15: **if** $\mu_P(x) > \mu_N(x)$ **then**
16:     Output P
17: **else**
18:     Output N
19: **end if**

---

4 Experiments and Analysis

In this section, we use 14 data sets from UCI. In order to make the results more interpretable, the degree of imbalance to these data sets are different. Experiments are carried out twice. The first time we use FRNN to make classification directly (FRNN). The second time we consider DB_SMOTE to modify data sets firstly, then we use the FRNN to make classification (DB_SOMTE + FRNN). We compare the result of this two experiments by a series of measures for algorithms performance. Finally, the comparison was illustrated in forms and histograms.

4.1 Evaluation

Considering the two-class classification of imbalanced data sets: Let P be the minority (positive) class, and N be the majority (negative) class. TN is the number of negative samples correctly classified (True Negatives), TP is the number of positive samples correctly classified (True Positives), FN is the number of negative samples incorrectly classified as positive (False Negatives), FP is the number of positive samples incorrectly classified as negative(False Positives). As the Eqs. (20)–(24) shown: TPR (True positive rate, also referred to as recall) is the percentage of positive samples that are correctly classified, and TNR (True negative rate) is the percentage of negative samples that are correctly classified. *Precision* is defined as the percentage of samples that are correctly labeled as positive. G value is the geometric mean of the true positive and true negative rates, and F value is the harmonic mean of minority classes accuracy and the precision [17].

$$TPR = \frac{TP}{TP + FN} \tag{20}$$

$$TNR = \frac{TN}{TN + FP} \tag{21}$$

$$Precision = \frac{TP}{TP + FP} \tag{22}$$

$$G = \sqrt{TPR * TNR} \tag{23}$$

$$F = \frac{2 * TPR * Precision}{TPR + Precision} \tag{24}$$

4.2   Description of Data Sets

There are 14 experimental data sets from UCI in our experiments, including Ionosphere, Pima, German, etc. The Car data set has four classes, then we separated it into 3 new data sets, they are Car2, Car3, and Car4. Similarly, Cleaveland-01, Cleaveland-02, Cleaveland-03, and Cleaveland-04 come from Cleaveland which has five classes. The number of the majority and minority samples, and the ratio of the majority to the minority from each data set are shown in Table 1.

Table 1. Description of the data sets used in the experiment

Data set name	Number of majority	Number of minority	Ratio of maj to min
Ionosphere	224	126	1.78
Pima	500	268	1.87
German	685	275	2.49
Haberman	225	81	2.78
Cleaveland-01	163	54	3.01
Car2	1210	384	3.15
Transfusion	570	178	3.20
Auto-mpg	247	70	3.53
Cleaveland-02	163	36	4.52
Cleaveland-03	163	35	4.65
Cleaveland-04	163	13	12.5
Car3	1210	69	17.54
Car4	1210	65	18.62
Shuttle-25	809	13	62.23

Table 2. FRNN VS DB_SMOTE + FRNN (TNR, TPR, $Precision$, G, and F)

UCI data sets	TNR	TPR	$Precision$	G	F
Ionosphere	1.000 vs 0.957	0.228 vs 0.544	0.644 vs 0.856	0.463 vs 0.639	0.796 vs 0.922
Pima	0.798 vs 0.078	0.532 vs 0.706	0.584 vs 0.766	0.648 vs 0.741	0.550 vs 0.733
German	0.815 vs 0.876	0.457 vs 0.685	0.490 vs 0.841	0.605 vs 0.766	0.467 vs 0.751
Haberman	0.819 vs 0.768	0.261 vs 0.637	0.361 vs 0.745	0.428 vs 0.698	0.287 vs 0.683
Cleaveland-01	0.835 vs 0.800	0.485 vs 0.737	0.488 vs 0.774	0.594 vs 0.766	0.477 vs 0.752
Car2	0.946 vs 0.946	0.247 vs 0.653	0.750 vs 0.983	0.387 vs 0.730	0.672 vs 0.879
Transfusion	0.847 vs 0.964	0.197 vs 0.668	0.285 vs 0.947	0.345 vs 0.801	0.229 vs 0.781
Auto-mpg	0.933 vs 0.964	0.363 vs 0.749	0.554 vs 0.946	0.527 vs 0.847	0.402 vs 0.831
Cleaveland-02	0.942 vs 0.878	0.648 vs 0.810	0.757 vs 0.839	0.766 vs 0.841	0.640 vs 0.823
Cleaveland-03	0.950 vs 0.920	0.633 vs 0.803	0.685 vs 0.896	0.724 vs 0.855	0.649 vs 0.840
Cleaveland-04	0.971 vs 0.952	0.450 vs 0.731	0.483 vs 0.837	0.503 vs 0.829	0.460 vs 0.769
Car3	0.986 vs 0.996	0.183 vs 0.907	0.169 vs 0.983	0.189 vs 0.944	0.172 vs 0.934
Car4	0.996 vs 0.996	0.200 vs 0.933	0.197 vs 0.980	0.199 vs 0.961	0.197 vs 0.953
Shuttle-25	1.000 vs 1.000	0.700 vs 0.990	0.285 vs 0.947	0.700 vs 0.995	0.700 vs 0.995

4.3   Experiments Results

From the information of Table 2 and the Fig. 2(a)–(e), we can see that the DB_SMOTE + FRNN approach can improve the performance in the imbalanced data sets classification, especially for minority class. And it performs as well as FRNN in TNR, the value only fluctuates slightly on different data sets, but it remains stable as Fig. 2.(a) shows. Then Fig. 2.(b) shows the change of

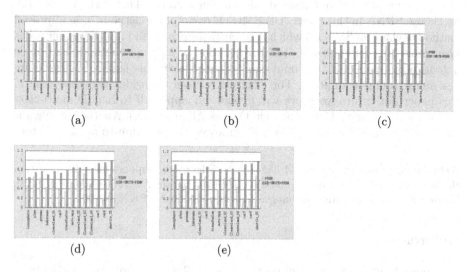

(a)   (b)   (c)

(d)   (e)

Fig. 2. Experiment comparisons of FRNN and DB_SMOTE + FRNN: (a) the TNR comparison, (b) the TPR comparison, (c) the $precision$ comparison, (d) the G value comparison and (e) the F value comparison

the TPR. The horizontal axis represents the fourteen test data sets, which are arranged in a ratio increased gradually from left to right. The vertical axis represents the value of TPR of the data sets. As we can see from this figure, from left to right, the more imbalance of the ratio of the majority to the minority, the more improve of the TPR. Such as the dataset Car3 and Car4, when we put to use our proposed method, TPR improved 70 % points. Similarly, changes in the Precision, G value and F value which shown as Fig. 2(c)–(e) have a similar law. There is a very large increase in the haberman, transfusion, car3, car4 and other data sets.

So we can conclude that the more imbalanced the data set is, the better performance our method shows. It can be concluded that for the classification of the imbalanced data, DB_SMOTE + FRNN is an effective combination which is worth to the further research and discussion. And we canS draw that, before classify the data using any classifier, it is necessary to preprocess our data first.

5   Conclusion

On the problem of imbalanced classification, there are plenty of methods to modify the imbalance data or to improve the classification accuracy. At the same time, the study about combining those methods can be found in many research literatures, which proved such study to be feasible. We consider that the key to this kind of attempt is that the researchers should discover the complementation of methods for combination based on grasping the basis mechanisms of those methods. Fuzzy Rough Nearest Neighbor (FRNN) algorithm is not only good at handling incomplete information system classification, also is effective in the classification of imbalanced information systems. This is the reason why we choose FRNN in this paper. DB_SMOTE resampling methods is an improved method of classical SMOTE which is recently proposed, and it's effectiveness has been proved by experiments. The combination of this two methods had made a good performance in the simulation experiment in this paper, but there are also much room to improve. For example, we only consider two-class case in this paper, for future work, we will extend our approach to handle multi-class problems (e.g. One-vs-One approach, One-vs-All approach). Another problem is the FRNN is time-consuming, we will improve the algorithm to make it faster.

Acknowledgements. We would like to acknowledge the support for this work from the National Natural Science Foundation of China (Grant Nos. 61403200, 61170180), Natural Science Foundation of Jiangsu Province (Grant No.BK20140800).

References

1. Cormack, R.M.: A review of classification. J. Roy. Stat. Soc. Ser. A (General), 321–367 (1971)
2. Kotsiantis, S.B., Zaharakis, I.: Supervised machine learning: a review of classification techniques. Emerg. Artif. Intell. Appl. Comput. Eng., 3–24 (2007)

3. Maloof, M.A.: Learning when data sets are imbalanced and when costs are unequal and unknown. In: ICML-2003 Workshop on Learning from Imbalanced Data Sets II, vol. 2, pp. 1–2 (2003)
4. Japkowicz, N.: Learning from imbalanced data sets: a comparison of various strategies. In: AAAI Workshop on Learning from Imbalanced Data Sets, pp. 10–15 (2000)
5. Duman, E., Ekinci, Y., Tanrıverdi, A.: Comparing alternative classifiers for database marketing: the case of imbalanced datasets. Expert Syst. Appl. **39**, 48–53 (2012)
6. Khreich, W., Granger, E., Miri, A., Sabourin, R.: Iterative boolean combination of classifiers in the ROC space. In: An Application to Anomaly Detection with HMMs. Pattern Recogn. **43**, 2732–2752 (2010)
7. Lee, Y., Hu, P., Cheng, T., Huang, T., Chuang, W.: A preclustering-based ensemble learning technique for acute appendicitis diagnoses. Artif. Intell. Med. **58**, 115–124 (2013)
8. Ramentol, E., Vluymans, S., Verbiest, N., Caballero, Y.: IFROWANN: imbalanced fuzzy-rough ordered weighted average nearest neighbor classification, pp. 1–15 (2014)
9. Weiss, G.M.: Mining with rarity: a unifying framework. ACM SIGKDD Explor. Newsl. **6**, 7–19 (2004)
10. Hwang, J., Park, S., Kim, E.: A new weighted approach to imbalanced data classification problem via support vector machine with quadratic cost function. Expert Syst. Appl. **38**, 8580–8585 (2011)
11. Sun, Y., Kamel, M.S., Wong, A.K., Wang, Y.: Cost-sensitive boosting for classification of imbalanced data. Pattern Recogn. **40**, 3358–3378 (2007)
12. Jensen, R., Cornelis, C.: Fuzzy-rough nearest neighbour classification. In: Peters, J.F., Skowron, A., Chan, C.-C., Grzymala-Busse, J.W., Ziarko, W.P. (eds.) Transactions on Rough Sets XIII. LNCS, vol. 6499, pp. 56–72. Springer, Heidelberg (2011)
13. Sarkar, M.: Rough-fuzzy functions in classification. Fuzzy Sets Syst. **132**(3), 353–369 (2002)
14. Liu, X., Liu, S.: New oversampling algorithm DB-SMOTE. Comput. Eng. Appl., 92–95 (2014)
15. Sáez, J., Luengo, J., Stefanowski, J., Herrera, F.: SMOTE-IPF: addressing the noisy and borderline examples problem in imbalanced classification by a resampling method with filtering. Inf. Sci. **291**, 184–203 (2015)
16. Barandela, R., Sánchez, J., Garcia, V., Rangel, E.: Strategies for learning in class imbalance problems. Pattern Recogn. **36**, 849–851 (2003)
17. Liu, J., Hu, Q.: A weighted rough set based method developed for class imbalance learning. Inf. Sci. **178**, 1235–1256 (2008)
18. Cover, T., Hart, P.: Nearest neighbor pattern classification. IEEE Trans. Inf. Theor. **13**, 21–27 (1967)

Three-Way Decisions

Region Vector Based Attribute Reducts in Decision-Theoretic Rough Sets

Guoshun Huang[1,2]([⊠]) and Yiyu Yao[2]

[1] School of Science, Foshan University, Foshan 528000, China
fshgs_72@163.com
[2] Department of Computer Science, University of Regina,
Regina, SK S4S 0A2, Canada
yyao@cs.uregina.ca

Abstract. When removing some attributes, the partition induced by a smaller set of attributes will be coarser and the decision regions may be changed. In this paper, we analyze the decision region changes when removing attributes and propose a new type of attribute reducts from the point of view of vector based three-way approximations of a partition. We also present a reduct construction method by using a discernibility matrix.

Keywords: Attribute reduction · Decision-theoretic rough sets · Discernibility matrix

1 Introduction

Decision-theoretic rough set (DTRS) models [15,16] are probabilistic generalizations of Pawlak rough sets [9]. One feature of DTRS models is the tolerance of decision errors. In contrast to Pawlak approximations [9], both probabilistic positive and negative regions are not certain and contain errors [1,8,13]. Consequently, the notion of attribute reducts becomes more complicated. In addition to the condition of positive region preservation required by a Pawlak attribute reduct [10], we must also consider boundary and negative regions. This leads to the notion of region vector based attribute reducts.

Compared with Pawlak's three regions, probabilistic regions of DTRS models are not monotonic with respect to the set inclusion relation on sets of attributes [5,18]. Definitions of, and methods for constructing, Pawlak attribute reducts may not be appropriate for DTRS models. To address the problem of non-monotonicity, many authors proposed different types of attribute reducts, such as minimum cost attribute reducts by Jia et al. [2,4], non-monotonic attribute reducts by Li et al. [5], cost-sensitive attribute reducts by Liao et al. [6], decision region distribution preservation reducts by Ma et al. [7], and so on.

Zhao et al. [18] proposed a general definition of an attribute reduct as a minimal set of condition attributes satisfying properties given in terms of a set

© Springer International Publishing Switzerland 2015
Y. Yao et al. (Eds.): RSFDGrC 2015, LNAI 9437, pp. 355–365, 2015.
DOI: 10.1007/978-3-319-25783-9_32

of evaluation measures. Jia et al. [3] studied systematically several specific classes
of attribute reducts based on a general definition of attribute reducts. Yao [14]
provided a more general conceptual definition of reducts of a set, embracing
notions of attribute reducts, attribute-value pair reducts, and rule reducts.

Based on results from these studies, in this paper we study attribute reducts
by considering changes of decision regions. By using region vector based three-
way approximations, we propose three new definitions of attribute reducts. The
new definitions require that an attribute reduct should perverse high confidence
rules. A method for constructing an attribute reduct is proposed by using the
notion of a discernibility matrix introduced by Skowron and Rauszer [11].

2   Three-Way Approximations of a Classification

An information table is defined by $IT = (U, AT, \{V_a | a \in AT\}, \{I_a | a \in AT\})$,
where U is a finite nonempty set of objects, AT is a finite nonempty set of
attributes, V_a is a nonempty set of values for $a \in AT$, and $I_a : U \rightarrow V_a$ is an
information function that maps an object in U to one value in V_a. If $AT = C \cup D$, where C is a finite set of condition attributes, D is a finite set of decision
attributes and $C \cap D = \emptyset$, the information table is called a decision table, denoted
by DT.

For a subset of attributes $A \subseteq AT$, one defines an equivalence relation on U
as follows:

$$xR_Ay \Leftrightarrow \forall a \in A(I_a(x) = I_a(y)). \tag{1}$$

The equivalence relation R_A induces a partition of U, denoted by $U/R_A = \{[x]_{R_A} \mid x \in U\}$ and U/A or π_A for short. The equivalence class containing x
is given by $[x]_A = \{y \mid \forall a \in A(I_a(x) = I_a(y))\}$. According to Pawlak [10], a
partition induced by a subset of attributes is called a classification of U. The
classification based on the set of decision attributes is given by U/D or π_D.

For a subset of objects $X \subseteq U$ and a subset of attributes $A \subseteq AT$, Let
$\Pr(X|[x]_A)$ denote the conditional probability of an object belonging to X given
that the object belongs to $[x]_A$. This probability may be simply estimated as
$\Pr(X|[x]_A) = |X \cap [x]_A| / |[x]_A|$, where $|\cdot|$ denotes the cardinality of a set. In
terms of conditional probability, the main results of DTRS models are approximations of a set through a pair of lower and upper approximations or three
pair-wise disjoint positive, boundary and negative regions. In this paper, we
adopt the formulation with three-way approximations [13,14] by using a slightly
different notional system.

Definition 1. *In a decision table $DT = (U, AT = C \cup D, \{V_a | a \in AT\}, \{I_a | a \in AT\})$, given a subset of attributes $A \subseteq C$ and a pair of thresholds $0 \leq \beta < \alpha \leq 1$, the following sets:*

$$\mathrm{POS}_{(\alpha,\beta)}(X|\pi_A) = \{x \in U \mid \Pr(X|[x]_A) \geq \alpha\},$$
$$\mathrm{BND}_{(\alpha,\beta)}(X|\pi_A) = \{x \in U \mid \beta < \Pr(X|[x]_A) < \alpha\},$$
$$\mathrm{NEG}_{(\alpha,\beta)}(X|\pi_A) = \{x \in U \mid \Pr(X|[x]_A) \leq \beta\}, \tag{2}$$

are called (α, β)-*positive, boundary and negative regions of a subset of objects* X *with respect to the partition* π_A.

The three regions are pair-wise disjoint and their union is U. Some of the three regions may be the empty set. Thus, the family of these regions may not necessarily be a partition of U.

One can extend the three-way approximations of a set to three-way approximations of a classification $\pi_D = \{D_1, D_2, \cdots, D_r\}$. A straightforward way is to define ¬way approximations component-wise.

Definition 2. *In a decision table* DT, *given a subset of attributes* $A \subseteq C$ *and a pair of thresholds* $0 \leq \beta < \alpha \leq 1$, *the vector based* (α, β)-*positive, boundary and negative regions of* π_D *with respect to the partition* π_A *are given by:*

$$\overrightarrow{\mathrm{POS}}_{(\alpha,\beta)}(\pi_D|\pi_A) = (\mathrm{POS}_{(\alpha,\beta)}(D_1|\pi_A), \cdots, \mathrm{POS}_{(\alpha,\beta)}(D_r|\pi_A)),$$

$$\overrightarrow{\mathrm{BND}}_{(\alpha,\beta)}(\pi_D|\pi_A) = (\mathrm{BND}_{(\alpha,\beta)}(D_1|\pi_A), \cdots, \mathrm{BND}_{(\alpha,\beta)}(D_r|\pi_A)),$$

$$\overrightarrow{\mathrm{NEG}}_{(\alpha,\beta)}(\pi_D|\pi_A) = (\mathrm{NEG}_{(\alpha,\beta)}(D_1|\pi_A), \cdots, \mathrm{NEG}_{(\alpha,\beta)}(D_r|\pi_A)). \tag{3}$$

By taking a component-wise union of regions in two vectors $\overrightarrow{\mathrm{POS}}_{(\alpha,\beta)}(\pi_D|\pi_A)$ *and* $\overrightarrow{\mathrm{BND}}_{(\alpha,\beta)}(\pi_D|\pi_A)$, *we obtain a non-negative region vector:*

$$\overrightarrow{\neg\mathrm{NEG}}_{(\alpha,\beta)}(\pi_D|\pi_A) = (\neg\mathrm{NEG}_{(\alpha,\beta)}(D_1|\pi_A), \cdots, \neg\mathrm{NEG}_{(\alpha,\beta)}(D_r|\pi_A)), \tag{4}$$

where $\neg\mathrm{NEG}_{(\alpha,\beta)}(D_j|\pi_A) = \mathrm{POS}_{(\alpha,\beta)}(D_j|\pi_A) \cup \mathrm{BND}_{(\alpha,\beta)}(D_j|\pi_A)$, $j = 1, \cdots, r$.

Another definition is based on the union of the three-way approximations of all decision classes, we call it a set based definition.

Definition 3. *In a decision table* DT, *given a subset of attributes* $A \subseteq C$ *and a pair of thresholds* $0 \leq \beta < \alpha \leq 1$, *the set based* (α, β)-*positive, boundary and negative regions of* π_D *with respect to the partition* π_A *are given by:*

$$\begin{aligned}
\mathrm{POS}_{(\alpha,\beta)}(\pi_D|\pi_A) &= \bigcup_{j=1}^{r} \mathrm{POS}_{(\alpha,\beta)}(D_j|\pi_A), \\
&= \{x \in U \mid \exists D_j \in \pi_D(\mathrm{Pr}(D_j|[x]_A) \geq \alpha)\}; \\
\mathrm{BND}_{(\alpha,\beta)}(\pi_D|\pi_A) &= \bigcup_{j=1}^{r} \mathrm{BND}_{(\alpha,\beta)}(D_j|\pi_A), \\
&= \{x \in U \mid \exists D_j \in \pi_D(\beta < \mathrm{Pr}(D_j|[x]_A) < \alpha)\}; \\
\mathrm{NEG}_{(\alpha,\beta)}(\pi_D|\pi_A) &= U - \mathrm{POS}_{(\alpha,\beta)}(\pi_D|\pi_A) \cup \mathrm{BND}_{(\alpha,\beta)}(\pi_D|\pi_A), \\
&= \{x \in U \mid \forall D_j \in \pi_D(\mathrm{Pr}(D_j|[x]_A) \leq \beta)\}. \tag{5}
\end{aligned}$$

A set based (α, β)-*non-negative region is defined as the union of positive and boundary regions:*

$$\neg\mathrm{NEG}_{(\alpha,\beta)}(\pi_D|\pi_A) = \bigcup_{j=1}^{r} \{x \in U | \exists D_j \in \pi_D(\mathrm{Pr}(D_j \mid [x]_A) > \beta)\}. \tag{6}$$

As $\text{POS}_{(\alpha,\beta)}(\pi_D|\pi_A)$ only depends on the threshold α, Li et al. [5] call it the α-positive region. For convenience, we still call it (α,β)-positive region. In the special case when $\alpha = 1$ and $\beta = 0$, $\text{POS}_{(1,0)}(\pi_D|\pi_A)$ and $\text{BND}_{(1,0)}(\pi_D|\pi_A)$ are, in fact, the positive and boundary regions in Pawlak rough set model, respectively. In this case, $\text{NEG}_{(1,0)}(\pi_D|\pi_A) = \emptyset$.

One can construct decision or classification rules from the three regions. In general, a rule in rough set theory can be expressed in the form of $[x]_A \rightarrow D_j$, stating that an object with description $[x]_A$ would be in the decision class D_j. Given a rule $[x]_A \rightarrow D_j$, $\text{Pr}(D_j|[x]_A)$ is its confidence and $\text{Pr}([x]_A|D_j)$ its coverage [17].

Skowron [12] introduced a concept of generalized decisions. For an object x, the generalized decision for x is the set all possible decisions for objects in the equivalence class $[x]_A$. For DTRS, Zhao et al. [18] introduced the notion of (α,β)-positive, boundary and non-negative decisions. Specifically, for an equivalence class $[x]_A \in \pi_A$, we have:

$$
\begin{aligned}
D_{\text{POS}(\alpha,\beta)}([x]_A) &= \{D_j \in \pi_D \mid (\text{Pr}(D_j|[x]_A) \geq \alpha)\}, \\
D_{\text{BND}(\alpha,\beta)}([x]_A) &= \{D_j \in \pi_D \mid (\beta < \text{Pr}(D_j|[x]_A) < \alpha)\}, \\
D_{\neg\text{NEG}(\alpha,\beta)}([x]_A) &= D_{\text{POS}(\alpha,\beta)}([x]_A) \cup D_{\text{BND}(\alpha,\beta)}([x]_A).
\end{aligned}
\tag{7}
$$

They are probabilistic versions of the generalized decisions.

3   Region Vector Based Attribute Reducts for DTRS Models

In existing studies, the set based regions are widely used in defining attribute reducts for DTRS models. The problem of searching for an attribute reduct is to remove redundant attributes. By successively removing attributes, the resulting partitions will become coarser. As a result, the decision regions will be changed. Compared with Pawlak rough set model, the monotonicity of decision region change no longer holds for DTRS models. It is therefore very important to analyze how decision region changes with the decrease of attributes.

3.1   An Analysis of Set Based Decision Regions

For the set based three regions, it is easy to verify the following properties:

(1) $\text{NEG}_{(\alpha,\beta)}(\pi_D|\pi_A)$ may not be empty,

(2) $\text{NEG}_{(\alpha,\beta)}(\pi_D|\pi_A) \cap \text{POS}_{(\alpha,\beta)}(\pi_D|\pi_A) = \emptyset$,

(3) $\text{NEG}_{(\alpha,\beta)}(\pi_D|\pi_A) \cap \text{BND}_{(\alpha,\beta)}(\pi_D|\pi_A) = \emptyset$,

(4) $\text{POS}_{(\alpha,\beta)}(\pi_D|\pi_A)$ and $\text{BND}_{(\alpha,\beta)}(\pi_D|\pi_A)$ may have an overlap,

(5) When $\alpha > 0.5, \forall D_i, D_j \in \pi_D, i \neq j$,

$$\text{POS}_{(\alpha,\beta)}(D_i|\pi_A) \cap \text{POS}_{(\alpha,\beta)}(D_j|\pi_A) = \emptyset,$$

(6) When $\alpha \leq 0.5$, there may exist decision classes such that

$$\text{POS}_{(\alpha,\beta)}(D_i|\pi_A) \cap \text{POS}_{(\alpha,\beta)}(D_j|\pi_A) \neq \emptyset, \quad i \neq j.$$

These properties are essential to our study of attribute reducts in DTRS models.

When we remove some attributes, the confidence of decision rules, as defined by the conditional probability may change. Consider a subset of condition attributes $A \subseteq C$. By definition, an equivalence class of π_A is the union of a family of equivalence classes of C. That is, $[x]_A = \cup_{y \in [x]_A} [y]_C = \cup_{k=1}^m [y_k]_C$, where $[y_1]_C, \ldots, [y_m]_C$ are the family of distinct equivalence classes of π_C such that their union is $[x]_A$. Based on this result, one can easily verify the next lemma.

Lemma 1. *For $D_j \in \pi_D$, we have:*

(1) $\Pr(D_j|[x]_A) = \sum_{k=1}^m \dfrac{|[y_k]_C|}{|[x]_A|} \Pr(D_j|[y_j]_C),$

(2) $\min\{\Pr(D_j|[y]_C) \mid y \in [x]_A\} \leq \Pr(D_j|[x]_A) \leq \max\{\Pr(D_j|[y]_C) \mid y \in [x]_A\}.$

By removing some attributes, we in fact combine a family rules into a single rule. According to Lemma 1, the confidence of the new rule $[x]_A \to D_j$ is the weighted sum of the confidence of individual rules $[y_k]_C \to D_j$, $k = 1, \ldots, m$.

By the set based definition of three regions, $[x]_C$ is put into the (α, β)-positive region $\mathrm{POS}_{(\alpha,\beta)}(\pi_D|\pi_C)$ if $\Pr(D_j|[x]_C) \geq \alpha$ for some $D_j \in \pi_D$. Equivalence classes in positive regions induced by π_C with different decision set may merge into a new equivalence class π_A when removing attributes and the new equivalence class may be in the negative region $\mathrm{NEG}_{(\alpha,\beta)}(\pi_D|\pi_C)$, as shown by the following example from [2].

Example 1. Consider a decision Table 1, where $U = \{x_1, \cdots, x_9\}$, $C = \{c_1, c_2\}$, and $D = \{d\}$. The partition induced by the set of condition attributes is given by $U/C = \{C_1, C_2, C_3, C_4\}$, where $C_1 = \{x_1, x_4\}$, $C_2 = \{x_2, x_6, x_8\}$, $C_3 = \{x_3\}$, and $C_4 = \{x_5, x_7, x_9\}$. The decision partition is given by $U/D = \{D_1, D_2, D_3\}$, where $D_1 = \{x_1, x_2\}$, $D_2 = \{x_3, x_4, x_5\}$, and $D_3 = \{x_6, x_7, x_8, x_9\}$. Let $A = \{c_1\}$, then $U/A = \{A_1, A_2\}$, where $A_1 = C_1 \cup C_2$ and $A_2 = C_3 \cup C_4$.

Assume that $\alpha = 0.6$ and $\beta = 0.5$. We have $\mathrm{POS}_{(0.6,0.5)}(\pi_D|\pi_C) = C_2 \cup C_3 \cup C_4$ and $\mathrm{NEG}_{(0.6,0.5)}(\pi_D|\pi_A) = U$. The equivalence classes C_3 and C_4 in partition U/C is merged into A_2 in partition U/A when we remove attribute c_2. The equivalence class A_2 is put into negative region $\mathrm{NEG}_{(0.6,0.5)}(\pi_D|\pi_A)$.

When $\alpha = 0.6$ and $\beta = 0.4$, the equivalence class A_2 is put into the boundary region $\mathrm{BND}_{(0.6,0.4)}(\pi_D|\pi_A)$. When $\alpha = 0.5$ and $\beta = 0.3$, A_2 is put into positive region $\mathrm{POS}_{(0.5,0.3)}(\pi_D|\pi_A)$.

Analogously, some equivalence classes in the positive or negative regions merge into a new equivalence class, which may be put into the positive, boundary or negative region. There is one exception, that is, the merge of equivalence classes in the negative region is still in the negative region according to the definition of negative region. Table 2 summarizes the main results, where the second column gives the regions of equivalences before the merge, and the third column gives the regions of the new equivalence class.

Table 1. A decision table

U	x_1	x_2	x_3	x_4	x_5	x_6	x_7	x_8	x_9
c_1	1	1	0	1	0	1	0	1	0
c_2	1	0	1	1	0	0	0	0	0
d	0	0	1	1	1	2	2	2	2

Table 2. Changes of decision regions after removing attributes

Cases	The original regions	The new regions after removing attributes
(1)	All positive	Positive, boundary or negative
(2)	Positive and boundary	Positive, boundary or negative
(3)	Positive and negative	Positive, boundary or negative
(4)	All boundary	Positive, boundary or negative
(5)	Boundary and negative	Positive, boundary or negative
(6)	All negative	negative
(7)	Positive,boundary and negative	Positive, boundary or negative

3.2 Region Vector Based Attribute Reducts

According to the previous analysis, a set based region may become larger, unchanged or smaller when we remove some condition attributes. This results in difficulties in constructing an attribute reduct for DTRS models. For example, when equivalence classes in positive and negative regions merge, the new equivalence class will be put into positive, boundary or negative regions. If it is put into the positive region, the new positive region will be larger. Such a move may not be desirable because it puts original lower confidence rules into positive region. On the other hand, if the merged equivalence class is put into the boundary or negative regions, it cannot guarantee that a positive rule is still a positive rule after removing attributes.

When equivalence classes with the same decision set are combined, the new equivalence class can put into the positive region according to Lemma 1. An original positive rule is still a positive rule after reducing attributes. The generality stay the same and there is no risk to misclassify a low confidence rule as a high confidence rule.

Similarly, if equivalence classes in boundary regions with the same decision set are merged, the merged equivalence class is still in the boundary region by Lemma 1. This will not change the positive, boundary and negative regions. When equivalence classes in negative region are merged, the merged equivalence classes must be put into negative region. This does not change the generality and nor increases the cost of decision risk.

Based on the above analysis, we propose a new type of attribute reducts based on a vector representation of three regions.

Definition 4. *In a decision table DT, given a subset of attributes $R \subseteq C$ and a pair of thresholds $0 \leq \beta < \alpha \leq 1$, R is an (α, β)-vregion attribute reduct of C with respect to the set of decision attributes D if it satisfies the following two conditions:*

(1) $\overrightarrow{\text{vregion}}_{(\alpha,\beta)}(\pi_D|\pi_R) = \overrightarrow{\text{vregion}}_{(\alpha,\beta)}(\pi_D|\pi_C);$

(2) *for any* $R' \subset R, \overrightarrow{\text{vregion}}_{(\alpha,\beta)}(\pi_D|\pi_{R'}) \neq \overrightarrow{\text{vregion}}_{(\alpha,\beta)}(\pi_D|\pi_C).$

Especially, if vregion is a positive, boundary, negative or non-negative region, then R is called an (α, β)-positive, boundary, negative or non-negative region attribute reduct of C, respectively. If R only satisfies condition (1), it is called an (α, β)-positive, boundary, negative or non-negative region consistent attribute subset of C.

Theorem 1. $\overrightarrow{\text{POS}}_{(1,0)}(\pi_D|\pi_R) = \overrightarrow{\text{POS}}_{(1,0)}(\pi_D|\pi_C)$ *if and only if* $\text{POS}_{(1,0)}(\pi_D|\pi_R) = \text{POS}_{(1,0)}(\pi_D|\pi_C)$.

Proof. ($\Rightarrow$) If $\overrightarrow{\text{POS}}_{(1,0)}(\pi_D|\pi_R) = \overrightarrow{\text{POS}}_{(1,0)}(\pi_D|\pi_C)$, by definitions of these regions, we immediately have $\text{POS}_{(1,0)}(\pi_D|\pi_R) = \text{POS}_{(1,0)}(\pi_D|\pi_C)$.
($\Leftarrow$) By definition, $\text{POS}_{(1,0)}(D_i|\pi_R) \cap \text{POS}_{(1,0)}(D_j|\pi_R) = \emptyset$ for any $D_i, D_j \in \pi_D, i \neq j$ and $\text{POS}_{(1,0)}(D_j|\pi_R) \subseteq \text{POS}_{(1,0)}(D_j|\pi_C)$. Hence, if $\text{POS}_{(1,0)}(\pi_D|\pi_R) = \text{POS}_{(1,0)}(\pi_D|\pi_C)$, we have $\text{POS}_{(1,0)}(D_j|\pi_R) = \text{POS}_{(1,0)}(D_j|\pi_C)$. It follows that $\overrightarrow{\text{POS}}_{(1,0)}(\pi_D|\pi_R) = \overrightarrow{\text{POS}}_{(1,0)}(\pi_D|\pi_C)$.

Theorem 1 shows that, in Pawlak rough set model, the two representation forms are equivalent in judging whether R is a positive region preservation consistent set of C. For boundary region, we can only get a one-way inference. That is, $\overrightarrow{\text{BND}}_{(\alpha,\beta)}(\pi_D|\pi_R) = \overrightarrow{\text{BND}}_{(\alpha,\beta)}(\pi_D|\pi_C) \Rightarrow \text{BND}_{(1,0)}(\pi_D|\pi_R) = \text{BND}_{(1,0)}(\pi_D|\pi_C)$, and the reverse is not necessarily true. For a decision-theoretic rough set model, we can only get one-way inference, i.e.,

$$\overrightarrow{\text{POS}}_{(\alpha,\beta)}(\pi_D|\pi_R) = \overrightarrow{\text{POS}}_{(\alpha,\beta)}(\pi_D|\pi_C) \Rightarrow \text{POS}_{(\alpha,\beta)}(\pi_D|\pi_R) = \text{POS}_{(\alpha,\beta)}(\pi_D|\pi_C),$$

$$\overrightarrow{\text{BND}}_{(\alpha,\beta)}(\pi_D|\pi_R) = \overrightarrow{\text{BND}}_{(\alpha,\beta)}(\pi_D|\pi_C) \Rightarrow \text{BND}_{(\alpha,\beta)}(\pi_D|\pi_R) = \text{BND}_{(\alpha,\beta)}(\pi_D|\pi_C).$$

The reverse is not necessarily true. The regions based on vector representation do not have the monotonicity, as shown by the next example.

Example 2. (continued from Example 1) Let $\alpha = 0.40, \beta = 0.35, R = \{c_1\}$, we have:

$$\text{POS}_{(0.40,0.35)}(\pi_D|\pi_R) = \text{POS}_{(0.40,0.35)}(\pi_D|\pi_C) = U,$$

$$\overrightarrow{\text{POS}}_{(0.40,0.35)}(\pi_D|\pi_C) = (C_1, C_1 \cup C_3, C_2 \cup C_4),$$

$$\overrightarrow{\text{POS}}_{(0.40,0.35)}(\pi_D|\pi_R) = (C_1 \cup C_2, C_3 \cup C_4, U).$$

They show that the monotonicity does not hold.

4   A Reduct Construction Method

For an (α, β)-positive region attribute reduct, the positive rule set does not change if we merge equivalence classes with the same decisions. We can construct a discernibility matrix $M_{D_{POS}}([x]_C, [y]_C)$ for positive region attribute reduct as follows:

$$M_{D_{POS}}([x]_C, [y]_C) = \begin{cases} \{a \in C | I_a(x) \neq I_a(y)\}, \\ \qquad \text{if } D_{POS(\alpha, \beta)}([x]_C) \neq D_{POS(\alpha, \beta)}([y]_C), \\ C, \quad \text{otherwise.} \end{cases} \qquad (8)$$

By the discernibility function proposed by Skowron and Rauszer [11]:

$$f(M_{D_{POS}}) = \wedge \vee M_{D_{POS}}([x]_C, [y]_C). \qquad (9)$$

We can get a decision reduct, which is a prime implicants of the reduced disjunctive form of the discernibility function.

If $\alpha > 0.5$, Eq. (8) degenerates into the discernibility matrix of a variable precision rough set (VPRS) model [8]. We can get an α-lower-approximation attribute reduct. Since VPRS is a special case of DTRS with $\alpha > 0.5$ and $\beta = 1 - \alpha$, we give a general conclusion for any pair of thresholds (α, β) with $0 \leq \beta < \alpha \leq 1$.

Theorem 2. *In a decision table DT, given a subset of attributes $A \subseteq C$ and a pair of thresholds $0 \leq \beta < \alpha \leq 1$, A is an (α, β)-positive region consistent set of C iff for any $x, y \in U$, $D_{POS(\alpha, \beta)}([x]_C) \neq D_{POS(\alpha, \beta)}([y]_C)$ implies $[x]_A \cap [y]_A = \emptyset$.*

Proof. ($\Rightarrow$) For any $x, y \in U$, if $[x]_A \cap [y]_A \neq \emptyset$, $[x]_A = [y]_A$. Thus, we have $D_{POS(\alpha, \beta)}([x]_A) = D_{POS(\alpha, \beta)}([y]_A)$. Since A is an (α, β)-positive region consistent set of C, $POS_{(\alpha, \beta)}(D_j | \pi_A) = POS_{(\alpha, \beta)}(D_j | \pi_C)$ for any $D_j \in \pi_D$. It follows that $x \in POS_{(\alpha, \beta)}(D_j | \pi_A) \Leftrightarrow x \in POS_{(\alpha, \beta)}(D_j | \pi_C)$, i.e., $D_{POS(\alpha, \beta)}([x]_A) = D_{POS(\alpha, \beta)}([x]_C)$. Similarly, we have $D_{POS(\alpha, \beta)}([y]_A) = D_{POS(\alpha, \beta)}([y]_C)$. Therefore, we can conclude that $D_{POS(\alpha, \beta)}([x]_C) = D_{POS(\alpha, \beta)}([y]_C)$, which conflicts with the assumption.

($\Leftarrow$) Since $A \subseteq C$, it is easy to verify that $T([x]_A) = \{[y]_C \mid [y]_C \in [x]_A/C\}$ is a partition of $[x]_A$.

(i) For any $D_j \in \pi_D$, if $x \in POS_{(\alpha, \beta)}(D_j | \pi_A)$, then $[x]_A \subseteq POS_{(\alpha, \beta)}(D_j | \pi_A)$. Since $[x]_A = [y]_A$ for any $y \in [x]_A$, according to the arbitrariness of y and the assumption that $D_{POS(\alpha, \beta)}([x]_C) = D_{POS(\alpha, \beta)}([y]_C)$, we have $D_j \in D_{POS(\alpha, \beta)}([x]_C)$, i.e., $x \in POS_{(\alpha, \beta)}(D_j | \pi_C)$.

(ii) If $x \in POS_{(\alpha, \beta)}(D_j | \pi_C)$, then $D_j \in D_{POS(\alpha, \beta)}([x]_C)$. Since $[x]_A = [y]_A$ for any $[y]_C \in T([x]_A)$, by the assumption, we have $D_{POS(\alpha, \beta)}([x]_C) = D_{POS(\alpha, \beta)}([y]_C)$. It follows that $D_j \in D_{POS(\alpha, \beta)}([y]_C)$ i.e., $\Pr(D_j | [y]_C) \geq \alpha$. Hence,

$$p(D_j | [x]_A) = \sum_{[y]_C \in T([x]_A)} \frac{|[y]_C|}{|[x]_A|} \frac{|[y]_C \cap D_j|}{|[y]_C|} > \alpha, \qquad (10)$$

i.e., $x \in POS_{(\alpha, \beta)}(D_j | \pi_A)$.

Based on (i) and (ii), we have $\text{POS}_{(\alpha,\beta)}(D_j|\pi_A) = \text{POS}_{(\alpha,\beta)}(D_j|\pi_C)$ for any $D_j \in \pi_D$, which follows that A is an (α,β)-positive region consistent set of C.

According to Theorem 2, when deleting some attributes, if only equivalence classes with same decision set are merged, it can guarantee that the deleted attributes are redundant and the remaining set of attributes is an (α,β)-positive region consistent set of C.

Example 3. Consider a decision Table 3, where $U = \{x_1, \cdots, x_{10}\}$, $C = \{c_1, c_2\}$, and $D = \{d\}$. For the set of condition attributes, we have $U/C = \{C_1, C_2, C_3\}$, where $C_1 = \{x_1, x_2, x_3, x_4\}$, $C_2 = \{x_5, x_6, x_7\}$, $C_3 = \{x_8, x_9, x_{10}\}$. For the set of decision attributes, we have $U/D = \{D_1, D_2\}$, where $D_1 = \{x_1, x_2, x_3, x_7, x_8, x_9\}$ and $D_2 = \{x_4, x_5, x_6, x_{10}\}$. Let $R = \{c_1\}$. We have $U/R = \{R_1, R_2\}$, where $R_1 = C_1 \cup C_3$ and $R_2 = C_2$. If $\alpha = 0.6$, then $\overrightarrow{\text{POS}}_{(\alpha,\beta)}(\pi_D|\pi_C) = (C_1 \cup C_3, C_2)$ and $\overrightarrow{\text{POS}}_{(\alpha,\beta)}(\pi_D|\pi_R) = (R_1, R_2) = \overrightarrow{\text{POS}}_{(\alpha,\beta)}(\pi_D|\pi_C)$. It is easy to verity that $\{c_1\}$ is an (α,β)-positive region attribute reduct of C, it guarantees that the positive rules unchanged although their confidence is different. Before removing attributes, the positive rules are $C_1 \rightarrow_P D_1$, $C_3 \rightarrow_P D_1$, with confidence 0.750 and 0.667, respectively. By deleting attribute set $\{c_2\}$, equivalence classes C_1 and C_3 merge into R_1. We get a new positive rule $R_1 \rightarrow_P D_1$, with a confidence of 0.714.

On the other hand, we have $D_{\text{POS}(\alpha,\beta)}(C_1) = \{D_1\}$, $D_{\text{POS}(\alpha,\beta)}(C_2) = \{D_2\}$, $D_{\text{POS}(\alpha,\beta)}(C_3) = \{D_1\}$. The corresponding discernibility matrix $M_{\text{POS}(0.6,\beta)}$ is showed in Table 4. The attribute reduct is $\{c_1\}$, which is consistent with Definition 4.

Similarly, we can present the discernibility matrix $M_{D_{\neg\text{NEG}}}([x]_C, [y]_C)$ for non-negative region attribute reducts as follows:

$$M_{D_{\neg\text{NEG}}}([x]_C, [y]_C) = \begin{cases} \{a \in C | I_a(x) \neq I_a(y)\}, \\ \quad \text{if } D_{\neg\text{NEG}(\alpha,\beta)}([x]_C) \neq D_{\neg\text{NEG}(\alpha,\beta)}([y]_C), \\ C, \quad \text{otherwise.} \end{cases} \quad (11)$$

Table 3. A decision table

U	x_1	x_2	x_3	x_4	x_5	x_6	x_7	x_8	x_9	x_{10}
c_1	1	1	1	1	0	0	0	1	1	1
c_2	1	1	1	1	0	0	0	0	0	0
d	1	1	1	2	2	2	1	1	1	2

Table 4. The discernibility matrix $M_{\text{POS}(0.6,\beta)}$ of Table 3

	C_1	C_2	C_3
C_1	C		
C_2	$\{c_1\}$	C	
C_3	C	$\{c_1, c_2\}$	C

Theorem 3. *Given a decision table DT, $A \subseteq C$ is an (α, β)-non-negative region consistent set of C iff for any $x, y \in U$, if $D_{\neg\mathrm{NEG}(\alpha,\beta)}([x]_C) \neq D_{\neg\mathrm{NEG}(\alpha,\beta)}([y]_C)$, then $[x]_A \cap [y]_A = \emptyset$.*

Proof. It is similar to the proof of theorem 2.

The discernibility function of the matrix is given by:

$$f(M_{D_{\neg\mathrm{NEG}}}) = \wedge \vee M_{D_{\neg\mathrm{NEG}}}([x]_C, [y]_C). \tag{12}$$

A prime implicant of the reduced disjunctive form of the discernibility function is an (α, β)-non-negative region attribute reduct.

5   Conclusion

In this paper, we analyze the decision region changes when deleting attributes. It is found that positive rules are unchanged when equivalence classes in positive region with the same decision set are merged. It is also true for the equivalence classes in non-negative region with the same decision set. The notion of vector based attribute reduct is proposed. One can use the standard method to construct a reduct based on a discernibility matrix.

Acknowledgements. This work was supported in part by the Project of Department of Education of Guangdong Province (no. 2014KTSCX152) and a Discovery Grant from NSERC, Canada.

References

1. Grzymala-Busse, J.W., Clarka, P.G., Kuehnhausena, M.: Generalized probabilistic approximations of incomplete data. Int. J. Approximate Reasoning **50**, 180–196 (2014)
2. Jia, X.Y., Liao, W.H., Tang, Z.M., Shang, L.: Minimum cost attribute reduction in decision-theoretic rough set models. Inf. Sci. **219**, 151–167 (2013)
3. Jia, X.Y., Shang, L., Zhou, B., Yao, Y.Y.: Generalized attribute reduct in rough set theory. Knowledge-Based Systems, 2015. http://dx.doi.org/10.1016/j.knosys.2015.05.017
4. Jia, X.Y., Tang, Z.M., Liao, W.H., Shang, L.: On an optimization representation of decision-theoretic rough set model. Int. J. Approximate Reasoning **55**, 156–166 (2014)
5. Li, H.X., Zhou, X.Z., Zhao, J.B., Liu, D.: Non-monotonic attribute reduction in decision-theoretic rough sets. Fundamenta Informaticae **126**, 415–432 (2013)
6. Liao, S.J., Zhu, Q.X., Min, F.: Cost-sensitive attribute reduction in decision-theoretic rough set models. Math. Prob. Eng. **2014**, 1–9 (2014)
7. Ma, X.A., Wang, G.Y., Yu, H., Li, T.R.: Decision region distribution preservation reduction in decision-theoretic rough set model. Inf. Sci. **278**, 614–640 (2014)
8. Mi, J.S., Wu, W.Z., Zhang, W.X.: Approaches to knowledge reduction based on variable precision rough set model. Inf. sci. **159**, 255–272 (2004)
9. Pawlak, Z.: Rough sets. Int. J. Comput. Inf. Sci. **11**, 341–356 (1982)

10. Pawlak, Z.: Rough Sets: Theoretical Aspects of Reasoning about Data. Kluwer Academic Publishers, Dordrecht (1991)
11. Skowron, A., Rauszer, C.: The discernibility matrices and functions in information systems. In: Slowinski, R. (ed.) Intelligent Decision Support, Handbook of Applications and Advances of the Rough Sets Theory, pp. 331–362. Kluwer, Dordrecht (1992)
12. Skowron, A.: Boolean reasoning for decision rules generation. In: Komorowski, J., Raś, Z.W. (eds.) Methodologies for Intelligent Systems. LNCS, vol. 689, pp. 295–305. Springer, Heidelberg (1993)
13. Yao, Y.Y.: Three-way decisions with probabilistic rough sets. Inf. Sci. **180**, 341–353 (2010)
14. Yao, Y.Y.: The two sides of the theory of rough sets. Knowl.-based Syst. **80**, 67–77 (2015)
15. Yao, Y.Y., Wong, S.K.M.: A decision theoretic framework for approximating concepts. Int. J. Man-Mach. Stud. **37**, 793–809 (1992)
16. Yao, Y.Y., Wong, S.K.M., Lingras, P.: A decision-theorectic rough set model. In: Ras, Z.W., Zemankova, M., Emrich, M.L. (eds.) Methodologies for Intelligent systems, vol. 5, pp. 17–24. Norh-Holland, New York (1990)
17. Yao, Y.Y., Zhao, Y.: Attribute reduction in decision-theoretic rough set models. Inf. Sci. **178**, 3356–3373 (2008)
18. Zhao, Y., Wong, S.K.M., Yao, Y.: A Note on Attribute Reduction in the Decision-Theoretic Rough Set Model. In: Peters, J.F., Skowron, A., Chan, C.-C., Grzymala-Busse, J.W., Ziarko, W.P. (eds.) Transactions on Rough Sets XIII. LNCS, vol. 6499, pp. 260–275. Springer, Heidelberg (2011)

How to Evaluate Three-Way Decisions Based Binary Classification?

Xiuyi Jia[1]($\boxtimes$) and Lin Shang[2]

[1] School of Computer Science and Engineering,
Nanjing University of Science and Technology, Nanjing, China
jiaxy@njust.edu.cn
[2] State Key Laboratory for Novel Software Technology,
Nanjing University, Nanjing, China
shanglin@nju.edu.cn

Abstract. Appropriate measures are important for evaluating the performance of a classifier. In existing studies, many performance measures designed for two-way decisions based classification are applied to three-way decisions based classification directly, which may result in an incomprehensive evaluation. However, there is a lack of systematically research on the performance measures for three-way decisions based classification. This paper introduces some numerical measures and graphical measures for three-way decisions based binary classification.

Keywords: Three-way decisions · Binary classification · Performance measure

1 Introduction

Three-way decisions theory proposed by Yao [14] is an extension of classical two-way decisions theory, in which a deferment decision is adopted when current information is limited or insufficient to make acceptance decision or rejection decision. Since three-way decisions is similar to human decision strategy in real world applications, it has drawn wide attention in recent years [6,10,12].

For three-way decisions based classification, current studies can be categorized as follows. One group concentrates on the theoretical analysis or model construction of three-way decisions based classification, such as analyzing the superiority of three-way decisions model through theoretical analysis [15], defining different three-way decisions models through different decision criteria [11], and dealing with the deferment objects [8,9]. The second group concentrates on the application of three-way decisions, such as email spam filtering [4,18], text classification [7], cluster analysis [17], and web-based support systems [16].

Many researchers reported that three-way decisions based classifiers can obtain a better result rather than two-way decisions based classifiers on several criteria [3–5,7,8,18]. By reviewing these studies, we find that most comparing

© Springer International Publishing Switzerland 2015
Y. Yao et al. (Eds.): RSFDGrC 2015, LNAI 9437, pp. 366–375, 2015.
DOI: 10.1007/978-3-319-25783-9_33

Table 1. The cost function matrix in DTRS.

	X: Actual positive	$\neg X$: Actual negative
a_P: Predicted positive	λ_{PP}	λ_{PN}
a_B: Predicted boundary	λ_{BP}	λ_{BN}
a_N: Predicted negative	λ_{NP}	λ_{NN}

criteria are precision, F-measure and misclassification cost. All these performance measures are defined directly based on two-way decisions based classification. However, since expressions of three-way decisions based classification result and two-way decisions based classification result are different, it is necessary to consider defining specific performance measures for three-way decisions based classification. To the best of our knowledge, there is a lack of systematically investigation on performance measures for three-way decisions based classification.

In this paper, we focus on investigating different performance measures for three-way decisions based binary classification. Besides the common numerical measures are defined, two graphical measures including ROC curve and AUC are also redefined in a three-dimensional space. The remainder of this paper is organized as follows: Sect. 2 gives an introduction of three-way decisions based classification and some basic performance measures. Section 3 introduces some performance measures for two-way decisions based binary classification. Section 4 defines the numerical measures and graphical measures for three-way decisions based classification. Section 5 concludes this paper.

2 Three-Way Decisions Based Classification

In this section, we give the basic notions of three-way decisions based classification followed by some basic performance measures for three-way decisions based binary classification.

As three-way decisions can be applied in many models or frameworks, we use decision-theoretic rough set (DTRS), a typical three-way decisions model, to briefly explain what three-way decisions theory is.

In DTRS, for binary classification, 2 states and 3 actions are considered. $\Omega = \{X, \neg X\}$ is the set of states indicating that an object x is in class X and not in X. $\mathcal{A} = \{a_P, a_B, a_N\}$ is the set of actions representing 3 different decisions. a_P is the acceptance decision: deciding $x \in \text{POS}(X)$, which means classifying x into X. a_B is the deferment decision: deciding $x \in \text{BND}(X)$, which means x needs further examination. a_N is the rejection decision: deciding $x \in \text{NEG}(X)$, which means classify x into $\neg X$. The cost function matrix regarding the cost of actions in different states is shown in Table 1.

Based on Bayesian decision theory, a pair of thresholds (α, β) can be derived from given cost functions. Let $p(X|x)$ be the conditional probability of x being in X, then the three-way decisions can be represented as follows:

Table 2. The confusion matrix for three-way decisions based classification.

	Actual positive	Actual negative
Predicted positive	n_{PP}	n_{PN}
Predicted boundary	n_{BP}	n_{BN}
Predicted negative	n_{NP}	n_{NN}

(P) If $p(X|x) > \alpha$, deciding $x \in \text{POS}(X)$;
(B) If $\beta \leq p(X|x) \leq \alpha$, deciding $x \in \text{BND}(X)$;
(N) If $p(X|x) < \beta$, deciding $x \in \text{NEG}(X)$.

The reader is referred to Ref [14] for more detailed explanation.

The confusion matrix of three-way decisions based classification is shown in Table 2.

In Ref [15], Yao compared probabilistic three-way decisions model with two-way decisions model and Pawlak rough set model on 16 basic performance measures. All these measures are summarized in Table 3. Based on the confusion matrix, first 3 measures represented the proportion of each row to all objects, second 6 measures represented the relationship between each element and its corresponding row, and third 6 measures represented the relationship between each element and its corresponding column. The last measure combined the cost function matrix and the confusion matrix to represent the total classification cost.

3   Performance Measures for Two-Way Decisions Based Binary Classification

In classical two-way decisions based binary classification, most performance measures are derived from a confusion matrix [1,13]. In the confusion matrix shown in Table 4, n_{PP} denotes the number of correctly classified positive objects, n_{PN} denotes the number of predicted positive objects which are actual negative, n_{NP} denotes the number of predicted negative objects which are actual positive, and n_{NN} denotes the number of correctly classified negative objects, respectively.

Based on the confusion matrix, basic performance measures for two-way decisions based binary classification are summarized in Table 5.

Positive precision, F-score, Accuracy are common adopted measures for two-way decisions based classification. Besides the basic performance measures, some other measures were also discussed in existing studies, such as *Positive likelihood, Negative likelihood, Youden's index, Discriminant power,* and so on [2]. All these combined measures can be computed from above basic measures directly.

In order to show the efficiency of classifiers intuitively, some graphical measures were also defined, including *ROC curve* and *AUC*. We will introduce them in Sect. 4.2.

Table 3. 16 basic measures for three-way decisions based binary classification defined by Yao [15]. $|U| = n_{PP} + n_{PN} + n_{BP} + n_{BN} + n_{NP} + n_{NN}$.

Measure	Formula	Description
three regions based measures:		
acceptance rate	$\frac{n_{PP}+n_{PN}}{|U|}$	Proportion of accepted objects which are identified by the classifier
deferment rate	$\frac{n_{BP}+n_{BN}}{|U|}$	Proportion of deferred objects which are identified by the classifier
rejection rate	$\frac{n_{NP}+n_{NN}}{|U|}$	Proportion of rejected objects which are identified by the classifier
row-wise normalization based measures:		
acceptance accuracy	$\frac{n_{PP}}{n_{PP}+n_{PN}}$	Proportion of actual positives which are predicted positive
acceptance error	$\frac{n_{PN}}{n_{PP}+n_{PN}}$	Proportion of actual negatives which are predicted positive
deferred positive error	$\frac{n_{BP}}{n_{BP}+n_{BN}}$	Proportion of actual positives which are deferred
deferred negative error	$\frac{n_{BN}}{n_{BP}+n_{BN}}$	Proportion of actual negatives which are deferred
rejection error	$\frac{n_{NP}}{n_{NP}+n_{NN}}$	Proportion of actual positives which are predicted negative
rejection accuracy	$\frac{n_{NN}}{n_{NP}+n_{NN}}$	Proportion of actual negatives which are predicted negative
column-wise normalization based measures:		
correct acceptance rate	$\frac{n_{PP}}{n_{PP}+n_{BP}+n_{NP}}$	Proportion of predicted positives which are actual positive
deferred positive rate	$\frac{n_{BP}}{n_{PP}+n_{BP}+n_{NP}}$	Proportion of deferred positives which are actual positive
incorrect rejection rate	$\frac{n_{NP}}{n_{PP}+n_{BP}+n_{NP}}$	Proportion of predicted negatives which are actual positive
incorrect acceptance rate	$\frac{n_{PN}}{n_{PN}+n_{BN}+n_{NN}}$	Proportion of predicted positives which are actual negative
deferred negative rate	$\frac{n_{BN}}{n_{PN}+n_{BN}+n_{NN}}$	Proportion of deferred negatives which are actual negative
correct rejection rate	$\frac{n_{NN}}{n_{PN}+n_{BN}+n_{NN}}$	Proportion of predicted negatives which are actual negative
cost related measure:		
total cost	$\frac{\lambda_{PP}\cdot n_{PP}+\lambda_{PN}\cdot n_{PN}+\lambda_{BP}\cdot n_{BP}+\lambda_{BN}\cdot n_{BN}+\lambda_{NP}\cdot n_{NP}+\lambda_{NN}\cdot n_{NN}}{|U|}$	

Table 4. The confusion matrix for two-way decisions based binary classification.

	Actual positive	Actual negative
Predicted positive	n_{PP}	n_{PN}
Predicted negative	n_{NP}	n_{NN}

Table 5. Basic measures for two-way decisions based binary classification.

Measure	Formula	Description
Positive related measures:		
Positive precision	$\frac{n_{PP}}{n_{PP}+n_{PN}}$	Proportion of actual positives which are predicted positive
Positive recall(Sensitivity)	$\frac{n_{PP}}{n_{PP}+n_{NP}}$	Proportion of predicted positives which are actual positive
Positive F-Score	$\frac{(\beta^2+1)\cdot n_{PP}}{(\beta^2+1)\cdot n_{PP}+\beta^2\cdot n_{NP}+n_{PN}}$	Weighted average of positive precision and positive recall
Negative related measures:		
Negative precision	$\frac{n_{NN}}{n_{NP}+n_{NN}}$	Proportion of actual negatives which are predicted negative
Negative recall(Specificity)	$\frac{n_{NN}}{n_{PN}+n_{NN}}$	Proportion of predicted negatives which are actual negative
Negative F-Score	$\frac{(\beta^2+1)\cdot n_{NN}}{(\beta^2+1)\cdot n_{NN}+\beta^2\cdot n_{PN}+n_{NP}}$	Weighted average of negative precision and negative recall
Overall measures:		
Accuracy	$\frac{n_{PP}+n_{NN}}{n_{PP}+n_{NN}+n_{PN}+n_{NP}}$	Overall effectiveness of a classifier
Error rate	$\frac{n_{PN}+n_{NP}}{n_{PP}+n_{NN}+n_{PN}+n_{NN}}$	Overall classification error

4   Performance Measures for Three-Way Decisions Based Classification

Although Ref [15] has introduced 16 measures, the purpose of that paper is to compare three models on these measures. For first 15 measures, in order to do theoretical analysis or derivation on them directly, each of them represented a local property of the confusion matrix only. Some overall or complex measures were not considered, such as accuracy, F-score and so on. Therefore, in order to evaluate a three-way decisions based classifier comprehensively, we give a systematically introduction and analysis of performance measures in the follows.

According to their representations, the measures can be classified into two groups. One group contains numerical measures, which produce a single number summarizing the classifier's performance. The other group contains graphical measures, which depict performance in a graph with three dimensions.

4.1 Numerical Measures

The numerical measures for three-way decisions based binary classification are summarized in Table 6. The first 3 measures describe the performance of the classifier on positive objects. The second 3 measures describe the performance of the classifier on negative objects. The last 6 measures describe the overall performance of the classifier on all objects. In these measures, F-score is a weighted mean of precision (accuracy) and recall (coverage), which represents a kind of tradeoff between them. A high F-score ensures a high precision and a high recall.

4.2 Graphical Measures

Since graphical measures can be applied to visualize the classifier's performance, they have drawn many attention in the machine learning and data mining communities. Typical graphical measures to evaluate two-way decisions based classifiers are ROC(Receiver operating characteristics) curve and AUC(Area Under ROC Curve). In this section, based on existing definitions of graphical measures for two-way decisions based classifiers, we define new ROC curve and AUC for three-way decisions based classifiers.

ROC. Typical ROC for two-way decisions based classifiers is a two-dimensional graph in which *true positive rate* is plotted on the Y axis and *false positive rate* is plotted on the X axis. The true positive rate and false positive rate based on Table 4 are defined as following:

$$true\ positive\ rate = \frac{n_{PP}}{n_{PP} + n_{NP}}, \tag{1}$$

$$false\ positive\ rate = \frac{n_{PN}}{n_{PN} + n_{NN}}. \tag{2}$$

In the ROC space shown in Fig. 1(a), the performance of a classifier is represented by a point. A classifier C_1 will be better than another classifier C_2 if C_1's point is on the northwest of C_2's, that means higher true positive rate and lower false positive rate.

Since two-way decisions based classifiers do not produce boundary region, the two-dimensional ROC graph can depict the tradeoff between true positives and false positives. However, for a three-way decisions based classifier, both the true positives and false positives can be simultaneously affected by the deferment

Table 6. Numerical measures for three-way decisions based binary classification.

Measure	Formula	Description
Positive related measures:		
Positive precision	$Pp = \frac{n_{PP}}{n_{PP}+n_{PN}}$	Proportion of actual positives which are predicted positive
Positive recall	$Pr = \frac{n_{PP}}{n_{PP}+n_{BP}+n_{NP}}$	Proportion of predicted positives which are actual positive
Positive F-Score	$Pf = \frac{(\beta^2+1)\cdot n_{PP}}{(\beta^2+1)\cdot n_{PP}+\beta^2\cdot(n_{BP}+n_{NP})+n_{PN}}$	Weighted average of positive precision and positive recall
Negative related measures:		
Negative precision	$Np = \frac{n_{NN}}{n_{NP}+n_{NN}}$	Proportion of actual negatives which are predicted negative
Negative recall	$Nr = \frac{n_{NN}}{n_{PN}+n_{BN}+n_{NN}}$	Proportion of predicted negatives which are actual negative
Negative F-Score	$Nf = \frac{(\beta^2+1)\cdot n_{NN}}{(\beta^2+1)\cdot n_{NN}+\beta^2\cdot(n_{BN}+n_{PN})+n_{NP}}$	Weighted average of negative precision and negative recall
Overall measures:		
Overall accuracy	$Oa = \frac{n_{PP}+n_{NN}}{n_{PP}+n_{NN}+n_{PN}+n_{NP}}$	Proportion of accepted objects identified by the classifier
Overall coverage	$Oc = \frac{n_{PP}+n_{NN}+n_{PN}+n_{NP}}{|U|}$	Proportion of objects being decided immediately
Overall deferment rate	$Od = \frac{n_{BP}+n_{BN}}{|U|}$	Proportion of deferred objects identified by the classifier
Overall error rate	$Oe = \frac{n_{PN}+n_{NP}}{n_{PP}+n_{NN}+n_{PN}+n_{NP}}$	Proportion of objects which are classified incorrectly
Overall F-score	$Of = \frac{(\beta^2+1)\cdot Oa\cdot Oc}{\beta^2\cdot Oa+Oc}$	Weighted average of overall accuracy and overall coverage
Total cost	$Tc = \lambda_{PP}\cdot n_{PP}+\lambda_{PN}\cdot n_{PN}+\lambda_{BP}\cdot n_{BP}$ $+\lambda_{BN}\cdot n_{BN}+\lambda_{NP}\cdot n_{NP}+\lambda_{NN}\cdot n_{NN}$	Total cost of classification

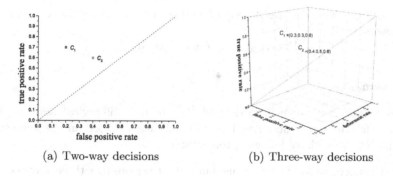

(a) Two-way decisions   (b) Three-way decisions

Fig. 1. ROC spaces for two-way decisions and three-way decisions based classifiers.

decision, then the tradeoff among three aspects should be considered in the ROC graph.

Figure 1(b) illustrates a new three-dimensional ROC graph for three-way decisions based classifiers, in which *true positive rate* is plotted on the Z axis, *false positive rate* is plotted on the X axis and *overall deferment rate* is plotted on the Y axis. The overall deferment rate Od has been defined in Table 6.

In the three-dimensional ROC graph, a good classifier will produce a point with a high Z value, a low X value and a low Y value.

AUC. AUC is a single value which represents the area under ROC curve. In Fig. 2(a), classifier C_1 has a greater area than C_2, which means C_1 has a better average performance. Similar to the AUC graph for two-way decisions based classifiers, we introduce a new AUC graph for three-way decisions based classifiers shown in Fig. 2(b).

In the three-dimensional AUC graph, it is not easy to compute the area under ROC curve directly. Therefore, we propose a new computation method. First, project three-dimensional points into XZ plane and YZ plane, respectively. Second, compute the AUC value in each two-dimensional space. Last, combine

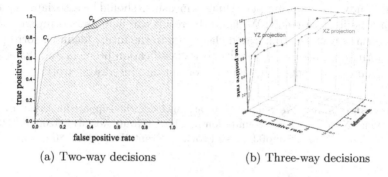

(a) Two-way decisions   (b) Three-way decisions

Fig. 2. AUC for two-way decisions and three-way decisions based classifiers.

the two AUC values in a specific way. A simple combination method can be used as following:

$$AUC_{XYZ} = AUC_{XZ} + \mu \cdot AUC_{YZ}. \tag{3}$$

4.3 Remarks

In above subsections, we have defined the common numerical and graphical measures for three-way decisions based classification. Based on the analysis on these measures, we have following remarks:

* Since the universe will be partitioned into three regions by a three-way decisions based classifier, in order to evaluate the classifier comprehensively, we have to consider all aspects of the three regions to construct appropriate measures. Therefore, the positive related measures, the negative related measures and the overall deferment rate are considered in numerical measures.
* For the calculation of three-way decisions based AUC, more other complex and reasonable weighting methods can be considered.
* Due to the space limit, some other measures including likelihood measures, Youden's index and Lift chart, and comparison experiments are not introduced in this paper.

5 Conclusions

Choosing appropriate performance measures plays an important role in classification tasks. A good measure can evaluate the classifier correctly and comprehensively. In this paper, we focus on the problem of how to evaluate a three-way decisions based classifier in binary classification tasks. By reviewing the basic measures for two-way decisions based classification, we introduce the corresponding numerical measures and graphical measures for three-way decisions based classification. Accuracy, coverage, deferment rate and other basic numerical measures are redefined according to the three-way decisions based confusion matrix. Three-dimensional ROC graph and AUC are also introduced to visualize the three-way decisions based result.

This study suggests that performance measures should be associated with the form of classification result. We hope the result can help users comprehensively evaluate three-way decisions based classifiers. In the future, defining corresponding performance measures for multi-class classification problem and doing some comparison experiments will be an important and interesting work.

Acknowledgements. We would like to acknowledge the support for this work from the National Natural Science Foundation of China (Grant Nos. 61403200, 61170180), Natural Science Foundation of Jiangsu Province(Grant No.BK20140800).

References

1. Fawcett, T.: An introduction to ROC analysis. Pattern Recogn. Lett. **27**, 861–874 (2006)
2. Gu, Q., Zhu, L., Cai, Z.H.: Evaluation measures of the classification performance of imbalanced data sets. In: Proceedings of the 4th International Symposium on Computational Intelligence and Intelligent Systems, pp. 461–471 (2009)
3. Jia, X.Y., Liao, W.H., Tang, Z.M., Shang, L.: Minimum cost attribute reduction in decision-theoretic rough set models. Inf. Sci. **219**, 151–167 (2013)
4. Jia, X., Shang, L.: Three-Way Decisions Versus Two-Way Decisions on Filtering Spam Email. In: Peters, J.F., Skowron, A., Li, T., Yang, Y., Yao, J.T., Nguyen, H.S. (eds.) Transactions on Rough Sets XVIII. LNCS, vol. 8449, pp. 69–91. Springer, Heidelberg (2014)
5. Jia, X.Y., Shang, L., Zhou, B., Yao, Y.Y.: Generalized attribute reduct in rough set theory. Knowl.-Based Syst. **80**, 67–77 (2015). doi:10.1016/j.knosys.2015.05.017
6. Jia, X.Y., Tang, Z.M., Liao, W.H., Shang, L.: On an optimization representation of decision-theoretic rough set model. Int. J. Approximate Reasoning **55**(1), 156–166 (2014)
7. Li, W., Miao, D.Q., Wang, W.L., Zhang, N.: Hierarchical rough decision theoretic framework for text classification. In: Proceedings of the 9th IEEE International Conference on Cognitive Informatics, pp. 484–489 (2010)
8. Li, W., Huang, Z., Jia, X.: Two-phase classification based on three-way decisions. In: Lingras, P., Wolski, M., Cornelis, C., Mitra, S., Wasilewski, P. (eds.) RSKT 2013. LNCS, vol. 8171, pp. 338–345. Springer, Heidelberg (2013)
9. Li, H., Zhou, X., Huang, B., Liu, D.: Cost-sensitive three-way decision: a sequential strategy. In: Lingras, P., Wolski, M., Cornelis, C., Mitra, S., Wasilewski, P. (eds.) RSKT 2013. LNCS, vol. 8171, pp. 325–337. Springer, Heidelberg (2013)
10. Li, H.X., Zhou, X.Z., Zhao, J.B., Liu, D.: Non-monotonic attribute reduction in decision-theoretic rough sets. Fundamenta Informaticae **126**(4), 415–432 (2013)
11. Liu, D., Liang, D.: An overview of function based three-way decisions. In: Miao, D., Pedrycz, W., Slezak, D., Peters, G., Hu, Q., Wang, R. (eds.) RSKT 2014. LNCS, vol. 8818, pp. 812–823. Springer, Heidelberg (2014)
12. Liu, D., Liang, D.C., Wang, C.C.: A novel three-way decision model based on incomplete information system. Knowl.-Based Syst. (2015). doi:10.1016/j.knosys. 2015.07.036
13. Sokolova, M., Lapalme, G.: A systematic analysis of performance measures for classification tasks. Inf. Process. Manag. **45**, 427–437 (2009)
14. Yao, Y.Y.: Three-way decisions with probabilistic rough sets. Inf. Sci. **180**(3), 341–353 (2010)
15. Yao, Y.Y.: The superiority of three-way decisions in probabilistic rough set models. Inf. Sci. **181**(6), 1080–1096 (2011)
16. Yao, J.T., Azam, N.: Web-based medical decision support systems for three-way medical decision making with game-theoretic rough sets. IEEE Trans. Fuzzy Syst. **23**(1), 3–15 (2015)
17. Yu, H., Zhang, C., Hu, F.: An incremental clustering approach based on three-way decisions. In: Cornelis, C., Kryszkiewicz, M., Ślęzak, D., Ruiz, E.M., Bello, R., Shang, L. (eds.) RSCTC 2014. LNCS, vol. 8536, pp. 152–159. Springer, Heidelberg (2014)
18. Zhou, B., Yao, Y.Y., Luo, J.G.: Cost-sensitive three-way email spam filtering. J. Intell. Inf. Syst. **42**(1), 19–45 (2014)

A Moderate Attribute Reduction Approach in Decision-Theoretic Rough Set

Hengrong Ju[1,2], Xibei Yang[2], Pei Yang[1,3,4], Huaxiong Li[1,4(✉)], and Xianzhong Zhou[1,4]

[1] School of Management and Engineering, Nanjing University, Nanjing 210093, Jiangsu, People's Republic of China
[2] School of Computer Science and Engineering, Jiangsu University of Science and Technology, Zhenjiang 212003, Jiangsu, People's Republic of China
[3] State Key Laboratory for Novel Software Technology, Nanjing University, Nanjing 210093, People's Republic of China
[4] Research Center for Novel Technology of Intelligent Equipments, Nanjing University, Nanjing 210093, People's Republic of China
huaxiongli@nju.edu.cn

Abstract. Attribute reduction is an important topic in Decision-Theoretic Rough Set theory. To overcome the limitations of lower-approximation-monotonicity based reduct and cost minimum based reduct, a moderate attribute reduction approach is proposed in this paper, which combines the lower approximation monotonicity criterion and cost minor criterion. Furthermore, the proposed attribute reduct is searched by solving an optimization problem, and a fusion fitness function is proposed in a generic algorithm, such that the reduct is computed in a low time complexity. Experimental analysis is included to validate the theoretic analysis and quantify the effectiveness of the proposed attribute reduction algorithm. This study indicates that **the optimality is not the best and sub-optimum may be the best choice.**

Keywords: Attribute reduction · Decision cost · DTRS · Lower-approximation-monotonicity

1 Introduction

As a representative of decision-cost-sensitive rough set model, Decision-Theoretic Rough Set theory (DTRS) [1,2] attracted more and more attention since it was firstly proposed by Yao in early 1990s. Following Yao's pioneer works, many theoretical and applied results related to decision-theoretic rough set have been obtained, For example, Jia et al. proposed an optimization representation of DTRS model [3]; Li et al. developed a cost-sensitive sequential three-way decision model [4]. More generalizations of DTRS model, see Refs. [5–14] for more details.

Attribute reduction is one of the most fundamental and important topics in rough set theory and has received much attention from many researchers [15,16].

© Springer International Publishing Switzerland 2015
Y. Yao et al. (Eds.): RSFDGrC 2015, LNAI 9437, pp. 376–388, 2015.
DOI: 10.1007/978-3-319-25783-9_34

As far as attribute reduction in DTRS, an important topic on attribute reduction has been discussed, i.e., the monotonicity property of the decision regions with respect to the set inclusion of attributes does not hold in DTRS [17,18]. In order to deal with such difficulty of non-monotonic probabilistic regions with respect to the monotonic variation of attributes, two main types of attribute reduction approaches have been considered in DTRS: one is the decision-based reduct [19,20] while the other is the cost-based reduct [19,21]. Presently, unremitting efforts by a lot of the researchers lead to great successes around these two mechanisms, respectively.

– Decision-based reduct: considering the decision in attribute reduction, Yao et al. [19] studied different definitions of the attribute reductions in DTRS and proposed decision-monotocity criterion and generality criterion based reducts; Li et al. [20] investigated the non-monotonic attribute reduction issue in DTRS.
– Cost-based reduct: As far as cost related reduct is considered, Yao and Zhao [19] proposed cost criterion based reduct; Jia et al. [21] studied the minimum decision cost attribute reduction in DTRS with three different algorithms.

In previous work, great achievements of attribute reductions in DTRS have been obtained by considering decision rule and decision cost, respectively. Nevertheless, few researchers pay attention to the attribute reduction with decision rule and decision cost simultaneously. The purpose of this paper is to further investigate the attribute reduction in DTRS when we consider the decision rule and decision cost at the same time. In this paper, we propose a moderate attribute reduction approach in DTRS, which is called Lower Approximation Monotonicity Fusion with Cost Minor Attribute Reduction (LAMFCMAR). Furthermore, we regard it as an optimization problem and apply the generic algorithm to compute the reduct.

2 Preliminary Knowledge of Yao's DTRS

Formally, a decision system [22–24] can be considered as a pair $S =< U, AT \cup D >$, in which universe U is a finite set of the objects; AT is a nonempty set of the conditional attributes, D is the set of decisional attribute, such that $\forall a \in AT \cup D$, V_a is the domain of a. $\forall x \in U$, $a(x)$ denotes the value of x on a. $\forall A \subseteq AT$, an indiscernibility relation $IND(A)$ can be defined as $IND(A) = \{(x,y) \in U^2 : \forall a \in A, a(x) = a(y)\}$, $[x]_A = \{y \in U : (x,y) \in IND(A))\}$ is the equivalence class of x in terms of $IND(A)$.

The Bayesian decision procedure deals with making decision with minimum risk based on observed evidence. Yao et al. introduced a more general rough set model called a Decision-Theoretic Rough Set (DTRS) model [25–27]. In this part, we briefly introduce original DTRS model. According to the Bayesian decision procedure, the DTRS model is composed of two states and three actions. The set of states is given by $\Omega = \{X, \sim X\}$ indicating that an object is in X or not, respectively. The probabilities for these two complement states can be denoted as $P(X|[x]_A) = \frac{|X \cap [x]_A|}{|[x]_A|}$ and $P(\sim X|[x]_A) = 1 - P(X|[x]_A)$. The set of actions is given by $\mathscr{A} = \{a_P, a_B, a_N\}$, where a_P, a_B and a_N represent the three actions in classifying an object x, namely, deciding x belongs to the positive region, deciding x belongs to the boundary region and deciding x belongs to the negative region, respectively. The loss functions regarding the risk or cost of actions in different states. Let λ_{PP}, λ_{BP} and λ_{NP} denote the cost incurred for taking actions a_P, a_B, a_N, respectively, when an object belongs to X, and λ_{PN}, λ_{BN} and λ_{NN} denote the cost incurred for taking the same actions when an object belongs to $\sim X$. The loss functions regarding the risks or costs of actions in different states is given by a 3×2 matrix [25, 26] in Table 1.

Table 1. The loss function in DTRS.

	X	$\sim X$
a_P	λ_{PP}	λ_{PN}
a_B	λ_{BP}	λ_{BN}
a_N	λ_{NP}	λ_{NN}

According to the loss functions, the expected costs associated with taking different actions for objects in $[x]_A$ can be expressed as:

$$R_P = R(a_P|[x]_A) = \lambda_{PP} \cdot P(X|[x]_A) + \lambda_{PN} \cdot P(\sim X|[x]_A);$$
$$R_B = R(a_B|[x]_A) = \lambda_{BP} \cdot P(X|[x]_A) + \lambda_{BN} \cdot P(\sim X|[x]_A); \qquad (1)$$
$$R_N = R(a_N|[x]_A) = \lambda_{NP} \cdot P(X|[x]_A) + \lambda_{NN} \cdot P(\sim X|[x]_A).$$

The Bayesian decision procedure leads to the following minimum-risk decision rules:

(P) If $R_P \leq R_B$ and $R_P \leq R_N$, then decides x belongs to the positive region;

(B) If $R_B \leq R_P$ and $R_B \leq R_N$, then decides x belongs to the boundary region;

(N) If $R_N \leq R_P$ and $R_N \leq R_B$, then decides x belongs to the negative region.

Considering a special kind of loss functions with $\lambda_{PP} \leq \lambda_{BP} < \lambda_{NP}$ and $\lambda_{NN} \leq \lambda_{BN} < \lambda_{PN}$. That is to say, the loss of classifying an object x belonging to X into the positive region is no more than the loss of classifying x into the boundary region, and both of these losses are strictly less than the loss of classifying x into the negative region. The reverse order of losses is used for classifying an object not in X. We further assume that the loss functions satisfy the condition:

$$(\lambda_{PN} - \lambda_{BN}) \cdot (\lambda_{NP} - \lambda_{BP}) > (\lambda_{BP} - \lambda_{PP}) \cdot (\lambda_{BN} - \lambda_{NN}). \qquad (2)$$

Based on the above two assumptions, we have the simplified rules:

(P1)   If $P(X|[x]_A) \geq \alpha$, then decides x belongs to the positive region;
(B1)   If $\beta < P(X|[x]_A) < \alpha$, then decides x belongs to the boundary region;
(N1)   If $P(X|[x]_A) \leq \beta$, then decides x belongs to the negative region;
where

$$\alpha = \frac{\lambda_{PN} - \lambda_{BN}}{(\lambda_{PN} - \lambda_{BN}) + (\lambda_{BP} - \lambda_{PP})};$$

$$\beta = \frac{\lambda_{BN} - \lambda_{NN}}{(\lambda_{BN} - \lambda_{NN}) + (\lambda_{NP} - \lambda_{BP})};$$

with $0 \leq \beta < \alpha \leq 1$.

Using these three decision rules, $\forall A \subseteq AT$ and $\forall X \subseteq U$, we get the probabilistic approximations:

$$\underline{A}_{(\alpha,\beta)}(X) = \{x \in U : P(X|[x]_A) \geq \alpha\}; \tag{3}$$

$$\overline{A}_{(\alpha,\beta)}(X) = \{x \in U : P(X|[x]_A) > \beta\}. \tag{4}$$

The pair $[\underline{A}_{(\alpha,\beta)}(X), \overline{A}_{(\alpha,\beta)}(X)]$ is referred to as the lower and upper approximations of X with respect to the set of attributes A in Yao's DTRS. Based on $\underline{A}_{(\alpha,\beta)}(X)$ and $\overline{A}_{(\alpha,\beta)}(X)$, the positive region, boundary region and negative region in DTRS can be expressed as follows:

$$POS_{(\alpha,\beta)}(X) = \underline{A}_{(\alpha,\beta)}(X); \tag{5}$$

$$BND_{(\alpha,\beta)}(X) = \overline{A}_{(\alpha,\beta)}(X) - \underline{A}_{(\alpha,\beta)}(X); \tag{6}$$

$$NEG_{(\alpha,\beta)}(X) = U - \overline{A}_{(\alpha,\beta)}(X). \tag{7}$$

By the definitions of decision rules and decision region, three important costs should be considered:

$-$ (P) cost : $COST_{POS} = \sum\limits_{x \in POS(X)} \left(\lambda_{PP} \cdot P(X|[x]_A) + \lambda_{PN} \cdot P(\sim X|[x]_A)\right);$

$-$ (N) cost : $COST_{NEG} = \sum\limits_{x \in NEG(X)} \left(\lambda_{NP} \cdot P(X|[x]_A) + \lambda_{NN} \cdot P(\sim X|[x]_A)\right);$

$-$ (B) cost : $COST_{BND} = \sum\limits_{x \in BND(X)} \left(\lambda_{BP} \cdot P(X|[x]_A) + \lambda_{BN} \cdot P(\sim X|[x]_A)\right).$

Considering a special case where we assume zero cost for a correct classification, i.e., $\lambda_{PP} = \lambda_{NN} = 0$, then the decision costs of decision rules can be simply expressed as:

$-$ (P) cost : $COST_{POS} = \sum\limits_{x \in POS(X)} \lambda_{PN} \cdot P(\sim X|[x]_A);$

$-$ (N) cost : $COST_{NEG} = \sum\limits_{x \in NEG(X)} \lambda_{NP} \cdot P(X|[x]_A);$

$-$ (B) cost : $COST_{BND} = \sum\limits_{x \in BND(X)} \left(\lambda_{BP} \cdot P(X|[x]_A) + \lambda_{BN} \cdot P(\sim X|[x]_A)\right).$

Therefore, $\forall A \subseteq AT$, the decision cost of A can be expressed as:

$$\textbf{COST}_A = COST_{POS} + COST_{NEG} + COST_{BND}.$$

3   Attribute Reduction in DTRS

Considering two subsets of attributes $A, B \subseteq AT$ such that $A \subseteq B$. For any $x \in U$, we can obtain $[x]_B \subseteq [x]_A$. In Pawlak rough set model, if $[x]_A$ belongs to the positive region, then its subset $[x]_B$ also is in the positive region. However, in Yao's DTRS, for a subset $[x]_B$ of a equivalence class $[x]_A$, no matter to which region $[x]_A$ belongs, we do not know to which region $[x]_B$ belongs. Therefore, we cannot obtain the monotonicity of the probabilistic regions with respect to set inclusion of attributes. The probabilistic positive and non-negative regions are monotonically increasing with respect to the decreasing of α and β. From this point of view, the positive region preservation based reduction may be not appropriate in DTRS, since we may obtain a subset $B \subseteq AT$, satisfying $\underline{B}_{(\alpha,\beta)}(X) \supseteq \underline{AT}_{(\alpha,\beta)}(X)$. With respect to different requirements, two popular types of the attribute reductions have been considered in DTRS, respectively, i.e., lower approximation monotonicity based reduction and cost minimum based reduction.

3.1   Lower Approximation Monotonicity Based Attribute Reduction (LAMAR)

In many practical applications, the main objective is to classify an object into a class other than exclude an object from a class. Classification of object is corresponding to the positive decision rule in DTRS while exclusion of object is corresponding to the negative decision rule in DTRS. In addition, in three-way decision, the boundary decision rule represents a delay or uncertainty decision [27]. Therefore, we argue that the lower-approximation-monotonicity reduct is a very important reduct in DTRS because it requires to keep or increase the number of positive decision rules in a decision system.

Definition 1. *Let* $S = <U, AT \cup D>$ *be a decision system,* $\forall A \subseteq AT$, $U/IND(D) = \{X_1, X_2, \ldots, X_n\}$ *is the set of the decision classes induced by the decisional attribute* D. *A is referred to as a lower-approximation-monotonicity reduct if and only if A is the minimal set of the conditional attributes, which preserves* $\underline{A}_{(\alpha,\beta)}(X_j) \supseteq \underline{AT}_{(\alpha,\beta)}(X_j)$ *for each* $X_j \in U/IND(D)$.

The LAMAR requires two things. Firstly, it requires that by reducing attributes, a positive decision rule is still a positive decision rule of the same decision. Secondly, the criterion requires that by reducing attributes a boundary decision rule is still a boundary decision rule, or is upgraded to a positive decision rule with the same decision.

However, it is noticed that Definition 1 is only the qualitative definition, in order to deal with this issue, it is suitable to regard the attribute reduction problem as the attribute optimization problem. Let S be a decision system, $\forall A \subseteq AT$, based on the lower-approximation-monotonicity criterion, the fitness function can be defined as follows:

$$f_A^L = \frac{\sum\limits_{X_j \in U/IND(D)} \left(\underline{A}_{(\alpha,\beta)}(X_j) \odot \underline{AT}_{(\alpha,\beta)}(X_j) \right)}{|U|};$$ (8)

where $|\cdot|$ denotes the cardinality of a set, and

$$P \odot Q = \begin{cases} |P - Q| & P \supseteq Q \\ -\infty & otherwise. \end{cases}$$

In above definition, it is not difficult to observe that $f_A^L \in (\infty, 1)$, $f_A^L > 0$ indicates that the subset A obtains more positive decision rules than original attribute set AT; $f_A^L = 0$ indicates that the positive decision rules induced by subset A are same to the original attribute set; $f_A^L < 0$ indicates the subset A obtains less positive decision rules than original attribute set. According to the definition of LAMAR, the main objective of this criterion is to increasing the value of f_A^L.

3.2 Cost Minimum Based Attribute Reduction(CMAR)

Cost is one of the important features of DTRS model. In the attribute reduction process, from the viewpoint of cost minimum criterion, the decision maker always want to obtain a reduct with lower or lowest cost [21].

Definition 2. *Let S be a decision system, $\forall A \subseteq AT$, A is referred to as a cost minimum reduct in S if and only if A is the minimal set of conditional attributes, which satisfies* $\mathbf{COST}_A \leq \mathbf{COST}_{AT}$ *and for each set* $B \subset A$, $\mathbf{COST}_B > \mathbf{COST}_A$.

In such definition, we want to find a subset of conditional attributes so that the overall decision cost will be decreased or unchanged based on the reduct. In most situations, it is better for decide to obtain a smaller or smallest cost in the decision procedure. The minimum cost can be denoted as [21]:

$$\min \mathbf{COST}_A.$$

Before we compute the cost minimum reduct, we can introduce the cost fitness function as follows:

$$f_A^C = \frac{\mathbf{COST}_A}{\mathbf{COST}_{AT}}.$$ (9)

Based on the definition of cost fitness function, we can observe that $f_A^C \in (0, 1]$, the main objective of cost minimum criterion is to decreasing the value of f_A^C.

3.3   Lower-Approximation-Monotonicity Fusion with Cost Minor Attribute Reduction(LAMFCMAR)

As we discussed above, the LAMAR and CMAR are two important mechanisms in DTRS's attribute reduction. However, in the present research, these
two approaches are mutual independent and both have several limits. In the
paper "Want more? Pay more!" [28], Yang et al. have explored the relationships
between approximation increasing and the variation of decision cost. Through
experimental analysis, they have drawn an important conclusion such that the
increasing of decision-theoretic rough approximation leads to the increasing
of decision cost. That is to say, on the one hand, the lower–approximation-
monotonicity criterion will result in a high decision cost of the reduct; on the
other hand, a cost minimum criterion ignores the effect of positive decision rule
and leads to few positive decision rules of the reduct. In order to deal with
this issue, we can define a fusion fitness function for pooling together the lower-
approximation-monotonicity criterion and cost minimum criterion, that is:

$$f_A^{LC} = g\left(f_A^L, f_A^C\right). \tag{10}$$

Many measures or strategies for the fusion functions have been studied, one
of the popular strategies is the weighted arithmetic average. It can be written
as follows:

$$f_A^{LC} = w_L \cdot f_A^L - w_C \cdot f_A^C. \tag{11}$$

where the weight $w_L > 0$, $w_C > 0$.

The setting of these two weights reflects the different statuses of these two
criteria. In different applications, the emphases are different. In this paper, for
simply, we suppose that these two weights satisfy the following relationships:
$w_L + w_C = 1$. We also suppose that the lower–approximation-monotonicity criterion and cost minimum criterion have equal statuses and set the weights to
$w_L = 0.5$, $w_C = 0.5$. Form this point of view, if we want to increasing the lower
approximation set and decreasing the decision cost, increasing the value of f_A^{LC}
can be a appropriate choice.

Definition 3. *Let S be a decision system, $\forall A \subseteq AT$, A is referred to as a
LAMFCM reduct in S if and only if A is the minimal set of conditional attributes,
which satisfies $f_A^{LC} \geq f_{AT}^{LC}$ and for each set $B \subset A$, $f_B^{LC} < f_A^{LC}$.*

In light of above discussions, the definition of lower-approximation-
monotonicity fusion with cost minor attribute reduction can be obtained.

3.4   Genetic Algorithm for Attrition Reduction

In many rough set literatures, two approaches have been widely adopted to
find an optimal reducts: exhaustive algorithm and optimization algorithm. An
exhaustive algorithm is to construct the set of all reducts by discernibility matrix.

Unfortunately, this algorithm is NP-hard. This is why optimization algorithm has attracted much attention. In what follows, we will present a genetic algorithm to compute these three types of reducts. In genetic algorithm, a chromosome is represented as a binary string of length $\#(AT)$ which is the number of condition attributes, where "1" means that the corresponding attribute is included in the subset, and "0" denotes that the corresponding attribute is not included. The selection method in our genetic approach is the roulette wheel selection. In the roulette wheel selection, individuals in the population are assigned probabilities of being selected that is directly proportionate to their fitness. For the parameters setting, in the crossover procedure, the mixing ratio is set as 0.9 and the mutation operation is applied with the probability 0.3. The details of genetic algorithm are described as follows.

Algorithm. Genetic approach for attribute reduction based on different criteria.

Input: A decision decision system S;
Output: An optimal reduct *red*.
1: Create an initial random population (number=70);
2: Evaluation the population;
3: While Number of generations < 100 **do**
　　　　Computer the fitness function f_A^L, f_A^C, f_A^{LC};
　　　　Select the fittest chromosomes in the population;
　　　　Perform crossover on the selected chromosomes to create offspring;
　　　　Perform mutation on the selected chromosomes;
　　　　Evaluate the new population;
　　End while
4: Selected fittest chromosome from current population and output it as *red*.

4　Experimental Analysis

In this section, by experimental analysis, we will illustrate the differences among these three attribute reductions, i.e., LAMAR, CMAR and LAMFCMAR. All the experiments have been carried out on a personal computer with Windows 7, Intel Core 2 DuoT5800 CPU (4.00 GHz) and 4.00 GB memory. The programming language is Matlab 2012b.

We download six public data sets from UCI Repository of Machine Learning Databases, which are described in Table 2. In this paper, for each data set, the loss functions are randomly generated. Their values are in [10,100] with following constraint conditions: $\lambda_{BP} < \lambda_{NP}$, $\lambda_{BN} < \lambda_{PN}$, $\lambda_{PN} < \lambda_{NP}$, $\lambda_{PP} = \lambda_{NN} = 0$. Note that for each data set, 10 different groups of loss functions are randomly generated.

Table 3 shows the experimental results of (P) rules and (B) rules. The number of these rules is equivalent to the number of objects in positive region and boundary region, respectively. This is mainly because each object in positive/boundary region can induce a (P)/(B) decision rule. In Table 3, the raw represents the decision rule numbers of original data set, LAMAR is short for attribute reduction

Table 2. Data sets description.

ID	Data sets	Samples	Features	Decision classes
1	Annealing	798	38	5
2	Dermatology	366	34	6
3	Ionosphere	351	34	2
4	Spect Heart	267	22	2
5	Wdbc	569	30	4
6	Zoo	101	17	7

based on lower-approximation-monotonicity criterion; similar explanations can be provided for CMAR and LAMFCMAR. Note that in our experiments, for each data set, 10 different groups of loss functions are randomly generated, Therefore, we recorded the average results. From Table 3, it is not difficult to draw the following conclusions:

1. From the viewpoint of (P) rules, the reducts based on LAMAR and LAMFC-MAR can generate more (P) decision rules than original data set and CMAR based reducts. This is mainly because the condition of LAMAR and LAM-FCMAR based reducts require that, a positive decision rule is still a positive decision rule, or a boundary decision rule is upgraded to a positive decision rule by reducing attributes. This mechanism not only keeps the original (P) decision rules unchanged, but also increases the numbers of (P) decision rules. From the perspective of (P) decision rule number, the reducts based on LAMAR generate the most numbers of (P) decision rules. In most cases, by comparing the original data sets, the reduct based on CMAR can induce the same numbers of (P) decision rules. The (P) decision rule numbers of LAMFCMAR based reducts are between LAMAR based reducts and CMAR based reducts.

2. From the viewpoint of (B) rules, the reducts based on LAMAR and LAMFC-MAR can generate less (B) decision rules than original data set and CMAR based reducts. This is mainly because the condition of LAMAR and LAMFC-MAR based reducts require that by reducing attributes, a boundary decision rule is still a boundary decision rule, or is upgraded to a positive decision rule, which results in the decreasing of (B) decision rules. From the perspective of (B) decision rule number, the reducts based on LAMAR generate the lest numbers of (B) decision rules. In most cases, by comparing the original data sets, the reduct based on CMAR can induce the same numbers of (B) decision rules, and the (B) decision rule numbers of LAMFCMAR based reducts are between LAMAR based reducts and CMAR based reducts.

Table 4 shows the experimental results of decision cost and attribute length comparisons between raw data and three criteria based reducts. Similar to Table 3, we also recorded the average results in Table 4. Based on a careful investigation of Table 4, it is not difficult to draw the following two conclusions:

Table 3. The decision rules comparison between raw data and reducts.

ID	(P) rules				(B) rules			
	Raw	LAMAR	CMAR	LAMFCMAR	Raw	LAMAR	CMAR	LAMFCMAR
1	578.2	737.9	578.2	726.4	359.4	115.4	359.1	184.8
2	336.1	353.9	336.1	353.5	50.4	34.7	50.4	32.5
3	297.8	303.6	297.8	303.3	82.8	74.1	82.8	61.8
4	228.9	238.1	228.9	235.7	66	50.4	66	54.3
5	534.6	566.9	534.6	566.5	43.9	3.4	43.9	3.8
6	90.7	91.8	90.7	91.5	16.1	14.2	16.1	14.2

1. From the viewpoint of the decision cost, by comparing the original data sets, the costs of reducts based on these three criteria have been decreased. Considering the differences of these three criteria based reducts, we have noticed that, the LAMAR based reducts obtain the most decision costs, which are can be validated by the results in [28]. The CMAR based reducts obtain the lowest costs and the decision costs of LAMFCMAR based reducts are between LAMAR based reducts and CMAR based reducts. This is mainly because that the goal of LAMFCMAR is to obtain a reduct with minor decision costs.
2. Considering the attribute length, we can observe that the LAMAR based reducts require the least amount of attributes. The CMAR based reducts require the most attributes and the attribute amount of LAMFCMAR based reducts are between LAMAR based reducts and CMAR based reducts.

Table 4. The cost and length comparisons between raw data and reducts.

ID	Decision cost				Attribute length			
	Raw	LAMAR	CMAR	LAMFCMAR	Raw	LAMAR	CMAR	LAMFCMAR
1	676.88	243.34	128.46	209.24	38	16.3	32.3	19.4
2	298.63	40.08	24.3	33.64	34	21.5	27.2	23.3
3	290.48	57.56	45.37	48.28	34	20.9	26.5	24
4	222.03	39.59	20.48	23.42	22	14.1	19	16.9
5	435.81	87.23	51.48	66.32	30	11.7	23.4	17.8
6	78.92	12.32	5.92	8.96	17	10.9	11.4	11.2

To sum up, we can obtain the following conclusions:

1. The LAMAR based reducts can generate the most amount of (P) decision rules and least amount of (B) decision rules on the premise of the least amount of attributes. However, it results in the increase of the decision costs.

2. The CMAR based reducts can obtain the minimum decision costs on the premise of the same numbers of (P) and (B) decision rules with original data sets. Unfortunately, it requires the most amount of conditional attributes.
3. The LAMFCMAR based reducts perform better in all comparisons. i.e., the amount of decision rules, decision costs and attributes are all between the amount of those of reducts based on LAMAR and CMAR. Even through it is not the best, it is a sub-optimal performance.

5   Conclusion

Decision rule and decision cost are two important aspects of DTRS. In this paper, we proposed a moderate attribute reduction, which is the fusion consideration of decision rule and decision cost. Through experimental analysis, we can observe that the LAMFCMAR based reducts perform better in all comparisons. Although the proposed approach is not the best, it is a sub-optimal choice for decision maker. The present study is the first step to fusion attribute reduction in decision-theoretic rough set. The following are challenges for further research.

1. Fusion attribute reduction approach to complicated data type, such as interval-valued data and incomplete data may be an interesting topic.
2. The weight learning of the fitness function in this paper is also a serious challenge.

Acknowledgment. This work is supported by the Natural Science Foundation of China (Nos. 61100116, 71201076, 61170105, 61473157,71171107), Qing Lan Project of Jiangsu Province of China, and the Ph.D. Programs Foundation of Ministry of Education of China (20120091120004).

References

1. Yao, Y.Y., Wong, S.K.M., Lingras, P.: A decision-theoretic rough set model. In: Ras, Z.W., Zemankova, M., Emrich, M.L. (eds.) Methodologies for Intelligent Systems, vol. 5, pp. 17–24. North-Holland, New York (1990)
2. Yao, Y.Y., Wong, S.K.M.: A decision theoretic framework for approximating concepts. Int. J. Man Mach. Stud. **37**, 793–809 (1992)
3. Jia, X.Y., Tang, Z.M., Liao, W.H., et al.: On an optimization representation of decision-theoretic rough set model. Int. J. Approx. Reason. **55**, 156–166 (2014)
4. Li, H., Zhou, X., Zhao, J., Huang, B.: Cost-sensitive classification based on decision-theoretic rough set model. In: Li, T., Nguyen, H.S., Wang, G., Grzymala-Busse, J., Janicki, R., Hassanien, A.E., Yu, H. (eds.) RSKT 2012. LNCS, vol. 7414, pp. 379–388. Springer, Heidelberg (2012)
5. Li, H., Zhou, X., Huang, B., Liu, D.: Cost-sensitive three-way decision: a sequential strategy. In: Lingras, P., Wolski, M., Cornelis, C., Mitra, S., Wasilewski, P. (eds.) RSKT 2013. LNCS, vol. 8171, pp. 325–337. Springer, Heidelberg (2013)
6. Liang, D.C., Liu, D., Pedrycz, W., Hu, P.: Triangular fuzzy decision-theoretic rough sets. Int. J. Approx. Reason. **54**, 1087–1106 (2013)

7. Liang, D.C., Liu, D.: Systematic studies on three-way decisions with interval-valued decision-theoretic rough sets. Inform. Sci. **276**, 186–203 (2014)

8. Liu, D., Li, T.R., Li, H.X.: A multiple-category classification approach with decision-theoretic rough sets. Fundam. Inform. **115**, 173–188 (2012)

9. Liu, D., Li, T.R., Liang, D.C.: Incorporating logistic regression to decision-theoretic rough sets for classification. Int. J. Approx. Reason. **55**(1), 197–210 (2014)

10. Yu, H., Liu, Z.G., Wang, G.Y.: An automatic method to determine the number of clusters using decision-theoretic rough set. Int. J. Approx. Reason. **55**, 101–115 (2014)

11. Qian, Y.H., Zhan, G.H., Sang, Y.L., et al.: Multigranulation decision-theoretic rough sets. Int. J. Approx. Reason. **55**, 225–237 (2013)

12. Li, W.T., Xu, W.H.: Double-quantitative decision-theoretic rough set. Inform. Sci. **316**, 54–67 (2015)

13. Li, W.T., Xu, W.H.: Multigranulation decision-theoretic rough set in ordered information system. Fundam. Inform. **139**, 67–89 (2015)

14. Li, W., Xu, W.: Probabilistic rough set model based on dominance relation. In: Miao, D., Pedrycz, W., Slezak, D., Peters, G., Hu, Q., Wang, R. (eds.) RSKT 2014. LNCS, vol. 8818, pp. 856–864. Springer, Heidelberg (2014)

15. Ju, H.R., Yang, X.B., Song, X.N., et al.: Dynamic updating multigranulation fuzzy rough set: approximations and reducts. Int. J. Mach. Learn. Cyber. **5**(6), 981–990 (2014)

16. Ju, H.R., Yang, X.B., Dou, H.L., et al.: Variable precision multigranulation rough set and attributes reduction. Trans. Rough Set **8**, 52–68 (2014)

17. Zhao, Y., Wong, S.K.M., Yao, Y.: A note on attribute reduction in the decision-theoretic rough set model. In: Peters, J.F., Skowron, A., Chan, C.-C., Grzymala-Busse, J.W., Ziarko, W.P. (eds.) Transactions on Rough Sets XIII. LNCS, vol. 6499, pp. 260–275. Springer, Heidelberg (2011)

18. Ma, X.A., Wang, G.Y., Yu, H., et al.: Decision region distribution preservation reduction in decision-theoretic rough set model. Inform. Sci. **278**, 614–640 (2014)

19. Yao, Y.Y., Zhao, Y.: Attribute reduction in decision-theoretic rough set models. Inform. Sci. **178**, 3356–3373 (2008)

20. Li, H.X., Zhou, X.Z., Zhao, J.B., et al.: Non-monotonic attribute reduction in decision-theoretic rough sets. Fundam. Inform. **126**(4), 415–432 (2013)

21. Jia, X.Y., Liao, W.H., Tang, Z.M., et al.: Minimum cost attribute reduction in decision-theoretic rough set models. Inform. Sci. **219**, 151–167 (2013)

22. Yang, X.B., Song, X.N., Chen, Z.H., et al.: On multigranulation rough sets in incomplete information system. Int. J. Mach. Learn. Cyb. **3**, 223–232 (2012)

23. Yang, X.B., Qi, Y.S., Song, X.N., et al.: Test cost sensitive multigranulation rough set: model and minimal cost selection. Inform. Sci. **250**, 184–199 (2013)

24. Yang, X.B., Song, X.N., She, Y.H., et al.: Hierarchy on multigranulation structures: a knowledge distance approach. Int. J. Gen. Syst. **42**(7), 754–773 (2013)

25. Yao, Y.Y.: Probabilistic rough set approximations. Int. J. Approx. Reason. **49**, 255–271 (2008)

26. Yao, Y.: Three-way decision: an interpretation of rules in rough set theory. In: Wen, P., Li, Y., Polkowski, L., Yao, Y., Tsumoto, S., Wang, G. (eds.) RSKT 2009. LNCS, vol. 5589, pp. 642–649. Springer, Heidelberg (2009)

27. Yao, Y., Zhou, B.: Naive Bayesian rough sets. In: Yu, J., Greco, S., Lingras, P., Wang, G., Skowron, A. (eds.) RSKT 2010. LNCS, vol. 6401, pp. 719–726. Springer, Heidelberg (2010)
28. Yang, X., Qi, Y., Yu, H., Yang, J.: Want more? Pay more!. In: Cornelis, C., Kryszkiewicz, M., Ślęzak, D., Ruiz, E.M., Bello, R., Shang, L. (eds.) RSCTC 2014. LNCS, vol. 8536, pp. 144–151. Springer, Heidelberg (2014)

Decisions Tree Learning Method Based on Three-Way Decisions

Yangyang Liu[1]($\boxtimes$), Jiucheng Xu[1,2,3], Lin Sun[1,2,3], and Lina Du[1]

[1] College of Computer and Information Engineering, Henan Normal University,
Xinxiang 453007, Henan, China
15903863829@163.com
[2] Engineering Technology Research Center for Computing Intelligence
and Data Mining, Xinxiang 453007, Henan, China
[3] Engineering Laboratory of Intellectual Business and Internet
of Things Technologies, Xinxiang 453007, Henan, China

Abstract. Aiming at the problems that traditional data mining methods ignore inconsistent data, and general decision tree learning algorithms lack of theoretical support for the classification of inconsistent nodes. The three-way decision is introduced to decision tree learning algorithms,and the decision tree learning method based on three-way decisions is proposed. Firstly, the proportion of positive objects in node is used to compute the conditional probability of the three-way decision of node. Secondly, the nodes in decision tree arepartitioned to generate the three-way decision tree. The merger and pruning rules of the three-way decision tree are derived to convert the three-way decision tree into two-way decision tree by considering the information around nodes. Finally, an exampleisimplemented. The results show that the proposed method reserves inconsistent information, partitions inconsistent nodes by minimizing the overall risk, not only generates decision tree with cost-sensitivity, but also makes the partition of inconsistent nodes more explicable. Besides, the proposed method reduces the overfitting to some extent and the computation problem of conditional probability of three-way decisions is resolved.

Keywords: Three-way decisions · Decision tree · Conditional probability · Boundary nodes · Minimizing the overall risk · Merger and pruning

1 Introduction

A lot of data, in the complex systems of reality fields, is often inconsistent. Usually, the decision table containing inconsistent objects is called inconsistent decision table. Inadequate information is one of the causes of inconsistent data [1]. As the computer technology and network technology being popular in various application fields, the data generated in applications is increasingly complicated in content and form, inconsistent data is more and more common.The theories, methods and technologies, which can directly deal with and analyze inconsistent

© Springer International Publishing Switzerland 2015
Y. Yao et al. (Eds.): RSFDGrC 2015, LNAI 9437, pp. 389–400, 2015.
DOI: 10.1007/978-3-319-25783-9_35

data, need to be developed [2]. Traditional data mining methods usually regard inconsistent data as dirty data, and clean up inconsistent data in the data pre-processing stage. This way can reduce the interference of inconsistent data, but the useful information in inconsistent data will be lost.

Decision tree [3–6] is a kind of typical induction learning algorithm, is an important research field in machine learning and data mining, and is used widely and successfully in practical applications, such as ID3 algorithm, CART algorithm, and C4.5 algorithm. The early decision tree classification algorithms are intended to improve the accuracy of classification, and minimize misclassification. Decision tree considers the more frequent class as the class of inconsistent nodes containing some objects that have identical values for each attribute but belong to different classes. This classification method lacks of reasonable theoretical bases,it not only can not differentiate the nodes in which the numbers of objects belonging to different classes are equals, but also will cause that the classifier internally biases towards the more frequent class and ignores the infrequent class that has less influence on accuracy and significant influence on classification results, namely ignores the cost brought by misclassification [7]. Therefore, the early decision tree learning algorithms are not cost-sensitive, and easily cause misclassification of inconsistent leaf nodes, which will affect the classification effect of decision tree.

The three-way decision is one of key ideas in decision-theoretic rough set (DTRS), and extends two-way decision semantics including positive region and negative region to three-way decision semantics including positive region, negative region and boundary region. In three-way decision, boundary region is considered as a feasible decision, which is consistent with the method that human intelligence deals with decision problems. Different with classical Pawlak rough set model, the three-way decision, which is a data analysis tool with cost sensitivity, tends to focus on risk cost resulting from the fault of decision classification, rather than maintain the consistency of decision knowledge and experience data. Besides, the three-way decision considers that how much misclassification rate can be allowed from the perspective of minimizing the overall risk, the risk of middle decision is lower than that of misclassification when the data information is insufficient [8]. Therefore, the three-way decision can help decision tree to deal with inconsistent nodes, and makes the partition of inconsistent nodes more explicable. Although the three-way decision has accomplished lots of achievements in some fields, it still faces some challenges including the calculations of conditional probability and threshold parameters [9]. For the calculation of conditional probability, Yao and Zhou [10] presented a naive Bayesian rough set model estimating the conditional probability based on the Bayesian theorem and the naive probabilistic independence assumption. Liu et al. [9] employed logistic regression to compute conditional probability in three-way decision. So the calculation methods of conditional probability are not limited to equivalence class method in rough set. For computing the thresholds α and β in DTRS, Jia et al. [11] derived the proper cost functions and thresholds without any prior knowledge by constructing an optimum problem based on the minimization of the decision cost, and proposed an adaptive learning parameters algorithm to solve the optimum problem [12,13].

In this paper, we focus on developing algorithms to address the problem that general decision tree learning algorithms lack of theoretical support for the classification of inconsistent nodes. The three-way decision is introduced into decision tree algorithm to partition nodes. Firstly, the proportion of positive objects in node is used to compute the conditional probability of the three-way decision of node. Secondly, an adaptive learning cost functions algorithm (*Alcofa*) proposed by Jia et al. [11] is used to search the optimal thresholds. On the basis of the optimal thresholds, the nodes in decision tree can be partitioned and the three-way decision tree with cost-sensitive can be generated. Then the merger and pruning rules of the three-way decision tree are derived. According to the merger and pruning rules, the three-way decision tree can be pruned to become two-way decision tree by considering the information around nodes. Finally, an example is implemented to prove the validity and rationality of the proposed method.

The remainder of this paper is organized as follows. Section 2 reviews the main ideas of decision tree and three-way decision. The three-way decision tree is constructed in Sect. 3. In Sect. 4, the merger and pruning rules of three-way decision tree are derived. The example is given in Sect. 5. The paper ends with conclusions and future research work in Sect. 6.

2 Preliminaries

Basic concepts, notations and results of the ID3 decision tree learning algorithm and the three-way decision are briefly reviewed in this section.

2.1 ID3 Decision Tree Learning Algorithm

In this subsection, we present ID3 decision tree learning algorithm [5,14]. The ID3 decision tree learning algorithm is described as Algorithm 1.

Algorithm 1. ID3 decision tree learning algorithm

Input: Training set $D = \{x_1, \ldots, x_n\}$, attribute set $A = \{A_1, \ldots, A_m\}$, threshold ε;

Output: Decision tree T.

Step 1: If all objects in D belong to same class C_k, then T is a single node tree. Let C_k be the class of this node, return T;

Step 2: If $A = \phi$, then T is a single node tree, Let the more frequent class C_k in D be the class of this node, return T;

Step 3: Otherwise, compute the information gain of each attribute of A with respect to D. Select the attribute A_g corresponding to maximal information gain;

Step 4: If the information gain of A_g is less than threshold ε, then let T be a single node tree, and let the more frequent class C_k in D be the class of this node, return T;

Step 5: Otherwise, aiming at each value a_i of A_g, divide D into some non-empty subsets D_i according to $A_g = a_i$. Let the more frequent class in D_i be the class label to construct children nodes, and the tree T consists of nodes and children nodes of this nodes;

Step 6: For ith children node, let D_i be training set, $A - \{A_g\}$ be attribute set. The tree T can be constructed by recursively using step 1 $\sim$ step 5, return T.

2.2　The Three-Way Decision

In this subsection, we present the three-way decision model [9,15–19].

For the Bayesian decision procedure and the thoughts of three-way decision, the DTRS model is composed of 2 states and 3 actions. The set of states is given by $\Omega = \{C, \neg C\}$ indicating that an object is in C and not in C, respectively. And the set of actions is given by $\mathcal{A} = \{a_P, a_B, a_N\}$ where a_P, a_B, and a_N represent the three actions in classifying an object x, namely, deciding $x \in POS(C)$, deciding x should be further investigated $x \in BND(C)$, and deciding $x \in NEG(C)$, respectively. Let λ_{PP}, λ_{BP} and λ_{NP} denote the losses incurred for taking actions of a_P, a_B and a_N , respectively, when an object belongs to C. Similarly, λ_{PN}, λ_{BN} and λ_{NN} denote the losses incurred for taking the same actions when the object belongs to $\neg C$. $P(C||[x])$ is the conditional probability of an object x belonging to C given that the object is described by its equivalence class $[x]$. For an object x, the expected loss $R(a_i||[x])$ associated with taking the individual actions can be expressed as:

$$
\begin{aligned}
R(a_P||[x]) &= \lambda_{PP}P(C||[x]) + \lambda_{PN}P(\neg C||[x]), \\
R(a_B||[x]) &= \lambda_{BP}P(C||[x]) + \lambda_{BN}P(\neg C||[x]), \\
R(a_N||[x]) &= \lambda_{NP}P(C||[x]) + \lambda_{NN}P(\neg C||[x]).
\end{aligned}
\tag{1}
$$

The Bayesian decision procedure suggests the following minimum-cost decision rules:

(P) If $R(a_P||[x]) \leq R(a_B||[x])$ and $R(a_P||[x]) \leq R(a_N||[x])$, decide $x \in POS(C)$;

(B) If $R(a_B||[x]) \leq R(a_P||[x])$ and $R(a_B||[x]) \leq R(a_N||[x])$, decide $x \in BND(C)$;

(N) If $R(a_N||[x]) \leq R(a_P||[x])$ and $R(a_N||[x]) \leq R(a_B||[x])$, decide $x \in NEG(C)$.

By considering a reasonable kind of loss functions with $\lambda_{PP} \leq \lambda_{BP} < \lambda_{NP}$ and $\lambda_{NN} \leq \lambda_{BN} < \lambda_{PN}$. Let

$$
\begin{aligned}
\alpha &= \frac{(\lambda_{PN} - \lambda_{BN})}{(\lambda_{PN} - \lambda_{BN}) + (\lambda_{BP} - \lambda_{PP})}, \\
\beta &= \frac{(\lambda_{BN} - \lambda_{NN})}{(\lambda_{BN} - \lambda_{NN}) + (\lambda_{NP} - \lambda_{BP})}, \\
\gamma &= \frac{(\lambda_{PN} - \lambda_{NN})}{(\lambda_{PN} - \lambda_{NN}) + (\lambda_{NP} - \lambda_{PP})}.
\end{aligned}
\tag{2}
$$

Based on $P(C||x]) + P(\neg C||x]) = 1$, if $\frac{(\lambda_{BP} - \lambda_{PP})}{(\lambda_{PN} - \lambda_{BN})} < \frac{(\lambda_{NP} - \lambda_{BP})}{(\lambda_{BN} - \lambda_{NN})}$, it implies $0 \le \beta < \gamma < \alpha \le 1$. In this case, the following simplified rules are obtained:

(P1) If $P(C||x]) \ge \alpha$, decide $x \in POS(C)$;
(B1) If $\beta < P(C||x]) < \alpha$, decide $x \in BND(C)$;
(N1) If $P(C||x]) \le \beta$, decide $x \in NEG(C)$.

In a lot of learning process, as the lack of preliminary knowledge, the accurate loss function values can not be obtained. Jia et al. [11] proposed an adaptive learning cost functions algorithm ($Alcofa$) to search the optimal thresholds.

3 The Construction of Three-Way Decision Tree

Suppose D is training data set, $D = \{(x_1, y_1), (x_2, y_2), \ldots, (x_n, y_n)\}$, assume that there are only two classes denoted $C = \{X, \neg X\}$, $x_i \in R^N$, $y_i \in C$, $i = 1, \ldots, n$. X indicates positive class, $\neg X$ indicates negative class. $A = \{a_1, a_2, \ldots, a_N\}$ is attribute set. Train data set D based on ID3 algorithm to obtain decision tree $T(E, B)$, E indicates the set of internal nodes (IE) and leaf nodes (LE), internal node denotes the value of attribute, leaf node denotes class of each path, B denotes all branch. Rank all nodes in T from top to down and from left to right, and let $E = \{E_1, E_2, \ldots, E_m\}$, $E_i \in A$. Assume $n(E_i)$ is the number of objects in node E_i, $n_p(E_i)$ is the number of objects belonging to X in node E_i, $n_n(E_i)$ is the number of objects belonging to $\neg X$ in node E_i.

Definition 1. *Let $D = \{(x_1, y_1), (x_2, y_2), \ldots, (x_n, y_n)\}$ be a training data set, $T(E, B)$ is decision tree with respect to D learned by ID3 algorithm. $\forall E_i \in A$, if $n_p(E_i) = n(E_i)$, $n_n(E_i) = 0$, or $n_p(E_i) = 0$, $n_n(E_i) = n(E_i)$, then define the E_i as consistent node; if $n_p(E_i) > 0$, $n_n(E_i) > 0$, $n(E_i) = n_p(E_i) + n_n(E_i)$, then define the E_i as inconsistent node.*

ID3 decision tree learning algorithm shows that $\forall E_i \in LE$ in T, the objects in E_i may belong to different class, namely E_i may be inconsistent node. The cause of inconsistent nodes is as follows:

(1) From Step2 and Step5 of ID3 algorithm, if there are inconsistent data in D, then $\forall E_i \in LE$, E_i may be inconsistent node;

(2) From Step4 of ID3 algorithm, $\forall E_i \in LE$, the threshold ε may cause E_i is an inconsistent node.

ID3 decision tree learning algorithm states that the class of inconsistent node is the more frequent class in this node. As we know, this method not only can not distinguish the nodes in which the numbers of objects belonging to different class are equals, but also will cause that the classifier internally biases towards the frequent class and ignores the infrequent class that has less influence on accuracy and significant influence on classification results, namely ignores the cost brought by misclassification. For this problem, the three-way decision is introduced to ID3 decision tree learning algorithm, which can decide how much misclassification rate can be allowed according to minimizing the overall risk.

3.1   The Construction of Conditional Probability in the Three-Way Decision

Definition 2. *Let $D = \{(x_1, y_1), (x_2, y_2), \ldots, (x_n, y_n)\}$ be a training data set, $T(E, B)$ is decision tree with respect to D learned by ID3 algorithm. Considering the three-way decision model of nodes in decision tree having two class X and $\neg X$, X denotes the positive class, $\neg X$ denotes the negative class, define the conditional probability in the three-way decision of $\forall E_i \in E$ is the proportion of positive objects in E_i, namely*

$$p(X|E_i) = \frac{n_p(E_i)}{n(E_i)} \tag{3}$$

From Definition 2, we can obtain that for $\forall E_i \in E$, the greater the conditional probability in the three-way decision of E_i is, the greater the probability of E_i belonging to X is.

3.2   The Construction Algorithm of the Three-Way Decision Tree

The conditional probability in three-way decision of nodes in decision tree can be computed based on Sect. 3.1, the optimal threshold (α, β) can be obtained by *Alcofa* algorithm proposed by Jia et al. [11]. The three-way decision partitions the nodes in decision tree according to the following rules:

(1) If $P(X|E_i) \geq \alpha$, then $E_i \in POS(X)$, namely E_i is positive node;
(2) If $\beta < P(X|E_i) < \alpha$, then $E_i \in BND(X)$, namely E_i is boundary node;
(3) If $P(X|E_i) \leq \beta$, then $E_i \in NEG(X)$, namely E_i is negative node.

Assume $T(E, B)$ is decision tree with respect to D learned by ID3 algorithm. Partition E based on the three-way decision, define the decision tree after partitioning as the three-way decision tree $THT(E, B, THL)$. For $E = \{E_1, E_2, \ldots, E_m\}$, the $THL = \{THL_1, THL_2, \ldots, THL_m\}$ is the three-way class label set with respect to $E = \{E_1, E_2, \ldots, E_m\}$. Namely for $\forall E_i \in E$, $THL_i \in \{positive\ node, negative\ node, boundary\ node\}$.

The construction algorithm of three-way decision tree is described as Algorithm 2.

Algorithm 2. The construction algorithm of three-way decision tree

Input: Training set $D = \{x_1, \ldots, x_n\}$;

Output: The three-way decision tree THT.

Step 1: Train data set D to generate decision tree $T(E, B)$ according to ID3 algorithm;

Step 2: Compute the conditional probability in three-way decision of nodes in decision tree according to Definition 2;

Step 3: Search the optimal threshold (α, β) in the three-way decision according to conditional probability in three-way decision of nodes obtained by Step 2 and *Alcofa* algorithm;

Step 4: Based on Sect. 2.2, partition T according to conditional probability in three-way decision of nodes obtained by Step 2 and the optimal threshold (α, β) obtained by Step 3. Let the decision three after partitioning be THT, return THT.

4 The Merger and Pruning Rules of the Three-Way Decision Tree

In order to decrease the number of leaf nodes and increase the generalization of decision tree, the three-way decision tree need to be pruned.

Definition 3. *Let $THT(E, B, THL)$ be a three-way decision tree with respect to data set D, the set of leaf nodes $LE \subseteq E$, $\forall E_i \in LE$. Assume FE_i is the parent node of E_i, $FE_i \in A$, FE_i has n different values $\{v_1, v_2, \ldots, v_n\}$. The children node corresponding to the value v_j of FE_i is FE_{ij}. The class FE_{ij} could be positive nodes, boundary nodes, or negative nodes. Then for any two leaf nodes FE_{ip} and FE_{iq} corresponding to v_p and v_q, v_p, $v_q \in \{v_1, v_2, \ldots, v_n\}$, if their classes are same, then this two leaf nodes can be merged into one node, and the class of this new node is same with the class of nodes before merger.*

The boundary node indicates that the decisions of acceptance or rejection cannot be determine immediately on account of lacking information and recognition [20]. To eliminate the boundary nodes in three-way decision tree, and transform three-way decision tree into two-way decision tree (TWT). We will further consider the information around boundary nodes of boundary nodes, namely the information of parent nodes and brother nodes, to merge and prune three-way decision tree.

Definition 4. *Let $THT(E, B, THL)$ be a three-way decision tree with respect to data set D. After the merger operation of Definition 3, let the new decision tree is $THT'(E', B', THL')$, the new set of nodes is $E' = \{E'_1, E'_2, \ldots, E'_m\}$, m is the new number of nodes. The set of leaf nodes $LE' \subseteq E'$, $\forall E'_i \in LE'$, $THL'_i ='$ boundarynode', assume BE'_i is the set of brother nodes of E'_i, FE'_i is the parent node of E'_i. Let the leaf nodes belonging to positive class in BE'_i is PBE'_i, the leaf nodes belonging to negative class in BE'_i is NBE'_i, merge E'_i and the nodes in BE'_i into a new node, namely ME'_i. $n_p(ME'_i) = n_p(E'_i) + n_p(NBE'_i) + n_p(PBE'_i)$, $n_n(ME'_i) = n_n(E'_i) + n_n(NBE'_i) + n_n(PBE'_i)$, $n_p(ME'_i) = n_p(ME'_i) + n_n(ME'_i)$. Based on the optimal threshold (α, β) and Definition 2, the classification rules of ME_i are as follows:*

(1) If $P(X|ME_i) \geq \alpha$, then ME_i is positive node;

(2) If $\beta < P(X|ME_i) < \alpha$, then ME_i is boundary node;

(3) If $P(X|ME_i) \leq \beta$, then ME_i is negative node.

Definition 5. *Let $THT(E, B, THL)$ be a three-way decision tree with respect to data set D. After the merger operations of Definitions 3 and 4, let the new decision tree is $THT''(E'', B'', THL'')$, the new set of nodes is $E'' = \{E''_1, E''_2, \ldots, E''_t\}$, t is the new number of nodes. LE'' is the set of leaf nodes in E'', $\forall E''_i \in LE''$, FE''_i is the parent node of E''_i. The pruning rules of E''_i are as follows:*

(1) If E''_i has not brother nodes, then prune E''_i, FE''_i becomes a new leaf node;

(2) If E''_i belongs to 'boundary node', and the brother node of E''_i is not leaf node, then let E''_i belongs to the more frequent class in D.

The merger and pruning algorithm of the three-way decision tree are described as Algorithm 3.

Algorithm 3. The merger and pruning algorithm of the three-way decision tree

Input: Training set $D = \{x_1, \ldots, x_n\}$;

Output: The two-way decision tree TWT.

Step 1: Train data set D to generate decision tree $T(E, B)$ according to ID3 algorithm;

Step 2: Partition the nodes in T to generate the three-way decision tree $THT(E, B, THL)$;

Step 3: Starting from the bottom of the tree, merge all leaf nodes meeting conditions of Definition 3 from left to right;

Step 4: Starting from the bottom of the tree, search the leaf nodes in boundary region to merge these leaf nodes from left to right by the rules in Definition 4;

Step 5: Prune decision tree obtained by Step 4 according to Definition 5;

Step 6: Recursively use Step3, Step4 and Step5 until there are not leaf nodes belonging to boundary node;

Step 7: On the basis of decision tree obtained by Step 6, split the branch including more than two attribute values, to generate two-way decision tree TWT, return TWT.

The objects in data set can be partitioned into X or $\neg X$ according to the classes of leaf nodes in two-way decision tree. Because of the cost-sensitive of the three-way decision tree, the two-way decision tree is also cost-sensitive.

5   Example Verification

To verify the validity of the method proposed by this paper, the liver-disorders data set is analyzed. The 15 rows of data are randomly extracted from liver-disorders data set, and extra 5 rows of inconsistent data are added to 15 rows of data extracted from liver-disorders data set to test the effect that the method proposed by this paper deal with inconsistent data. Data discretization is implemented and denoted as D. The result is listed in Table 1.

From Table 1, we can know that the decision classes are "2" and "1". Let X denotes "2", $\neg X$ denotes "1". The decision tree T with respect to D can be generated by weka 3.6.12 according to ID3 algorithm, showed in Fig. 1.

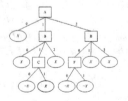

Fig. 1. The decision tree T

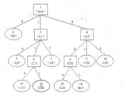

Fig. 2. The decision tree T containing classification information

Table 1. The discretization result of liver-disorders data set

Objects number	A	B	C	D	E	F	Dec
1	0	0	0	0	0	0	2
2	1	1	0	0	0	0	1
3	2	0	0	0	0	0	1
4	2	0	0	0	0	0	1
5	1	1	1	1	1	0	2
6	0	0	0	0	0	0	2
7	0	0	0	0	0	0	2
8	1	0	0	0	0	0	2
9	2	1	0	0	0	0	2
10	1	2	0	0	0	1	2
11	0	0	0	0	0	1	2
12	2	0	0	0	0	1	1
13	2	2	0	0	1	1	2
14	0	2	0	0	0	0	2
15	1	0	0	0	0	1	2
16	1	1	0	0	0	0	2
17	2	0	0	0	0	0	1
18	2	0	0	0	0	0	1
19	2	0	0	0	0	0	2
20	2	0	0	0	0	0	2

In Fig. 1, the leaf nodes are illustrated as ellipses, other nodes are illustrated as rectangles.

Starting from the root of T, rank the nodes in T from left to right, and number them. Denote the nodes in T by nodes number and the numbers of positive objects and negative objects in nodes, illustrated in Fig. 2.

The conditional probability in the three-way decision of nodes can be compute according to Fig. 2 and Definition 2, according to conditional probability in the three-way decision of nodes and $Alcofa$ algorithm, the optimal threshold (α, β) can be obtain as $(0.5, 0.3)$, the three-way decision tree can be generated by algorithm 2, illustrated in Fig. 3.

Finally, merge and prune the three-way decision tree in Fig. 3 to remove all boundary nodes, a decision tree having not boundary nodes can be obtained as Fig. 4.

Split all branch including more than two values to generate two-way decision tree TWT, illustrated as Fig. 5.

Figure 2 illustrates that the number of positive objects and the number of negative objects in node 11 is equal. The traditional ID3 decision tree learning algorithm in Algorithm 1 is difficult to determine the class of node 11.

| Fig. 3. The three-way decision tree *THT* | Fig. 4. The decision tree having not boundary nodes |

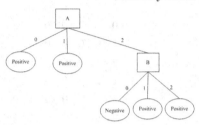

Fig. 5. The two-way decision tree *TWT*

However, the method proposed by this paper calculates the optimal threshold (α, β) by minimizing overall risk namely *Alcofa* algorithm, the optimal threshold is $(0.5, 0.3)$. The three-way decision determines the most appropriate misclassification rate by the optimal threshold, and partitions node 11 into positive class. For node 1, 8 and 13, the traditional ID3 decision tree learning algorithm considers the more frequent class as the class of these inconsistent nodes. However, the three-way decision considers that the risk of boundary decision is lower than that of misclassification at this moment. Consequently, the three-way decision does not determine immediately the decisions of acceptance or rejection, and partitions these nodes into boundary region. The classes of these nodes will be decided by considering the information around these nodes, namely the merger and prune of the three-way decision tree. Compare Figs. 1 and 5, the conclusion can be obtained that the method proposed by this paper decreases the depth of decision tree and the number of leaf nodes in decision tree.

From the above analysis, the following conclusions can be obtained immediately.

(1) For the inconsistent node in which the numbers of objects belonging to positive class and the numbers of objects belonging to negative class are equal. The traditional ID3 decision tree learning algorithm can not decide the class of this node, but the method proposed by this paper can determine the class of this node by minimizing the overall decision;

(2) For the inconsistent node in which the numbers of objects belonging to positive class and the numbers of objects belonging to negative class are unequal. The traditional ID3 decision tree learning algorithm considers the more frequent class as the class of this inconsistent node. However, the three-way decision

considers how much misclassification rate can be allowed from the perspective of minimizing the overall risk, and makes the partition of inconsistent nodes is more explicable;

(3) The method proposed by this paper can reduce the overfitting of ID3 decision tree learning algorithm.

6   Conclusions and Future Work

Aiming at the problems that traditional data mining methods easily ignore inconsistent data, and general decision tree learning algorithms lack of theoretical support for the classification of inconsistent nodes,as well as the computation problem of conditional probability in three-way decision. This paper introduces a decisions tree learning method based on three-way decisions. This method has the following advantages: firstly, the computation problem of conditional probability in three-way decision is solved. Secondly, this method reserves inconsistent data in data set, deals with these inconsistent data by using the three-way decisions to avoid the loss of useful information. Thirdly, the three-way decision considers how much misclassification rate can be allowed from the perspective of minimizing the overall risk, and makes the partition of inconsistent nodes more explicable. Fourthly, for the boundary nodes caused by the insufficient information, this method converts the three-way decision tree into the two-way decision tree by considering the surroundings information of boundary nodes. Fifthly, the decision tree with cost-sensitive is generated by introducing the three-way decision. It should be pointed out that the boundary nodes can also be removed by incremental learning method more than considering the information around boundary nodes, so our future research work will focus on the research of the incremental learning method based on the three-way decision tree.

Acknowledgements. The authors wish to thank the anonymous reviewers and Editor-in-Chief for their valuable comments and hard work. This work was supported by the National Natural Science Foundation of China (Nos. 61370169, 61402153,60873104), the Key Project of Science and Technology Department of Henan Province (Nos. 142102210056, 112102210194), the Science and Technology Research Key Project of Educational Department of Henan Province (Nos.12A520027, 13A520529), the Key Project of Science and Technology of Xinxiang Government (No. ZG13004), the Education Fund for Youth Key Teachers of Henan Normal University, and the 2014 Henan Normal University Youth Science Fund(No. 2014QK28).

References

1. Qian, W.B., Yang, B.R., Xu, Z.Y., Xie, Y.H.: Rule extraction algorithm based on discernibility matrix in inconsistent decision table. Comput. Sci. **40**(6), 215–218 (2013)
2. Meng, Z.Q., Zhou, S.Q.: Research method of generalized decision rule acquisition based on GrC in inconsistent decision systems. Comput. Sci. **39**(1), 198–202 (2012)

3. Diogo, R., Ferreira, E.V.: Using logical decision trees to discover the cause of process delays from event logs. Comput. Ind. **70**, 194–207 (2015)
4. Hong, K.S., Melanie, P.O., Ye, C.K.: Sparse alternating decision tree. Pattern Recogn. Lett. **60**, 57–64 (2015)
5. Mistikoglu, G., Gerek, I.H., Erdis, E., Mumtaz Usmen, P.E., Cakan, H., Kazan, E.E.: Decision tree analysis of construction fall accidents involving roofers. Expert Syst. Appl. **42**(4), 2256–2263 (2015)
6. Chen, J.K., Wang, X.Z., Gao, X.H.: Improved ordinal decisions trees algorithms based on rank entropy. Pattern Recogn. Artif. Intell. **27**(2), 134–140 (2014)
7. Ruan, X.H., Huang, X.M., Yuan, D.R., Duan, Q.L.: Classification algorithm based on heterogeneous cost-sensitive decision tree. Comput. Sci. **40**(11A), 140–142 (2013)
8. Jia, X.Y., Shang, L., Zhou, X.Z., Liang, J.Y., Miao, D.Q., Wang, G.Y., Li, T.R., Zhang, Y.P.: The Theory and Application of Three-way Decision. Nanjing University Press, Nanjing (2012)
9. Liu, D., Li, T., Liang, D.: A new discriminant analysis approach under decision-theoretic rough sets. In: Yao, J.T., Ramanna, S., Wang, G., Suraj, Z. (eds.) RSKT 2011. LNCS, vol. 6954, pp. 476–485. Springer, Heidelberg (2011)
10. Yao, Y., Zhou, B.: Naive bayesian rough sets. In: Yu, J., Greco, S., Lingras, P., Wang, G., Skowron, A. (eds.) RSKT 2010. LNCS, vol. 6401, pp. 719–726. Springer, Heidelberg (2010)
11. Jia, X.Y., Tang, Z.M., Liao, W.H., Shang, L.: On an optimization representation of decision-theoretic rough set model. Int. J. Approximate Reasoning **55**(1), 156–166 (2014)
12. Li, H.X., Zhou, X.Z., Zhao, J.B.: Non-nonotonic attribute reduction indecision-theoretic rough sets. Fundamenta Informaticae **126**(4), 415–432 (2013)
13. Liu, D., Li, T.R., Liang, D.C.: Incorporating logistic regression to decisiontheoretic rough sets for classifications. Int. J. Approximate Reasoning **55**(1), 197–210 (2014)
14. Li, H.: Statistical Learning Method. Tsinghua University Press, Beijing (2012)
15. Yao, Y.Y.: Three-way decisions with probabilistic rough sets. Inf. Sci. **180**(3), 341–353 (2010)
16. Liu, D., Li, T.R., Liang, D.C.: Incorporating logistic regression to decision-theoretic rough sets for classifications. Int. J. Approximate Reasoning **55**, 197–210 (2014)
17. Liu, D., Yao, Y.Y., Li, T.R.: Three-way decision-theoretic rough sets. Comput. Sci. **38**(1), 246–250 (2011)
18. Yao, Y.: Three-way decision: an interpretation of rules in rough set theory. In: Wen, P., Li, Y., Polkowski, L., Yao, Y., Tsumoto, S., Wang, G. (eds.) RSKT 2009. LNCS, vol. 5589, pp. 642–649. Springer, Heidelberg (2009)
19. Liu, D., Li, T.R., Liang, D.C.: Incorporating logistic regression to decision-theoretic rough sets for classifications. Int. J. Approximate Reasoning **55**(1), 197–210 (2013)
20. Liu, D., Li, T.R., Li, H.X.: Rough set theory: a three-way decisions perspective. J. Nanjing Univ. (Natural Sciences) **49**(5), 574–581 (2013)

Multi-decision-makers-based Monotonic Variable Consistency Rough Set Approach with Multiple Attributes and Criteria

Wenbin Pei, He Lin[(⊠)], and Li Li

School of Information Science and Engineering, LanZhou University, Lanzhou
730000, People's Republic of China
{Peiwb12,linhe}@lzu.edu.cn

Abstract. The paper separates decision expression system into three parts: the relation system, the decision-making system and the causal system by the perspective of Pansystems theory. In these three separated systems, the extended approach involves multiple types of attributes and many decision-makers, and it aims at modelling data expressed by monotonic variable consistency measures. Furthermore, the two referred thresholds, according to Bayes decision procedure that is applied by Decision Theoretic Rough Set, can be calculated directly. So the paper proposes Multi-decision-makers-based Monotonic Variable Consistency Rough Set Approach with Multiple Attributes and Criteria, and its properties are proposed and proved.

Keywords: Monotonicity · Multiple types of attributes and criteria · Many decision-makers · Rough set · Pansystems theory

1 Introduction

Dominance-based Rough Set Approach (referred to as DRSA) has been introduced for dealing with multiple criteria sorting problems [4,6]. According to the literature entitled Rough approximation of a preference relation by dominance relations, the decision express system in its original form does not allow the representation of preference binary relations between actions. In order to do with preference-ordered domains and decision classes, Greco et al. have proposed the dominance relation to operate on pairwise comparison table [6].

The paper regards decision expression system as a kind of Pansystems generalized systems or complex systems. Generally, it is not very easy to figure out the entire relationships among all the factors of a complex system. In order to analyze the complex system and to describe the objects in it better, the system need to be decomposed into several subsystems that is relatively simple because its functional and structural complexities are greatly reduced. So decision expression system should be divided into the relation system, the decision-making system and the causal system by the thoughts of aggregation-dispersion [5] from

© Springer International Publishing Switzerland 2015
Y. Yao et al. (Eds.): RSFDGrC 2015, LNAI 9437, pp. 401–413, 2015.
DOI: 10.1007/978-3-319-25783-9_36

Pansystems Theory. Each factor of the relation system can be expressed as granules of knowledge. Similarly, for the decision-making system, each factor can be described as decision classes. Furthermore, the factors of the causal system can be formed by connections between the relation system and the decision-making system, so it can be interpreted as relationships between granules of knowledge and decision classes.

It was acknowledged that the objects are usually estimated by multiple types of attributes and criteria in the relation system, and multiple types of relations should be adopted to construct the granules of knowledge. We focus on the situation that the relation system contains qualitative attributes, quantitative attributes, ordinal criteria and cardinal criteria. In this scenario, what kinds of relations should be applied to calculate granules of knowledge reasonably? According to the previous literature [4], the indiscernibility relation is defined on qualitative attributes, nevertheless, the similarity relation is defined on quantitative attributes. Based on the definition of cardinal criteria [7], although the intensity grades are numerically valued, they may be interpreted in terms of linguistic ordinal qualifiers, for example, "very weak preference", "weak preference", "strict preference", "strong preference". As a result, the graded preference relation is adopted to define on Universe evaluated by cardinal criteria [7]. However, in the case of ordinal criteria, the dominance relation is defined directly on Universe [7]. Taken together, the granules of knowledge are defined by applying four types of relations simultaneously: the indiscernibility relation, the similarity relation, the graded preference relation, and the dominance relation.

To obtain more reasonable analyses, the decision-making system always refer to the set of decision-makers, so the preferences of many decision-makers need to be merged together.

Related to the causal system, some consistent objects might be treated as inconsistent objects because of inevitable interference, so conditions of the lower approximation might be weakened. Therefore some extensions of rough set model have been proposed, such as Variable Precision Rough Set [8,9], Parameterized Rough Set Models [10,11], Variable Consistency Model of Dominance-based Rough Set Approach [25], Probabilistic Rough Set Model Based on Dominance Relation, etc. However, the semantics explanations for the parameters requirement of these models are not clear enough. To overcome these weaknesses, Yao et al. introduced the Bayesian decision procedure to Probabilistic Rough Sets and proposed Decision Theoretic Rough Set(referred to as DTRS)[21]. The thresholds can be calculated directly and systematically by minimizing the decision costs with Bayes decision procedure, which gives a brief semantics explanation in practical applications with minimum decision risks [21].

As is explicated above, we propose Multi-decision-makers-based Monotonic Variable Consistency Rough Set Approach with Multiple Attributes and Criteria. The extended approach involves multiple types of attributes and many decision-makers, and it aims at modelling data expressed by monotonic variable consistency measures. Furthermore, the two referred thresholds, according to Bayes decision procedure, can be directly calculated.

The paper is organized as follows. Section 2 is devoted to introduce the reason that the decision expression system can be divided into the relation system, the decision-making system and the causal system. In Sect. 3, we define granules of knowledge referred to the relation system and decision classes referred to the decision-making System. Section 4 is dedicated to propose Monotonic variable consistency measures referred to the causal system. Section 5 proposes Multi-decision-makers-based Monotonic Variable Consistency Rough Set Approach with Multiple Attributes and Criteria. Section 6 concludes the paper.

2   Analysis of Decision Expression System from Views of the Pansystems Theory

Formally, the 4-tuple $S =< U, A, V, f >$ denotes a decision expression system, where U is the universe; $A = C \cup D$, C is the set of condition attributes, D is the set of decision attributes; $V = \bigcup_{a \in A} V_a$, V_a is the value domain of attribute a; for every $a \in A$ there exists an information function $f : U \Rightarrow V_a$, which specifies the attribute values for each x in U.

Mathematical expression of Pansystems generalized systems(or complex systems) are defined as $S = (A, B)$, the generalized hardware A is the set of objects or the generalized system which has been defined previously, and the generalized software B is a relation defined on A or pansystems-weight composite of relations [5]. For the decision expression system S, U can be described as the generalized hardware A in the Pansystems generalized system. Similarly, the set of attributes C and D can be described as the generalized software B in the Pansystems generalized system. The analysis indicates that decision expression system is a kind of the Pansystems generalized systems(or complex systems). Generally, it is not very easy to figure out the entire relationships among all the factors of a complex system. In order to analyze a complex system and to describe the objects in it, the system needs to be decomposed into several subsystems that is relatively simple according to the thoughts of aggregation-dispersion [5] from Pansystems Theory. The thoughts of aggregation-dispersion are the extension and generalization about relations between centralization and decentralization, clustering and group, etc.

Based on the above analyses, decision expression system should be decomposed into several systems: the relation system $R_C = \{U, C, \cup, \cap\}$, the decision-making system $R_D = \{U, D, \cup, \cap\}$ and the causal system $R_{CD} = \{U, C, D, \cup, \cap, \Rightarrow\}$. The relation system is composed of granules of knowledge, and the decision-making system may be described as the set of decision classes. Furthermore, the factors of the causal system can be interpreted as relationships between granules of knowledge and decision classes. According to Rough Set, the approximated sets are analyzed by granules of knowledge, so the separation of the approximated sets and knowledge may make distinct analyses for a complex system. The Fig. 1 outlines a frame about these three separated systems. Then we investigate the related methods in order to describe and analyze better in these three separated systems.

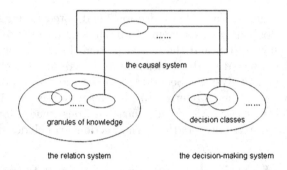

Fig. 1. A frame about these three separated systems

The cognitive complicacy of the things stems from their uncertainties and diversities. The uncertainties mainly express in the uncertainties of knowledge discernment, denotation description (vague definition of some attribute values), and denotation numbers. However, the diversity is frequently noted as using multiple attributes of various types to describe objects. As a result, for relation system R_C, set C contains several types of condition attributes. Therefore, multiple types of relations should be adopted to construct granules of knowledge. In fact, the importance degree of each attribute is variable at various angles or under the different situations and conditions. For simplicity, the importance degree of every attribute is assumed into constant in this paper.

For decision-making system R_D, D is the set of decision-makers. The fact can not be denied that the preferences of a decision-maker are expressed as a generalized software, and preferences of every decision-maker require to be brought together if many decision-makers have been taken into account.

The definition of the generalized system indicates that the system changes can be resulted from the hardware changes or the software changes or both of them. In this paper, we assume that there are only software changes.

3   Granules of Knowledge Referred to the Relation System and Decision Classes Referred to the Decision-Making System

Generally, values of the qualitative attributes exist by the unordered symbols or concepts, so they are analyzed by the indiscernibility relation since the concepts and symbols are usually stable or accurate. However, values of the quantitative attributes are numbers, so the objects described by quantitative attributes are analyzed by the similarity relation [3,4]. It is acknowledged, however, that the domain of a criterion has to be ordered according to a decreasing or increasing preference [4]. The criteria set is distinguished between ordinal criteria and cardinal criteria by the literature [7]. As was mentioned before, the graded preference relation is adopted to define on cardinal criteria, while the dominance relation is defined directly on ordinal criteria.

For a relation system $R_C = \{U, C, \cup, \cap\}$, given $P \subseteq C$, P can be divided into $P_o^>$, $P_N^{\succeq h}$, $P^=$ and $P^\frown$, and $P_o^>$ is the set of ordinal criteria, $P_N^{\succeq h}$ is the set of cardinal criteria, $P^=$ is the set of qualitative attributes, $P^\frown$ is the set of quantitative attributes. It is worth noting that $P_o^>$, $P_N^{\succeq h}$, $P^=$ and $P^\frown$ need to satisfy the equivalence of $P = P_o^> \cup P_N^{\succeq k} \cup P^= \cup P^\frown$, and the intersection of any two subsets is an empty set.

The preference relation, denoted by S_g^h, assumes different grades h of intensity for a cardinal criterion $g \in P_N^{\succeq k}$. Let $H_i \subset [-1, 1]$ be the finite set of all admitted values of h with respect to $g_i \in P_N^{\succeq k}$, and H_i^+(resp. H_i^-) is the subset of strictly positive(resp., strictly negative) values of h[6]. So a finite set $H_i = (H_i^- \cup \{0\} \cup H_i^+)$ is constituted by marginal intensity grades, and $\forall h \in H_i, \forall x, y \in U,$

(1) $xS_{g_i}^h y$, $h \in H_i^+$, means that action x is preferred to action y in grade h on criterion g_i,

(2) $xS_{g_i}^h y$, $h \in H_i^-$, means that action x is not preferred to action y in grade h on criterion g_i,

(3) $xS_{g_i}^0 y$ means that action x is similar (asymmetrically indifferent) to action y on criterion g_i.

In this paper, we define the set of possible degrees of preference $H_q = \{-3, -2, -1, 0, 1, 2, 3\}$. For each $q \in P_N^{\succeq k}, \forall x, y \in U$, the preference relations S_q^h have the following interpretation [6]:

(1) $xS_q^3 y$(or $yS_q^{-3}x$) means that x is strongly preferred to y with respect to q,

(2) $xS_q^2 y$(or $yS_q^{-2}x$) means that x is preferred to y with respect to q,

(3) $xS_q^1 y$(or $yS_q^{-1}x$) means that x is weakly preferred to y with respect to q,

(4) $xS_q^0 y$(or $yS_q^0 x$) means that x is indifferent to y with respect to q.

Then, $\forall q \in P_N^{\succeq k}, \forall x, y \in U$, let $\sigma_q = max_{\forall x, y \in U}\{C_q(x) - C_q(y)\}$ [6], and the following values of the thresholds are adopted: $\Delta_q^4 = -\Delta_q^{-4} = \sigma_q$, $\Delta_q^3 = -\Delta_q^{-3} = 0.75\sigma_q$, $\Delta_q^2 = -\Delta_q^{-2} = 0.5\sigma_q$, $\Delta_q^1 = -\Delta_q^{-1} = 0.25\sigma_q$.

In the $R_C, \forall x, y \in U$, symbol $xS_P^o y$ indicates that x dominates y on $P_o^>$; symbol $xS_P^h y$ indicates that x dominates y in grade h on $P_N^{\succeq h}$; symbol $xI_P y$ indicates that x is indiscernible from y on $P^=$; symbol $xR_P y$ indicates that x is similar to y on $P^\frown$.

It is acknowledged that the similarity of x to y with respect to $q \in P^\frown$ may be calculated by the similarity of value $f(x, q) = r$ to value $f(y, q) = s$. For $\forall x, y \in U$, $\forall q \in P^\frown$, $xR_P y$ if and only if $\frac{|f(x,q)-f(y,q)|}{f(y,q)} \leq t$, where t is a similarity threshold [4]. Similarily, $\forall q \in P^=$, $xI_P y$ if and only if $f(x, q) = f(y, q)$. For $\forall x, y \in U, \forall q \in P_o^>$, $xS_P^o y$ if and only if $f(x, q) \geq f(y, q)$, however, $\forall q \in P_N^{\succeq h}$, $xS_P^h y$ if and only if $f(x, q) - f(y, q) \geq \Delta_q^h$.

Two binary relations are defined on U, denoted by H_{hP} and H_{hP}^*, $\forall x, y \in U$
• $xH_{hP}y$ if and only if $xS_P^o y$ for each $q \in P_o^>$, $xS_P^h y$ for each $q \in P_N^{\succeq h}$, $xI_P y$ for any $q \in P^=$, $yR_P x$ for any $q \in P^\frown$.
• $xH_{hP}^* y$ if and only if $xS_P^o y$ for each $q \in P_o^>$, $xS_P^h y$ for each $q \in P_N^{\succeq h}$, $xI_P y$ for any $q \in P^=$, $xR_P y$ for any $q \in P^\frown$.

From H_{hP} and H_{hP}^*, four types of granules of knowledge can be generated, $H_{hP}^{L+}(x)$, $H_{hP}^{U+}(x)$, $H_{hP}^{L-}(x)$, $H_{hP}^{U-}(x)$ respectively:

(1) $H_{hP}^{L+}(x) = \{y \in U | yH_hPx\}$ is a set of objects y dominating x on $P_o^>$, dominating x in grade h on $P_N^{\geq h}$, indiscernible from x on $P^=$, and x is similar to y on $P^\sim$.

(2) $H_{hP}^{U+}(x) = \{y \in U | yH_h^*Px\}$ is a set of objects y dominating x on $P_o^>$, dominating x in grade h on $P_N^{\geq h}$, indiscernible from x on $P^=$, and y is similar to x on $P^\sim$.

(3) $H_{hP}^{L-}(x) = \{y \in U | xH_h^*Py\}$ is a set of objects y dominated by x on $P_o^>$, dominated by x in grade h on $P_N^{\geq h}$, indiscernible from x on $P^=$, and x is similar to y on $P^\sim$.

(4) $H_{hP}^{U-}(x) = \{y \in U | xH_hPy\}$ is a set of objects y dominated by x on $P_o^>$, dominated by x in grade h on $P_N^{\geq h}$, indiscernible from x on $P^=$, and y is similar to x on $P^\sim$.

In the paper, h has four different preference grades, therefore four groups of granules of knowledge can be acquired.

Then an example is contributed to explain the necessity of many decision-makers. About the communication quality assessment, users from the security department consider reliability as a key element to evaluate the quality of communication, while users from the monitoring department regard real-time as the main foundation. It is acknowledged that taking the views from the security department and monitoring department into account would be better than considering separately. So it is reasonable to contain multiple decision-makers.

It is assumed, for a decision-making system $R_D = \{U, D, \cup, \cap\}$, that a decision-maker corresponds to a decision criterion. If only a decision-makers is included in the set $D = \{d\}$. Set D makes a partition of the set of reference objects U into a finite number of decision classes $Cl_1, Cl_2, \cdots, Cl_n$. Let $Cl = \{Cl_t, t \in T\}$, $T = \{1, 2, \cdots, n\}$, Cl is a set of these classes such that for each $x \in U$ belongs to one and only one $Cl_t \in Cl$. The classes are ordered, i.e. for all $u, v \in T$, if $u > v$, then the objects from Cl_u are preferred to the objects from Cl_v. And the sets to be approximated are t-upward union $Cl_t^\geq = \bigcup_{s \geq t} Cl_s$ and t-downward union $Cl_t^\leq = \bigcup_{s \leq t} Cl_s$. To take the preferences of many decision-makers into account, some new concepts are introduced as following.

Definition 1. [14] For $R_D = \{U, D, \cup, \cap\}$, the set of decision-makers is assumed to be $D = \{d_1, d_2, \cdots, d_n\}$. If preferences of all the decision-makers need to be satisfied in the D, then the sets to be approximated are upward-union $Cl_D^{\geq t} = \bigcap_{d \in D} Cl_d^{\geq t}$ and downward-union of decision classes $Cl_D^{\leq t} = \bigcap_{d \in D} Cl_d^{\leq t}$, where $Cl_d^\geq = \bigcup_{\substack{s \geq t \\ d \in D}} Cl_s$, $Cl_d^\leq = \bigcup_{\substack{s \leq t \\ d \in D}} Cl_s$.

Definition 2. [14] For $R_D = \{U, D, \cup, \cap\}$, the set is assumed to be $MD = \{D_1, D_2, \cdots, D_n\}$, where $D_i \in D$. If preferences of decision-makers need to be satisfied in at least one subset D_i, then the sets to be approximated are the upward-union $Cl_{MD}^{\geq t} = \bigcup_{D_i \in MD} Cl_{D_i}^{\geq t} = \bigcup_{D_i \in MD}(\bigcap_{d \in D} Cl_d^{\geq t})$ and downward-union of decision classes $Cl_{MD}^{\leq t} = \bigcup_{D_i \in MD} Cl_{D_i}^{\leq t} = \bigcup_{D_i \in MD}(\bigcap_{d \in D} Cl_d^{\leq t})$.

4 Monotonic Variable Consistency Measures Referred to the Causal System

Since some consistent objects might be treated as inconsistent objects because of the inevitable interference, conditions of the lower approximation might be weakened. Literature [9] proposed two types of consistency measures, that is, the gain-type consistency measure $f_x^p(y)$ and the cost-type consistency measure $g_x^p(y)$. For a gain-type consistency measure $f_x^p(y)$, the higher the value, the more consistent is the given action. However, for a cost-type measure $g_x^p(y)$, the lower the value, the more consistent is the given action. It is reasonable for the consistency measures to require several properties of monotonicity.

Four types of properties of monotonicity are given by literature [9]:
(m_1) Extension of the set of attributes.
(m_2) Extension of the set of objects.
(m_3) Extension of the union of ordered classes.
(m_4) Improvement of evaluation of an object.

In this paper, the gain-type consistency measures are defined when decision classes are $Cl_t^{\geq}$ and $Cl_t^{\leq}$:

$$\eta_{Cl_t^{\geq}}^{hP}(x) = \max_{\substack{T \subseteq P \\ z \in H_{hT}^{U-}(x) \cap Cl_t^{\geq}}} \left\{ \frac{card(H_{hT}^{L+}(z) \cap Cl_t^{\geq})}{card(H_{hT}^{L+}(z))} \right\} \tag{1}$$

$$\sigma_{Cl_t^{\geq}}^{hP}(x) = \max_{\substack{T \subseteq P \\ z \in H_{hT}^{L-}(x) \cap Cl_t^{\geq}}} \left\{ \frac{card(H_{hT}^{U+}(z) \cap Cl_t^{\geq})}{card(H_{hT}^{U+}(z))} \right\} \tag{2}$$

$$\eta_{Cl_t^{\leq}}^{hP}(x) = \max_{\substack{T \subseteq P \\ z \in H_{hT}^{U+}(x) \cap Cl_t^{\leq}}} \left\{ \frac{card(H_{hT}^{L-}(z) \cap Cl_t^{\leq})}{card(H_{hT}^{L-}(z))} \right\} \tag{3}$$

$$\sigma_{Cl_t^{\leq}}^{hP}(x) = \max_{\substack{T \subseteq P \\ z \in H_{hT}^{L+}(x) \cap Cl_t^{\leq}}} \left\{ \frac{card(H_{hT}^{U-}(z) \cap Cl_t^{\leq})}{card(H_{hT}^{U-}(z))} \right\} \tag{4}$$

Theorem 1. If $R \subseteq P \subseteq C, \forall x \in U$, then $\eta_{Cl_t^{\geq}}^{hP}(x) \geq \eta_{Cl_t^{\geq}}^{hR}(x)$, $\sigma_{Cl_t^{\geq}}^{hP}(x) \geq \sigma_{Cl_t^{\geq}}^{hR}(x)$, $\eta_{Cl_t^{\leq}}^{hP}(x) \geq \eta_{Cl_t^{\leq}}^{hR}(x)$, $\sigma_{Cl_t^{\leq}}^{hP}(x) \geq \sigma_{Cl_t^{\leq}}^{hR}(x)$.

Proof. For all $R \subseteq P \subseteq C, \forall x \in U$, $\max\limits_{\substack{T \subseteq P \\ z \in H_{hT}^{U-}(x) \cap Cl_t^{\geq}}} \left\{ \frac{card(H_{hT}^{L+}(z) \cap Cl_t^{\geq})}{card(H_{hT}^{L+}(z))} \right\} \geq$

$\max\limits_{\substack{T \subseteq R \\ z \in H_{hT}^{U-}(x) \cap Cl_t^{\geq}}} \left\{ \frac{card(H_{hT}^{L+}(z) \cap Cl_t^{\geq})}{card(H_{hT}^{L+}(z))} \right\}$, so $\eta_{Cl_t^{\geq}}^{hP}(x) \geq \eta_{Cl_t^{\geq}}^{hR}(x)$.

The same can be proved for measure $\sigma_{Cl_t^{\geq}}^{hP}(x)$, $\eta_{Cl_t^{\leq}}^{hP}(x)$, $\sigma_{Cl_t^{\leq}}^{hP}(x)$. □

Theorem 2. If $R \subseteq C, \forall x \in U, Cl_t^{\geq} \subseteq U, Cl_t^{*\geq} = Cl_t^{\geq} \cup Cl_t^{\circ\geq}$, $Cl_t^{\circ\geq}$ is the set of objects and satisfy $Cl_t^{\circ\geq} \cap U \neq \varnothing$, then $\eta_{Cl_t^{*\geq}}^{hP}(x) \geq \eta_{Cl_t^{\geq}}^{hP}(x)$, $\sigma_{Cl_t^{*\geq}}^{hP}(x) \geq \sigma_{Cl_t^{\geq}}^{hP}(x)$; if $Cl_t^{\leq} \subseteq U, Cl_t^{*\leq} = Cl_t^{\leq} \cup Cl_t^{\circ\leq}$, $Cl_t^{\circ\leq}$ is the set of objects and satisfy $Cl_t^{\circ\leq} \cap U \neq \varnothing$, then $\eta_{Cl_t^{*\leq}}^{hP}(x) \geq \eta_{Cl_t^{\leq}}^{hP}(x)$, $\sigma_{Cl_t^{*\leq}}^{hP}(x) \geq \sigma_{Cl_t^{\leq}}^{hP}(x)$.

Proof. For any $T \subseteq C$, $\forall x \in U$, $Cl_t^{\geq} \subseteq U$, $Cl_t^{*\geq} = Cl_t^{\geq} \cup Cl_t^{\circ\geq}$, and $Cl_t^{\circ\geq} \cap U \neq \varnothing$,
$\frac{card(H_{hT}^{L+}(x) \cap Cl_t^{\geq})}{card(H_{hT}^{L+}(x))} \leq \frac{card(H_{hT}^{'L+}(x) \cap Cl_t^{*\geq})}{card(H_{hT}^{'L+}(x))}$, where $H_{hT}^{'L+}(x)$ denotes the set of the
objects which can dominate x when considering the set of attributes T and
$Cl_t^{*\geq} = Cl_t^{\geq} \cup Cl_t^{\circ\geq}$. Thus, $max\{\frac{card(H_{hT}^{L+}(x) \cap Cl_t^{\geq})}{card(H_{hT}^{L+}(x))}\} \leq max\{\frac{card(H_{hT}^{'L+}(x) \cap Cl_t^{*\geq})}{card(H_{hT}^{'L+}(x))}\}$, that
is, $\eta_{Cl_t^{*\geq}}^{P}(x) \geq \eta_{Cl_t^{\geq}}^{P}(x)$.
In the same way, $\sigma_{Cl_t^{*\geq}}^{hP}(x) \geq \sigma_{Cl_t^{\geq}}^{hP}(x)$, $\eta_{Cl_t^{*\leq}}^{hP}(x) \geq \eta_{Cl_t^{\leq}}^{hP}(x)$, $\sigma_{Cl_t^{*\leq}}^{hP}(x) \geq \sigma_{Cl_t^{\leq}}^{hP}(x)$ can
be proved.   □

Theorem 3. If $P \subseteq C$, $\forall x \in U$, $Cl_j^{\geq} \subseteq Cl_i^{\geq} \subseteq U$, then $\eta_{Cl_i^{\geq}}^{hP}(x) \geq \eta_{Cl_j^{\geq}}^{hP}(x)$,
$\sigma_{Cl_i^{\geq}}^{hP}(x) \geq \sigma_{Cl_j^{\geq}}^{hP}(x)$. If $Cl_t^{\leq} \subseteq Cl_k^{\leq} \subseteq U$, then $\eta_{Cl_t^{\leq}}^{hP}(x) \geq \eta_{Cl_k^{\leq}}^{hP}(x)$, $\sigma_{Cl_t^{\leq}}^{hP}(x) \geq \sigma_{Cl_t^{\leq}}^{hP}(x)$.
The proof is omitted because the theorem can be easily proved.   □

Theorem 4. If $P \subseteq C$, $Cl_i^{\geq} \subseteq U$, $\forall x, y \in U$, if $xD_P y$, then $\eta_{Cl_i^{\geq}}^{hP}(x) \geq \eta_{Cl_i^{\geq}}^{hP}(y)$,
$\eta_{Cl_i^{\leq}}^{hP}(x) \geq \eta_{Cl_i^{\leq}}^{hP}(y)$; if $xD_P^* y$, then $\sigma_{Cl_i^{\geq}}^{hP}(x) \geq \sigma_{Cl_i^{\geq}}^{hP}(y)$, $\sigma_{Cl_i^{\leq}}^{hP}(x) \geq \sigma_{Cl_i^{\leq}}^{hP}(y)$.
The proof is omitted because the theorem can be easily proved.   □

5   Multi-decision-makers-based Monotonic Variable Consistency Rough Set Approach with Multiple Attributes and Criteria

The semantics explanation for the parameters requirement of some models(such as Variable Precision Rough Set, Parameterized rough set models, etc.) is hoped to be clear, which is, however, may not be all that clear. To overcome these weaknesses, DTRS is proposed. The thresholds can be calculated directly and systematically by minimizing the decision costs with Bayes decision procedure.

For the Bayesian decision procedure, the DTRS use two states and three actions to describe the decision process. The set of states is given by $\Omega = \{X, -X\}$ indicating that an element is in X and not in X. The set of actions is given by $\Lambda = \{a_P, a_b, a_N\}$, where a_P, a_B, and a_N represent the three actions in classifying an object x, namely, deciding $x \in POS(X)$, deciding $x \in BND(X)$, and deciding $x \in NEG(X)$, respectively [21]. In addition, $\lambda_{\bullet\bullet}$ describes the loss functions. The first place of the subscript of $\lambda_{\bullet\bullet}$ denotes the action of the decisions, and the second place of the subscript of $\lambda_{\bullet\bullet}$ denotes the state of the decisions. Hence, we use λ_{PP}, λ_{BP} and λ_{NP} to represent the losses incurred for taking actions a_P, a_B, and a_N, respectively, when an object belongs to X[21]. Similarly, λ_{PN}, λ_{BN} and λ_{NN} represent the losses incurred for taking the same actions when the object does not belong to X[21].

Let $\lambda_{PP} = [\lambda_{PP}^L, \lambda_{PP}^U]$, $\lambda_{BP} = [\lambda_{BP}^L, \lambda_{BP}^U]$ and $\lambda_{NP} = [\lambda_{NP}^L, \lambda_{NP}^U]$, $\lambda_{PN} = [\lambda_{PN}^L, \lambda_{PN}^U]$, $\lambda_{BN} = [\lambda_{BN}^L, \lambda_{BN}^U]$ and $\lambda_{NN} = [\lambda_{NN}^L, \lambda_{NN}^U]$. $\lambda_{\bullet\bullet}^L$ represents minimum value of $\lambda_{\bullet\bullet}$, which is provided by optimistic decision-makers. However, $\lambda_{\bullet\bullet}^U$ represents maximum value of $\lambda_{\bullet\bullet}$, which is given by pessimistic decision-makers. And they must satisfy the conditions of $\lambda_{PP}^L \leq \lambda_{PP}^U < \lambda_{BP}^L \leq \lambda_{BP}^U < \lambda_{NP}^L \leq \lambda_{NP}^U$,

$\lambda_{NN}^L \leq \lambda_{NN}^U < \lambda_{BN}^L \leq \lambda_{BN}^U < \lambda_{PN}^L \leq \lambda_{PN}^U$. So, the consistency coefficients can be calculated directly and systematically by minimizing the decision costs with Bayes decision procedure.

For the optimistic decision-makers, consistency coefficients ∂_1, β_1 are defined as [20]:

$$\partial_1 = \frac{\lambda_{PN}^L - \lambda_{BN}^L}{(\lambda_{PN}^L - \lambda_{BN}^L) + (\lambda_{BN}^L - \lambda_{PP}^L)} \qquad \beta_1 = \frac{\lambda_{BN}^L - \lambda_{NN}^L}{(\lambda_{BN}^L - \lambda_{NN}^L) + (\lambda_{NP}^L - \lambda_{BP}^L)}$$

For the pessimistic decision-makers, consistency coefficients ∂_2, β_2 are defined as [20]:

$$\partial_2 = \frac{\lambda_{PN}^U - \lambda_{BN}^U}{(\lambda_{PN}^U - \lambda_{BN}^U) + (\lambda_{BN}^U - \lambda_{PP}^U)} \qquad \beta_2 = \frac{\lambda_{BN}^U - \lambda_{NN}^U}{(\lambda_{BN}^U - \lambda_{NN}^U) + (\lambda_{NP}^L - \lambda_{BP}^L)}$$

So the two thresholds, that is, consistency coefficients of lower approximation ∂_X and consistency coefficients of upper approximation β_X, can be defined as [20]:

$$\partial_X \in [\frac{\lambda_{PN}^L - \lambda_{BN}^U}{(\lambda_{PN}^U - \lambda_{BN}^L) + (\lambda_{BN}^U - \lambda_{PP}^L)}, min\{\frac{\lambda_{PN}^U - \lambda_{BN}^L}{(\lambda_{PN}^L - \lambda_{BN}^U) + (\lambda_{BP}^L - \lambda_{PP}^U)}, 1\}] \qquad (5)$$

$$\beta_X \in [\frac{\lambda_{BN}^L - \lambda_{NN}^U}{(\lambda_{BN}^U - \lambda_{NN}^L) + (\lambda_{NP}^U - \lambda_{BP}^L)}, min\{\frac{\lambda_{BN}^U - \lambda_{NN}^L}{(\lambda_{BN}^L - \lambda_{NN}^U) + (\lambda_{NP}^L - \lambda_{BP}^U)}, 1\}] \qquad (6)$$

Then there are two alternatives of many decision-makers:

(1) The extended models satisfies preferences of all the decision-makers in the set D.

(2) The extended models satisfies preferences of all the decision-makers in at least one subset D_i.

When preferences of all the decision-makers are satisfied in the set D, the approximated sets are the upward union set $Cl_D^{\geq t}$ and the downward union set $Cl_D^{\leq t}$. However, if preferences of all the decision-makers are satisfied in at least one subset D_i, the sets to be approximated are the upward union set $Cl_{MD}^{\geq t}$ and downward union set $Cl_{MD}^{\leq t}$. In both cases, granules of knowledge are used for approximation are $H_{hT}^{L+}(x)$, $H_{hT}^{U+}(x)$, $H_{hT}^{L-}(x)$ and $H_{hT}^{U-}(x)$.

In the following definitions or properties, if preferences of all the decision-makers are satisfied in the D, then $Cl_B^{\geq t}$ needs to be replaced by $Cl_D^{\geq t}$. However, preferences of all the decision-makers are satisfied in at least one subset D_i, $Cl_B^{\geq t}$ needs to be replaced by $Cl_{MD}^{\geq t}$.

According to monotonicity properties (m_1), (m_2), (m_3), (m_4), the gain-type consistency measures are defined as follows:

$$\eta_{Cl_B^{\geq t}}^{hP}(x) = \max_{\substack{T \subseteq P \\ z \in H_{hT}^{U-}(x) \cap Cl_B^{\geq t}}} \{\frac{card(H_{hT}^{L+}(z) \cap Cl_B^{\geq t})}{card(H_{hT}^{L+}(z))}\} \qquad (7)$$

$$\sigma_{Cl_B^{\geq t}}^{hP}(x) = \max_{\substack{T \subseteq P \\ z \in H_{hT}^{L-}(x) \cap Cl_B^{\geq t}}} \{\frac{card(H_{hT}^{U+}(z) \cap Cl_B^{\geq t})}{card(H_{hT}^{U+}(z))}\} \qquad (8)$$

$$\eta_{Cl_B^{\leq t}}^{hP}(x) = \max_{\substack{T \subseteq P \\ z \in H_{hT}^{U+}(x) \cap Cl_B^{\leq t}}} \{\frac{card(H_{hT}^{L-}(z) \cap Cl_B^{\leq t})}{card(H_{hT}^{L-}(z))}\} \qquad (9)$$

$$\sigma^{hP}_{Cl^{\leq t}_B}(x) = \max_{\substack{T \subseteq P \\ z \in H^{L+}_{hT}(x) \cap Cl^{\leq t}_B}} \{ \frac{card(H^{U-}_{hT}(z) \cap Cl^{\leq t}_B)}{card(H^{U-}_{hT}(z))} \} \qquad (10)$$

Definition 3. If $P \subseteq C$, $t \in T$, the impact of environmental noise and preferences of many decision-makers are considered, then the P-lower approximation of $Cl^{\geq t}_B$ is

$$P^{hx}_-(Cl^{\geq t}_B) = \{ x \in U \mid \max_{\substack{T \subseteq P \\ u = max\{h\} \\ z \in H^{U-}_{uT}(x) \cap Cl^{\geq t}_B}} \{ \frac{card(H^{L+}_{uT}(z) \cap Cl^{\geq t}_B)}{card(H^{L+}_{uT}(z))} \} \geq \partial_X \} \qquad (11)$$

the P-upper approximation of $Cl^{\geq t}_B$ is

$$P^-_{hx}(Cl^{\geq t}_B) = \{ x \in Cl^{\geq t}_B \mid \max_{\substack{T \subseteq P \\ v = min\{h\} \\ z \in H^{L-}_{vT}(x) \cap Cl^{\geq t}_B}} \{ \frac{card(H^{U+}_{vT}(z) \cap Cl^{\geq t}_B)}{card(H^{U+}_{vT}(z))} \} > \beta_X \} \qquad (12)$$

and the P-boundary of $Cl^{\geq t}_B$ is defined as

$$Bn^{hx}_P(Cl^{\geq t}_B) = P^-_{hx}(Cl^{\geq t}_B) - P^{hx}_-(Cl^{\geq t}_B) \qquad (13)$$

In the above definitions, ∂_X, β_X are calculated according to formula (5) and formula (6). By the same token, the P-lower approximation of $Cl^{\leq t}_B$ is $P^{hx}_-(Cl^{\leq t}_B) = \{ x \in U \mid \max_{\substack{T \subseteq P \\ u = max\{h\} \\ z \in H^{U+}_{uT}(x) \cap Cl^{\leq t}_B}} \{ \frac{card(H^{L-}_{uT}(z) \cap Cl^{\leq t}_B)}{card(H^{L-}_{uT}(z))} \} \geq \partial_X \}$, the P-upper approxima-

tion of $Cl^{\leq t}_B$ is $P^-_{hx}(Cl^{\leq t}_B) = \{ x \in Cl^{\leq t}_B \mid \max_{\substack{T \subseteq P \\ v = min\{h\} \\ z \in H^{L+}_{vT}(x) \cap Cl^{\leq t}_B}} \{ \frac{card(H^{U-}_{vT}(z) \cap Cl^{\leq t}_B)}{card(H^{U-}_{vT}(z))} \} >$

$\beta_X \}$, and the P-boundary of $Cl^{\leq t}_B$ is defined as $Bn^{hx}_P(Cl^{\leq t}_B) = P^-_{hx}(Cl^{\leq t}_B) - P^{hx}_-(Cl^{\leq t}_B)$. In the above definition, ∂_X, β_X are calculated according to formula (5) and (6).

Property 1. If $P \subseteq C$, $t \in T$, then $P^{hx}_-(Cl^{\geq t}_B) \subseteq Cl^{\geq t}_B \subseteq P^-_{hx}(Cl^{\geq t}_B)$, $P^{hx}_-(Cl^{\leq t}_B) \subseteq Cl^{\leq t}_B \subseteq P^-_{hx}(Cl^{\leq t}_B)$.

The proof is omitted because the property can be easily proved.   □

Property 2. If $R \subseteq P \subseteq C$, $t \in T$, then

$$P^-_{hx}(Cl^{\geq t}_B) \subseteq R^-_{hx}(Cl^{\geq t}_B) \quad R^{hx}_-(Cl^{\geq t}_B) \subseteq P^{hx}_-(Cl^{\geq t}_B) \quad Bn^{hx}_P(Cl^{\geq t}_B) \subseteq Bn^{hx}_R(Cl^{\geq t}_B)$$

$$P^-_{hx}(Cl^{\leq t}_B) \subseteq R^-_{hx}(Cl^{\leq t}_B) \quad R^{hx}_-(Cl^{\leq t}_B) \subseteq P^{hx}_-(Cl^{\leq t}_B) \quad Bn^{hx}_P(Cl^{\leq t}_B) \subseteq Bn^{hx}_R(Cl^{\leq t}_B)$$

The proof is omitted because the property can be easily proved.   □

Definition 4. If $P \subseteq C$, $t \in T$, the impact of environmental noise and preferences of many decision-makers are considered, then the accuracy of approximation of $Cl^{\geq t}_B$ is defined as $\partial^{hx}_P(Cl^{\geq t}_B) = \frac{card(P^{hx}_-(Cl^{\geq t}_B))}{card(P^-_{hx}(Cl^{\geq t}_B))}$, and the accuracy of approximation of $Cl^{\leq t}_B$ is defined as $\partial^{hx}_P(Cl^{\leq t}_B) = \frac{card(P^{hx}_-(Cl^{\leq t}_B))}{card(P^-_{hx}(Cl^{\leq t}_B))}$.

Definition 5. If $P \subseteq C$, $t \in T$, the impact of environmental noise and preferences of many decision-makers are considered, then the quality of approximation of partition Cl is defined as $\lambda_P^{hx}(Cl) = \frac{card(U - (\bigcup_{t \in T} Bn_P^{hx}(Cl_B^{\geq t})) \cup (\bigcup_{t \in T} Bn_P^{hx}(Cl_B^{\leq t})))}{card(U)}$.

Definition 6. The impact of environmental noise and preferences of many decision-makers are considered, each minimal subset $R \subseteq C$ such that $\lambda_R^{hx}(Cl) = \lambda_C^{hx}(Cl)$, then R is called a reduction of Cl. It was acknowledged that a decision express table can have more than one reduction. The intersection of all the reductions is called the core and denoted by $CORE_{Cl}(C)$.

6   Conclusion

The paper raised Multi-decision-makers-based Monotonic Variable Consistency Rough Set Approach with Multiple Attributes and Criteria. The extended approach is based on perspective of Pansystems to divide decision expression system into the relation system, the decision-making system and the causal system. In the relation system, because the objects are usually estimated by multiple types attributes and criteria, the attribute set is distinguished between qualitative attributes and quantitative attributes, and among criteria we distinguish between ordinal criteria and cardinal criteria. In that case four types of relations are applied jointly: the indiscernibility relation, the similarity relation, the dominance relation, and the graded preference relation. For the decision-making system, the preferences of the set of decision-makers are considered. It, owing to noise, aimes at modelling data expressed in terms of monotonic variable consistency measures rather than by full inclusion relations to define the lower and upper approximation sets.

Acknowledgement. The paper is supported by the Fundamental Research Funds for the Central Universities(lzujbky-2012-43). The authors thank valued amendments which are raised by Professor Yongli Li.

References

1. Pawlak, Z.: Rough sets. Int. J. Comput. Inf. Sci. **11**, 341–356 (1982)
2. Pawlak, Z.: Rough sets: Theoretical Aspects of Reasoning About Data. Kluwer Academic Publishers, Dordrecht (1991)
3. Slowinski, R., Vanderpooten, D.: A generalized definition of rough approximations based on similarity. IEEE Trans. Knowl. Data Eng. **12**, 331–336 (2000)
4. Greco, S., Matarazzo, B., Slowinski, R.: Rough sets methodology for sorting problems in presence of multiple attributes and criteria. Eur. J. Oper. Res. **138**, 247–259 (2002)
5. Wu, X.M.: Views to World from Pansystem. The Renmin University of China Press, Beijing (1990)
6. Greco, S., Matarazzo, B., Slowinski, R.: Rough approximation of a preference relation by dominance relations. Eur. J. Oper. Res. **117**, 63–83 (1999)

7. Fortemps, P., Greco, S., Slowinski, R.: Multicriteria decision support using rules that represent rough-graded preference relation. Eur. J. Oper. Res. **188**, 206–223 (2008)

8. Ziarko, W.: Variable precision rough sets model. J. Comput. Syst. Sci. **46**, 39–59 (1993)

9. Blaszczynski, J., Greco, S., Slowinski, R., Szelag, M.: Monotonic variable consistency rough set approaches. Int. J. Approximate Reasoning **50**, 979–999 (2009)

10. Greco, S., Pawlak, Z., Slowinski, R.: Can bayesian confirmation measures be useful for rough set decision rules? Eng, Appl. Artif. Intell. **17**, 345–361 (2004)

11. Greco, S., Matarazzo, B., Slowinski, R.: Rough membership and bayesian confirmation measures for parameterized rough sets. Int. J. Approximate Reasoning **49**, 285–300 (2008)

12. Slowinski, R., Stefanowski, J., Greco, S., Matarazzo, B.: Rough sets based processing of inconsistent information in decision analysis. Control Cybern. **29**, 379–404 (2000)

13. Yao, Y.Y.: Relational interpretations of neighborhood operators and rough set approximation operators. Inf. Sci. **111**, 239–259 (1998)

14. Greco, S., Matarazzo, B., Słowiński, R.: Dominance-based rough set approach on pairwise comparison tables to decision involving multiple decision makers. In: Yao, J.T., Ramanna, S., Wang, G., Suraj, Z. (eds.) RSKT 2011. LNCS, vol. 6954, pp. 126–135. Springer, Heidelberg (2011)

15. Greco, S., Mousseau, V., Slowinski, R.: Ordinal regression revisited: multiple criteria ranking with a set of additive value functions. Eur. J. Oper. Res. **191**, 415–435 (2008)

16. Slowinski, R.: Rough set learning of preferential attitude in multi-criteria decision making. In: Komorowski, J. (ed.) Methodologies Intell. Syst., vol. 689, pp. 642–651. Springer, Heidelberg (1993)

17. Yao, Y.Y.: Granular computing. Comput. Sci. **31**, 1–5 (2004)

18. Yao, Y.Y.: Relational interpretations of neighborhood operators and rough set approximation operators. Inf. Sci. **111**, 239–259 (1998)

19. Miao, D.Q., Li, D.T., Yao, Y.Y.: Uncertainty and Granular Computing. Science Press, Beijing (2011)

20. Liu, D., Li, T.R., Li, H.X.: Interval-valued decision-theoretic rough sets. Comput. Sci. **39**, 179–182 (2012)

21. Yao, Y.: Decision-theoretic rough set models. In: Yao, J.T., Lingras, P., Wu, W.-Z., Szczuka, M.S., Cercone, N.J., Ślęzak, D. (eds.) RSKT 2007. LNCS (LNAI), vol. 4481, pp. 1–12. Springer, Heidelberg (2007)

22. Pedrycz, W.: Granular Computing: Analysis and Design of Intelligent Systems. CRC Press/Francis Taylor, Boca Raton (2013)

23. Li, W., Xu, W.: Probabilistic rough set model based on dominance relation. In: Miao, D., Pedrycz, W., Slezak, D., Peters, G., Hu, Q., Wang, R. (eds.) RSKT 2014. LNCS, vol. 8818, pp. 856–864. Springer, Heidelberg (2014)

24. Li, W.T., Xu, W.H.: Multigranulation decision-theoretic rough set in ordered information system. Fundamenta Informaticae **139**, 67–89 (2015)

25. Greco, S., Matarazzo, B., Słowiński, R., Stefanowski, J.: Variable consistency model of dominance-based rough sets approach. In: Ziarko, W.P., Yao, Y. (eds.) RSCTC 2000. LNCS (LNAI), vol. 2005, pp. 170–181. Springer, Heidelberg (2001)

26. Greco, S., Matarazzo, B., Slowinski, R.: Dominance-based rough set approach as a proper way of handling graduality in rough set theory. Trans. Rough Sets **4400**, 36–52 (2007)

27. Greco, S., Słowiński, R., Yao, Y.: Bayesian decision theory for dominance-based rough set approach. In: Yao, J.T., Lingras, P., Wu, W.-Z., Szczuka, M.S., Cercone, N.J., Ślęzak, D. (eds.) RSKT 2007. LNCS (LNAI), vol. 4481, pp. 134–141. Springer, Heidelberg (2007)

28. Qian, Y., Zhang, H.H., Sang, Y.L., Liang, J.Y.: Multigranulation decision-theoretic rough sets. Int. J. Approximate Reasoning 55(1), 225–237 (2014)

29. Li, Y.L., Lin, Y., Wang, X.Y., Lin, H.: An initial comparision of fuzzy sets and rough sets from the view of pansystem theory. In: Hu, X.H., Liu, Q., Skowron, A. (eds) IEEE International Conference on Granular Computing, vol. 2, pp. 520–525 (2005)

30. Pei, W.B., Lin, H., Li, L.Y.: Optimal-neighborhood statistics rough set approach with multiple attributes and criteria. In: Miao, D., Pedrycz, W., Slezak, D., Peters, G., Hu, Q., Wang, R. (eds.) RSKT 2014. LNCS, vol. 8818, pp. 683–692. Springer, Heidelberg (2014)

Determining Three-Way Decision Regions by Combining Gini Objective Functions and GTRS

Yan Zhang[(⊠)] and JingTao Yao

Department of Computer Science, University of Regina, Regina, SK S4S 0A2, Canada
{zhang83y,jtyao}@cs.uregina.ca

Abstract. Game-theoretic rough set model (GTRS) is a recent advancement in determining decision regions by formulating competition or cooperation between multiple measures of decision regions. Different competitions can be formulated with GTRS to gain optimal and balanced decision regions. In three-way decisions, there are some remaining issues where GTRS may be employed to reach a compromise between conflicting measures. When Gini coefficient is used to measure impurity of decision regions, Gini objective functions may be formulated to optimize impurities of multiple decision regions. We aim to examine the problem of minimizing the impurities of immediate and non-commitment decision regions simultaneously. In particular, we consider using GTRS to determine three-way decision regions by finding a solution to Gini objective functions. A compromise solution from various Pareto optimal strategies is obtained with GTRS. The game formulation, Pareto optimal strategies, Nash equilibrium of games, as well as iteration learning mechanism are investigated in detail. An example to demonstrate that compromise decision regions can be obtained by using GTRS to formulate competitions between decision regions is presented.

Keywords: Game-theoretic rough sets · Three-way decisions · Impurity · Gini objective function · Game theory

1 Introduction

Three-way decisions are constructed based on the notions of acceptance, rejection and non-commitment [16]. Given U as a finite nonempty set of objects and C as a finite set of criteria, the aim of three-way decisions is to partition U based on C into three disjoint decision regions, acceptance, rejection, and non-commitment regions [16]. We may make decisions and induce rules from acceptance and rejection regions. In many cases, we may have to defer decisions or make non-commitment decisions from non-commitment region. The acceptance and rejection regions can be called immediate decision regions [4]. Determining three-way decision regions is one of key issues in three-way decisions. Game-theoretic rough set model (GTRS) employs game mechanisms to determine three

© Springer International Publishing Switzerland 2015
Y. Yao et al. (Eds.): RSFDGrC 2015, LNAI 9437, pp. 414–425, 2015.
DOI: 10.1007/978-3-319-25783-9_37

decision regions [5,13]. It formulates competition or cooperation between multiple measures of decision regions to reach agreements [1]. The essential idea of GTRS is to implement a game to obtain rough thresholds that improves the rough sets based decision making. In the existing formulations, GTRS is applied in probabilistic rough sets to determine and interpret the probabilistic thresholds that define the decision regions. It implements games between two conflicting criteria to optimize probabilistic thresholds. Herbert and Yao proposed that two probabilistic thresholds (α, β) compete against each other to directly reduce the boundary regions [5]. They also formulated competition games between two classification approximation measures, i.e., accuracy and precision, to improve the classification ability of rough set model [5]. Azam and Yao applied GTRS to formulate competition games between two measures for evaluating positive rules, i.e., confidence and coverage, to solve multiple criteria decision making problems in rough sets [1]. Azam and Yao optimized the probabilistic thresholds with GTRS model by considering the competition between two properties of rough set model, accuracy and generality [2]. They also proposed a competition game between immediate and deferred decision regions to improve the overall uncertainty level of the rough set classification [2]. Zhang and Yao summarize the rule measures that can be set as game players in GTRS [17]. These studies not only provide a good beginning for GTRS research, but also build up a solid foundation for future GTRS research.

Gini coefficient is a kind of entropy calculation, and it can be adopted to measure the impurity of three-way decision regions defined by rough set model [18]. Gini coefficient provides a tradeoff between applicability and accuracy of decision regions, and it has been used to generate optimal and balanced decision regions [18]. We formulated Gini objective functions to interpret a balance between decision regions. There are two types of objective functions, i.e., single-objective optimization and multi-objective optimization. The Gini objective function that minimizes the overall impurity is a single-objective optimization problem, and it aims to minimize the summation of three impurities of decision regions. Decision-theoretic rough sets (DTRS) [14] and information-theoretic rough sets (ITRS) [3] are both based on single optimization and they aim to minimize cost and uncertainty, respectively. Deng and Yao used gradient-descent approach to search optimal pair of thresholds [3]. Azam and Yao applied GTRS to find three-way decision regions with minimal overall uncertainty [2].

The Gini objective function that minimizes impurities of immediate and non-commitment decision regions simultaneously is a multi-objective optimization problem [19]. When three-way decision regions changes, impurities of immediate and non-commitment decision regions changes as well. In particular, impurities of immediate decision regions decrease at expense of the increase of impurity of non-commitment region, and vice versa [18,19]. Lower impurity levels of immediate decision regions mean the three-way decisions have a high level of accuracy. A lower impurity level of non-commitment region means the three-way decisions have a high level of coverage. In other words, both regions compete each other to decrease their own impurities [18,19]. In this paper, we will use GTRS to find the solution of the Gini objective function. The competitive games between immediate and non-commitment decision regions will be formulated. Since impurities of immediate and non-commitment decision regions change in the opposite direction, all possible threshold pairs are Pareto optimal. we will focus on how to apply GTRS to find equilibrium from all Pareto optimal strategies. Compromise decision regions can be obtained by repeating competitive games. The results in this study may enhance our understanding of GTRS and make it practical in applications.

2   Background Knowledge

In this section, we briefly introduce the background concepts about three-way decisions and game theory.

2.1   Three-Way Decisions

The three-way decision theory was outlined by Yao in [16]. It classifies the objects of U into acceptance, rejection and non-commitment decision regions. When formulating three-way decisions we should define evaluation functions and designed values [16]. Three-way decisions can be formulated from many models or approaches, such as such as rough sets, fuzzy sets, shadow sets, interval sets [16]. These models provide different definitions of evaluation functions, as well as determinations of designed values for acceptance and rejection.

In this paper, we use probabilistic rough sets to formulate three-way decisions. Suppose the universe U is a finite nonempty set. Let $E \subseteq U \times U$ be an equivalence relation on U, where E is reflexive, symmetric, and transitive [10]. For an element $x \in U$, the equivalence class containing x is given by $[x] = \{y \in U | xEy\}$. The family of all equivalence classes defines a partition of the universe and is denoted by $U/E = \{[x] | x \in U\}$, that is the intersection of any two elements is an empty set and the union of all elements are the universe U [10]. For an indescribable target concept $C \subseteq U$, probabilistic rough sets utilize conditional probability and thresholds (α, β) to define three decision regions, i.e., acceptance, rejection and non-commitment regions of C [15]:

$$POS_{(\alpha,\beta)}(C) = \bigcup \{[x] \mid [x] \in U/E, Pr(C|[x]) \geq \alpha\},$$

$$NEG_{(\alpha,\beta)}(C) = \bigcup \{[x] \mid [x] \in U/E, Pr(C|[x]) \leq \beta\},$$

$$BND_{(\alpha,\beta)}(C) = \bigcup \{[x] \mid [x] \in U/E, \beta < Pr(C|[x]) < \alpha\}. \tag{1}$$

Here, the conditional probability of an equivalence class $[x]$ in C given the equivalence class $[x]$, i.e., $Pr(C|[x]) = \frac{|[x] \cap C|}{|[x]|}$ is used as an evaluation function. Thresholds α and β are used as designed values for acceptance and rejection respectively [15]. Intuitively speaking, given an equivalence class $[x]$, if $Pr(C|[x]) \geq \alpha$, we consider that all objects in $[x]$ belong to the concept C, i.e., accept $[x]$ as C. If $Pr(C|[x]) \leq \beta$, we consider that all objects in $[x]$ do not belong to the concept

C, i.e., reject $[x]$ as C. If $\beta < Pr(C|[x]) < \alpha$, we defer to make decisions. So far, three-way decision model formulated by rough sets has been used in many applications [7,8].

2.2   Concepts in Game Theory

Game theory is a study of strategic decision-making where several players make choices that potentially affect the payoffs of other players [11]. It can represent multi-objective optimization with multiple decision-makers, each controlling certain design variables [12]. In game theory, a game can be any situation of conflict or cooperation between two or more players each with multiple possible strategies [9]. There are different representations of games. The normal form, also known as the strategic or matrix form, is the most familiar representation of strategic interactions in game theory [6].

Definition 1 *Normal-form game*. *A normal-form game is a tuple (O, S, u):*

- *O is a finite set of n players, indexed by i;*
- *$S = S_1 \times ... \times S_n$, where S_i is a finite set of all possible strategies for player i. The n-tuple $s = (s_1, s_2, ..., s_n) \in S$ is called a strategy profile where one strategy for each player;*
- *$u = (u_1, ..., u_n)$ where $u_i : S \mapsto \mathbb{R}$ is a real-valued utility (or payoff) function for player i. Utility $u_i(s)$ means the payoff for player i with the strategy profile s.*

It is natural to represent games with an n-dimensional matrix which shows the players, strategies, and payoffs. A two-player game is a 2-dimensional matrix, where each row denotes a possible strategy for player 1, each column denotes a possible strategy for player 2, and each cell denotes one possible outcome. Each player's payoff for an outcome is in the cell corresponding to that outcome, with player 1's utility listed first.

In the multi-player games, all players are hoping to maximize their payoffs. The optimal strategy depends on the choice of every player. We use two fundamental solution concepts, Pareto optimality and Nash equilibrium. Pareto optimality explains which outcomes of a game are better from the point of view of an outside observer [6]. Nash equilibrium observes what features the optimal strategy should have from an individual player's point of view [6].

Definition 2 *Pareto domination*. *Strategy profile s Pareto dominates strategy profile s' if for all $i \in P, u_i(s) \geq u_i(s')$, and there exists some $j \in O$ for which $u_j(s) > u_j(s')$.*

Definition 3 *Pareto optimality*. *Strategy profile s is Pareto optimal, if there does not exist another strategy profile $s' \in S$ that Pareto dominates s.*

In a Pareto dominant strategy profile no player can be made better without making any other player worse off.

Suppose $s_{-i} = (s_1, s_2, ..., s_{i-1}, s_{i+1}, ..., s_n)$ represent an opposing strategy profile without player i's strategy, it can be denoted as $s = (s_i, s_{-i})$.

Definition 4 *Best response. Player i's best response to the strategy profile s_{-i} is a mixed strategy $s_i^* \in S_i$ such that $u_i(s_i^*, s_{-i}) \geq u_i(s_i, s_{-i})$ for all strategies $s_i \in S_i$.*

Definition 5 *Nash equilibrium. A strategy profile $s = (s_1, ..., s_n)$ is a Nash equilibrium, if for all players $i \in O$, s_i is a best response to s_{-i}.*

Intuitively speaking, a Nash equilibrium is a strategy profile in which no player can better their payoffs by changing their strategies if they knew what strategies the other players were performing.

3   Solving Gini Objective Functions with GTRS

Given a criterion C and acceptance and rejection thresholds (α, β), the three-way decisions divide U into three disjoint decision regions, i.e., acceptance, rejection, and non-commitment regions which are denoted as $POS_{(\alpha,\beta)}(C)$, $NEG_{(\alpha,\beta)}(C)$ and $BND_{(\alpha,\beta)}(C)$, respectively. Balanced decision regions are necessary conditions for making effective decisions. In this section, we will discuss how to use game-theoretic rough set model to determine three decision regions which can simultaneously minimize impurities of immediate and non-commitment regions.

3.1   Gini Objective Function

When we use Gini coefficients to evaluate impurities of three-way decision regions, the impurities of three regions can be computed by the following formulas [18],

$$
\begin{aligned}
G_P(\alpha, \beta) =& Pr(POS_{(\alpha,\beta)}(C)) \times \\
& \left(1 - Pr(C|POS_{(\alpha,\beta)}(C))^2 - Pr(C^c|POS_{(\alpha,\beta)}(C))^2\right), \\
G_N(\alpha, \beta) =& Pr(NEG_{(\alpha,\beta)}(C)) \times \\
& \left(1 - Pr(C|NEG_{(\alpha,\beta)}(C))^2 - Pr(C^c|NEG_{(\alpha,\beta)}(C))^2\right), \\
G_B(\alpha, \beta) =& Pr(BND_{(\alpha,\beta)}(C)) \times \\
& \left(1 - Pr(C|BND_{(\alpha,\beta)}(C))^2 - Pr(C^c|BND_{(\alpha,\beta)}(C))^2\right).
\end{aligned}
\tag{2}
$$

In these equations, the probabilities of decision regions $Pr(\Delta_{(\alpha,\beta)}(C))$ can be calculated as,

$$
Pr(\Delta_{(\alpha,\beta)}(C)) = \frac{|\Delta_{(\alpha,\beta)}(C)|}{|U|},
\tag{3}
$$

where Δ can be replaced by POS, NEG or BND, and $|\cdot|$ denotes the cardinality of a set. The probability $Pr(C|\Delta_{(\alpha,\beta)}(C))$ denotes the conditional probability of an object x in C given that the object is in the decision region $\Delta_{(\alpha,\beta)}(C)$. The conditional probabilities $Pr(C|\Delta_{(\alpha,\beta)}(C))$ and $Pr(C^c|\Delta_{(\alpha,\beta)}(C))$ are computed as,

$$
Pr(C|\Delta_{(\alpha,\beta)}(C)) = \frac{|C \cap \Delta_{(\alpha,\beta)}(C)|}{|\Delta_{(\alpha,\beta)}(C)|},
$$

$$
Pr(C^c|\Delta_{(\alpha,\beta)}(C)) = \frac{|C^c \cap \Delta_{(\alpha,\beta)}(C)|}{|\Delta_{(\alpha,\beta)}(C)|}.
\tag{4}
$$

The impurities of immediate decision regions are the summation of impurities of acceptance and rejection decision regions [18],

$$G_I(\alpha, \beta) = G_P(\alpha, \beta) + G_N(\alpha, \beta). \tag{5}$$

One of Gini objective functions is to minimize impurities of immediate and non-commitment decision regions simultaneously, that is,

$$(\alpha, \beta) = \{(\alpha, \beta)| min\, (G_I(\alpha, \beta), G_B(\alpha, \beta))\}. \tag{6}$$

The goal is to obtain balanced three-way decision regions with minimal impurities of immediate and non-commitment decision regions. In fact, the two sub-objectives contained in Eq. (6) are conflicting. The decrease of one inevitably causes the increase of the other. The Eq. (6) is a typical multi-objective optimization problem, and we will use GTRS to find a compromise solution to this problem.

3.2   Game Formulation and Analysis

We represent games with the normal form. We will discuss game players, strategies, payoffs, equilibrium, and learning mechanism in detail in this section.

Players. In a competitive game, the players are immediate decision regions and non-commitment decision region, i.e., $O = \{I, B\}$. Here I denotes immediate decision regions and B denotes non-commitment decision region.

Initial Values and Possible Strategies. The strategy set are made up of all possible strategies or actions performed by all players. The strategy set is $S = \{S_I, S_B\}$. Let us investigate what actions could be included in the strategy set of each player. We change decision regions by tuning the values of acceptance and rejection thresholds (α, β). Two players are both trying to minimize their impurities according to Gini objective function shown in Eq. (6). The threshold α controls which objects can be classified in acceptance region. The greater the value of α, the stricter the condition of admission to acceptance regions. That means the impurity of acceptance region is lower, or the region is purer and smaller. The threshold β controls which objects can be classified in rejection region. The lower the value of β, the stricter the condition of admission to rejection regions. That means the impurity of rejection region is lower, or the region is purer. Player I can obtain lower impurity with the increase of α and decrease of β. The actions that player I preferred can be increasing α, decreasing β, or both of them. Since the universe U is fixed, the three-way decisions with smaller immediate decision regions may have a larger non-commitment region. Player B and I desire contradictory situations. For player B, the decrease of α and increase of β can result in lower impurity. The actions that player B preferred can be decreasing α, increasing β, or both.

The initial values of (α, β) have an effect on the actions. The thresholds (α, β) should satisfy the constraint $0 \leq \beta \leq \alpha \leq 1$ in order to keep decision regions disjoint. When initial threshold $\alpha = 1$, the players are not able to increase α. If we set initial values (α, β) as $(1, 0.5)$, player I can decrease β and player B can decrease α. The strategy set of player I can be $S_I = \{\beta, \beta - 1c_I, \beta - 2c_I, ...\}$. The strategy set of player B can be $S_B = \{\alpha, \alpha - 1c_B, \alpha - 2c_B, ...\}$. Here c_I and C_B denote the step values of

Table 1. The payoff table

		B		
		α	$\alpha - 0.1$	$\alpha - 0.2$
	β	$\langle u_I(\frac{\alpha+\alpha}{2}, \frac{\beta+\beta}{2}),$ $u_B(\frac{\alpha+\alpha}{2}, \frac{\beta+\beta}{2})\rangle$	$\langle u_I(\frac{2\alpha-0.1}{2}, \frac{\beta+\beta}{2}),$ $u_B(\frac{2\alpha-0.1}{2}, \frac{\beta+\beta}{2})\rangle$	$\langle u_I(\frac{2\alpha-0.2}{2}, \frac{\beta+\beta}{2}),$ $u_B(\frac{2\alpha-0.2}{2}, \frac{\beta+\beta}{2})\rangle$
I	$\beta - 0.1$	$\langle u_I(\frac{\alpha+\alpha}{2}, \frac{2\beta-0.1}{2}),$ $u_B(\frac{\alpha+\alpha}{2}, \frac{2\beta-0.1}{2})\rangle$	$\langle u_I(\frac{2\alpha-0.1}{2}, \frac{2\beta-0.1}{2}),$ $u_B(\frac{2\alpha-0.1}{2}, \frac{2\beta-0.1}{2})\rangle$	$\langle u_I(\frac{2\alpha-0.2}{2}, \frac{2\beta-0.1}{2}),$ $u_B(\frac{2\alpha-0.2}{2}, \frac{2\beta-0.1}{2})\rangle$
	$\beta - 0.2$	$\langle u_I(\frac{\alpha+\alpha}{2}, \frac{2\beta-0.2}{2}),$ $u_B(\frac{\alpha+\alpha}{2}, \frac{2\beta-0.2}{2})\rangle$	$\langle u_I(\frac{2\alpha-0.1}{2}, \frac{2\beta-0.2}{2}),$ $u_B(\frac{2\alpha-0.1}{2}, \frac{2\beta-0.2}{2})\rangle$	$\langle u_I(\frac{2\alpha-0.2}{2}, \frac{2\beta-0.2}{2}),$ $u_B(\frac{2\alpha-0.2}{2}, \frac{2\beta-0.2}{2})\rangle$

decreasing β and increasing α and $0 \leq c_I, c_B \leq 1$. When we set $c_I = 0.1$ and $C_B = 0.1$, $S_I = \{\beta, \beta - 0.1, \beta - 0.2, ...\}$ and $S_B = \{\alpha, \alpha - 0.1, \alpha - 0.2, ...\}$. The steps to decrease or increase thresholds depends on data set. If the data set is big, a smaller step is recommended. However, a very small step will cause execution time increased and a very large step will influence the precision of thresholds. The number of strategies or actions is not fixed. Normally, we set 3 or 4 actions for each player in one game iteration. If there are too many actions in an iteration, more strategy profiles will be contained in one payoff table, which will increase the amount of calculation of payoffs.

Payoffs. The payoffs of players are $U = \{u_I, u_B\}$, and functions $u_I(\alpha, \beta)$ and $u_B(\alpha, \beta)$ are defined by the impurities of immediate and non-commitment decision regions. As in game-theoretic analysis we are interested in measuring profits, we use 1 minus impurities of decision regions for calculating payoff functions:

$$u_I(\alpha, \beta) = 1 - G_I(\alpha, \beta), \ u_B(\alpha, \beta) = 1 - G_B(\alpha, \beta), \tag{7}$$

where $G_I(\alpha, \beta)$ and $G_B(\alpha, \beta)$ are defined as in Eqs. (2) and (5). Both players are trying to maximize their own payoffs. We use payoff tables to represent games, as shown in Table 1. On each cell, the final threshold values are the average of corresponding thresholds adopted by two players. For example, the cell on the second row and second column in Table 1, player I chooses the strategy decreasing β by 0.1 and player B chooses the strategy decreasing α by 0.1, the threshold values after affected by two players are $(\frac{\alpha+\alpha-0.1}{2}, \frac{\beta+\beta-0.1}{2}) = (\frac{2\alpha-0.1}{2}, \frac{2\beta-0.1}{2})$. Here, α and β are initial values of thresholds.

Solutions and Equilibria. We have defined the players and the strategies available to players in games, as well as payoff table, now the question is how to find solutions to this competitive game. In multi-objective problems, all players are hoping to maximize their own payoffs. Thus an optimal strategy for one player is not meaningful. The best strategy depends on the choices of all players.

Generally speaking, when thresholds (α, β) change, the decrease of one impurity inevitably causes the increase of the other impurity. When α decreases or β increases, the conditions of admission to immediate decision regions become much looser. More objects belonging to non-commitment decision region are classified to immediate decision regions. The impurity levels of immediate decision regions are increased. Meanwhile, the impurity level of non-commitment decision region is decreased. Similarly, when α increases or β decreases, the impurity levels of immediate decision regions will be decreased, and that of non-commitment decision region will be increased. For any two strategy profiles $s, s' \in S$, $u_I(s) > u_I(s')$ and $u_B(s) > u_B(s')$ can not hold simultaneously. In other words, there does not exist any strategy profile $s \in S$ that dominates all others strategy profiles. All strategy profiles are Pareto optimal according to Definitions 2 and 3.

Based on the payoff table shown in Table 1, player I can always better its payoff when decreasing β more no matter what strategy player B performs, i.e., $u_I(\beta-0.2, s) \geq u_I(s^*, s)$, and $s^* \in S_I$, $s \in S_B$. Player I's strategy $\beta - 0.2$ is the best response to player B's strategies. Similarly, player B can gain more profit if α decreass no matter what strategy player I chooses. Player B's strategy $\alpha - 0.2$ is the best response to player I's strategies. According to Definition 5, we can see that the strategy profile $(\beta-0.2, \alpha-0.2)$ corresponding to the bottom right cell in the Table 1 is Nash equilibrium.

Repetition of Games. GTRS aims to optimize a suitable thresholds pair by formulating games. If the values of thresholds obtained in the current game are not good enough to apply in decision making, we will repeat the game with updated thresholds. Repetition of game formulation means that we may be able to find suitable thresholds with repetition of thresholds modification. Assuming the initial thresholds are (α, β), equilibrium analysis shows that the result thresholds are (α', β'). In the subsequent iteration of the game, the initial thresholds will be (α', β'), the strategies are the change of (α', β'), i.e., $S_I = \{\beta', \beta'-step, \beta'-step\times 2, ...\}$ and $S_B = \{\alpha', \alpha'-step, \alpha'-step\times 2, ...\}$. The game may be repeated until the stop conditions are satisfied.

Stop Conditions. There are many possible stop conditions, for example, the acceptance region and rejection region are not overlapped $(\alpha \leq \beta)$, the payoffs of players are beyond some specific values, subsequent iteration does not improve previous configurations. In this paper, we set the stop condition as the gain of one player's payoff is less than the loss of the other player's payoff in the current game. The final decision regions can be defined by the initial thresholds used in the current game. A concrete example will be given in the next section.

4   An Example

In this section, we present an example to demonstrate that suitable decision regions can be obtained by formulating a GTRS game between immediate and non-commitment decision regions. Table 2 summarizes probabilistic data about a concept C. There are 16 equivalence classes denoted by $X_i(i = 1, 2, ..., 16)$, which are listed in a decreasing order of the conditional probabilities $Pr(C|X_i)$ for convenient computations.

When $(\alpha, \beta) = (1, 0)$, the immediate decision regions are

$$POS_{(1,0)}(C) \cup NEG_{(1,0)}(C) = X_1 \cup X_{16},$$

the non-commitment decision region is

$$BND_{(1,0)}(C) = X_2 \cup X_3 \cup ... \cup X_{15}.$$

The impurities of immediate decision regions are

$$G_I(1,0) = G_P(1,0) + G_N(1,0) = 0.$$

Table 2. Summary of the experimental data

	X_1	X_2	X_3	X_4	X_5	X_6	X_7	X_8
$P(X_i)$	0.093	0.088	0.093	0.089	0.069	0.046	0.019	0.015
$P(C\|X_i)$	1	0.978	0.95	0.91	0.89	0.81	0.72	0.61
	X_9	X_{10}	X_{11}	X_{12}	X_{13}	X_{14}	X_{15}	X_{16}
$P(X_i)$	0.016	0.02	0.059	0.04	0.087	0.075	0.098	0.093
$P(C\|X_i)$	0.42	0.38	0.32	0.29	0.2	0.176	0.11	0

For the non-commitment decision region, the probability is

$$Pr(BND_{(1,0)}(C)) = \sum_{i=2}^{15} Pr(X_i) = 0.814.$$

The conditional probability of C is

$$Pr(C|BND_{(1,0)}(C)) = \frac{\sum_{i=2}^{15} Pr(C|X_i)Pr(X_i)}{\sum_{i=2}^{15} Pr(X_i)} = \frac{0.4631}{0.814} = 0.5689.$$

The impurity of the non-commitment decision region is

$$G_B(1,0) = Pr(BND_{(1,0)}(C)) \times (1 - (0.5689)^2 - (1 - 0.5689)^2)$$
$$= 0.814 \times 0.4905 = 0.3993.$$

When formulating games between immediate and non-commitment decision regions, the players are immediate and non-commitment decision regions, i.e., $O = \{I, B\}$. The strategy sets are $S = \{S_I, S_B\}$. The initial thresholds are $(\alpha, \beta) = (1, 0.5)$. Player I tries to decrease β, so its strategy set is $S_I = \{(1, 0.5), (1, 0.4), (1, 0.3)\}$. Player B tries to decrease α, so its strategy set is $S_B = \{(1, 0.5), (0.9, 0.5), (0.8, 0.5)\}$. All possible strategy profiles of payoff table are shown in Table 3. The payoff table is shown in Table 4.

The strategy profile $((1, 0.3), (0.8, 0.5))$ is the equilibrium, then we will repeat the game by setting $(0.9, 0.4)$ as initial thresholds. In the second iteration of the game, the initial thresholds is $(\alpha, \beta) = (0.9, 0.4)$, two players' strategy sets are $S_I = \{(0.9, 0.4), (0.9, 0.3), (0.9, 0.2)\}$ and $S_B = \{(0.9, 0.4), (0.8, 0.4), (0.7, 0.4)\}$, respectively.

Table 3. All possible strategy profiles for the payoff table

		B		
		$(1, 0.5)$	$(0.9, 0.5)$	$(0.8, 0.5)$
	$(1, 0.5)$	$\langle u_I(1, 0.5),$ $u_B(1, 0.5)\rangle$	$\langle u_I(0.95, 0.5),$ $u_B(0.95, 0.5)\rangle$	$\langle u_I(0.9, 0.5),$ $u_B(0.9, 0.5)\rangle$
I	$(1, 0.4)$	$\langle u_I(1, 0.45),$ $u_B(1, 0.45)\rangle$	$\langle u_I(0.95, 0.45),$ $u_B(0.95, 0.45)\rangle$	$\langle u_I(0.9, 0.45),$ $u_B(0.9, 0.45)\rangle$
	$(1, 0.3)$	$\langle u_I(1, 0.4),$ $u_B(1, 0.4)\rangle$	$\langle u_I(0.95, 0.4),$ $u_B(0.95, 0.4)\rangle$	$\langle u_I(0.9, 0.4),$ $u_B(0.9, 0.4)\rangle$

Table 4. The payoff table of competition between decision regions

	B		
	$(1, 0.5)$	$(0.9, 0.5)$	$(0.8, 0.5)$
I $(1, 0.5)$	$< 0.8581, 0.9243 >$	$< 0.8452, 0.9396 >$	$< 0.8301, 0.9551 >$
$(1, 0.4)$	$< 0.8581, 0.9243 >$	$< 0.8452, 0.9396 >$	$< 0.8301, 0.9551 >$
$(1, 0.3)$	$< 0.8678, 0.9094 >$	$< 0.8550, 0.9262 >$	$< \mathbf{0.8399, 0.9428} >$

We set the stop condition as the gain of one player's payoff is less than the loss of the other player's payoff in the current game. The competition will be repeated four times. The result is shown in Table 5. In the forth iteration, we can see that the gain of the payoff values of player I is 0.0630 which is less than the loss of payoff values of player B 0.0647. The repetition of game is stopped and the final result is the initial thresholds of the forth competitive game $(\alpha, \beta) = (0.7, 0.2)$.

Table 5. The repetition of games

	Initial (α, β)	Result (α, β)	Payoffs	Difference of payoffs change
1	$(1, 0.5)$	$(0.9, 0.4)$	$< 0.8399, 0.9428 >$	
2	$(0.9, 0.4)$	$(0.8, 0.3)$	$< 0.8505, 0.9366 >$	$0.0206 - 0.0062 = 0.0144$
3	$(0.8, 0.3)$	$(0.7, 0.2)$	$< 0.8597, 0.9309 >$	$0.0092 - 0.0057 = 0.0035$
4	$(0.7, 0.2)$	$(0.6, 0.1)$	$< 0.9227, 0.8652 >$	$0.0630 - 0.0647 = -0.0017$

The impurities of immediate and non-commitment decision regions are 0.1403 and 0.0691, respectively. At this time, two types of regions reach the compromise on simultaneously minimizing their impurity levels. Now we will use accuracy and coverage to measure the three-way decisions defined by thresholds $(\alpha, \beta) = (0.7, 0.2)$. The accuracy represents the proportion of objects that can be correctly classified by three-way decisions. It is the ratio of the number of correctly classified objects by the immediate decision regions and the number of objects covered by these regions. The coverage is intended to express the applicability of three-way decisions. It is the ratio of the number of objects in the immediate decision regions to the number of all objects in the universe. The definition formulas can be referred in [15, 17]. When $(\alpha, \beta) = (0.7, 0.2)$, the three-way decisions have accuracy of 90.98 % and coverage of 85 %. This means 85 % objects are able to be classified with accuracy of 90.98 %. The remain 15 % objects can not be classified without providing more information. The three-way decision regions defined by the initial thresholds $(\alpha, \beta) = (1, 0.5)$ have accuracy of 85.34 % and coverage of 58.1 %. Compared with the initial three-way decision regions, the decision regions defined by $(0.7, 0.2)$ increases the accuracy by 5.64 % and coverage by 25.9 % which are more accurate and more applicable.

5   Conclusion

Game-theoretic rough sets provide a tradeoff mechanism by simultaneously considering multiple measures for an effective determination of three-way decision regions.

In this paper, we examine the issues about using game-theoretic rough set model to find solutions to a multi-objective optimization problem, i.e., the Gini objective function minimizing impurities of immediate and non-commitment decision regions simultaneously. The competitive games between immediate and non-commitment decision regions are formulated. The strategies are the change of thresholds which can define decision regions. The payoff functions are 1 minus impurities of corresponding decision regions. All strategy profiles are Pareto optimal and game mechanism is adopted to obtain a compromise solution which is represented by Nash equilibrium from these Pareto optimizations. We employ iterative learning mechanism so that the decision regions with the minimal impurities of immediate and non-commitment decision regions can be obtained.

Acknowledgements. This work is partially supported by a Discovery Grant from NSERC Canada, the University of Regina Gerhard Herzberg Fellowship and Verna Martin Memorial Scholarship.

References

1. Azam, N., Yao, J.T.: Multiple criteria decision analysis with game-theoretic rough sets. In: Li, T., Nguyen, H.S., Wang, G., Grzymala-Busse, J., Janicki, R., Hassanien, A.E., Yu, H. (eds.) RSKT 2012. LNCS, vol. 7414, pp. 399–408. Springer, Heidelberg (2012)
2. Azam, N., Yao, J.T.: Analyzing uncertainties of probabilistic rough set regions with game-theoretic rough sets. Int. J. Approximate Reasoning **55**(1), 142–155 (2014)
3. Deng, X.F., Yao, Y.Y.: A multifaceted analysis of probabilistic three-way decisions. Fundamenta Informaticae **132**(3), 291–313 (2014)
4. Herbert, J.P., Yao, J.T.: Criteria for choosing a rough set model. Comput. Math. Appl. **57**(6), 908–918 (2009)
5. Herbert, J.P., Yao, J.T.: Game-theoretic rough sets. Fundamenta Informaticae **108**(3–4), 267–286 (2011)
6. Leyton-Brown, K., Shoham, Y.: Essentials of game theory: a concise multidisciplinary introduction. Synth. Lect. Artif. Intell. Mach. Learn. **2**(1), 1–88 (2008)
7. Li, H.X., Zhang, L., Huang, B., Zhou, X.Z.: Sequential three-way decision and granulation for cost-sensitive face recognition. Knowledge-Based Systems (2015). http://dx.doi.org/10.1016/j.knosys.2015.07.040
8. Liu, D., Liang, D.C., Wang, C.C.: A novel three-way decision model based on incomplete information system. Knowledge-Based Systems (2015). http://dx.doi.org/10.1016/j.knosys.2015.07.036
9. Marler, R.T., Arora, J.S.: Survey of multi-objective optimization methods for engineering. Struct. Multi. Optim. **26**(6), 369–395 (2004)
10. Pawlak, Z.: Rough Sets: Theoretical Aspects of Reasoning About Data. Kluwer Academic Publishers, Boston (1991)
11. Turocy, T., Von Stengel, B.: Game theory. Encycl. Inf. Syst. **2**, 403–420 (2002)
12. Vincent, T.L.: Game theory as a design tool. J. Mech. Design **105**(2), 165–170 (1983)
13. Yao, J.T., Herbert, J.P.: A game-theoretic perspective on rough set analysis. J. Chongqing Univ. Posts Telecommun. **20**(3), 291–298 (2008)

14. Yao, Y.Y.: Decision-theoretic rough set models. In: Yao, J.T., Lingras, P., Wu, W.-Z., Szczuka, M.S., Cercone, N.J., Ślęzak, D. (eds.) RSKT 2007. LNCS (LNAI), vol. 4481, pp. 1–12. Springer, Heidelberg (2007)
15. Yao, Y.Y.: The superiority of three-way decisions in probabilistic rough set models. Inf. Sci. **181**(6), 1080–1096 (2011)
16. Yao, Y.Y.: An outline of a theory of three-way decisions. In: Yao, J.T., Yang, Y., Słowiński, R., Greco, S., Li, H., Mitra, S., Polkowski, L. (eds.) RSCTC 2012. LNCS, vol. 7413, pp. 1–17. Springer, Heidelberg (2012)
17. Zhang, Y., Yao, J.T.: Rule measures tradeoff using game-theoretic rough sets. In: Zanzotto, F.M., Tsumoto, S., Taatgen, N., Yao, Y. (eds.) BI 2012. LNCS, vol. 7670, pp. 348–359. Springer, Heidelberg (2012)
18. Zhang, Y., Yao, J.T.: Determining three-way decision regions with gini coefficients. In: Cornelis, C., Kryszkiewicz, M., Ślęzak, D., Ruiz, E.M., Bello, R., Shang, L. (eds.) RSCTC 2014. LNCS, vol. 8536, pp. 160–171. Springer, Heidelberg (2014)
19. Zhang, Y., Yao, J.T.: Gini objective functions for three-way decisions (manuscript)

IJCRS 2015 Data Challenge

Mining Data from Coal Mines:
IJCRS'15 Data Challenge

Andrzej Janusz[1]([✉]), Marek Sikora[2], Łukasz Wróbel[2,3], Sebastian Stawicki[1,4],
Marek Grzegorowski[1], Piotr Wojtas[3], and Dominik Ślęzak[1,4]

[1] Institute of Mathematics, University of Warsaw, Banacha 2, 02-097 Warsaw, Poland
janusza@mimuw.edu.pl, m.grzegorowski@mimuw.edu.pl
[2] Institute of Computer Science, Silesian University of Technology,
Akademicka 16, 44-100 Gliwice, Poland
marek.sikora@polsl.pl
[3] Institute of Innovative Technologies EMAG, Leopolda 31, 40-189 Katowice, Poland
lukasz.wrobel@ibemag.pl
[4] Infobright Inc., Krzywickiego 34, lok. 219, 02-078 Warsaw, Poland
{stawicki,slezak}@mimuw.edu.pl

Abstract. We summarize the data mining competition associated with
IJCRS'15 conference – IJCRS'15 Data Challenge: Mining Data from Coal
Mines, organized at Knowledge Pit web platform. The topic of this competition was related to the problem of active safety monitoring in underground corridors. In particular, the task was to design an efficient method
of predicting dangerous concentrations of methane in longwalls of a
Polish coal mine. We describe the scope and motivation for the competition. We also report the course of the contest and briefly discuss a
few of the most interesting solutions submitted by participants. Finally,
we reveal our plans for the future research within this important subject.

Keywords: Data mining competitions · Time series data · Attribute
engineering · Feature extraction · High dimensional data.

1 Introduction

Coal mining requires working in hazardous conditions. Miners in an underground
coal mine can face several threats, such as, e.g. methane explosions or rock-
burst. To provide protection for people working underground, systems for active
monitoring of a production processes are typically used. One of their fundamental
applications is screening dangerous gas concentrations (methane in particular)
in order to prevent spontaneous explosions [1]. Therefore, for that purpose the
ability to predict dangerous concentrations of gases in the nearest future can be
even more important than monitoring the current sensor readings [2].

Partially supported by the Polish National Science Centre – grant 2012/05/B/ST6/-
03215 and by the Polish National Centre for Research and Development (NCBiR) –
grant PBS2/B9/20/2013 in frame of the Applied Research Programs.

© Springer International Publishing Switzerland 2015
Y. Yao et al. (Eds.): RSFDGrC 2015, LNAI 9437, pp. 429–438, 2015.
DOI: 10.1007/978-3-319-25783-9_38

Typically, a monitoring system in a coal mine is conjugated with a power supply system and may automatically cut the power off if it detects any viable threat. Such action is necessary to prevent accidents. However, every break in the mining process results in costly losses for the mine. If the monitoring system could foresee dangerous concentrations of methane in the nearest future, dispatchers that control the mining process would be able to make some adjustments in order to decrease the transmission of the gas (e.g. by reducing the currents consumed by the cutter loader and thus lowering the intensity of the coal drilling which causes the release of methane from the soil). By doing so, it would be possible to avoid the necessity of shutting down the whole heavy machinery working in the mine. In this way, the monitoring system coupled with a decision support system could not only increase the safety of miners working underground, but also significantly reduce the overhead of the mining process.

From a data processing point of view, a decision support system which could aid in controlling the coal mining process requires efficient methods for handling continuous streams of data [3]. Such methods have to be able to handle large volumes of data from multiple sensors. They also need to be robust with regard to missing or corrupted data. Moreover, a good decision support system should be easy to comprehend by the experts and end-users who need to have access not only to its outcomes, but also to arguments or causes that were taken into account. A few practical studies have been already conducted with this respect, relying on rule-based models for predicting the methane level [4]. However, the literature on this important subject is still very scarce.

One of very few research initiatives in that field is DISESOR - a Polish national R&D project aimed at creation of an integrated decision support system for monitoring of the mining process and early detection of viable threats to people and equipment working underground. The system developed in the frame of DISESOR integrates data from different monitoring tools. It contains an expert system module that can utilize specialized domain knowledge and an analytical module which can be applied to make a diagnosis of the mining processes. When combined, these modules are capable of reliable prediction of natural hazards. The idea to popularize this topic among the data mining community by organizing an open data challenge originated within this project.

In IJCRS'15 Data Challenge we aimed to address the problem of active monitoring and prevention of methane outbreaks. We decided on the formula of an on-line and open to a whole data mining community competition due to the fact that it allows to conveniently review and test performance of the available state-of-the-art approaches. It is also an objective way of verifying the viability of not only the predictive models but also the whole analytic processes which include preprocessing methods, feature extraction, model construction and post processing of predictions (ensemble approaches). As the host we used the Knowledge Pit platform [5] which we designed to support the organization of data mining competitions associated with data science-related conferences.

The following sections give more details about the competition and its results. Section 2 briefly describes the above-mentioned Knowledge Pit platform and highlights its main functionalities. Section 3 explains our motivation to organize the competition and shows details of the utilized data set and the chosen evaluation method. Section 4 summarizes the competition results and gives some insights regarding the most successful approaches. Finally, Sect. 5 concludes the paper and indicates directions for our future research on this topic.

2   Knowledge Pit Data Challenge Platform

Knowledge Pit[1] is a web platform created to support organization of data mining challenges. On the one hand, this platform is appealing to members of the machine learning community for whom competitive challenges can be a source of new interesting research topics. Solving real-life complex problems can also be an attractive addition to academic courses for students who are interested in practical data mining. On the other hand, setting up a publicly available competition can be seen as a form of outsourcing the task to the community. This can be highly beneficial to the organizers who define the challenge, since it is an inexpensive way to solve the problem which they are investigating. Moreover, an open data mining competition can bring together domain experts and data analysts, which in a longer perspective may leverage a cooperation between the industry and academic researchers.

The Knowledge Pit platform is designed in a modular way, on top of an open-source e-learning platform *Moodle.org* and as such, it follows the best practices of a software development. The current modules of the platform include user accounts management system, competition management subsystems, time and calendar functionalities, communications features (i.e. forums and messaging subsystems), and a flexible interface for connecting automated evaluation services prepared to assess contestants' submissions.

Figure 1 shows an architecture schema of the Knowledge Pit platform. Its two main parts are the platform's engine located at a dedicated server and the evaluation subsystems. Currently, Knowledge Pit is hosted on a server belonging to Polish Information Processing Society[2].

The two main parts of the platform are the platform's engine and the evaluation subsystems. The first one provides interfaces for defining and maintaining of data challenges, management of user's profiles, submissions and private files, maintaining *Leaderboards* and the internal messaging systems (competition forums, chats, email and notification sending services, etc.). It is based on a very popular solution stack, i.e. Apache, MySQL and PHP. Together they constitute a bridge between the platform and different groups of users (guests, participants of competitions, moderators and organizers of particular challenges, managers and administrators of the system).

The second part of the platform is responsible for assessment of solutions submitted by participants of particular competitions. Due to a flexible communication mechanism, this service may be distributed among several independent

---

[1] https://knowledgepit.fedcsis.org.
[2] http://pti.org.pl/English-Version.

Knowledge Pit server

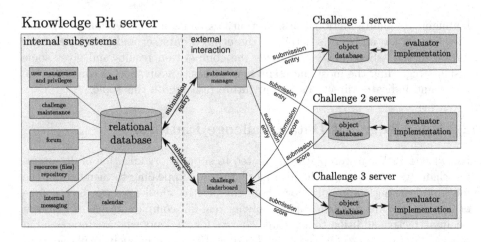

Fig. 1. A system architecture of the Knowledge Pit web platform.

workstations, which guarantees the scalability of the evaluation process. Since evaluating submissions for some competitions may require a lot of resources (e.g. memory, CPU time, disc I/O or database connections), this is a very important aspect of system's architecture. For example, the assessment of a single submission to AAIA'14 Data Mining Competition[3] required constructing several Naive Bayes classification models for a data table consisting of 50,000 objects and testing their performance on a different table with 50,000 objects described by 11,852 conditional attributes [5]. In that case, distribution of the required computations allowed for nearly real-time evaluation, even during the most busy moments when the rapid system load peak was observed.

Another advantage of separating the evaluation subsystems from the platform's engine is that it may be implemented in any suitable programming language, as a script or a standalone compiled application that can use any external libraries. In this way, the responsibility for preparation of a suitable evaluation procedure can be delegated to organizers of individual competitions. In such a case, the only requirement for the implementation of the evaluator is that it should maintain a correct protocol of information exchange with the platform's engine. The proposed flow of responsibilities releases the Knowledge Pit platform from the specific requirements and responsibilities which could not be processed in a generic way. It also gives competition organizers a very flexible method of expressing their data mining task in a form of a fully customizable evaluation procedures. For instance, the solutions submitted to IJCRS'15 Data Challenge[4], that is described in this paper, were evaluated using a script written in R language [6], which was running on an external server.

---

[3] https://knowledgepit.fedcsis.org/contest/view.php?id=83.
[4] https://knowledgepit.fedcsis.org/contest/view.php?id=109.

3   Scope of the IJCRS'15 Data Challenge

One of the main goals that we wanted to achieve was to increase the involvement of the community in devising efficient methods for assessing natural hazards in coal mines. As was explained in Sect. 1, a good solution to this issue could not only save human lives but it would also help to minimize financial losses caused by suspensions in the mining process. The task in our competition was to come up with a prediction model which could be efficiently integrated with a safety monitoring system of a coal mine, in order facilitate foreseeing warning levels of methane concentrations at three crucial methane meters located in vulnerable points of a longwall. In this section we describe the data which we provided for the competition and discuss the chosen evaluation criteria.

3.1   The Competition Data

The data used in the competition came from an active Polish coal mine. They correspond to a mining period between March 2, 2014 and June 16, 2014. The main data file for the competition consisted of multivariate time series corresponding to readings of sensors used for monitoring the conditions at the longwall. It was provided in a tabular format. In total, in the training data set there were series from $51,700$ time periods, each 10 min long, with measurements taken every second (600 values in a single series for every sensor). The values for each time period were stored in a different row of the data file. Each of the rows contained readings from 28 different sensors thus, in total, the data consisted of $16,800$ numerical attributes. The time periods in the training data were overlapping and given in a chronological order.

Data labels indicated whether a warning threshold had been reached in a period between three and six minutes after the end of the training period, for three methane meters: $MM263$, $MM264$ and $MM256$. If a given row corresponds to a period between t_{-599} and t_0, then the label for a methane meter MM in this row is "warning" if and only if $\max(MM(t_{181}), ..., MM(t_{360})) \geq 1.0$. The labels for the training data were provided in separate files. The test data file was in the same format as the training data set, however, the labels for the test series were not available for participants during the course of the competition. It is important to note that time periods in the test data did not overlap and they were given in a random order. This temporal disjunction between the training and test data makes the common assumption regarding i.i.d. data unfulfilled [7] and constitutes the biggest difficulty in the considered task.

In the data files provided for participants of the competition we also included a mining scheme utilized in the mine (Fig. 2), which explains the placement of all sensors used for monitoring the mining process. This file, combined with the provided description of sensors, constituted an important source of domain knowledge which was necessary to understand the dependencies between readings of different sensors. The cutter loader moves along the longwall between the sensors $MM262$ and $MM264$. The bigger are the currents consumed by the cutter loader, the more efficient is its mining work, which in theory results

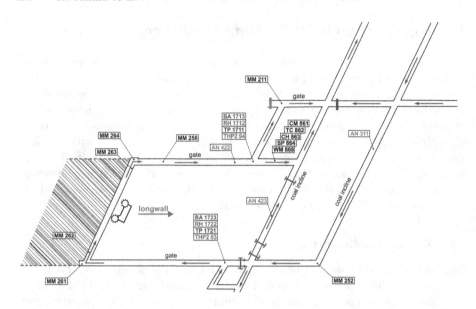

Fig. 2. A scheme of the mining process that generated the competition data.

in more methane emitted to the air. The arrows on the provided scheme show the directions of the air flow (along with methane flow) in the corridors. If the methane concentration measured by any of the sensors reaches the alarm level, the cutter loader is switched off automatically. However, if we were able to predict ahead the warning methane concentrations, we could reduce the speed of the cutter loader and give the methane more time to spread out – before the necessity of switching off the whole production line.

All the data for the competition were provided by Research and Development Centre EMAG[5] which was also the main sponsor of the competition.

3.2   Evaluation of Submissions

The evaluation procedure in the competition was two-fold. During the course of the contest an on-line evaluation system was providing constant feedback for the participants in a form of a publicly available Leaderboard — a dynamic ranking of participants' best results. However, the scores displayed on the Leaderboard were only preliminary assessments of the solutions' quality. They were computed using only 20 % of the available test data. The second evaluation round was conducted after completion of the competition. It was available only for those teams which had provided a short description of their solution in a form of a competition report. The final evaluation was carried out independently from the preliminary one, using the remaining part of the test data set.

---

[5] http://www.ibemag.pl/index.php?l=ang.

Table 1. Proportions of examples from the "warning" class for each of the target sensors.

Data set\sensor ID	$MM263$	$MM264$	$MM256$
Training set	0.009	0.026	0.025
Preliminary test set	0.007	0.023	0.025
Final test set	0.007	0.025	0.027

The quality criterion chosen for the evaluation of submissions was based on a well known Area Under ROC Curve (AUC) measure. This measure was selected due to the sparsity of positive examples from the "warning" decision class for each of the considered target sensors. Table 1 shows proportions of the cases from the "warning" class for the methane meters $MM263$, $MM264$ and $MM256$ in the training and test parts of the data.

A correctly formatted submission consisted of a single text file in the *csv* format. Consecutive rows of this file corresponded to cases from the test set. They ought to contain three real numbers indicating likelihoods of the label "warning" for the three target methane meters. The values did not have to be in a particular range, however, higher numerical values should have indicated higher chances of the label "warning". For each of the submitted file, AUC values were computed independently for each of the target sensors. The score received by a submission was a simple arithmetical mean of the obtained AUC values.

4   Results of the Competition

IJCRS'15 Data Challenge attracted many skilled data mining practitioners who managed to submit a variety of interesting solutions. There were 90 registered teams with members from 18 different countries. In total they submitted 1,676 solutions. Additionally, 40 teams provided a brief report describing their approach. These reports turned out to be a valuable source of knowledge regarding the state-of-the-art in the predictive analysis of time series data.

In order to stimulate the competitiveness among participants and to give a good reference point to their results we provided a simple baseline solution. It was based on a technique that derives from the rough set theory [8]. The likelihood predictions for the target decision classes were made using an ensemble of 10 sets of decision rules. The sets were obtained by applying *LEM2* rule induction algorithm [9] to the training data discretized using the greedy discernibility-based heuristic [10,11]. The whole prediction model, including preprocessing of data, discretization, computation of decision rules and prediction of the likelihoods, was implemented in R System using the *RoughSets* package [12].

The top-ranked participants exceeded the score of the baseline solution by nearly 6 percentage points. Table 2 shows names of teams that achieved the best results in the competition.

Table 2. The final and preliminary results of the top-ranked teams.

Rank	Team name	Preliminary	Final score
1	Zagorecki	0.9666259	0.95926715
2	Marcb	0.94607103	0.94392893
3	Dymitrruta	0.93371596	0.94369948
4	Moomean	0.92862237	0.94280921
5	Trzewior	0.94691336	0.94134706
6	Kkurach_kp7	0.96853101	0.94002446
...	...	...	...
36	Baseline	0.8930246	0.90044912

The submitted solutions proved to be a valuable source of knowledge. They not only provided an insightful view on the state-of-the-art in multidimensional time series analysis but also contained inspiring ideas, designed specifically for the considered problem. The most interesting of these ideas are described by their authors in separate papers submitted to the competition track of IJCRS'15.

Investigation of the results revealed big differences between scores obtained by the participants on the preliminary and test sets. Since every team could submit more than one solution, it seems that some participants tried fine-tune their algorithms using the feedback from the preliminary test data. In theory, this approach could lead to a strong overfitting to that small set, which could explain the differences in results. However, a statistical test did not reveal a significant divergence between the differences in the preliminary and final results of the three top-ranked and the lower-ranked participants. The average difference was lower for the top-ranked teams but the p-value of the test was 0.3573.

We decided to compare submission histories of some of teams with the most interesting results. Figure 3 shows scores (both preliminary and final) of top 10 and last 10 solutions submitted by the winning team and a team from the third ten of the ranking. This team had a similar number of submissions as the winner and its best score (on the final test set) was also close to the maximum. It is important to notice that participants were not aware of final scores received by their submissions until the end of the competition and only one solution was taken into account for the final ranking. The final solution was either manually selected by participants or was automatically chosen based on its preliminary score. This comparison shows an interesting phenomenon – in a vast majority of cases the top-ranked teams achieved better results of the final test set than on the preliminary one, whereas for the lower-ranked teams there was an opposite tendency. Finding an explanation to this fact could be very important for the future development of predictive models for multivariate time series data.

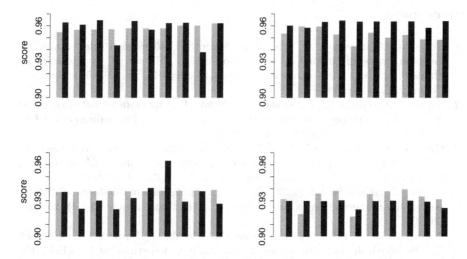

Fig. 3. The top left plot shows the preliminary (gray bars) and final scores (black bars) of top 10 solutions submitted by the team "zagorecki" (selected based on the preliminary results). The top plot on the right shows the scores of the 10 latest submissions from this team. For comparison we also present similar plots for a team placed in the third ten of the ranking.

5 Summary and the Future Research

IJCRS'15 Data Challenge was an on-line data mining competition hosted by the Knowledge Pit platform. It was held between April 13 and June 25, 2015. The task was to come up with a prediction model which could be applied to foresee warning levels of methane concentrations at a longwall of a Polish coal mine.

In this paper we explained our motivation for organizing this event and also provided details regarding the competition data sets and the task. We believe that our effort will result in stimulating research within the machine learning community in this very important field of study. We are planning to make available all the competition data, including labels for the test cases. This will make it possible to facilitate a collaboration between research teams that did not participate in the competition but are interested in this subject.

We are also planning to organize a continuation of IJCRS'15 Data Challenge. This time the data will be about a seismic activity and its impact on the safety in coal mines. The task will be to devise a classification model that can accurately predict energy of seismic shocks that will be perceived at longwalls of a mine. We hope that the success of IJCRS'15 Data Challenge will help us in attracting even more participants to the new competition.

References

1. Kozielski, M., Skowron, A., Wróbel, Ł., Sikora, M.: Regression rule learning for methane forecasting in coal mines. In: Kozielski, S., Mrozek, D., Kasprowski, P., Malysiak-Mrozek, B., Kostrzewa, D. (eds.) BDAS 2015, pp. 495–504. Springer, Cham (2015)
2. Krasuski, A., Jankowski, A., Skowron, A., Ślęzak, D.: From sensory data to decision making: a perspective on supporting a fire commander. In: Proceedings of WI-IAT 2013 Workshops, pp. 229–236. IEEE (2013)
3. Grzegorowski, M., Stawicki, S.: Window-based feature extraction framework for multi-sensor data: a posture recognition case study. In Ganzha, M., Maciaszek, L.A., Paprzycki, M. (eds.) Proceedings of FedCSIS 2015. IEEE (2015)
4. Kabiesz, J., Sikora, B., Sikora, M., Wróbel, Ł.: Application of rule-based models for seismic hazard prediction in coal mines. Acta Montanist. Slovaca 18(4), 262–277 (2013)
5. Janusz, A., xc, A., Stawicki, S., Rosiak, M., Ślęzak, D., Nguyen, H.S.: Key risk factors for polish state fire service: a data mining competition at knowledge pit. In: Ganzha, M., Maciaszek, L.A., Paprzycki, M., (eds.) Proceedings of FedCSIS 2014, pp. 345–354. IEEE (2014)
6. R Development Core Team: R: A Language and Environment for Statistical Computing. R Foundation for Statistical Computing, Vienna, Austria (2008)
7. Boullé, M.: Tagging fireworkers activities from body sensors under distribution drift. In: Ganzha, M., Maciaszek, L.A., Paprzycki, M. (eds.) Proceedings of FedCSIS 2015. IEEE (2015)
8. Pawlak, Z., Skowron, A.: Rudiments of rough sets. Inf. Sci. 177(1), 3–27 (2007)
9. Grzymała-Busse, J.W.: A new version of the rule induction system LERS. fundamenta Informaticae 31(1), 27–39 (1997)
10. Nguyen, H.S.: On efficient handling of continuous attributes in large data bases. Fundamenta Informaticae 48(1), 61–81 (2001)
11. Janusz, A.: Algorithms for similarity relation learning from high dimensional data. In: Peters, J.F., Skowron, A. (eds.) Transactions on Rough Sets XVII. LNCS, vol. 8375, pp. 174–292. Springer, Heidelberg (2014)
12. Riza, L.S., Janusz, A., Bergmeir, C., Cornelis, C., Herrera, F., Ślęzak, D., Benítez, J.M.: Implementing algorithms of rough set theory and fuzzy rough set theory in the R package 'roughsets'. Inf. Sci. 287, 68–89 (2014)

Prediction of Methane Outbreak in Coal Mines from Historical Sensor Data under Distribution Drift

Marc Boullé[(✉)]

Orange Labs, 2 Avenue Pierre Marzin, 22300 Lannion, France
marc.boulle@orange.com
http://www.marc-boulle.fr

Abstract. We describe our submission to the IJCRS'15 Data Mining Competition, where the objective is to predict methane outbreaks from multiple sensor readings. Our solution exploits a selective naive Bayes classifier, with optimal preprocessing, variable selection and model averaging, together with an automatic variable construction method that builds many variables from time series records. One challenging part of the challenge is that the input variables are not independent and identically distributed (i.i.d.) between the train and test datasets, since the train data and test data rely on different time periods. We suggest a methodology to alleviate this problem, that enabled to get a final score of 0.9439 (team marcb), second among the 50 challenge competitors.

Keywords: Multi-Relational Data Mining · Supervised classification · Feature selection · Drift detection

1 Introduction

The IJCRS'15 Data Mining Competition[1] is related to a problem of prediction of methane outbreaks in a coal mine. The coal mine is equipped with 28 sensors of different types (barometer, anemometer, temperature meter, humidity meter, methane meter...). Sensor readings are available as time series for time periods of 10 min long with measurements taken every second. The train data consists of 51,700 samples (time periods of 10 min long), whereas the test data contains 5,076 samples with time periods that do not overlap with those in the train data. The objective is to predict whether a warning threshold has been reached in a delay between three and six minutes after the end of the time period, for three methane meters. The evaluation criterion is the mean AUC of the three target classes. In this paper, we present our submission to the challenge. It exploits a Selective Naive Bayes classifier together with an automatic variable construction method (Sect. 2). We motivate the choice of this classification framework and describe its application to the challenge in Sect. 3. A good classifier trained on the

---

[1] https://knowledgepit.fedcsis.org/contest/view.php?id=109.

© Springer International Publishing Switzerland 2015
Y. Yao et al. (Eds.): RSFDGrC 2015, LNAI 9437, pp. 439–451, 2015.
DOI: 10.1007/978-3-319-25783-9_39

train data obtained a poor leaderboard score. This is not caused by over-fitting, but by a severe distribution drift between the train and test data. We suggest in Sect. 4 a methodology to alleviate this problem. Finally, Sect. 5 summarizes the paper.

2   Supervised Classification Framework

We summarize the Selective Naive Bayes (SNB) classifier introduced in [4]. It extends the Naive Bayes classifier [16] using an optimal estimation of the class conditional probabilities, a Bayesian variable selection and a Compression-based Model Averaging. We also describe the automatic variable construction framework presented in [5], used to get a tabular representation from the times series.

2.1   Optimal Discretization

The Naive Bayes (NB) classifier has proved to be very effective in many real data applications [10,16]. It is based on the assumption that the variables are independent within each class, and solely relies on the estimation of univariate conditional probabilities. The evaluation of these probabilities for numerical variables has already been discussed in the literature [8,18]. Experiments demonstrate that even a simple equal width discretization brings superior performance compared to the assumption using a Gaussian distribution per class. Using a discretization method, each numerical variable is recoded as a categorical variable, with a distinct value per interval. Class conditional probabilities are assumed to be piecewise constant per interval, and obtained by counting the number of instances per class in each interval. These class conditional probabilities are used as inputs for the naive Bayes classifier.

In the MODL approach [3], the discretization is turned into a model selection problem and solved in a Bayesian way. First, a space of discretization models is defined. The parameters of a specific discretization model M are the number of intervals, the bounds of the intervals and the class frequencies in each interval. Then, a prior distribution is proposed on this model space. This prior exploits the hierarchy of the parameters: the number of intervals is first chosen, then the bounds of the intervals and finally the class frequencies. The choice is uniform at each stage of the hierarchy. Finally, the multinomial distributions of the class values in each interval are assumed to be independent from each other. A Bayesian approach is applied to select the best discretization model, which is found by maximizing the maximum a posteriori (MAP) model. Owing to the definition of the model space and its prior distribution, the Bayes formula is applicable to derive an exact analytical criterion to evaluate the posterior probability of a discretization model. The optimized criterion is $p(M)p(D|M)$, where $p(M)$ is the prior probability of a preprocessing model and $p(D|M)$ the conditional likelihood of the data given the model.

Efficient search heuristics allow to find the most probable discretization given the data sample. Extensive comparative experiments report high performance.

Univariate Informativeness Evaluation A 0-1 normalized version of the optimized criterion provides a univariate informativeness evaluation of each input variable. Taking the negative log of the MAP criterion, $c(M) = -(\log p(M) + \log p(D|M))$, the approach receives a Minimim Description Length (MDL) [21] interpretation, where the objective is to minimize the coding length of the model plus that of the data given the model. The null model $M_\emptyset$ is the preprocessing model with one single interval, which represents the case with no correlation between the input and output variables. We then introduce the $I(V)$ criterion in Eq. 1 to evaluate the informativeness of a variable V.

$$I(V) = 1 - \frac{c(M)}{c(M_\emptyset)}.$$ (1)

The value of $I(V)$ grows with the informativeness of an input variable. It is a between 0 and 1, 0 for irrelevant variables uncorrelated with the target variable and 1 for variables that perfectly separate the target values.

2.2 Bayesian Approach for Variable Selection

The naive independence assumption can harm the performance when violated. In order to better deal with highly correlated variables, the Selective Naive Bayes approach [17] exploits a wrapper approach [13] to select the subset of variables which optimizes the classification accuracy. Although the Selective Naive Bayes approach performs quite well on datasets with a reasonable number of variables, it does not scale on very large datasets with hundreds of thousands of instances and thousands of variables, such as in marketing applications or text mining. The problem comes both from the search algorithm, whose complexity is quadratic in the number of variables, and from the selection process which is prone to overfitting. In [4], the overfitting problem is tackled by relying on a Bayesian approach, where the best model is found by maximizing the probability of the model given the data. The parameters of a variable selection model are the number of selected variables and the subset of variables. A hierarchic prior is considered, by first choosing the number of selected variables and second choosing the subset of selected variables. The conditional likelihood of the models exploits the Naive Bayes assumption, which directly provides the conditional probability of each class. This allows an exact calculation of the posterior probability of the models. Efficient search heuristic with super-linear computation time are proposed, on the basis of greedy forward addition and backward elimination of variables.

2.3 Compression-Based Model Averaging

Model averaging has been successfully exploited in bagging [6] using multiple classifiers trained from re-sampled datasets. In this approach, the averaged classifier uses a voting rule to classify new instances. Unlike this approach, where each classifier has the same weight, the Bayesian Model Averaging (BMA) approach [11] weights the classifiers according to their posterior probability. In the case of the Selective Naive Bayes classifier, an inspection of the optimized models reveals that their posterior distribution is so sharply peaked that averaging them according to the BMA approach almost reduces to the MAP model. In this situation, averaging is useless. In order to find a trade-off between equal weights as in bagging and extremely unbalanced weights as in the BMA approach, a logarithmic smoothing of the posterior distribution, called Compression-based Model Averaging (CMA), is introduced in [4]. The weighting scheme on the models reduces to a weighting scheme on the variables, and finally results in a single Naive Bayes classifier with weights per variable. Extensive experiments demonstrate that the resulting Compression-based Model Averaging scheme clearly outperforms the Bayesian Model Averaging scheme. In the rest of the paper, the classifier resulting from model averaging is called Selective Naive Bayes (SNB).

2.4   Automatic Variable Construction for Multi-Table

In a data mining project, the data preparation phase aims at constructing a data table for the modeling phase [7,20]. The data preparation is both time consuming and critical for the quality of the mining results. It mainly consists in the search of an effective data representation, based on variable construction and selection. Variable construction [19] has been less studied than variable selection [9] in the literature. However, learning from relational data has recently received an increasing attention. The term Multi-Relational Data Mining (MRDM) was initially introduced in [12] to address novel knowledge discovery techniques from multiple relational tables. The common point between these techniques is that they need to transform the relational representation. Methods named by propositionalisation [1,14,15] try to flatten the relational data by constructing new variables that aggregate the information contained in non target tables in order to obtain a classical tabular format.

In [5], an automatic variable construction method is proposed for supervised learning, in the multi-relational setting using a propositionalisation-based approach. Domain knowledge is specified by describing the multi-table structure of the data and choosing construction rules. The formal description of the data structure relies on a root table that contains the main statistical units and secondary tables in 0 to 1 or 0 to n relationship with the root table. For example, Fig. 1 describes the structure of the data for the challenge. The construction rules available for automatic construction of variables are detailed below:

- *Selection(Table, Num)→Table*: selection of records from a secondary table according to a conjunction of selection terms (membership in a numerical interval of a variable *Num* in the secondary table),
- *Count(Table)→Num*: count of records in a table,
- *Mean(Table, Num) →Num*: mean value of variable *Num*,
- *Median(Table, Num)→Num*: median value,
- *Min(Table, Num)→Num*: min value,
- *Max(Table, Num)→Num*: max value,
- *StdDev(Table, Num)→Num*: standard deviation,
- *Sum(Table, Num) →Num*: sum of values.

The space of variables that can be constructed is virtually infinite, which raises both combinatorial and over-fitting problems. When the number of constructed variables increases, the chance for a variable to be wrongly considered as informative becomes critical. A prior distribution over all the constructed variables is introduced. This provides a Bayesian regularization of the constructed variables, which allows to penalize the most *complex* variables. An effective algorithm is introduced as well to draw samples of constructed variables from this prior distribution. Experiments show that the approach is robust and efficient.

3   Applying the Framework for the Challenge

We motivate our choice of the classification framework[2], then describe how we apply it on the challenge dataset.

3.1   Choice of the Classification Framework

In all our challenge submissions, we exploit the framework described in Sect. 2 to train a selective naive Bayes classifier, with optimal discretization, variable selection and model averaging. The classifier is trained on a flat data representation, obtained using the automatic variable construction method (Sect. 2.4) that builds many variables from the time series records data. Once the data schema is specified, the only parameter is the number of variables to construct. The method is fully automatic, scalable and highly robust, with test performance mainly equivalent to train performance.

The SNB classifier is resilient to noise and to redundancies between the input variables, but it is blind to non-trivial interactions between the variables. This can be leveraged by feature engineering, relying on domain expertise rather than on statistical expertise. More accurate classification methods are available, such as random forests, gradient boosting methods, support vector machines or neural networks. However, these methods require intensive feature engineering to get a flat input data table representation, are prone to over-fitting, are mainly black-box, not suitable for an easy interpretation of the models and finally require fine parameter tuning, both time consuming and expertise intensive. In an industrial context like the Orange telecommunication operator, the major issue is to quickly provide an accurate, robust and interpretable solution to many data mining problems, rather than a very accurate solution to few problems. In this context, the generic framework described in Sect. 2 and used in this challenge offers a good solution.

3.2   Application to the Challenge Dataset

For the IJCRS'15 Data Mining Competition, coal mines are described using a root table that contains the three class variables and a secondary table for the sensor readings. An identifier variable *Id* is added in each record of both tables, to enable the join between the root and secondary tables.

---

[2] Available as a shareware at http://www.khiops.com.

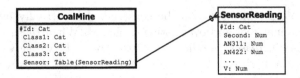

Fig. 1. Multi-table representation for the data of the IJCRS'15 challenge

The multi-table representation of the challenge data is presented in Fig. 1. The root table (*CoalMine*) contains 51,700 train instances, with 4 variables: *Id* and the three class variables. The secondary table (*SensorReading*) contains 51,700 × 600 records, with 30 variables: *Id* as a join key, the time variable *Second* and the 28 time series variables for the sensor readings (*AN311, AN422,..., V*). Using the data structure presented in Fig. 1 and the construction rules introduced in Sect. 2.4, one can for example construct the following variables ("name" = *formula*: comment) to enrich the description of a *CoalMine*:

- "Mean(Sensor.TP1721)" =
  Mean(Sensor, TP1721):
  mean of the temperature sensor TP1721 readings,
- "Count(Sensors) where AN422 ∈]1.75, 1.85]" =
  Count(Selection(Sensor, AN422 ∈]1.75, 1.85])):
  number of sensor readings where the anemometer AN422 value is between 1.75 and 1.85,
- "Max(Sensors.MM263) where Second > 300" =
  Max(Selection(Sensor, Second > 300), MM263):
  max of the methane meter sensor MM263 readings in the last five minutes.

The number of variables to construct is the only user parameter. An input flat data table representation is then obtained from the set of all automatically constructed variables. All these variables are then preprocesses using the optimal discretization method (cf. Sect. 2.1) to assess their informativeness and evaluate their class conditional probabilities, before training the SNB classifier.

4   Challenge Submissions

In this section, we describe our submissions to the challenge and suggest a methodology to alleviate the problem of the drift between the train and test distributions of the challenge dataset.

4.1   Preliminary Trials

To get familiar with the challenge evaluation protocol, we made preliminary trials, using all the sensor variables and constructing 100, 1000 and 10000 variables to summarize the sensors times series. We collect in Table 1 the train and

Table 1. Performance per number of constructed variables

Variables	Train AUC	Test AUC	Leaderboard score
100	0.9760	0.9684	0.8067
1000	0.9835	0.9766	0.7419
10000	0.9885	0.9802	0.7876

test AUC (averaged over the three target classes) using a 70 %–30 % split of the train dataset as well as the leaderboard score obtained using the corresponding submissions.

We obtained surprisingly robust and high train AUC. With only 100 variables constructed from the 28 sensors, the train AUC is about 0.97, with less than 1 % difference between the train and test splits. However, these promising performance dropped down on the challenge leaderboard, with a non monotonous behavior w.r.t. the number of constructed variables. This large drop of performance was not cause by over-fitting, but by a drift between the train and test (based on two distinct time periods).

4.2 A Methodology to Reduce the Drift Problem

Let us consider two tasks: classification of the methane outbreaks and detection of the drift. The drift detection task can be turned into a classification task as in [2], by merging the train and test datasets and using the dataset label ('train' or 'test') as the target variable. Using the initial input representation (cf. Sect. 4.1) with 10000 constructed variables, the drift detection task achieved an almost perfect performance with an AUC of 0.9999. This means that the train and test distributions can be well separated using the sensors data. As the data is not i.i.d, obtaining good classification performance on the train data does not guarantee good performance on the test data. Intuitively, if we are able to select an input representation with good classification performance on the train data but poor drift detection, we expect that our classifier will be less sensitive to drift and its performance drop on the test dataset will be reduced.

The objective is then to explore varying input representations and select the one with the best classification performance together with the poorest drift detection. To do so, we represent in Fig. 2 the informativeness (cf. Formula 1) of the 10,000 constructed variables for the classification and drift detection tasks. The results show that there are variables with large drift informativeness and small classification informativeness (top-left of the figure), or on the contrary variables with small drift informativeness and large classification informativeness (bottom-right). The interesting variables are those on the right and close to the X axis, with small drift informativeness.

To gain further insights, we collect the mean informativeness (cf. formula 1) per input sensor (gathering all constructed variables involving each sensor) and per target class.

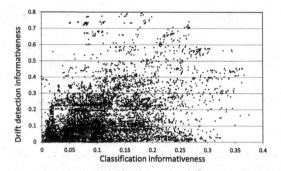

Fig. 2. Informativeness of 10,000 variables

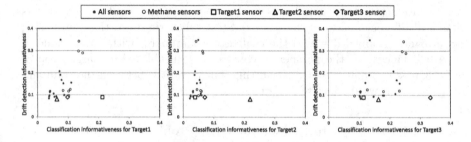

Fig. 3. Mean informativeness per sensor for each target

The results, shown in Fig. 3, suggest to consider the following subsets of variables, by decreasing number of variables with high drift informativeness.

1. all the 28 input sensors,
2. only the 11 sensors related to methane,
3. only the 3 sensors related to the three target classes,
4. only one sensor per target class.

We then build classifiers using only 100 constructed variables (which seems enough from Sect. 4.1), and obtain the classification and drift detection train AUC as well as the leaderboard scores, reported in Table 2.

Table 2. Performance per number of input sensors

Input sensors	Drift AUC	Train AUC	Leaderboard AUC
28	0.9998	0.9684	0.8067
11	0.9996	0.9723	0.8978
3	0.7625	0.9742	0.9225
1	0.6210	0.9675	0.9304

Table 2 shows that the drift detection AUC rapidly decreases with smaller number of sensors while the train classification AUC remains almost the same. Meanwhile, the leaderboard score increases, from 0.8067 using all the sensors to 0.9304 when only the target sensor is used for the prediction.

4.3 Simplification of the Solution

As only 100 variables and one single sensor per target variable where enough to get a good leaderboard score, we made several trials and errors to simplify the solution, improve its interpretability and performance. We finally kept four time periods (full 10 min period, last 5 min, last 2 min 30 s, and last minute) and two construction rules (*Mean* and *Max*), representing the input data by only 8 variables per target class.

4.4 Further Improvement

According to the challenge organizers, the time periods in the training data are overlapping and are given in a chronological order. This raises an additional issue, where instances in short time periods are non i.i.d. and over-sampled: this might cause additional over-fitting. Inspecting the data, we estimated that the over-sampling factor was about 10, and decided to train again the solution described in Sect. 4.3, using 10 % of the train data. To improve the robustness, we divided the train data into 10 folds, trained the solution on each fold and averaged the predictions. Using this method, our final chosen submission obtained a leaderboard score of 0.9461, very close to our final score of 0.9439.

Interestingly, many time series problems suffer from the two same kind of problems: different time periods that cause drift between the distributions in case of non stationary data, and over-sampled data when the sampling rate is large compared to the typical size of the time windows that govern the behavior of the time series. For example, for climate time series (temperature, humidity, pressure...), different data collection periods (e.g. winter and summer) involve distinct distributions of the data. And for a given period, a dataset with one record every second is clearly over-sampled, compared to the typical change rate in climate times series. This results in dramatic over-fitting, since predicting future value to be the same as the last past value is likely to be very accurate for a prediction windows of one second, but valueless for longer windows.

4.5 Insights on Relevant Variables

The IJCRS'2015 challenge comes with abundant data: 28 sensors with records every seconds during 10 min, which amounts to 28*600=16,800 values per train instance. Remarkably, our final solution exploits only 8 variables per class: mean and max of the target methane meter readings during the last 10 min, 5 min, 2 min 30 s and last min.

To get further insights on these variables, we inspect their discretizations, obtained from the first fold in our ten fold process described in Sect. 4.4. The

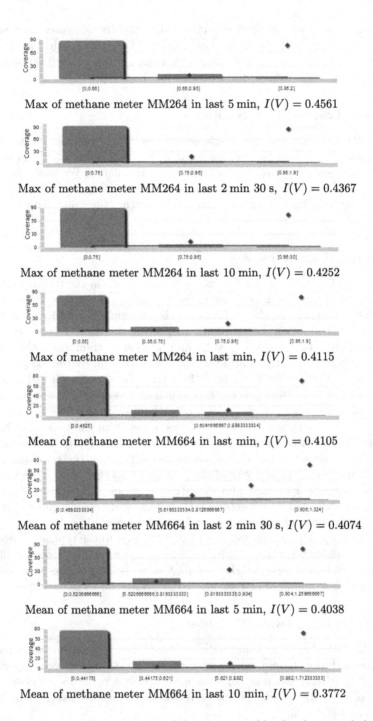

Fig. 4. Optimal discretizations of the input variables for the second class

overall behavior is of the same kind for the three classes, with slight differences. The variables are all discretized into 3 intervals for the first class, into 3 to 5 intervals for the second class, and into 3 or 4 intervals for the third class. Figure 4 displays the optimal discretizations for the 8 variables related to the second class (methane meter *MM264*, with two class values: *normal* and *warning*). For example, the first histogram is related to the variable "Max of methane meter MM264 in last 5 min", which have the highest informativeness ($I(V) = 0.4561$) among the 8 variables. The three intervals cover respectively 85 %, 12 % and 3 % of the instances. The overall mean coverage of the minority class value *warning* is represented by the horizontal line (2.6 %). The coverage of *warning* per interval is represented by the diamonds: 0.16 % in the first interval, 6 % in the second one and 54 % in the last one.

The optimal discretizations consist of 3 to 5 intervals, but they all represent the same kind of information. Most of the instances belong to the first interval, with very small value of the input variable and very small proportion of the *warning* target value. The second interval comes beyond a first threshold (typically between 0.5 and 0.75) and the last one is generally beyond a threshold of about 0.9. The proportion of *warning* quickly grows with the value of input variable, with between 40 % and 80 % of *warning* in the last interval.

Overall, the probability of a methane outbreak in the next few minutes quickly increases with the value of the related methane meter in the last minutes. This behavior, represented for the second class, is the same for the other classes (not displayed in this paper). The methodology presented in this paper allowed to retrieve this interpretable behavior and to quantify it precisely.

5   Conclusion

Whereas most data mining methods rely on i.i.d. data, this is not the case in IJCRS'15 Data Mining Competition, where the train and test data where collected from two different time periods. In this case, a robust classifier was able to achieve 0.98 AUC in a 70 %–30 % split of the train data, with a severe drop of the test performance down to 0.80. This is not an overfitting problem, but a problem of distribution drift between the train and test data. In this paper, we have suggested a methodology to alleviate this problem by evaluating the informativeness of each variable for the classification and drift detection tasks. We follow the intuition that the classifiers that exploit input variables with high class informativeness and low drift informativeness are more likely to be resilient to drift. We explored several axis for choosing representations that are robust to drift: selection of sensors, selection of construction rules that summarize sensor readings and number of constructed variables. In the end, we kept only one sensor per target class, summarized by 8 input variables. We were then able to build a classifier with 0.9439 final score, which is a large improvement compared to our initial solution.

References

1. Blockeel, H., De Raedt, L., Ramon, J.: Top-Down Induction of Clustering Trees. In: Proceedings of the Fifteenth International Conference on Machine Learning, pp. 55–63. Morgan Kaufmann (1998)
2. Bondu, A., Boullé, M.: A supervised approach for change detection in data streams. In: Proceedings of International Joint Conference on Neural Networks, pp. 519–526 (2011)
3. Boullé, M.: MODL: a Bayes optimal discretization method for continuous attributes. Mach. Learn. **65**(1), 131–165 (2006)
4. Boullé, M.: Compression-based averaging of selective naive Bayes classifiers. J. Mach. Learn. Res. **8**, 1659–1685 (2007)
5. Boullé, M.: Towards automatic feature construction for supervised classification. In: Calders, T., Esposito, F., Hüllermeier, E., Meo, R. (eds.) ECML PKDD 2014, Part I. LNCS, vol. 8724, pp. 181–196. Springer, Heidelberg (2014)
6. Breiman, L.: Bagging predictors. Mach. Learn. **24**(2), 123–140 (1996)
7. Chapman, P., Clinton, J., Kerber, R., Khabaza, T., Reinartz, T., Shearer, C., Wirth, R.: CRISP-DM 1.0 : step-by-step data mining guide. Technical report, The CRISP-DM consortium (2000)
8. Dougherty, J., Kohavi, R., Sahami, M.: Supervised and unsupervised discretization of continuous features. In: Proceedings of the 12th International Conference on Machine Learning, pp. 194–202. Morgan Kaufmann, San Francisco (1995)
9. Guyon, I., Gunn, S., Nikravesh, M., Zadeh, L. (eds.): Feature Extraction: Foundations And Applications. Studies in Fuzziness and Soft Computing, 1st edn. Springer, Heidelberg (2006)
10. Hand, D., Yu, K.: Idiot's bayes ? not so stupid after all? Int. Stat. Rev. **69**(3), 385–399 (2001)
11. Hoeting, J., Madigan, D., Raftery, A., Volinsky, C.: Bayesian model averaging: a tutorial. Stat. Sci. **14**(4), 382–417 (1999)
12. Knobbe, A.J., Blockeel, H., Siebes, A., Van Der Wallen, D.: Multi-Relational Data Mining. In: Proceedings of Benelearn 1999 (1999)
13. Kohavi, R., John, G.: Wrappers for feature selection. Artif. Intell. **97**(1–2), 273–324 (1997)
14. Kramer, S., Flach, P.A., Lavrač, N.: Propositionalization approaches to relational data mining. In: Džeroski, S., Lavrač, N. (eds.) Relational data mining, chap. 11, pp. 262–286. Springer-Verlag, Heidelberg (2001)
15. Krogel, M.-A., Wrobel, S.: Transformation-based learning using multirelational aggregation. In: Rouveirol, C., Sebag, M. (eds.) ILP 2001. LNCS (LNAI), vol. 2157, p. 142. Springer, Heidelberg (2001)
16. Langley, P., Iba, W., Thompson, K.: An analysis of Bayesian classifiers. In: 10th National Conference on Artificial Intelligence, pp. 223–228. AAAI Press (1992)
17. Langley, P., Sage, S.: Induction of selective Bayesian classifiers. In: Proceedings of the 10th Conference on Uncertainty in Artificial Intelligence, pp. 399–406. Morgan Kaufmann (1994)
18. Liu, H., Hussain, F., Tan, C., Dash, M.: Discretization: an enabling technique. Data Min. Knowl. Disc. **4**(6), 393–423 (2002)
19. Liu, H., Motoda, H.: Feature Extraction: A Data Mining Perspective, Construction and Selection. Kluwer Academic Publishers, Boston (1998)

20. Pyle, D.: Data Preparation for Data Mining. Morgan Kaufmann Publishers, Inc., San Francisco (1999)
21. Rissanen, J.: Modeling by shortest data description. Automatica **14**, 465–471 (1978)

Window-Based Feature Engineering
for Prediction of Methane Threats
in Coal Mines

Marek Grzegorowski[✉] and Sebastian Stawicki

Faculty of Mathematics, Informatics and Mechanics, University of Warsaw,
Banacha 2, 02-097 Warsaw, Poland
{M.Grzegorowski,Stawicki}@mimuw.edu.pl

Abstract. We present our results of experiments concerning the
methane threats prediction in coal mines obtained during IJCRS'15 Data
Challenge. The data mining competition task poses the problem of active
monitoring and early threats detection which is essential to prevent spontaneous gas explosions. This issue is very important for the safety of
people and equipment as well as minimization of production losses. The
discussed research was conducted also to verify the effectiveness of the feature engineering framework developed in the DISESOR project. The utilized framework is based on a sliding window approach and is designed to
handle numerous streams of sensor readings.

1 Introduction

We introduce a framework for automated feature engineering for numeric data
streams which has been developed for the purpose of processing multiple time
series of readings generated by various sensors within coal mines. The main
objective of participation in the data mining competition IJCRS'15 Data Challenge[1] was to assess the quality and usefulness of created features by comparing
the result achieved with the results of other participants.

The main goal of the competition was to verify whether warning conditions
associated with an elevated level of methane concentration in the atmosphere of
the mine shafts has been reached in a period between three and six minutes after
the end of the training period. During the conducted experiments the possibility
to predict the warning conditions recorded by any of three methane sensors
located in the immediate vicinity of the cutter loader was verified.

A decision to analyze sensors located in a close proximity to the workplace of
the shearer is justified by the fact that in connection with the ongoing extraction
of coal, the dynamics of the release of methane gas from the ground is the
greatest. Additionally, it is a place that requires the presence of a relatively

Partially supported by Polish National Science Centre - grant DEC-2012/05/B/
ST6/03215 and by the Polish National Centre for Research and Development
(NCBiR) – grant PBS2/B9/20/2013.

[1] https://knowledgepit.fedcsis.org/contest/view.php?id=109.

Y. Yao et al. (Eds.): RSFDGrC 2015, LNAI 9437, pp. 452–463, 2015.
DOI: 10.1007/978-3-319-25783-9_40

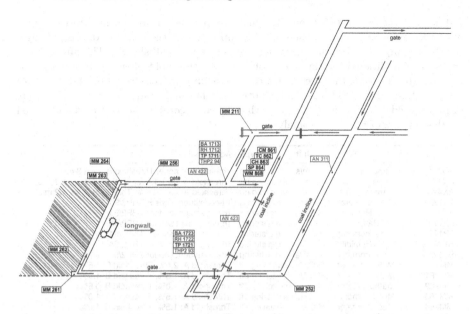

Fig. 1. A shearer moves along the wall of coal extraction between the sensors MM261 and MM264. The progress of the coal extraction is unveiled by an arrow described with "longwall". Thin arrows depict flow direction of the air in the mine sidewalks which is enforced by a ventilation system.

large number of miners, hence the highest level of safety should be ensured, e.g. by providing possibility of early risk detection. However, if we were able to predict the alarming methane concentrations, we could in turn reduce the speed of the cutter loader while maintaining the same effectiveness of the ventilation system what should bring a reduction in the methane concentration before the necessity of switching off the whole production line.

Another issue which was of the competition organizers interest was to discover correlation between work of a shearer expressed in the intensity of the electric energy consumption and the release rate of the methane gas from the soil in the mine. The organizers have staked the proposition that the bigger the currents consumed by the cutter loader are, the more intense its mining work is, which in theory, results in more methane emitted to the atmosphere. This correlation is very important, since according to legal regulations in Poland the high methane concentration causes the electricity cut off in a large part of the mine in case any of the sensors reaches the warning level.

The paper is organized as follows. In Sect. 2 the original data set is presented. In Sect. 3 the competition problem and an evaluation score function are described. Section 4 introduces a window-based feature extraction mechanism. In Sect. 5 a feature selection approach and a course of experiments are described. In Sect. 6 experimental results can be found. In Sect. 7 we provide information of the DISESOR project. In Sect. 8 the summary of the research, conclusions and plans for the nearest future are presented.

Table 1. Three groups of sensors. The first group is responsible for monitoring condition of the mine atmosphere, the second group monitors the methane drainage flange, the third group monitors the operating status of a longwall shearer. The fifth column provides an additional information about security thresholds assigned to the selected sensors. After crossing the threshold A the "switching off" sensors cut off the electricity supply. After crossing the threshold B both the "alarming" and "switching off" sensors display a predefined warning message. All of sensor recordings are collected and stored for the purpose of the further analysis.

Sensor	Type	Unit	Type	Additional Info
AN311	Anemometer	m/s	alarming	Threshold A: none, Threshold B: <= 0.3 m/s
AN422	Anemometer	m/s	switching off	Threshold A: <= 1.1 m/s, Threshold B: <= 1.3 m/s
AN423	Anemometer	m/s	switching off	Threshold A: <= 1.0 m/s, Threshold B: <= 1.2 m/s
TP1721	Thermometer	$°C$	registering	Tri-constituent sensor THP2/93
RH1722	Humidity	%RH	registering	Tri-constituent sensor THP2/93
BA1723	Barometer	hPa	registering	Tri-constituent sensor THP2/93
TP1711	Thermometer	$°C$	registering	Tri-constituent sensor THP2/94
RH1712	Humidity	%RH	registering	Tri-constituent sensor THP2/94
BA1713	Barometer	hPa	registering	Tri-constituent sensor THP2/94
MM252	Methanometer	$\%CH_4$	switching off	Threshold A: 2.0%, Threshold B: 1.5%
MM261	Methanometer	$\%CH_4$	switching off	Threshold A: 1.5%, Threshold B: 1.0%
MM262	Methanometer	$\%CH_4$	switching off	Threshold A: 1.0%, Threshold B: 0.6%
MM263	Methanometer	$\%CH_4$	switching off	Threshold A: 1.5%, Threshold B: 1.0%
MM264	Methanometer	$\%CH_4$	switching off	Threshold A: 1.5%, Threshold B: 1.0%
MM256	Methanometer	$\%CH_4$	switching off	Threshold A: 1.5%,Threshold B: 1.0%
MM211	Methanometer	$\%CH_4$	switching off	Threshold A: 2.0%, Threshold B: 1.5%
CM861	Methanometer	$\%CH_4$	registering	Measures high concentrations of methane
CR863	Pressure difference	Pa	registering	Sensor is placed on the demethanisation orifice
P_864	Barometer	kPa	registering	Pressure inside the pipeline for methane drainage
TC862	Temperature	$°C$	registering	Temperature inside the pipeline for methane drainage
WM868	Methane expense	m^3/min	registering	Methane expense calculated by CM, CR, P, TC
AMP1	Ammeter	A	registering	Current in the motor in the left arm of the shearer
AMP2	Ammeter	A	registering	Current in the motor in the right arm of the shearer
DMP3	Ammeter	A	registering	Current in the motor in the left tractor of the shearer
DMP4	Ammeter	A	registering	Current in the motor in the right tractor of the shearer
AMP5	Ammeter	A	registering	Current in the hydraulic pump motor of the shearer
F_SIDE	Drive direction	left, right	registering	The driving direction of the shearer
V	Shearer speed	Hz	registering	Work frequency, 100Hz means ca 20 m/min

2   The Data Set and the Problem

The data sets for this competition were provided in a tabular format in CSV files by R&D Centre EMAG and were prepared based on actual readings of sensors located in an active Polish coal mine. Among the thousands of sensors located on tens of kilometers of underground tunnels of the mine, 28 sensors monitoring the work in the immediate vicinity of the shearer workplace were selected. The detailed information about all the sensors can be found in Table 1. In Fig. 1 a detailed location of all sensors, as well as a workplace of a longwall shearer, on a fragment of the coal mine plan is shown.

In total, the provided files contain sensor readings for 51,700 time periods. The sensor readings took place with the frequency of 1 second, what gives us 600 columns corresponding to data from a single sensor in every row of both the training and test sets. Therefore, each row contains 16,800 values corresponding

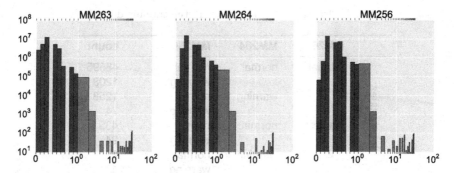

Fig. 2. Frequency distributions for sensors MM263, MM264 and MM256 built on top of the whole training data set. The majority of readings are in the $[0, 1.2]$ range and a relatively small number is spread in the $(1.2, 30]$ range. The dark blue bars drawn on a linear scale with a step of 0.1 correspond to the readings below the warning threshold of 1.0. The light blue bars drawn on a logarithmic scale with a step of 1 represent hazardous and outlier readings (Color figure online).

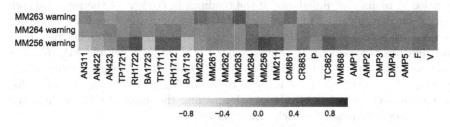

Fig. 3. Pearson's correlation coefficients between the label columns and the mean of 600 readings of a given sensor, computed for each row. The strongest correlation is visible for the sensors that correspond to warning labels on the sensors under consideration, however on MM264 it is only slightly better than on other columns (Color figure online).

to 10-minutes long series of readings from 28 sensors that were recorded in the same time interval. The names of columns were granted based on the following convention: <the sensor name>_value_<the number of the reading>, e.g. $MM256\_value\_1$, $MM256\_value\_2$, $MM256\_value\_3$, etc. The periods in the training data are overlapping and are given in a chronological order.

Each row of the training data has been tagged with three labels, each from the set $\{0, 1\}$, where 0 and 1 correspond to "normal" and "warning" labels, respectively. The warning is defined as the level of methane concentration which exceeds the threshold established at 1 %. The labels in the data indicate whether a warning threshold has been reached in a period between three and six minutes after the end of the training period, for three methane meters: MM263, MM264 and MM256. In particular, if a given row corresponds to a period between t_{-599} and t_0, then the label for a methane meter MM256 in this row is 1 (that is warning) if and only if $max(MM256(t_{181}), ..., MM256(t_{360})) \geq 1$. In Table 2 a

Table 2. Occurrences of the labels in the training data set.

	MM263	MM264	MM256	count
	normal	normal	normal	48695
			warning	1208
		warning	normal	1258
			warning	74
	warning	normal	normal	435
			warning	24
		warning	normal	2
			warning	4

(row label: label values)

summary of labels occurrences is presented. It is important to note that time periods in the test data do not overlap and are given in a randomized order.

The raw readings provided in the training data set may be used as an another premise to realize the imbalance of the readings in context of the methane concentration monitoring. The vast majority of the methane sensor values stored in the data sets are below the warning threshold. Figure 2 presents frequency distributions of values for the three sensors MM263, MM264 and MM256. The maximal value for all of the MM263, MM264 and MM256 sensor readings that occur in the training data set is equal to 30. In Fig. 3 the Pearson's correlation coefficients between the three label columns and the mean of 600 readings of a given sensor, computed for each row is shown.

3   Evaluation of Results

The task of the competition was to predict the likelihood of the label 1 (warning) for each time series from the test set. The final result consists of three real numbers corresponding to the target methane meter sensors. The predicted numbers do not need to fit in a particular range, however, a higher numerical value indicates a higher chance of the label 1. The assigned values should be provided with respect to the following order: MM263, MM264, MM256 for each of the 5,076 consecutive lines of the test set.

The preliminary scores were computed after each submission of the solution on a fixed subset of roughly 20 % of the test set. Partial results were published on the leaderboard on the Knowledge Pit data challenge platform[2] for the insight and further motivation of the other competitors. The final evaluation was performed using the remaining part of the test data. The assessment of solutions was done using the Area Under the ROC Curve (AUC) measure which was computed separately for each of the target sensors. The final score corresponds to the average AUC. For a submitted solution "s":

---

[2] See https://knowledgepit.fedcsis.org/.

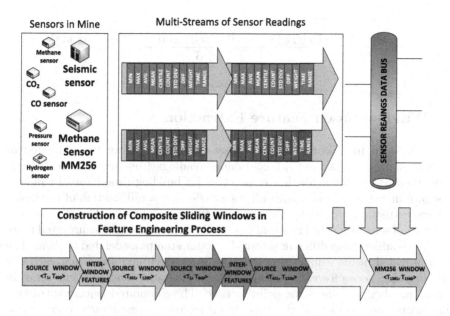

Fig. 4. The time series are divided into fragments and then processed separately.

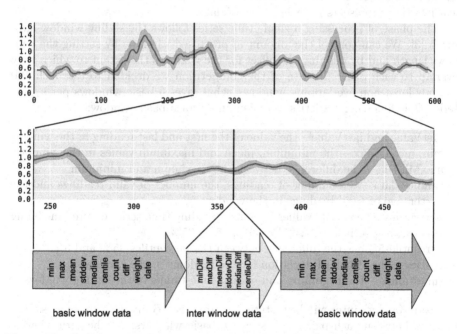

Fig. 5. The readings are split into five non-overlapping basic windows of equal length. The statistics are computed for each window separately. Inter-window features express the dynamics of window statistics changes.

$$score(s) = \frac{AUC_{MM263}(s) \ + \ AUC_{MM264}(s) \ + \ AUC_{MM256}(s)}{3}$$

4 Window-Based Feature Extraction

The developed framework for the feature extraction [1] can be configured to accept on its input a data set containing readings from multiple sensors. A general outline of the algorithm responsible for building time interval windows is shown in Fig. 4. The approach allows an effective parallelization of the whole feature extraction process by means of the framework proposed in [2].

A set of sensors located in a coal mine collect the data in a synchronized manner, i.e., readings from different sensors have timestamps associated to them. The data provided in the competition follow this approach in a sense that each row contains time series for each sensor under consideration and the time series in a data row begin at the same point in time. The elaborated feature extraction framework process each of the time series separately, producing a set of features that aggregate the information from raw readings. Afterwards, the sets of features computed for different sensors are composed to form one large set of features that represents the whole row of data.

The phase of processing a single time series follows the sliding window approach [3]. We can control the amount of processed windows by setting the windows length and by defining the offset for the windows the extent to which they overlap to each other. In case of the competition we decided to establish 120 seconds long, non-overlapping windows, what gives 5 basic windows per a time series. The following statistics were computed for a basic window:

- firstValue and lastValue – the value of the first and last reading in the window
- min, mean, max – the minimum, mean and maximum values in the window
- maxMinDiff – the difference between the maximum and minimum
- validity data – the number of readings, the number of valid readings and the percentage of valid readings in the window
- percentiles – percentile values for sensor reading time series covered in sliding window, respectively 5 %, 25 %, 50 %, 75 %, 95 %
- percentiles5Diff – the difference between the percentiles 95 % and 5 %
- stdDev – the standard deviation for the readings in the window
- timestamps – the beginning and ending timestamps in the sliding window.

Additionally, inter-window statistics are also computed to express the changes between each pair of consecutive basic windows. A schematic view of extracted window-based features and inter-window differential statistics as well as their relation to the processed sensor streams is illustrated in Fig. 5.

5   Feature Selection and the Course of Experiments

In the proposed solution the original problem of assignment the likelihood of
the occurrence of the label 1 was reduced to the problem of regression [4] for
the particular methane sensor. The higher the sensor indication, the greater the
likelihood of exceeding the warning threshold. The analogy is very natural and
follows directly the definition of the label 1, which occurs after exceeding the
threshold. Since, the labels were provided in a numerical form of 0 or 1, in the
case of regression algorithms there was no need to modify the labels.

The experimental environment was built upon R programming language
which has a lot of additional data mining packages [5]. To determine the optimal
set of attributes the wrapper approach [6] was utilized. The preliminary experiments have shown that a data mining approach to predict indications of methane
meters in order to determine the risk associated with raised methane concentration has a great potential and gives very promising results. The first attempts to
analysis using regression trees [7] implemented with the rPart package yielded
the results at the level of the competition's baseline - almost 90 %.

In Fig. 3 correlation between sensors and the "warning" label treated as a
number is shown. The strongest correlation is observed between the label and
the mean value of the same methane detector. For each label, the efficacy of
application the mean value of the last sliding window as a likelihood of the
warning label for the same sensor was explored. As observed, the mean value
retains similar efficacy to simple models generated by rPart trees (run without
tuning). Such behavior is justified by the relatively small dynamics of changes

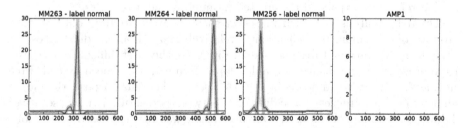

Fig. 6. Examples of outliers in methane concentration time series.

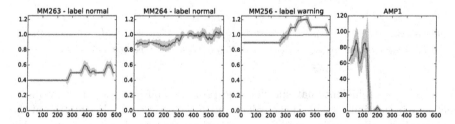

Fig. 7. Methane indications oscillating near the warning threshold. On the rightmost
plot the current cut off after exceeding the methane concentration threshold.

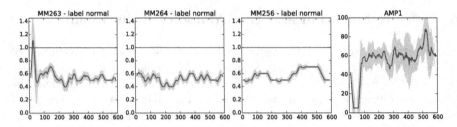

Fig. 8. Relatively small dynamics of changes in methane concentration for three methane detectors. On the rightmost plot we can observe current consumption of the cutter loader that corresponds to ongoing coal mining.

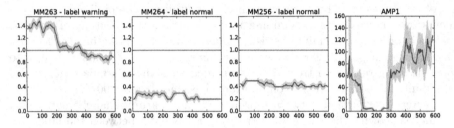

Fig. 9. Three methane detectors and ammeter located on the cutter loader. The sensor MM263 has exceeded the warning threshold.

in the atmosphere of the mine. On the other hand, the mean is susceptible to outliers (see Fig. 6) that have left no negligible impact on the results in the test set. Therefore, we expanded the created formula by the percentiles.

The data analysis described in Sect. 2 shows that, apart from outliers, the variation of methanometer indications is relatively small. Hence, indication oscillating near the warning threshold, like in Fig. 7, are the most difficult to classify. To improve the performance of the developed formula, it has been extended to include short-term trend of the investigated sensor. For this purpose, the inter-windowed features reflecting the difference between percentiles 75 in consecutive sliding windows have been used. Our best result was reached based on the following manually constructed formulas:

$$label_{MM263} = mean_{MM263} + percentile75_{MM263} + C * percentile75Diff_{MM263},$$
$$label_{MM264} = mean_{MM264} + percentile75_{MM264} + C * percentile75Diff_{MM264},$$
$$label_{MM256} = mean_{MM256} + percentile75_{MM256} + C * percentile75Diff_{MM256},$$

where C is a coefficient that specifies the proportional increase in the value of methane indications based on the calculated short term trend. Estimation of the C coefficient takes into account the length of created windows and assumption that the trend is a linear function within the investigated time range.

6   Results and Remarks

The final result obtained in the competition is 0.947 %. It is the second highest score. The effectiveness of the proposed approach with respect to the inductive learning (e.g. regression trees) is justified by the following observations:

- As shown in Fig. 2, warning labels constituted a small percentage of the test data. Thus, it was difficult to train regression models.
- Table 2 compare the amount of warnings observed for each sensor. Warnings were rarely indicated by more than one sensor, see example in Fig. 9.
- Relatively small dynamics of changes in methane concentration allows to utilize the short time trend. See Fig. 8 as an example.

The following discussion identifies some shortcomings of the proposed approach for the future consideration. First of all, the proposed approach does not take into account neither the observations of other sensors nor historical readings for the investigated sensor. The correlation between the label and the most recent indications of the same sensor was dominant, but the correlation between other characteristics of the same or other sensors were high enough to take advantage of them and enhance the approach by models for untypical events observed in the training data.

Another issue that has not been considered is a relationship between particular labels and information of the airflow direction. It may be important, as it may give a hint on the sequence of warning occurrences implied by methane flow, e.g. high methane concentration on sensor X may imply same observation on sensor Y after a particular time. This type of probabilistic conditional dependence between the various sensors can be represented with the notion of Bayesian networks [8]. Another way to model the approximate conditional dependence between labels could be expressed by the application of classifier chains [9] or the multi-label classification [10] technique.

The last of the noticed shortcomings which pose a potential for improvement in the rating score is unused correlation between the trend in the change of methane concentration and the shearer current consumption, suggested by the organizers of the competition. Eventually, it should be noted that in experiments neither SVM [11] models nor classifier ensembles [12,13] have been studied. Considering the high efficiency of simple models we expect that those techniques could give significant improvement.

7   DISESOR

The developed feature engineering framework for streams of readings from multiple sensors was constructed in order to be used in the analysis of data from Polish mines. The research in this area is being carried out within the DISESOR project in which issues of methanometer regression, analysis of missing values, outliers and seismic phenomena are studied. DISESOR aims to build a decision support system for threats monitoring and early warnings in coal mines.

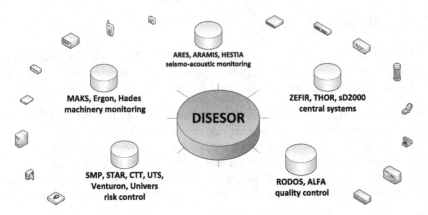

Fig. 10. Examples of computer platforms and applications that support different aspects of the mine operation. The systems are grouped by the area of control. Depending on the domain, the systems gather readings from specific sensors placed in mines, e.g. methanometers, CO and CO_2 detectors, seismic sensor, ammeters, thermometers, barometers, etc. Besides data integration, the high level design of DISESOR takes into account the data cleaning process, feature engineering, training of data mining models and on-line predictive reasoning.

Figure 10 illustrates the location and importance of the DISESOR in the ecosystem of computer applications supporting the operation of coal mines in Poland. DISESOR integrates data from many other systems responsible for various issues such as seismic or seismoacoustic monitoring, machine monitoring, control of the mine atmosphere, etc., which in turn aggregate readings from multiple sensors distributed in the mines. The systems displayed in Fig. 10 are provided by many different corporations and are developed in various technologies what causes many problems with data integration. On the other hand, this situation gives a great opportunity to use the collected data to improve the safety and increase the work efficiency. However, for this purpose, a central decision support system is indispensable.

8   Summary

The article summarizes results achieved in the IJCRS15 Data Challenge which was related to the problem of monitoring and prediction of dangerous concentrations of methane in longwalls of the coal mine. The competition was supposed to help us to verify the quality of developed framework for multi-stream feature engineering. In this paper we explained in details our motivation for participation in a data mining contest and presented: the challenge task, provided data, the elaborated feature engineering framework and conducted experiments.

The prepared solution achieved the second highest score, reaching 94.73 %, while the highest score was 95.92 %, the third highest score was 94.43 % and the competition's baseline reached 90 %. This shows that features prepared by the elaborated mechanism are suitable for the purpose of forecasting indications of methane detectors. Thanks to the developed framework one can achieve very promising results of data analysis without much attention and long lasting manual data cleaning and classifier tuning.

References

1. Grzegorowski, M., Stawicki, S.: Window-based feature extraction framework for multi-sensor data: a posture recognition case study. In Ganzha, M., Maciaszek, L.A., Paprzycki, M. (eds.) Proceedings of FedCSIS 2015. IEEE (2015)
2. Grzegorowski, M.: Scaling of complex calculations over big data-sets. In: Ślęzak, D., Schaefer, G., Vuong, S.T., Kim, Y.-S. (eds.) AMT 2014. LNCS, vol. 8610, pp. 73–84. Springer, Heidelberg (2014)
3. Wieczorkowska, A., Wróblewski, J., Ślęzak, D., Synak, P.: Problems with automatic classification of musical sounds. In: Klopotek, M.A., Wierzchoń, S.T., Trojanowski, K. (eds.) IIPWM 2003, pp. 423–430. Springer, Heidelberg (2003)
4. Kozielski, M., Skowron, A., Wróbel, L., Sikora, M.: Regression rule learning for methane forecasting in coal mines. In: Kozielski, S., Mrozek, D., Kasprowski, P., Malysiak-Mrozek, B., Kostrzewa, D. (eds.) BDAS 2015, pp. 495–504. Springer, Cham (2015)
5. Riza, L.S., Janusz, A., Bergmeir, C., Cornelis, C., Herrera, F., Ślęzak, D., Benítez, J.M.: Implementing algorithms of rough set theory and fuzzy rough set theory in the R package "roughsets". Inf. Sci. **287**, 68–89 (2014)
6. Janusz, A., Stawicki, S.: Applications of approximate reducts to the feature selection problem. In: Yao, J.T., Ramanna, S., Wang, G., Suraj, Z. (eds.) RSKT 2011. LNCS, vol. 6954, pp. 45–50. Springer, Heidelberg (2011)
7. Breiman, L., Friedman, J.H., Olshen, R.A., Stone, C.J.: Classification and Regression Trees. Wadsworth, Monterey (1984)
8. Ślęzak, D.: Degrees of conditional (in)dependence: a framework for approximate bayesian networks and examples related to the rough set-based feature selection. Inf. Sci. **179**(3), 197–209 (2009)
9. Read, J., Pfahringer, B., Holmes, G., Frank, E.: Classifier chains for multi-label classification. Mach. Learn. **85**(3), 333–359 (2011)
10. Park, S.H., Fürnkranz, J.: Multi-label classification with contraints. In: Proceedings of the Workshop on Preference Learning at ECML PKDD 2008, Antwerp, Belgium (2008)
11. Boser, B.E., Guyon, I.M., Vapnik, V.N.: A training algorithm for optimal margin classifiers. In: Proceedings of the Fifth Annual Workshop on Computational Learning Theory, COLT 1992, pp. 144–152. ACM, New York (1992)
12. Zhou, S., Zhang, S., Karypis, G. (eds.): ADMA 2012. LNCS, vol. 7713. Springer, Heidelberg (2012)
13. Ślęzak, D., Janusz, A.: Ensembles of bireducts: towards robust classification and simple representation. In: Kim, T., Adeli, H., Slezak, D., Sandnes, F.E., Song, X., Chung, K., Arnett, K.P. (eds.) FGIT 2011. LNCS, vol. 7105, pp. 64–77. Springer, Heidelberg (2011)

SVM Parameter Tuning with Grid Search and Its Impact on Reduction of Model Over-fitting

Petre Lameski[1], Eftim Zdravevski[1(✉)], Riste Mingov[2], and Andrea Kulakov[1]

[1] Faculty of Computer Science and Engineering, Saints Cyril and Methodius University, Skopje, Macedonia
{petre.lameski,eftim.zdravevski,andrea.kulakov}@finki.ukim.mk
[2] NI TEKNA - Intelligent Technologies, Negotino, Macedonia
riste.mingov@ni-tekna.com

Abstract. In this paper we describe our submission to the IJCRS'15 Data Mining Competition, which is concerned with prediction of dangerous concentrations of methane in longwalls of a Polish coalmine. We address the challenge of building robust classification models with support vector machines (SVMs) that are built from time series data. Moreover, we investigate the impact of parameter tuning of SVMs with grid search on the classification performance and its effect on preventing over-fitting. Our results show improvements of predictive performance with proper parameter tuning but also improved stability of the classification models even when the test data comes from a different time period and class distribution. By applying the proposed method we were able to build a classification model that predicts unseen test data even better than the training data, thus highlighting the non-over-fitting properties of the model. The submitted solution was about 2 % behind the winning solution.

Keywords: Support Vector Machines · SVM · Grid search · Over-fitting · Parameter tuning · Time series · Coalminig

1 Introduction

In general, mining is associated with work in hazardous conditions. Miners in an underground coalmine can face many threats, such as, methane explosions or rock-burst. The coal mining industry is the leading cause of fatal injuries in the United States [1]. Furthermore, not only accidents but also exposure to lethal gases can lead to long term diseases of miners [2]. According to the National Institute for Occupational Safety and Health, the fatality rate for coal mining in 2006 was 49.5 per 100,000 workers, more than 11 times greater than the fatality rate in all private industry. Inhaling coal dust also causes black lung disease in

P. Lameski—This work was partially financed by the Faculty of Computer Science and Engineering at the Ss.Cyril and Methodius University, Skopje, Macedonia.

© Springer International Publishing Switzerland 2015
Y. Yao et al. (Eds.): RSFDGrC 2015, LNAI 9437, pp. 464–474, 2015.
DOI: 10.1007/978-3-319-25783-9_41

coal mine workers. Aiming to provide protection for miners, systems for active monitoring of production processes are usually used. One of their fundamental applications is screening dangerous gas concentrations (methane in particular) in order to prevent spontaneous explosions [3]. Therefore, for that purpose the ability to predict dangerous concentrations of gases in the nearest future can be even more important then monitoring the current sensor readings [4]. In that context, the IJCRS'15 Data Mining Competition [5] is concerned with prediction of dangerous concentrations of methane in longwalls of a Polish coal mine. Being able to predict dangerous concentrations of gases in mines can save human lives, which stresses the importance of the topic of the competition.

Analysis of time series data can is a key to: pattern discovery, clustering, classification and rule discovery [6]. Due to the sampling rates of the sensors and the nature of the time series data, there are three main tasks that need to be defined in order to be able to automatically learn from the data [7]: dimensionality reduction and data representation, distance measurement and indexing. Dimensionality reduction and data representation is one of the most important and time-consuming tasks that are performed when processing the time series data. This step is crucial in order to be able to learn from the data. The data representation needs to take into account the alignment of the data in relation to the beginning of the series [8]. The distance metric should be invariant to many transformations of the time series data like amplitude or time shifting, uniform amplification, additive noise, time scaling, etc. [7]. The indexing problem is related to improving the retrieval speed for a given series when searching trough a database of time series data.

The remaining of the paper is structured as follows. In Sect. 2 we describe the challenge because its specifications were the driver of our research presented here. Then, in Sect. 3 we describe the Support Vector Machine (SVM) algorithm with accent of its parameters. Next, in Sect. 4 we describe the algorithm for optimization and tuning of SVM parameters. Thereupon, in Sect. 5 we describe the proposed methods for feature selection and modeling. Afterwards, in Sect. 6 we present the our results on the competition after applying the proposed methods. Finally, in Sect. 7 we discuss the benefits and limitations of the proposed methods and make some conclusions.

2   Challenge Description

The competition dataset came from an active Polish coal mine [5]. The task was to come up with a prediction model, which could predict warning levels of methane concentrations at three methane meters placed in the longwall of the mine. The organizers provided a mining scheme utilized in the mine, which explains all sensors used for monitoring the mining process and their placement. The cutter loader moves along the longwall between the sensors MM262 and MM264. The bigger the currents consumed by the cutter loader are, the more efficient is its mining work, which in theory results in more methane emitted in the air. If the methane concentration measured on any of the sensors reaches the alarm level the cutter loader is switched off automatically. However, if one is able to predict ahead the warning methane concentrations, the speed of the cutter loader could be reduced and give methane more time to spread out. This in turn might make the switching off of the whole production line unnecessary.

The training dataset is consisted of sensor readings for 51700 time periods, each 10 min long, with measurements taken every second (600 values for every sensor in a single series). Values for each time period are stored in different rows of the data. The data include readings from 28 different sensors thus, every row in the data consists of 16800 values stored in consecutive columns and separated by commas. One very important property of the training dataset is that its instances are chronologically ordered and overlapping. These properties will have important impact on the feature modeling. The test dataset conforms the same format and has 5076 instances that do not have any particular chronological ordering in the dataset.

Labels in the data indicate whether a warning threshold has been reached in a period between three and six minutes after the end of the training period, for three methane meters: MM263, MM264 and MM256. In particular, if a given row corresponds to a period between t_{-599} and t_0, then the label for a methane meter (MM) in this row is 'warning' if and only if $max(MM(t_{181}), ..., MM(t_{360})) \geq 1.0$. The goal is to predict the likelihood of the label 'warning' for particular time series from the test set. The predictions for each test instance should contain the likelihoods for the three target sensors.

The submitted solutions were evaluated on-line and the preliminary results were published on the competition leader-board. The preliminary score was computed on a random subset of the test set, fixed for all participants. It corresponded to approximately 10 % of the test data. The final evaluation was performed after completion of the competition using the remaining part of the test data. Those results was also published on-line. The assessment of solutions was done using the Area Under the ROC Curve (AUC) measure. It was computed separately for each of the three target sensors. The final score in the competition corresponds to the average AUC for those three sets of predictions. Interestingly, the organizers also provided a baseline solution that derives from the theory of rough sets. Namely, they used a discretization method based on maximum discernibility heuristic [9] in a combination with LEM2 algorithm [10] for decision rule induction, which were implemented in the RoughSets package for R System [11].

The main challenges related to the competition tasks are feature modeling and building classification models. Additionally to the difficulty of the task contribute the highly unbalanced distribution of classes for all three labels in the training dataset, as it can be seen from Table 1. In the following sections we describe how these tasks were addressed in our submissions.

3   Building Classification Models with SVMs

Before training any classifier we have defined the competition challenge as three binary classification problems. In order to be able to calculate the AUC the classification predictions should be probabilistic.

Table 1. Class distribution for all three labels (i.e. L1 (MM263), L2 (MM264), L3 (MM256)) in the training dataset

Class	L1 counts	L1 ratio	L2 counts	L2 ratio	L3 counts	L3 ratio
Normal	51235	99.1%	50362	97.4%	50390	97.5%
Warning	465	0.9%	1338	2.6%	1310	2.5%

The SVM divides the feature space by finding the support vectors with the highest distance between the nearest points of each class. Let the pairs $(x_i, y_i), i = 1, ..., k$, where $x_i \in R^n$, are such that $y_i \in \{-1, 1\}$. We need to find the solution to the problem stated with Eqs. (1) and (2) in order to generate the SVM classification model based on the training data.

$$\min_{w,b,\xi} \frac{1}{2} W^T W + C \sum_0^k (\xi i) \tag{1}$$

where (1) is subject to:

$$y_i(W^T \phi(x_i) + b) \geq 1 - \xi_i, \xi_i \geq 0 \tag{2}$$

The space is separated by the hyper-lines so all instances that are on one side belong to one class, and instances on the other side belong to the other class. If it is not able to find the exact hyper-lines that separate the space ideally based on the training data, the algorithm tries to separate the space in such a way that the classification error is minimal. In general, most of the classification problems are not linearly separable. Therefore, it is impossible to find a line that separates the space in a way that the features of both classes can be distinguished with a reasonable error in the classification. For that purpose we are using the Gaussian SVM, which uses parameters C and gamma to transform the feature vector space to a higher dimension where the separation can be executed with greater accuracy. The transformation is done using the kernel function $K(x_i, x_j) = \phi(x_i)^T \phi(x_j)$, defined for the Gaussian SVM as $K(x_i, x_j) = exp(-\gamma ||x_i - x_j||^2), \gamma > 0$. For building the SVM model, we used the scikit-learn [12] implementation of SVM. It contains a probabilistic SVM implementation that provides probabilistic output for the predictions.

4   SVM Parameter Tuning with Grid Search

The parameters C and gamma are provided as input to the SVM and influence the optimization process that divides the hyper-plane in an optimal way based on the training data [13]. The parameter C defines the penalty for errors in the data. Increased value of C increases the penalty and lowers the number of points that are allowed in the error margin. Smaller number of C usually allows a larger error margin when the hyper-plane is separated. The gamma parameter is specific for the Gaussian SVM and influences the hyper-line flexibility. For smaller values of gamma, the line that separates the hyper-plane is nearly linear, and for larger numbers it becomes more curved. Increasing the value of gamma too much might allow over-fitting on the train data [13].

The optimal parameters on the given data set need to be determined in order to make the best classification model. For this purpose we use grid search in the C and gamma parameter space as suggested in [14]. To increase the speed and lower the number of steps that we need to search, we propose using a two-step grid search. First, using exponentially increasing steps for the grid search, we can find good values for C and gamma. Then using smaller increments we search for other values of C and gamma near the values determined in the previous step. The grid search algorithm we propose is illustrated in Algorithm 1. The interval that we used for these experiments for the values of C and gamma is $(10^{-6}, 10^6)$ which means we needed to evaluate a total of $12 \times 12 = 144$ combinations. After that we have performed the finer-grained search near the best parameters for the dataset. Depending on the different training datasets we have obtained different optimal values, and those are reported in Sect. 6.

Algorithm 1. Grid search in parameter space for SVM with finer tuning

---

for $k = a \to b$ do
    for $l = c \to d$ do
        Set $C=10^k$
        Set $Gamma=10^l$
        Train SVM with C and $Gamma$ on $TrainingSet$
        Evaluate SVM classification on $ValidationSet$
        if precision is better than $MaxPrecision$ then
            Save $MaxC=C$ and $MaxGamma=Gamma$
        end if
    end for
end for
$IncrementC = MaxC/4$
$C = IncrementC$
while $C \leq MaxC \times 4$ do
    $IncrementGamma = MaxGamma/4$
    $Gamma = IncrementGamma$
    while $Gamma \leq MaxGamma \times 4$ do
        Train SVM with C and $Gamma$ on $TrainingSet$
        Evaluate SVM classification on $ValidationSet$
        if precision is better than $MaxPrecision$ then
            Set $OptimalC=C$ and $OptimalGamma=Gamma$
        end if
        $Gamma = Gamma + IncrementGamma$
    end while
    $C = C + IncrementC$
end while
return $OptimalC$ and $OptimalGamma$

---

5   Feature Modeling and Selection

For any data mining task, often the first challenge in to preprocess the data in order to create a feature set which will be later on used for building classification models. For the task at hand, the initial number of features is 16800, which is quite large for most machine learning algorithms. Moreover, they represent 28 different time series from different sensor types, so prior performing feature selection the time series need to be properly modeled. We have tried several different approaches to address this and they are described in the following subsections. All code used for the research in this paper was implemented in Python 3 using the SciKit Learn library [12] on a virtual machine with Intel Xeon X5680 at 3.33 Ghz with 8 GB RAM running Windows 7 Professional SP1 64 bit. All execution times mentioned in the remaining of the paper are for these hardware and software specifications.

5.1   Histogram-Based Modeling of Time Series Data

The first approach we tried was based on a robust histogram-based feature engineering of time series data, as described in [15]. We have applied this approach for modeling time series from other domains and it produced excellent results. Therefore our initial goal was to test its general applicability out-of-the-box without any modifications. Furthermore, our intent was to get initial understanding of the predictability of the problem. This method first adds first derivatives in order to model the speed of change, then discretizes and normalizes the original time series and the time series of first derivatives. Afterwards in order to perform feature reduction it generates histograms based on the discretized values of all time series. Before calculating the histograms, the dataset consisted of the 28 normalized time series having 16800 values and also 28 time series of first derivatives, each of them having 599 values. As a result, the unreduced dataset had a total of $28 \times 600 + 28 \times 599 = 33572$ values. With this method for each time series we keep 4 statistics: the minimum, maximum, standard deviation and mean values. When using 50 bins for discretization, we reduced the dataset to $2 \times 28 \times (50 + 4) = 3024$ features. This was still a large number of features but we decided to train a classification model in order to get a sense of how good the model is. The obtained AUC on the leaderboard dataset was about 0.78. These results did not include any parameter tuning for the SVMs, rather the default values for C and gamma.

The training time for this experiment was about one hour. Even though the number of features is significantly reduced than the original set of 16800 or the modified set of 33572, we needed to further reduce them so we can perform other experiments in timely manner. Performing grid search with cross-validation or with a random train/test split on a dataset of that size would take a lot of time. To address the feature selection we had two alternatives. The first was to do it in a systematic way with a filtering or a wrapper approach for feature selection. For a large dataset like this one a filtering approach, such as the one used in [16], would be more suitable. Also using PCA [17] would be applicable. From experience we knew that it would probably give good results, nevertheless the execution time would be considerable. The second approach was doing something more specific for the particular challenge. Aiming to better understand the dataset we have decide to experiment with this approach before trying any systematic feature selection, and our findings are presented in the next subsection. Due to the fact that those approaches turned out to be quite simpler, while offering good performance, and due to the time constraints of the competition we did not eventually return to exploring this approach in more details.

5.2   Using Time Series of the Target Sensors

From the problem description it was inherent that a 'warning' signal in the future for a particular methane sensor significantly depends on the past readings of that sensor. All other 27 sensors in the environment could potentially help understanding the circumstances that lead to a particular warning signal. If this assumption is valid, then the prediction model could be based only on this time series, with significantly less features. Other time series could be used as additional features, but the main predictive power of the model should come from the target time series.

Another key point that was discovered when analyzing the unprocessed time series data is that it was very slowly changing. As a result, many consecutive samples in the time series had the same value. On the other hand, the task description stated that the training examples are overlapping in time. Important to realize is that these two facts are related to the alignment problem, discussed in Sect. 1. Owing to the fact that the data in the time series is slowly changing and that the training dataset contains time-shifted data, the sensor readings in a time series could be directly used as features. Discovering that the data is slowly changing data also helped us understand the reason why the time series of first derivatives in the first approach, described in Subsect. 5.1, are useless features.

With this in mind, this approach attempts to use the original values of the target time series as features. To test it we have created three different datasets for the three classification problems. Each of the dataset contained only the corresponding time series with 600 readings of the particular sensor (i.e. the MM263, MM264 and MM256 sensor readings). With this approach each dataset had 600 different features. Important to realize is that the values of all 600 features are from the same domain, so data normalization is not needed. Indeed, our experiments showed that with proper parameter tuning the performance of the classifier built from the original time series and the normalized dataset is essentially the same. The AUC performance on the leaderboard dataset on both versions of the dataset prior performing parameter tuning were about 0.82 and after the parameter tuning with grid search the AUC score was about 0.92. The optimal values for the parameters were $C = 1$ and $gamma = 0.01$ for all three classification problems.

5.3   Using Samples from the Time Series of the Target Sensors

Being inspired by the good results of the previous approach, we wanted to reduce the number of features further to prevent over-fitting. We have tried discarding features that have low variance, selecting random subset of features, and selecting sequential subsets of features. As expected, the time for training, prediction and grid-search was substantially reduced. We have tried limiting the number of features to different values. The experiments showed that around 180 features gave good generalization performance. Somewhat counterintuitive, after performing SVM parameter tuning with grid search of various subsets of features the AUC score on the leaderboard was very similar. It was in the interval 0.91 to 0.925. We attribute this behavior to the fact that the data in the time series is slowly changing, which in turn results to redundant features. However, due to time limitations of the contest we did not have time to focus on finding and discarding redundant features to support this claim. Another thing to point out is that these reduced feature sets were leading to classifiers that are better than ones that use the original datasets of 600 features. As expected, for this or the previous approach data normalization did not help due to the fact that the domain of all features is the same. Similarly to the previous approach, the SVMs with tuned parameters outperformed the ones without tuned parameters by up to 0.15 AUC.

5.4   Using Time Series of the Target Sensors and Their Predecessors

The fourth approach we tried was based on the previous one. Here we have taken into account the configuration and placement of sensors in the mine. In addition to the 180 values that come from the target time series, we have also used 180 features from the methane sensor that is placed before the target sensor, in respect to the air flow direction. The datasets were consisted of the readings of MM262 and MM263 for the first classification problem, from MM263 and MM264 for the second, and from MM264 and MM256 for third. The logic behind this approach is that if the preceding sensor records high concentrations of methane then eventually the next sensor in the airflow line would record the increased concentration. For selecting the 180 features from each of two time series we have tried similar strategies as in the previous approach. As it turned out, this approach produced slightly worse results that the previous approach, but were also about 0.92 for the leaderboard dataset. Again, we explain this behavior by the fact that the time series are slowly changing, so the values of the time series of the previous sensor appear after some delay in the time series of the current sensor. This, in turn, makes the features that reflect the values of the previous sensor redundant in the prediction model. Consistently with the previous approaches, the SVMs without tuned parameters performed up to 0.15 AUC worse than the ones with tuned parameters.

6   Results

After analysis of the obtained results by the different approaches for feature modeling, we decided to use the simplest solution, described in the Subsect. 5.3, in our final submission. When comparing the leaderboard score and the final score of all teams, shown on Fig. 1, we can notice several things. Interestingly, our leaderboard rank was 38, while our final rank was 12, making a huge leap of 26 places. We attribute this to the robustness of our classification model. It was evident that many teams had significantly over-fitted their models to the leaderboard dataset. Therefore when testing on the final dataset, there is evident decrease of the performance. On the other hand, only few teams, among which was our, did not over-fit the models and therefore registered improved performance on the final dataset.

Fig. 1. Leaderboard vs final score of teams that achieved better performance than the baseline

7   Conclusion

During coal mining life-threatening gases are being produced and being able to predict dangerous concentrations, can save human lives. One way to achieve that is through analysis of time series data that originates from various sensors placed in a coalmine. In this paper we described our analysis of the dataset originating from a Polish coalmine provided in the IJCRS'15 data mining competition. We were able to identify important properties of the time series, such as the slowly changing nature of the data and the most predictive time series. More importantly, we have demonstrated that by using gird-search to tune the parameters of the SVMs, their performance can be significantly improved. In fact, this was evident for all types of datasets that we have tested. Furthermore, the combination of simple feature set and properly trained SVMs is notably resistant to over-fitting.

References

1. Finkelman, R.B.: Health impacts of coal: facts and fallacies. AMBIO J. Hum. Environ. **36**(1), 103–106 (2007)
2. Hendryx, M., Ahern, M.M., Nurkiewicz, T.R.: Hospitalization patterns associated with appalachian coal mining. J. Toxicol. Environ. Health Part A **70**(24), 2064–2070 (2007)
3. Kozielski, M., Skowron, A., Wrbel, L., Sikora, M.: Regression rule learning for methane forecasting in coal mines. In: Kozielski, S., Mrozek, D., Kasprowski, P., Malysiak-Mrozek, B., Kostrzewa, D. (eds.) Beyond Databases, Architectures and Structures. Communications in Computer and Information Science, vol. 521, pp. 495–504. Springer, Cham (2015)
4. Krasuski, A., Jankowski, A., Skowron, A., Slezak, D.: From sensory data to decision making: a perspective on supporting a fire commander. In: 2013 IEEE/WIC/ACM International Joint Conferences on Web Intelligence (WI) and Intelligent Agent Technologies (IAT), pp. 229–236. IEEE (2013)
5. Janusz, A., Ślęzak, D., Sikora, M., Wróbel, Ł., Stawicki, S., Marek, G., Slezak, D.: Mining data from coal mines: IJCRS'15 data challenge. In: Yao, Y., Hu, Q., Yu, H. Grzymala-Busse, J. (eds.) RSFDGrC 2015. LNCS, vol. 9437, pp. 429–438. Springer, Heidelberg (2015). https://knowledgepit.fedcsis.org/contest/view.php?id=109. Accessed 29 Jun 2015
6. Fu, T.C.: A review on time series data mining. Eng. Appl. Artif. Intell. **24**(1), 164–181 (2011)
7. Esling, P., Agon, C.: Time-series data mining. ACM Comput. Surv. **45**(1), 12:1–12:34 (2012)
8. Hu, B., Chen, Y., Keogh, E.: Classification of streaming time series under more realistic assumptions. Data Min. Knowl. Disc. 1–35 (2015)
9. Nguyen, H.S.: On efficient handling of continuous attributes in large data bases. Fundam. Inf. **48**(1), 61–81 (2001)
10. Grzymala-Busse, J.W.: A new version of the rule induction system lers. Fundam. Inf. **31**(1), 27–39 (1997)
11. Riza, L.S., Janusz, A., Bergmeir, C., Cornelis, C., Herrera, F., Slezak, D., Bentez, J.M.: Implementing algorithms of rough set theory and fuzzy rough set theory in the R package "roughsets". Information Sciences **287**, 68–89 (2014)
12. Pedregosa, F., Varoquaux, G., Gramfort, A., Michel, V., Thirion, B., Grisel, O., Blondel, M., Prettenhofer, P., Weiss, R., Dubourg, V., Vanderplas, J., Passos, A., Cournapeau, D., Brucher, M., Perrot, M., Duchesnay, E.: Scikit-learn: machine learning in Python. J. Mach. Learn. Res. **12**, 2825–2830 (2011)
13. Ben-Hur, A., Weston, J.: A users guide to support vector machines. In: Carugo, O., Eisenhaber, F. (eds.) Data Mining Techniques for the Life Sciences. Methods in Molecular Biology, vol. 609, pp. 223–239. Humana Press, New York (2010)
14. Hsu, C.W., Chang, C.C., Lin, C.J., et al.: A practical guide to support vector classification
15. Zdravevski, E., Lameski, P., Mingov, R., Kulakov, A., Gjorgjevikj, D.: Robust histogram-based feature engineering of time series data. In Ganzha, M., Maciaszek, L.A., Paprzycki, M., (eds.) Proceedings of the 2015 Federated Conference on Computer Science and Information Systems (2015, in print)

16. Zdravevski, E., Lameski, P., Kulakov, A., Gjorgjevikj, D.: Feature selection and allocation to diverse subsets for multi-label learning problems with large datasets. In: 2014 Federated Conference on Computer Science and Information Systems (FedCSIS), pp. 387–394, September 2014
17. Jolliffe, I.: Principal component analysis. In: Balakrishnan, N., Colton, T., Everitt, B., Piegorsch, W., Ruggeri, F., Teugels, J.L. (eds.) Wiley StatsRef: Statistics Reference Online. Wiley, Chichester (2014)

Detecting Methane Outbreaks from Time Series Data with Deep Neural Networks

Krzysztof Pawłowski[✉] and Karol Kurach

Faculty of Mathematics, Informatics and Mechanics, University of Warsaw,
Banacha 2, 02-097 Warsaw, Poland
{kpawlowski236,kkurach}@gmail.com

Abstract. Hazard monitoring systems play a key role in ensuring people's safety. The problem of detecting dangerous levels of methane concentration in a coal mine was a subject of IJCRS'15 Data Challenge competition. The challenge was to predict, from multivariate time series data collected by sensors, if methane concentration reaches a dangerous level in the near future. In this paper we present our solution to this problem based on the ensemble of Deep Neural Networks. In particular, we focus on Recurrent Neural Networks with Long Short-Term Memory (LSTM) cells.

Keywords: Machine learning · Recurrent neural networks · Ensemble methods · Time series forecasting · Hazard monitoring systems

1 Introduction

Working in a coal mine historically has been a very hazardous occupation. Over time the conditions and safety improved substantially, in part thanks to advances in technology. Despite remarkable progress, it is still one of the most dangerous professions [6]. One of the dangers present is a possibility of explosion caused by high concentration of methane in the air. Therefore, it is of utmost importance to monitor methane concentration levels and to ensure they are within a safe range. If the concentration levels reach a critical threshold, the production line needs to be shut down [23], which is a costly interruption.

On the other hand, mining effectiveness positively depends on the pace of methane emissions. Therefore, in order to maximize the efficiency, a fine balance needs to be made between mining effectiveness and the safety. If one could predict methane concentration in the future and tell when it is likely to be dangerously high, one could reduce the speed of the operation and try to avoid reaching dangerous levels. Design of such an efficient prediction algorithm is a goal of IJCRS'15 Data Challenge: Mining Data from Coal Mines competition [16]. The problem is an example of supervised learning classification task, with data given in a form of non-stationary multivariate time series. Between the training and test data sets there is a significant concept drift.

K. Pawłowski and K. Kurach—Both authors contributed equally.

© Springer International Publishing Switzerland 2015
Y. Yao et al. (Eds.): RSFDGrC 2015, LNAI 9437, pp. 475–484, 2015.
DOI: 10.1007/978-3-319-25783-9_42

Different methods of tackling similar problems have been proposed in the literature. One of them is an application of Deep Neural Networks. While artificial neural networks have been known for a very long time, in recent years they have achieved spectacular results in areas such as computer vision [17,18] or speech recognition [10,14], in part due to advances in computing hardware, such as Graphics Processing Units. Recurrent Neural Network is an architecture well-suited for the processing of time series data [4,8]. We aim to test how well such methods perform in the competition. In this paper we present our solution which uses Recurrent Neural Network with Long Short-Term Memory cells [15,27], Deep Feedforward Neural Network and ensembling [5] techniques.

The rest of this paper is organized as follows. In Sect. 2 we present the problem in detail. Section 3 describes the Recurrent Neural Network model we used. In Sect. 4 we present ensembling technique, detail Deep Feedforward Neural Network model and analyse the results. Finally, in Sect. 5 we conclude the paper and propose future work.

2   Problem Statement

In this section we describe the data used in the competition. Then, we document the evaluation procedure, including the target measure to be optimized. Finally, we review the most important challenges.

2.1   Data

The goal of *IJCRS'15 Data Challenge* competition is to predict dangerous level of methane concentration in coal mines based on the readings of 28 sensors. It is an example of supervised learning classification task. The data is split into training and test set, where the training set contains 51700 records and test set contains 5076 records.

Each record is a collection of 28 time series – corresponding to 28 sensors that are installed in the coal mine. The sensors record data such as level of methane concentration, temperature, pressure, electric current of the equipment etc. Each of the time series contains 600 readings, taken every second, for a total of 10 minutes of the same time period for each sensor. The time periods described in the training data set overlap and are given in a chronological order. For the test data, however, the time periods do **not** overlap and are given in random order.

For each record in the training set, three labels are given. The test set is missing the labels – it is the goal of the competition to predict those values. Each label instance can be either *normal* or *warning*. Those levels signify the amount of methane concentration, as recorded by the three known sensors, named *MM263*, *MM264* and *MM256*. The second-by-second readings of those sensors are described in time series mentioned in the previous paragraph. The predictions are to be made about the methane level in the future - that is during the period between three and six minutes after the end of the training (time series) period. If the level of methane concentration reaches or exceeds 1.0, then the corresponding label should be *warning*. Otherwise, it should be *normal*.

2.2   Evaluation

The submissions consist of three predictions of label values, made for each of 5076 records in the test set. Each prediction is a number – a higher value denotes a higher likelihood that the true label value is *warning*. The score is defined as a mean of *area under the ROC curve*, averaged over the three labels.

Participants may submit their predictions during the course of the competition. Until the finish of the competition, the participants are aware only of the score computed over *preliminary test set* – a subset of the whole test set that contains approximately 20 % of the records. This subset is picked at random by the organizers and is fixed for all competitors but it is not revealed to the participants which of the test records belong to it. The participants may choose a single final solution, possibly taking into the account the scores obtained on the preliminary test set. However, the final score is computed over the *final test set* – remaining approximately 80 % of the test data. This score is revealed only after the end of the competition and is used to calculate the final standings – the team with the highest score is declared the winner.

2.3   Challenges

The problem presents the following challenges.

Imbalanced Data. Only about 2 % of the labels in the training set belong to the *warning* class, while the remaining belong to the *normal* class. A trivial solution that predicts *normal* for every label achieves 98 % accuracy, obviously without having any practical significance (and with a bad 0.5 mean-ROC score). Unless special precautions are taken, methods that heavily optimize just the prediction accuracy can have significant problems with this task.

Overlapping Training Periods. Almost all adjacent training records overlap by 9 out of the total 10 minutes recorded in the time series. It clearly violates the assumption of i.i.d. that underpins the theoretical justification of many learning algorithms. In addition, due to overlap, a classical cross-validation approach may result in splits very "similar" data across different folds and in turn yield over-optimistic estimates of the model performances.

Noisiness. Seemingly small "meaningful" changes in sensor readings happen at the 1-second resolution. Most changes at this interval appear to be random fluctuations. That, combined with a large amount (16800) of readings per record, poses a severe danger of overfitting.

Large Data Size. The whole training set consists of over $868,560,000$ values. Just storing it in a computer memory requires 3.5 gigabytes of memory, when using 32-bit floating point representation. Thus storage and computational costs can be a significant constraint.

Concept Drift. Training and test data come from different time periods. The records in the training set are sorted by time, so it's easy to notice that there are very significant trends in the data that change along with the time. With test data samples taken at times belonging to a different interval than training samples, one can expect a severe concept drift - and indeed exploratory tests showed that classifier performance degrades on the test set, as compared to the same classifier's performance when it is evaluated on the interval of training data that was not used for its learning.

3   Long Short-Term Memory Model

This section describes the Recurrent Neural Network with Long Short-Term Memory (LSTM) [15] model, which is a crucial part of our final solution.

First, we give some background about the recurrent networks and formally define the dynamics of LSTM. Next, we describe how the data was preprocessed. Finally, we present the network architecture and describe in more detail the training procedure.

3.1   Overview

Recurrent Neural Network (RNN) is a type of artificial neural network in which dependencies between nodes form a directed cycle. This allows the network to preserve a state between subsequent time steps. This kind of network is particularly suited for modeling sequential data, where the length of the input is not fixed or can be very long. Parameters in RNNs are shared between different parts of the model, which allows better generalization.

LSTM is an RNN architecture designed to be better at storing and accessing information than standard RNN [11]. LSTM block contains memory cells that can remember a value for an arbitrary length of time and use it when needed. It also has a special *forget gate* that can erase the content of the memory when it is no longer useful. All the described components are built from differentiable functions and trained during backpropagation step.

The LSTM networks recently achieved state of the art performance in many tasks, including language modeling [21], handwriting [12] or speech [10] recognition, and machine translation [22]. There are several variants of LSTM that slightly differ in connectivity structure and activation functions. Definition 1 describes the architecture that we implemented based on the equations from [27].

Definition 1. *Let $h_t^l \in \mathbb{R}^n$ be a hidden state in layer l of the network at step t. We assume that h_t^0 is the input at time t. Similarly, let $c_t^l \in \mathbb{R}^n$ be a vector of long term memory cells in layer l at step t. We define $T_{n,m} : \mathbb{R}^n \to \mathbb{R}^m$ to be an affine transform ($x \to Wx + b$ for some W and b) and $\odot$ be a element-wise multiplication. Then, LSTM is a transformation that takes 3 inputs (h_{t-1}^l, h_t^{l-1}, c_{t-1}^l) and computes 2 outputs (h_t^l and c_t^l) as follows:*

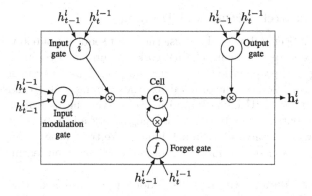

Fig. 1. A graphical representation of LSTM memory cells used in [27] and in our solution.

$$\begin{pmatrix} i \\ f \\ o \\ g \end{pmatrix} = \begin{pmatrix} \text{sigm} \\ \text{sigm} \\ \text{sigm} \\ \text{tanh} \end{pmatrix} T_{2n,4n} \begin{pmatrix} h_t^{l-1} \\ h_{t-1}^{l} \end{pmatrix}$$

$$c_t^l = f \odot c_{t-1}^l + i \odot g$$

$$h_t^l = o \odot \tanh(c_t^l)$$

In these equations, i, f, o, g are *input, forget, output* and *input modulation* gates respectively. The sigm and tanh functions are applied element-wise. This relationship is presented in Fig. 1.

3.2   Data Preprocessing

To address problems mentioned in Sect. 2.3, we apply several transformations to the original data.

First, for every sensor we normalize all of its readings to mean 0 and variance 1. This is a standard technique that improves convergence of gradient descent methods. To reduce overfitting and prevent algorithm from getting stuck in local minima, we shuffle training data after every epoch (one full pass over the data). We also address the problem of unbalanced classes by up-sampling with repetition from the *warning* class. It was done to ensure that at least 10 % of training examples are from the *warning* class.

Finally, we address the problem of large data. We modify the 600 input values from every sensor by grouping every 10 consecutive values and replacing them with their average.

3.3   Network Architecture and Training

As described in Sect. 3.2, we did not use the input sequence of 600 values directly but instead we grouped them. The network was unfolded to 60 time steps and trained using backpropagation through time [25]. This was mostly because of the practical purposes: for the original size the training took 10 times longer, we often exceeded the GPU memory and it was preventing us from testing bigger embedding or batch sizes. It could also negatively affect backpropagation, as the gradient was propagated for much longer. We tried other window sizes, but the value of 10 seemed to achieve a good trade-off between learning speed and quality.

The sensor values go through the hidden layer, which in this case consist solely of LSTM cells. At time step $t \in 1, .., 60$, the input for RNN are 28 average sensor values from seconds $[(t-1) * 10, t * 10]$.

After processing the whole sequence, the network's hidden state h_n^1 encodes all sensor averages in the same order in which they were seen. On top of this we build a standard supervised classifier (Multi-Layer Perceptron in this case) that predicts the binary outcome. The *warning* class is assigned a value of 1.0 and *normal* class is assigned a value of 0.0. The loss function used in the final model was Mean Squared Error. It performed better than Binary Cross Entropy loss which is typically used for binary classification

The training is done using standard Stochastic Gradient Descent algorithm and backpropagation through time. We initialize all the parameters by sampling from uniform distribution. To avoid *exploding gradient* problem, the gradients are scaled globally during training, so that their norm is not grater than 1 % of parameters' norm. All models were trained using Torch [3] on a machine with a GPU card.

4   Final Ensemble Model

We improve the quality of our prediction by making an ensemble model. In this section we give an overview of the ensemble technique, describe the extra base learner that was used in conjunction with the method described in Sect. 3 and document the procedure we used to obtain the final prediction.

4.1   Ensembling Methods

Ensemble methods combine results of multiple "base" learning algorithms, to form a single prediction that often achieves a better performance than any of individual base algorithm alone [5]. To give an example, one of the simplest forms of ensembling is to average the predictions of base algorithms. Ensembling works the best when base algorithms are accurate and, importantly, diverse. Diversity can mean that errors produced by the base algorithms are slightly correlated. Intuitively, in the mentioned example of ensembling by averaging, the prediction quality increases because uncorrelated errors "average out". The more sophisticated ensembling algorithms include bagging [2], boosting [7] or stacking [26].

4.2   Deep Feedforward Neural Network as a Base Learner

Deep feedforward neural network (DFNN) is an *artificial neural network* with multiple layers of hidden neurons. One notable difference between DFNN and LSTM network (described in Sect. 3) is that DFNN architecture does not contain recursive connections – instead, every neuron of the previous layer is connected with every neuron of the next layer. We use DFNN, together with LSTM, as base learners in an ensemble model. We train the DFNN with a *backpropagation* algorithm [24] that uses *stochastic gradient descent* (with *mini-batch* and *momentum*) as an optimization procedure to minimize the *root mean squared error* between numeric predictions and the target values. To avoid overfitting to the training set we use two regularization [9] methods: ad-hoc *early stopping* [19] and *dropout* [20].

Feature Engineering. To balance the impact of different features, further reduce overfitting and decrease the computational cost we preprocess the training and test data in the following way:

1. we scale the readings (separately for each sensor) across all the records to have *mean* equal 0 and *standard deviation* equal 1,
2. we transform the values with $x \rightarrow log(1 + x)$ function,
3. we compute *mean* and *standard deviation* for every sensor, taken over the last 30 readings (30-second period),
4. we keep the last 20 readings for the sensor that corresponds to the target label,
5. we discard all the original features.

Such preprocessing reduces the number of features from 16800 ($28 * 600$) to just 76 ($28 * 2 + 20$).

We also preprocess the training target values in the way described below. As mentioned in Subsect. 2.1, there are three labels for each training record with the *normal* class for future methane concentration levels under 1.0 and the *warning* class for methane concentration levels at or above 1.0. Recall from Subsect. 2.3, that an overwhelming majority of adjacent training periods overlap by exactly 9 minutes. That two facts together allow us to reconstruct the exact amount of methane concentration for a given record – to that end, we simply peek sensors values a few records in advance. We take reconstructed values as targets instead of the original label values, while also discarding all the training records for which such reconstruction is not possible (it turns out that less than 1000 records out of 51700 training records are discarded).

Training and Parameter Tuning. For each target label we train a different DFNN model and tune its parameters independently, to optimize the performance on:

– (initially) the validation set – created by us as 20 % of original training data. We train the model on the remainder of training data,

Table 1. Tuned parameter values for DFNN

Target label	MM263	MM264	MM256
activation	-	sigmoid	ReLU
layer sizes	-	76-15-5-2-1	76-25-7-3-1
dropout	-	none	.5
learning rate	-	0.1	0.1
mini-batch size	-	30	30
number of epochs	-	550	233

Table 2. Model scores

AUC score\label	MM263	MM264	MM256
LSTM	**0.9599**	0.9560	0.9605
DFFN	-	**0.9773**	0.9602
ensemble	-	0.9722	**0.9683**

- (finally) the preliminary test set. See Subsect. 4.4 for more discussion of model selection challenges.

We describe the final values of the most important parameters in Table 1. Parameters for the first target label (*MM263*) are omitted because for that label, the DFNN model fails to generate quality predictions for all the parameter combinations we tried.

4.3   Ensembling in Our Solution

Our final solution is an ensemble of two base models - LSTM described in Sect. 3 and DFNN described in Subsect. 4.2.

We decide to forgo the complex ensembling schemes that would require retraining of the models and perhaps additional parameter tuning. Instead, we consider a few simple averaging methods and finally we choose the method that gives the best AUC score – that is averaging the ranks of the base models' predictions. Table 2 illustrates the scores of particular models that are achieved on the preliminary test set. As the final submission we take the model that is a combination of the best-performing methods for each target label value. That is, for label *MM263* we use LSTM, for label *MM264* we use DFNN and for label *MM256* we use the ensemble of LSTM and DFNN.

4.4   Results and Discussion

Our final ensemble model achieved a score of 0.94 and the 6th place in the competition. It is interesting to mention that on the preliminary test set, our method obtained much higher score of 0.9685 which, if not decreased, would correspond to the 1st place.

Such a significant drop in score can be explained by overfitting. It was not a surprise, as we deliberately chose to perform model selection on the preliminary test set – that is, we submitted the model that achieved the best score on that set. Usually one would perform a cross-validation on the training set to perform model selection. The reason we decided not to, was because of the significant differences between the training and the test distributions (concept drift), as mentioned in Subsect. 2.3. We wanted to avoid a situation when the model is overly tuned to the training test and as a result does not generalize well on the test set. As we had only one shot for a final submission, it is not clear if the traditional (cross-validation) approach would have worked better – it can be an interesting topic for further research.

5 Conclusion

In this paper we presented our Deep Neural Network-based solution to *IJCRS'15 Data Challenge: Mining Data from Coal Mines* contest. It achieved a competitive score of 0.94 and the 6th place. The approach we developed should generalize well to other multivariate time series prediction problems. The obtained results confirm that methods based on Deep Neural Networks are not only effective for processing time series data, but also do not require extensive feature engineering to perform well. As expected, ensembling improves the quality of the prediction.

It would be interesting to see, if more advanced artificial neural network architectures, such as LSTMs with attention mechanism [1] or Neural Turing Machines [13], could achieve even better results. Another topic worth exploring are methods of handling the concept drift in context of parameter tuning and model selection, which was one of the main challenges in this task.

References

1. Bahdanau, D., Cho, K., Bengio, Y.: Neural machine translation by jointly learning to align and translate. CoRR abs/1409.0473 (2014). http://arxiv.org/abs/1409.0473
2. Breiman, L.: Bagging predictors. Mach. Learn. **24**(2), 123–140 (1996)
3. Collobert, R., Kavukcuoglu, K., Farabet, C.: Torch7: A matlab-like environment for machine learning. In: BigLearn, NIPS Workshop, No. EPFL-CONF-192376 (2011)
4. Connor, J.T., Martin, R.D., Atlas, L.E.: Recurrent neural networks and robust time series prediction. IEEE Trans. Neural Netw. **5**(2), 240–254 (1994)
5. Dietterich, T.G.: Ensemble methods in machine learning. In: Kittler, J., Roli, F. (eds.) MCS 2000. LNCS, vol. 1857, p. 1. Springer, Heidelberg (2000)
6. Donoghue, A.: Occupational health hazards in mining: an overview. Occup. Med. **54**(5), 283–289 (2004)
7. Freund, Y., Schapire, R., Abe, N.: A short introduction to boosting. J.-Jpn. Soc. Artif. Intell. **14**(771–780), 1612 (1999)
8. Giles, C.L., Lawrence, S., Tsoi, A.C.: Noisy time series prediction using recurrent neural networks and grammatical inference. Mach. Learn. **44**(1–2), 161–183 (2001)

9. Girosi, F., Jones, M.B., Poggio, T.: Regularization theory and neural networks architectures. Neural comput. **7**(2), 219–269 (1995)

10. Graves, A., Mohamed, A.R., Hinton, G.: Speech recognition with deep recurrent neural networks. In: IEEE International Conference on Acoustics, Speech and Signal Processing (ICASSP) 2013, pp. 6645–6649. IEEE (2013)

11. Graves, A.: Generating sequences with recurrent neural networks, CoRR abs/1308.0850 (2013). http://arxiv.org/abs/1308.0850

12. Graves, A., Liwicki, M., Fernández, S., Bertolami, R., Bunke, H., Schmidhuber, J.: A novel connectionist system for unconstrained handwriting recognition. IEEE Trans. Pattern Anal. Mach. Intell. **31**(5), 855–868 (2009)

13. Graves, A., Wayne, G., Danihelka, I.: Neural turing machines, CoRR abs/1410.5401 (2014). http://arxiv.org/abs/1410.5401

14. Hinton, G., Deng, L., Yu, D., Dahl, G.E., Mohamed, A.R., Jaitly, N., Senior, A., Vanhoucke, V., Nguyen, P., Sainath, T.N., et al.: Deep neural networks for acoustic modeling in speech recognition: the shared views of four research groups. IEEE Signal Process. Mag. **29**(6), 82–97 (2012)

15. Hochreiter, S., Schmidhuber, J.: Long short-term memory. Neural comput. **9**(8), 1735–1780 (1997)

16. Janusz, A., Ślęzak, D., Sikora, M., Wróbel, L., Stawicki, S., Grzegorowski, M., Wojtas, P.: Mining data from coal mines: IJCRS 2015 data challenge. In: Proceedings of IJCRS 2015. LNCS, Springer (2015), in print November 2015

17. Krizhevsky, A., Sutskever, I., Hinton, G.E.: Imagenet classification with deep convolutional neural networks. In: Advances in neural information processing systems, pp. 1097–1105 (2012)

18. Le, Q.V.: Building high-level features using large scale unsupervised learning. In: IEEE International Conference on Acoustics, Speech and Signal Processing (ICASSP) 2013, pp. 8595–8598. IEEE (2013)

19. Prechelt, L.: Early stopping - but when? In: Orr, G.B., Müller, K.-R. (eds.) NIPS-WS 1996. LNCS, vol. 1524, p. 55. Springer, Heidelberg (1998)

20. Srivastava, N., Hinton, G., Krizhevsky, A., Sutskever, I., Salakhutdinov, R.: Dropout: a simple way to prevent neural networks from overfitting. J. Mach. Learn. Res. **15**(1), 1929–1958 (2014)

21. Sundermeyer, M., Schlüter, R., Ney, H.: Lstm neural networks for language modeling. In: INTERSPEECH (2012)

22. Sutskever, I., Vinyals, O., Le, Q.V.: Sequence to sequence learning with neural networks. In: Advances in Neural Information Processing Systems, pp. 3104–3112 (2014)

23. Szlązak, N., Obracaj, D., Borowski, M., Swolkień, J., Korzec, M.: Monitoring and controlling methane hazard in excavations in hard coal mines. AGH J. Min. Geoengineering **37**, 105–116 (2013)

24. Werbos, P.: Beyond regression: new tools for prediction and analysis in the behavioral sciences, Ph.D. thesis, Harvard University, Cambridge (1974)

25. Werbos, P.J.: Generalization of backpropagation with application to a recurrent gas market model. Neural Netw. **1**(4), 339–356 (1988)

26. Wolpert, D.H.: Stacked generalization. Neural Netw. **5**(2), 241–259 (1992)

27. Zaremba, W., Sutskever, I., Vinyals, O.: Recurrent neural network regularization, CoRR abs/1409.2329 (2014). http://arxiv.org/abs/1409.2329

Self-Organized Predictor of Methane Concentration Warnings in Coal Mines

Dymitr Ruta$^{(\boxtimes)}$ and Ling Cen

Etisalat British Telecom Innovation Center,
Khalifa University of Science, Technology and Research, Abu Dhabi, UAE
{dymitr.ruta,cen.ling}@kustar.ac.ae

Abstract. Coal mining operation continuously balances the trade-off between the mining productivity and the risk of hazards like methane explosion. Dangerous methane concentration is normally a result of increased cutter loader workload and leads to a costly operation shutdown until the increased concentrations abate.

We propose a simple yet very robust methane warning prediction model that can forecast imminent high methane concentrations at least 3 minutes in advance, thereby giving enough notice to slow the mining operation, prevent methane warning and avoid costly shutdowns.

Our model is in fact an instance of the generic prediction framework able to rapidly compose a predictor of any future events upon the aligned time series big data. The model uses fast greedy backward-forward search applied subsequently upon the design choices of the machine learning model from the data granularity, feature selection, filtering and transformation up to the selection of the predictor, its configuration and complexity.

We have applied such framework to the methane concentration warning prediction in real coal mines as a part of the IJCRS'2015 data mining competition and scored 3rd place with the performance just under 85 %. Our top model emerged as a result of the rapid filtering through the large amount of sensors time series and eventually used only the latest 1 minute of aggregated data from just few sensors and the logistic regression predictor. Many other model setups harnessing multiple linear regression, decision trees, naive Bayes or support vector machine predictors on slightly altered feature sets returned nearly equally good performance.

Keywords: Big data · Events prediction · Time series forecasting · Feature selection · Classification · Regression

1 Introduction

Coal mining operation is under continuous exposure to all kinds of risks and threats. Methane explosion is probably the most common such threat that still accounts globally for thousands of miners' lives and losses of millions of dollars in mining equipment every year.

© Springer International Publishing Switzerland 2015
Y. Yao et al. (Eds.): RSFDGrC 2015, LNAI 9437, pp. 485–493, 2015.
DOI: 10.1007/978-3-319-25783-9_43

The methane that is adsorbed in the coal is released as the coal is mined or it migrates from surrounding sources above or below the coal seam through fractures created by the coal extraction process. At high concentrations ($> 5\,\%$) methane could cause explosions capable to collapse multiple shafts and bury or trap miners and their equipment. Good ventilation and methane concentration monitoring are therefore of critical importance to avoid such disasters.

Usual practise in the modern coal mines are installations of the early warning systems triggering alert when methane concentration exceeds the threshold of typically $1\,\%$ and automatically shutting down the mining operation until the methane concentration recedes to the normal levels. Since such shutdowns absorb great deal of cost there is a natural interest in better understanding the process of methane concentration buildup. Specifically, since methane release pace is attributed to the mining intensity it might be possible to prevent costly alerts, shutdowns and further reduce the risks of methane explosions by slowing down the mining processes whenever the methane concentration was on course to reach dangerously high levels. Such intelligent control requires, however, accurate predictions of the expected high methane concentrations and this is the objective of this work as well as the IJCRS'15 Data Mining Competition.

Accurate and reliable prediction of explosion or alarming situations in the coal mines has not been intensively explored in the literature. In [2], a hybrid system was proposed to predict gas concentration in hard-coal mines, which combined the advantage of a linear model and Regression and Classification Trees (CART) in a linear fashion that adaptively tuned the system parameters according to its efficiency in real-time operation. A power efficient Early Warning System (EWS) was presented in [1], which used a wireless gas-sensor network to detect concentrations of methane gas for the prediction of methane outbursts in coal mines. The major advantages of a wireless system over traditional wired safety systems is that it can reduce the cost of long wires and avoid destroying the wired communication links in high risk environment. Other works focussed on predictions of the conditions which trigger the gas explosion in the environments with insufficient oxygen [6]. In [3] and [4], the lower and upper explosion thresholds of multi-component gas mixture, e.g. mixture of hydrogen and methane, was estimated using a Least Square Support Vector Regression (LS-SVR) model and generalized regression neural network, respectively, which aimed at establishing a non-linear model between the composition of the explosive mixture and the explosion thresholds. Concentration of coal dusts is another cause of explosion in coal mines. In [5–7], mine explosion risks were linked to monitoring coal dust in conjunction with methane concentration and temperature data collected using multiple sensors. The joint impact of the coexistence of both methane and the dust at different concentrations on the explosion risks was analyzed and resulted in a single predictive model utilising both sources of evidence merged using fuzzy information fusion theory underpinnings.

1.1 Problem Formulation

The data and the problem specification comes from one of the Polish coal mines, scheme of which is presented in Fig. 1. The shaft covers a system of ventilated

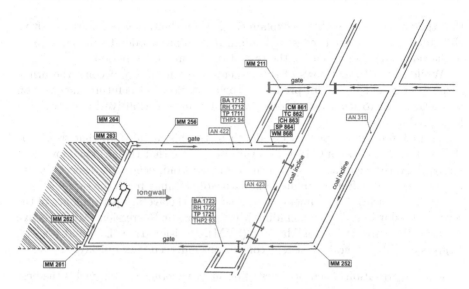

Fig. 1. Mining process scheme

corridors equipped with 28 sensors measuring air flow, temperatures, humidity, air pressure, methane concentrations and cutter loader characteristics including speed, consumed current and driving direction. The mining process is in fact represented by the cutter loader moving and extracting the coal along the corridor longwalls as shown in Fig. 1.

The sensor data is provided in a form of 10 time series slots each containing sequences of measures at the 1s time resolution. 51700 of such slots were provided for training along with the 3 target variables representing a binary indicators if methane sensors MM263, MM264 and MM256 will trigger an alert in between 3 to 6 minutes from the end of the slot.

The objective was to design a classification model that based on the provided sensor time series would most accurately predict alerts on the target methane sensors. More specifically the evaluation criterion embodying model performance was defined as an average area under the receiver operator curves (AUC) constructed as a result of testing the 3 classifiers over the unseen testing set of also 51700 10-min time slot cases.

1.2   Data Summarization and Filtering

The input data of 51700 cases each represented by 28 sensor series of 600s each amounted to almost 1 billion numerical measures. The target class variables, however, were compressed to indicate any alert presence in the period between 3 to 6 minutes after the end of the individual case time slot. Clearly there was a dissonance of the temporal data resolutions or complexity between the sensor data and the target variables, that needed adjustment in a form of aggregation of the sensor series. On the other hand, the difficulty with the aggregated data is that they might not be able to explain cases when the target becomes a positive class due to only short lived signal spikes remaining at odds to the large mass of the remaining signal flow in the rest of the 3 min active period.

While the historical flows of the target sensors naturally become the prime suspects of good features for predicting their own flow in the future the question is whether and to what extent other variables are contributive to the warning level predictions.

Third interesting aspect is the depth of the historical data that would be the most predictive about the target variables. Note that the air flow movement imposed by the ventilation system may further complicate this aspect due to introduced delays after which methane saturated air may reach the target sensor.

After rather brief experimentations guided by the average performance of the several standard predictors including Multiple Linear Regression (MLR), Naive Bayes (NB), Support Vector Machine (SVM), Decision Tree (DT) and Logistic Regression (LR) several consistent conclusions emerged very quickly:

1. 1-min aggregation of sensor series using simple average (mean) yields the best predictive results.
2. For all 28 sensors the most recent section of the input 10-min slot appears the most informative about the target variables.
3. Target variables' own histories appear to be by far the best predictors of their futures.

First conclusion emerges in response to the predictive time horizon that matches the target variables at the 1-min aggregation level. Interestingly, further aggregations including complete aggregation of the 10-min slot into a single point yield consistently worse results. Second conclusion is a reflection of the series temporal continuity, although separated by the 3 min directly following the input series. Finally, the third conclusion is the expected reflection of the series autocorrelation. Given these three points the task was simplified massively as the original inputs of 868 million sensor readings collapsed to at most thousands of numerical values.

2   Hierarchical Greedy Model Composition

After initial data filtering and observations of the data characteristics informing predictability of the target variables the objective now was to design a robust predictor that would incorporate this evidence.

However, apart from the sensor series quantisation at the 1-min intervals, we decided to leave all the other model design choices open and instead construct a generic predictor composition framework which could decide automatically upon the predictor design choices. Specifically, these design choices were narrowed down to the selection of features, selection of the depth of the historical input and the selection of the predictor from the choice of: MLR, NB, SVM, DT and LR models. For simplicity we decided to split the task into 3 independent classification tasks and simply aggregate the AUC performance results at the end.

The algorithm we chose as an engine for the design decisions selection is diversified backward-forward search (BFS) applied hierarchically upon the three layers of design choices, invariably guided by the AUC evaluation criterion. First we apply BFS for the feature selection choosing all 10 1−min aggregated sensor measures as features. Then starting from the top selected feature subsets we applied the very same BFS search to find the optimal depth of the historical sensor data, i.e. decide how many recent minutes of the data for each sensor give the best prediction results upon the target variables. Once this is settled the algorithm steps back to the feature selection task, this time on the updated historical depth and the result is considered to provide the optimised subset of refined features for the given initial classifier, chosen to be MLR. Then for the same feature configuration the algorithm traverses forward through other predictors, while also trying to step back to adjust feature configurations if they lead to improved performance for these new predictors. The process continues forward and backward until no further improvement is found. The terminal model represents the predictor hierarchically optimised at different levels of design choices.

2.1   Diversified Backward-Forward Feature Selection

The process of feature selection was dominated by the target variables own history in line with the conclusion from the previous section. Exploiting these observations we have initialised the BFS selection with the three groups of features corresponding to the 10-min aggregated data points from sensors: MM263, MM264 and MM256. For this search we have fixed all the other higher level choices of the depth of the historical evidence to the whole 10 minutes and using MLR predictor trained against the binary target variables. Such search attempts to scan through all the features and greedily add them to the pool of selected features if the inclusion improves the predictive performance measure. Next the BFS proceeds to the backward search also in a greedy fashion trying to remove features one at the time and validating such elimination if the predictive performance improves. The reason why backward search keeps improving the performance following the forward search is that the earlier addition choices may have been improved by the better latter choices such that some earlier selected suboptimal features are no longer needed or even lead to the degradation of performance. The rounds of additions and removals continue until no performance improvements are observed over the validation set, and given the very strong initial feature selection set such algorithm almost instantly converged to the very short solutions that introduced only very small deviations to the 3 target variable features.

The following stage involved optimising the feature historical depth in a form of choosing the best number of the last k minutes of each 10−min sensor slot features selected in the first stage. The same BFS search was applied to this task yet the result very clearly converged to the single last minute as the best choice of feature historical depth for all the selected features and all 3 classification tasks, that in fact were later also verified for all other not selected features. After such 10-fold reduction of the features set in line with the hierarchical BFS search model presented above the search reverts back to the feature selection stage with the newly optimised higher level design choice of historical depth set as fixed and the best feature subset found so far retained as the initial set for the search. This step resulted in a few single adjustments of the top features.

2.2 Predictive Model Selection

The initial choice of the predictor was fixed to the multiple linear regression (MLR) trained against binary target variables and evaluated using AUC performance criterion. With all the design layers optimised for MLR the same BFS selection was applied at the higher hierarchy level of predictor selection. In fact all the optimisations carried out at the bottom layers of the design constitute a single step in the same BFS search at the higher (predictor) level selection. The only difference here is that for simplicity we do not allow for multiple classifier/regression models to be selected, however this could be an open option if we further add predictor combiners. For the hierarchical predictor design composition presented in this paper we limit the design at the top of abstraction continuum to the choice of the prediction model. The BFS search proceeds therefore to test other predictors: LR, SVM, DT and NB, at each time attempting to adjust the bottom layers' design choices, which really come down to feature selection updates. Since the historical depth choice has consistently proven to point at the last minute as the optimal choice we have fixed this and taken away this selection from the design choices to speed up the search process.

Figure 2 depicts the concept of the hierarchical backward-forward search applied to automate the design of the machine learning model tasked to solve the IJCRS'15 problem, but equally applicable to any other automated predictor composition task with the number of hierarchy levels corresponding to the design choices at different abstraction levels.

3   Experimental Results

The experiments were carried out in line with the automated predictor design guided by the hierarchical BFS selection model described above. Enumerating the last average minutes of the 10−min slots from the 28 sensor variables by $f_1, ..., f_{28}$, the top feature selection solutions for the 3 sensor target variables resulted with the following design solutions:

- target MM_{263}: features set $\{f_4, f_7, f_{13}, f_{15}\}$ or $\{TP_{1721}, TP_{1711}, MM_{263}, MM_{256}\}$, predictor: LR
- target MM_{264}: features set $\{f_5, f_{14}, f_{19}, f_{20}\}$ or $\{RH_{1722}, MM_{264}, CR_{863}, P_{864}\}$, predictor: LR
- target MM_{256}: features set $\{f_7, f_{15}, f_{20}, f_{27}\}$ or $\{TP_{1711}, MM_{256}, P_{864}, MM_{256}\}$, predictor: LR

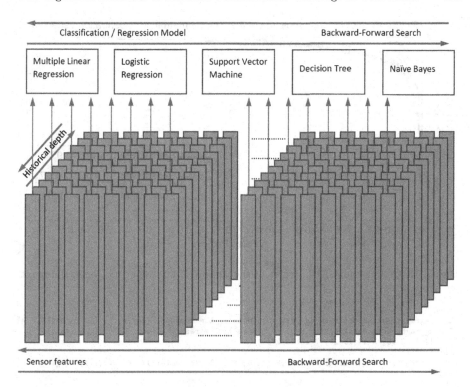

Fig. 2. Hierarchical Backward-Forward Search architecture for automated predictor design for IJCRS'15 Data Mining Competition task

First target sensor (MM_{263}) was enriched by the temperatures in the same corridor further along the airflow. Second target sensor (MM_{264}) was enriched by the humidity in the corridor preceding the sensor location along the airflow and the methane drainage pressure and its flange pressure difference, while the third sensor (MM_{256}) added a combination of the temperature and the methane drainage pipeline pressure as well as the current of the hydraulic pump engine in the cutter loader.

Interestingly, for all three prediction tasks logistic regression has been chosen as the top model beating much more complex models like support vector machines.

The ROC curves along with the AUC scores for the three top prediction models composed by our automated ML design system over the training set are presented in Fig. 3.

What is also rather interesting, an automated hierarchical design of the multi-target classification system resulted in a very simplistic yet very robust prediction model, for which the complete process of data preparation, model training and predictions over testing data, all can be written in just few lines of code as shown in the following Matlab script:

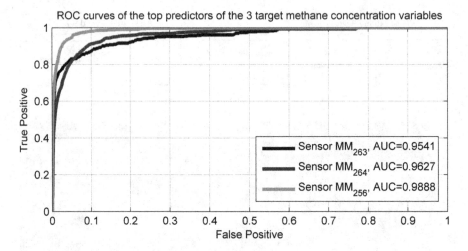

Fig. 3. ROC curves and AUC scores of the top predictors reported over training set

```
L = aggregate_sequence(learning_set, 10, 'mean'); L = reshape(L(end,:,:), [], n_feat)';
T = aggregate_sequence(testing_set, 10, 'mean'); T = reshape(T(end,:,:), [], n_feat)';
F = [4 7 13 15; 5 14 19 20; 7 15 20 27]';
for i = 1:3
    Y_pred(:,i) = glmval(glmfit(L(:,F(:,i)), Y(:,i), 'binomial'), T(:,F(:,i)), 'logit');
end
```

4   Evaluation and Conlcusions

Summarizing, for the predictive task of forecasting elevated methane concentration events, the collected high resolution sensor data proved unnecessarily complex compared to the aggregated target variables. Aggregation and selection of the most recent minute of the input sensor series turned out to be the most explanative about the subsequent methane concentration warning events. The target sensors' own history expectedly proved to provide the bulk of the explanative power, complemented by temperature, humidity and methane drainage pipeline pressures to only slightly improve the baseline predictive performance. The top results were achieved with the logistic regression predictor which was selected along with the set of the only 4 optimized features for each task on the course of hierarchical greedy backward-forward search algorithm carried out in a fully automated fashion. The presented self-organized composition of the robust predictive model resulted in the very high predictive performance measured by means of AUC metric in excess of 0.95 for all three predictive tasks over training set. The presented model results were submitted as an entry to the IJCRS'15 Data Mining Competition and scored 3rd place based on the evaluation on the testing set not available during the model design and training process. The AUC testing performance in the order of 0.85 gives a significant predictive power to forecast increased methane concentrations that would justify mining process slowdown and thereby prevent the effects of costly methane warning shutdowns and foremost eliminate catastrophic explosion events.

References

1. Unnikrishna Menon, K.A., Maria, D., Thirugnanam, H.: Power optimization strategies for wireless sensor networks in coal mines. In: International Conference on Wireless and Optical Communications Networks, pp. 1–5 (2012)
2. Sikora, M., Krzystanek, Z., Bojko, B., Spiechowicz, K.: Hybrid adaptative system of gas concentration prediction in hard-coal mines. In: International Conference on Systems Engineering, pp. 159–164 (2008)
3. Zheng, L., Xiao, Z., Yu, M., Jia, H.: Prediction of explosion limits of multi-component gas mixture using LS-SVR. In: International Conference on SE-Product E-Service and E-Entertainment, pp. 1–4 (2010)
4. Zheng, L., Jiang, L., Zheng, K., Yu, M.: Estimation of explosion limits of gas mixture using a single spread GRNN. In: International Conference on Artificial Intelligence, Management Science and Electronic Commerce, pp. 1113–1115 (2011)
5. Ma, F., Zang, H.: Improved coal mine dust monitoring system based on fuzzy information fusion. In: International Conference on Natural Computation, vol. 8, pp. 4430–4434 (2010)
6. Ma, F.: Sensor networks-based monitoring and fuzzy information fusion system for underground gas disaster. In: International Conference on Fuzzy Systems and Knowledge Discovery, pp. 596–600 (2012)
7. Ma, F.: Collaborative monitoring of underground gas disaster based on fuzzy information fusion. In: World Congress on Intelligent Control and Automation, pp. 4230–4234 (2012)

Prediction of Methane Outbreaks in Coal Mines from Multivariate Time Series Using Random Forest

Adam Zagorecki[1,2](✉)

[1] Defence Academy of the United Kingdom, Shrivenham SN6 8LA, UK
[2] Centre for Simulation and Analytics, Cranfield University, Bedford, UK
a.zagorecki@cranfield.ac.uk

Abstract. In recent years we have experienced unprecedented increase of use of sensors in many industrial applications. Examples of such are Health and Usage Monitoring Systems (HUMS) for vehicles, so-called intelligent buildings, or instrumentation on machinery in order to monitor performance, detect faults and gain insights in operational aspects. Modern sensors are capable of not only generating large volumes of data but as well transmitting that data through network and storing it for further analysis. Unfortunately, that collected data requires further analysis in order to provide useful information to the decision makers who want to reduce costs, improve safety, etc. Such analysis proved to be a challenge, as there are no generic methodologies that allow for automating data analysis and in practice costs required to analyze data are prohibitively high for many practical applications. This paper is a step in a direction of developing generic methods for sensor data analysis – it describes an application of a generic method that can be applied to arbitrary set of multivariate time series data in order to perform classification or regression tasks. The presented application relates to prediction of methane concentrations in coal mines based on time series data from various sensors. The method was tested within the framework of IJCRS'15 data mining competition and resulted in the winning model outperforming other solutions.

1 Introduction

In this paper I present an application of a generic approach to classification of multivariate time series data proposed in [1]. This approach was developed and evaluated in the context of the 2015 AAIA Data Mining Competition, where it led to the second highest score. In this paper I present the application of this approach to another data mining competition involving multivariate data series: the *IJCRS'15 Data Challenge: Mining Data from Coal Mines*. The presented solution resulted with the winning entry.

During the recent decade affordable and reliable sensors capable of collecting large amounts of data has become popular in many applications including

Y. Yao et al. (Eds.): RSFDGrC 2015, LNAI 9437, pp. 494–500, 2015.
DOI: 10.1007/978-3-319-25783-9_44

industrial, commercial and everyday life. One of most popular types of data collected by those sensors is time series data. This kind of data typically consist of sequences of measurements taken over time. With affordable sensors capable of transmitting data over network, multivariate time series data sets are becoming common in many domains. Examples can include vehicle or machinery monitoring, sensors from smartphones or sensor suites installed on human body. Because of the nature of time series, the collected measurements typically not directly exploitable – as the measurements consists of typically a large number of data points, the data is noisy, and requires further analysis in order to identify or discover interesting patterns that can be exploited by users. It is well recognized that the process of processing the raw measurements and transforming the data into knowledge useful for the users is a challenging and costly task. It is particularly true with multivariate time series data as time series are characterized by large volume and often need to undergo transformations (such as Fourier transforms, various filtering, etc.) to reveal potentially useful patterns. On the other hand, if generic methods for transforming multivariate time series data are developed, they can lead to rapid advances in utilization of sensor data in many areas. In this paper we present an application of a method for classification of multivariate time series data that was developed for a data mining competition involving motion sensors installed on human body and subsequently was successfully applied to different problem involving sensors in coal mines.

In the application presented in this paper the data consisted of measurements taken by various sensors installed in a coal mine and machinery operating in that coal mine. They involved different types of measurements, mostly environmental such as humidity, temperature, air pressure, methane concentration, etc. and some related to the state of operating machinery such as cutter loader speed, direction and currents at different parts of machinery. The task was to predict if methane level exceeding certain thresholds would occur in next 3 to 6 min. This knowledge would potentially enable extra warning time before methane warning level is exceeded and can be used to take preventive actions.

The rest of the paper is composed as follows: in the next section the competition task will be introduced with details of the sensors, available data and the evaluation. In the following section I will discuss the proposed approach to classification of multivariate time series data. Consequently each step in of the proposed approach will be discussed in more detail: feature engineering, and actual classification. I will finish the paper with a short discussion.

2   The Competition Task

This paper describes a solution to the IJCRS'15 data mining competition which was organized using the Knowledge Pit competition platform [3]. The objective of the competition was to gain insight into dependencies between cutter loader (mining machinery) performance methane level measured by several sensors distributed in the coal mine.

The basic task of the competition was to create a numeric model to predict exceedance of threshold levels at three methane sensors in the short future

(3 to 6 min) based on sensors readings from multiple sensors. For this purpose a commercial off-the-shelf body sensor suite was used to generate the data.

2.1 Data

The data made available for this competition consisted solely of time series. It consisted of 51,700 records, which corresponded to time periods.

Each record consisted of 28 time series. Each time series was composed of exactly 600 data points and corresponded to a 10 min worth measurements. The measurements included: anemometers, temperature sensors, methane sensors, barometers, humidity sensors, pressure and pressure difference sensors, current sensors (machinery), direction and speed of the cutter. The task was to predict methane level exceedance in the future at three sensors. The target variables were three binary variables (threshold exceeded or not) for three selected methane sensors. The true state of a target variable indicated that a methane warning level was exceeded within 3 to 6 min after the end of corresponding the time series.

The data was split into two sets: the training and test set. For the training set the target variables were provided. The task was to predict probability of the warning threshold exceedance for each of three target variables.

2.2 Evaluation

The evaluation of the results was performed using the Area Under the ROC Curve (AUC) measure concept. It was possible, because the target variables were defined in for of probability of threshold exceedance for each of three variables.

For each of three target variables the separate AUC score was first computed. Let us define the AUC score for the ith target variables as AUC_i. The final score was an average of three individual scores:

$$AUC_{total} = \frac{1}{3} \sum_{i=1}^{3} AUC_i.$$

During the competition only preliminary score was available to competitors. The preliminary score was based on a subset of the final test set, and it corresponded to approximately 20 % of the test data. The final evaluation was performed after completion of the competition.

3 Solution Overview

In this section I present an overview of the solution to the competition task. The method I used was based on the method developed for the AAIA'15 data mining competition that is described in [1]. The only difference is that for IJCRS'15 competition I did not use feature selection step. I decided not to use feature selection step as for the this competition feature selection resulted in inferior

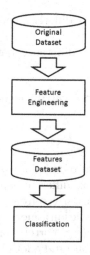

Fig. 1. The outline of the basic steps used during the competition.

results and proved unnecessary. The basic steps in the used method are presented in Fig. 1.

The first, and probably the most critical step was the feature engineering step. At this step the original data set was converted to a secondary data set that consisted of the features generated from the time series data. This step is discussed in detail in the Sect. 4.

For each of the three target variables I decided to create a separate classifier that made a binary decision. In this way I had to learn three separate classifiers. It is important to note, that no information was shared between three classifiers and all three of them were learned using the same features data set. The task called for assignment of probability (rather than hard decision) for presence of the threshold exceedance. As the basic classifier I used Random Forest, which allowed to compute the probability of class assignment.

3.1 Derived Time Series

I decided to expand the original set of time series used for feature selection by creating additional time series that were derived from the original time series. The derived time series were generated from a pair of the original times series. Let us assume that $x(t)$ and $y(t)$ are two original time series, then the derived time series were generated if:

- Both $x(t)$ and $y(t)$ were methane sensors (their names started with M)
- Both $x(t)$ and $y(t)$ started with BA, RH, TP, and AN – all of them corresponded to particular type of environmental sensors: pressure, humidity, temperature, and wind speed.
- For each of the pair of signals $x(t)$ and $y(t)$, I produced two derived time series $d_1(t)$ and $d_2(t)$ which were:

- $d_1(t) = x(t) - y(t)$ – simple difference between corresponding measurements
- $d_2(t) = \frac{x(t)-y(t)}{x(t)}$ – relative difference between corresponding measurements

The number of derived time series created was 52.

4   Feature Engineering

The next step was transformation of the data from time series form into a set of numerical values that summarize different aspects of the time series data.

The most basic features can be derived from individual time series are simple statistics (e.g. mean, standard deviation), more complex features can be derived from more than one time series (e.g. correlation coefficient between two time series). In the course of competition I did a lot of experimentation with different features. I was using feature selection algorithms implemented in Weka software [4] to identify most informative features. As the result of this analysis I put special emphasis on features related to maximal or minimal values, as those seemed to be most informative, at least according to feature selection algorithms. I would like to note, that I used the feature selection to inform feature engineering only – I did not use feature selection to actually select features for classification – I used all generated features for classification task.

4.1   Generated Features

For each of the time series (either original or derived) the following features were extracted:

- the mean value
- the standard deviation
- the minimal value
- the maximal value
- the average of top 5 minimal values
- the average of top 5 maximal values
- the minimal value expressed in standard deviations from the mean
- the maximal value expressed in standard deviations from the mean
- the average of top 5 minimal values expressed in standard deviations from the mean
- the average of top 5 maximal values expressed in standard deviations from the mean
- the maximal difference between minimal and maximal values taken over non-decreasing sequences of measurements
- the maximal difference between maximal and minimal values taken over non-increasing sequences of measurements
- the maximal values (frequency and power) for the fast Fourier transform with ignoring first three frequencies

- the parameters for linear regression: slope, intercept, the mean square error, and the absolute value of slope
- the parameters for polynomial fitting (done only for parabolic fitting): a_0, a_1 and a_2
- the parameters for polynomial fitting taken over the first half of the signal (done only for parabolic fitting): a_0, a_1 and a_2
- the parameters for polynomial fitting taken over the second half of the signal (done only for parabolic fitting): a_0, a_1 and a_2

Each of the above features generated a single number that was used as an individual feature for further analysis. This produced a total of 2214 features – 756 from the original time series and 1458 from derived time series.

4.2   Correlations

Finally, I decided to add correlation coefficients between time series. Additional parameters were derived from cross-correlations (those included auto-correlations) between selected pairs of signals:

- cross-correlations for the signal taken at $t=0$ and the same signal taken at $t=0$, 100, 200, and 300 using Pearsons' correlation coefficient
- cross-correlations for the signal taken at $t=0$ and the same signal taken at $t=0$, 100, 200, and 300 using Spearmans' correlation coefficient
- cross-correlations for the signal taken at $t=0$ and the same signal taken at $t=0$, 100, 200, and 300 using Kendalls' correlation coefficient

The pairs of signals $x(t)$ and $y(t)$ included:

- any methane sensors measurements MM taken pair-wise
- pairing signals starting with the same prefixes that were BA, RH, and AN – pairs only if two signals had the same prefix – for example BA with BA, but not with any other

This effectively lead to include auto-correlation as I allowed $x(t)=y(t)$. The total number of features in the winning set was 4914.

5   Classification

I used Random Forest [6] implemented in Weka software [4] as the basic classifier. I did experimented with other classifiers such as Neural Networks, Logistic Regression, Support Vector Machines, and others, however the Random Forest seemed to perform consistently better, One of the challenges with applying Random Forest effectively is selection of optimal number of features used for each tree. In the case of competitions it is typically done by trial and error approach. I experimented with different numbers of features per tree and for the particular feature set the numbers between 60 and 100 features seemed to work well. For the best score I could achieve, each of three Random Forest classifiers had 1000 trees. The number of features for each tree was limited to 80.

6   Conclusions

In this paper I presented winning solution for the IJCRS'15 Data Mining Competition. The approach is a slightly customized approach to classification of multivariate time series developed for other data mining competition that involved multivariate time series data and allowed to achieved very good score. This result seems to validate versatility of the proposed approach, as claimed in the original paper.

As suggested earlier, different features seemed to achieve better results comparing to the previous application of the method. Surprisingly, the same basic classifier, namely Random Forest seemed to perform consistently better over other classifiers – the same result was observed in the previous competition.

I believe that the result presented here provides empirical evidence that the developed approach can be easily generalized to similar problems for which multiple measurements in form of time series are available.

References

1. Zagorecki, A.: A Versatile Approach to Classification of Multivariate Time Series Data. In: The Proceedings of the 2015 Federated Conference on Computer Science and Information Systems (2015, to appear)
2. Meina, M., Janusz, A., Rykaczewski, K., Ślęzak, D., Celmer, B., Krasuski, A.: Tagging firefighter activities at the emergency scene: summary of AAIAâĂŹ15 data mining competition at knowledge pit. In: Proceedings of the 2015 Federated Conference on Computer Science and Information Systems (2015)
3. Janusz, A., Krasuski, A., Stawicki, S., Rosiak, M., Slezak, D., Nguyen, H.S.: Key risk factors for Polish State Fire Service: a data mining competition at knowledge pit. Federated Conference on Computer Science and Information Systems (FedCSIS) 2014, pp. 345–354 (2014) doi:10.15439/2014F507
4. Hall, M., Frank, E., Holmes, G., Pfahringer, B., Reutemann, P., Witten, I.H.: The WEKA data mining software: an update. SIGKDD Explorations, vol. 11(1), pp. 10–18 (2009)
5. Hall, M.A.: Correlation-based Feature Subset Selection for Machine Learning. Hamilton, New Zealand (1998)
6. Breiman, L.: Random forests. Mach. Learn. 45(1), 5–32 (2001)

Correction to: Building Granular Systems - from Concepts to Applications

Marcin Szczuka, Andrzej Jankowski, Andrzej Skowron,
and Dominik Ślęzak

Correction to:
Chapter "Building Granular Systems - from Concepts
to Applications" in Y. Yao et al. (Eds.):
Rough Sets, Fuzzy Sets, Data Mining, and Granular Computing,
LNAI 9437, https://doi.org/10.1007/978-3-319-25783-9_22

The acknowledgement section of this paper originally referred to grant DEC-2013/09/B/ST6/01568. The reference to this grant has been removed from the acknowledgement section at the request of one of the authors.

The updated version of this chapter can be found at
https://doi.org/10.1007/978-3-319-25783-9_22

Author Index

Printed in the United States
by Baker & Taylor Publisher Services